Automobile Electrical and Electronic Systems

Second edition

Automobile Electrical and Electronic Systems

Second edition

Tom Denton *BA, AMSAE, AMITRE, Cert.Ed.*

Associate Senior Lecturer in Motor Vehicle Engineering, Colchester Institute, UK
Associate Lecturer, Open University

ARNOLD

A member of the Hodder Headline Group
LONDON

First published in Great Britain in 1995 by
Arnold, a member of the Hodder Headline Group,
338 Euston Road, London NW1 3BH

Second edition published 2000

http://www.arnoldpublishers.com

British Library Cataloguing in Publication Data
A catalogue record for this book is available from the British Library

ISBN 0 340 73195 8

Typeset by J&L Composition Ltd, Filey, North Yorkshire
Printed and bound in Great Britain by The Bath Press, Bath

What do you think about this book? Or any other Arnold title?
Please send your comments to feedback.arnold@hodder.co.uk

Contents

Preface		ix
Introduction to the second edition		x
Acknowledgements		xi
1	**Development of the automobile electrical system**	**1**
1.1	A short history	1
1.2	Where next?	9
1.3	Self-assessment	10
2	**Electrical and electronic principles**	**11**
2.1	Safe working practices	11
2.2	Basic electrical principles	11
2.3	Electronic components and circuits	17
2.4	Digital electronics	25
2.5	Microprocessor systems	30
2.6	Measurement	35
2.7	Sensors and actuators	36
2.8	New developments	51
2.9	Diagnostics – electronics, sensors and actuators	53
2.10	Self-assessment	54
3	**Tools and test equipment**	**56**
3.1	Basic equipment	56
3.2	Multimeters	57
3.3	Specialist equipment	59
3.4	Dedicated equipment	65
3.5	On-board diagnostics	66
3.6	New developments in testing	67
3.7	Diagnostic procedures	71
3.8	Self-assessment	76
4	**Electrical systems and circuits**	**77**
4.1	The systems approach	77
4.2	Electrical wiring, terminals and switching	78
4.3	Multiplexed wiring systems	87
4.4	Circuit diagrams and symbols	92
4.5	Case study	93
4.6	Electromagnetic compatibility (EMC)	95
4.7	New developments in systems and circuits	98
4.8	Self-assessment	99
5	**Batteries**	**101**
5.1	Vehicle batteries	101
5.2	Lead-acid batteries	102
5.3	Maintenance and charging	103
5.4	Diagnosing lead-acid battery faults	105
5.5	Advanced battery technology	107
5.6	Developments in electrical storage	110
5.7	Self-assessment	116

6	**Charging systems**	**117**
6.1	Requirements of the charging system	117
6.2	Charging system principles	118
6.3	Alternators and charging circuits	119
6.4	Case studies	125
6.5	Diagnosing charging system faults	127
6.6	Advanced charging system technology	128
6.7	New developments in charging systems	132
6.8	Self-assessment	133
7	**Starting systems**	**134**
7.1	Requirements of the starting system	134
7.2	Starter motors and circuits	136
7.3	Types of starter motor	140
7.4	Case studies	147
7.5	Diagnosing starting system faults	148
7.6	Advanced starting system technology	148
7.7	New developments in starting systems	150
7.8	Self-assessment	152
8	**Ignition systems**	**153**
8.1	Ignition fundamentals	153
8.2	Electronic ignition	155
8.3	Programmed ignition	163
8.4	Distributorless ignition	168
8.5	Direct ignition	169
8.6	Spark-plugs	170
8.7	Case studies	173
8.8	Diagnosing ignition system faults	174
8.9	Advanced ignition technology	176
8.10	New developments in ignition systems	177
8.11	Self-assessment	178
9	**Electronic fuel control**	**180**
9.1	Combustion	180
9.2	Engine fuelling and exhaust emissions	186
9.3	Electronic control of carburation	190
9.4	Fuel injection	191
9.5	Diesel fuel injection	198
9.6	Case studies	201
9.7	Diagnosing fuel control system faults	209
9.8	Advanced fuel control technology	211
9.9	New developments	212
9.10	Self-assessment	215
10	**Engine management**	**217**
10.1	Combined ignition and fuel management	217
10.2	Exhaust emission control	221
10.3	Control of diesel emissions	225
10.4	Complete vehicle control systems	226
10.5	Case study – Mitsubishi GDI	228
10.6	Case study – Bosch Motronic M3	237
10.7	Diagnosing engine management system faults	245
10.8	Advanced engine management technology	248
10.9	New developments in engine management	256
10.10	Self-assessment	259

11 Lighting **260**
11.1 Lighting fundamentals 260
11.2 Lighting circuits 268
11.3 Gas discharge and LED lighting 269
11.4 Case studies 272
11.5 Diagnosing lighting system faults 275
11.6 Advanced lighting technology 276
11.7 New developments in lighting systems 278
11.8 Self-assessment 282

12 Auxiliaries **284**
12.1 Windscreen washers and wipers 284
12.2 Signalling circuits 289
12.3 Other auxiliary systems 290
12.4 Case studies 291
12.5 Diagnosing auxiliary system faults 294
12.6 Advanced auxiliary systems technology 295
12.7 New developments in auxiliary systems 296
12.8 Self-assessment 298

13 Instrumentation **299**
13.1 Gauges and sensors 299
13.2 Driver information 303
13.3 Visual displays 305
13.4 Case studies 310
13.5 Diagnosing instrumentation system faults 311
13.6 Advanced instrumentation technology 312
13.7 New developments in instrumentation systems 313
13.8 Self-assessment 314

14 Air conditioning **315**
14.1 Conventional heating and ventilation 315
14.2 Air conditioning 317
14.3 Other heating systems 319
14.4 Case studies 320
14.5 Diagnosing air conditioning system faults 324
14.6 Advanced temperature control technology 325
14.7 New developments in temperature control systems 326
14.8 Self-assessment 326

15 Chassis electrical systems **328**
15.1 Anti-lock brakes 328
15.2 Active suspension 332
15.3 Traction control 334
15.4 Automatic transmission 336
15.5 Other chassis electrical systems 338
15.6 Case studies 339
15.7 Diagnosing chassis electrical system faults 346
15.8 Advanced chassis systems technology 347
15.9 New developments in chassis electrical systems 350
15.10 Self-assessment 353

16 Comfort and safety **354**
16.1 Seats, mirrors and sun-roofs 354
16.2 Central locking and electric windows 356
16.3 Cruise control 358
16.4 In-car multimedia 360

16.5   Security   367
16.6   Airbags and belt tensioners   369
16.7   Other safety and comfort systems   373
16.8   Case studies   375
16.9   Diagnosing comfort and safety system faults   381
16.10   Advanced comfort and safety systems technology   383
16.11   New developments in comfort and safety systems   384
16.12   Self-assessment   386

17   Electric vehicles   387
17.1   Electric traction   387
17.2   Hybrid vehicles   390
17.3   Case studies   390
17.4   Advanced electric vehicle technology   395
17.5   New developments in electric vehicles   396
17.6   Self-assessment   398

18   World Wide Web   399
18.1   Useful contacts   399
18.2   Programs and information to download   401
18.3   Self-assessment   403

Appendix   404
Example program   404

Index   407

Preface

In the beginning, say 110 years ago, a book on vehicle electrics would have been very small. A book on vehicle electronics would have been even smaller!

As we move into the new millennium, the subject has become so large that some aspects of the vehicle electrical and electronic system have had to be glossed over or even missed out. Not many though, so there is still plenty to read.

This second edition of *Automobile Electrical and Electronic Systems* has been updated and extended, particularly by the inclusion of a whole chapter on basic principles and sections in each chapter for assessment. Subject coverage soon gets into a good depth; however, the really technical bits are kept in a separate section of each chapter so you can miss them out if you are new to the subject.

The nature of this subject is such that, as I produce this book, other innovations are taking place. To this end I have concentrated where possible on the underlying electrical and electronic principles. Many current and older systems are included to aid the reader with an understanding of these principles.

To set the whole automobile electrical subject in context, the first chapter covers some of the significant historical developments and even dares once again to speculate on the future!

What will be the next major step in automobile electronic systems? I think the 'auto-PC' and 'telematics' will be key factors – read on to find out more...

Tom Denton
2000

PS

Comments, questions and contributions are always welcome at my web site: http://www.automotive-technology.co.uk.

You will also find lots of useful information, updates and news about new books as well as automotive software and web links.

Introduction to the second edition

The book has grown! But then it was always going to, because the automobile electrical and electronic systems have grown. Another reason though, is that I have included more coverage of basic electrical technology. This can be used to learn the basics of electrical and electronic theory if you are new to the subject, or as a reference source for the more advanced user.

Significant case studies are included, some very new and others well tried and tested, but all have important aspects.

The main thing I would like to stress is that the book in general has become more 'web aware'. A significant number of web addresses are included, particularly in the final chapter. However, my main hope is that 'we' will become more interactive by using the power of the Internet.

Let me know when you find the odd mistake, but also let me know about new and interesting technology as well as interesting web sites. I will be doing the same on my web site so keep dropping in, the support will grow during the life of this book, more questions and assignments for example, revision and learning material as well as updates and software packages.

Keep in touch.

Tom Denton
2000

Acknowledgements

I am very grateful to the following companies who have supplied information and/or permission to reproduce photographs and/or diagrams, figure numbers are as listed:

AA Photo Library 1.8; AC Delco Inc. 7.26; Alpine Audio Systems Ltd. 13.27; Autodata Ltd. 10.1 (table); Autologic Data Systems Ltd.; BMW UK Ltd. 10.6; C&K Components Inc. 4.17; Citroën UK Ltd. 4.29, 4.31, 7.31; Clarion Car Audio Ltd. 16.21, 16.24; Delphi Automotive Systems Inc. 8.5; Eberspaecher GmbH. 10.13; Fluke Instruments UK Ltd. 3.5; Ford Motor Company Ltd. 1.2, 7.28, 11.4a, 12.18, 16.37; General Motors 11.24, 11.25, 15.20, 17.7; GenRad Ltd. 3.11, 3.18, 3.19; Hella UK Ltd. 11.19, 11.22; Honda Cars UK. Ltd. 10.5, 15.19; Hyundai UK Ltd. 11.4d; Jaguar Cars Ltd. 1.11, 11.4b, 13.24, 16.47; Kavlico Corp. 2.79; Lucas Ltd. 3.14, 5.5, 5.6, 5.7, 6.5, 6.6, 6.23, 6.34, 7.7, 7.10, 7.18, 7.21, 7.22, 8.7, 8.12, 8.37, 9.17, 9.24, 9.25, 9.26, 9.32, 9.33, 9.34, 9.46, 9.47, 9.48, 9.49, 9.51, 10.43; LucasVarity Ltd. 2.67, 2.81, 2.82, 2.83, 9.38, 9.60, 9.61; Mazda Cars UK Ltd. 9.57, 9.58, 9.59; Mercedes Cars UK Ltd. 5.12, 5.13, 11.4c, 16.14; Mitsubishi Cars UK Ltd. 10.21 to 10.38; NGK Plugs UK Ltd. 8.28, 8.30, 8.31, 8.32, 8.38, 9.41; Nissan Cars UK Ltd. 17.8; Peugeot UK Ltd. 16.28; Philips UK Ltd. 11.3; Pioneer Radio Ltd. 16.17, 16.18, 16.19; Porsche Cars UK Ltd. 15.12, 15.23; Robert Bosch GmbH. 2.72, 4.30, 5.2, 6.24, 7.19, 7.24, 7.25, 8.1, 8.9, 9.28, 10.10, 10.42, 10.53, 10.55a, 11.21; Robert Bosch Press Photos 1.1, 2.57, 2.58, 2.63, 2.69, 3.16, 4.21, 4.24, 4.25, 4.26, 6.35, 8.26, 9.18, 9.19, 9.27, 9.29, 9.30, 9.31, 9.35, 9.42, 9.43, 9.44, 9.45, 9.52, 9.53, 9.54, 10.7, 10.8, 10.9, 10.14, 10.15, 10.18, 10.19, 10.20, 10.59, 10.61, 11.7, 12.15, 12.19, 15.3, 15.8, 16.16, 16.33, 16.36, 16.52; Robert Bosch UK Ltd. 3.7, 6.28, 7.30, 8.34, 8.39; Rover Cars Ltd. 4.10, 4.11, 4.28, 8.19, 8.20, 10.3, 11.20, 12.17, 13.11, 14.9, 14.12, 14.13, 14.14, 14.15, 14.16, 14.17, 15.21, 16.2, 16.46; Saab Cars UK Ltd. 18.18, 13.15; Scandmec Ltd. 14.10; Snap-on Tools Inc. 3.1, 3.8; Sofanou (France) 4.8; Sun Electric UK Ltd. 3.9; Thrust SSC Land Speed Team 1.9; Toyota Cars UK Ltd. 7.29, 8.35, 8.36, 9.55; Tracker UK Ltd. 16.51; Unipart Group Ltd. 11.1; Valeo UK Ltd. 6.1, 7.23, 11.23, 12.2, 12.5, 12.13, 12.20, 14.4, 14.8, 14.19, 15.35, 15.36; VDO Instruments 13.16; Volvo Cars Ltd. 4.22, 10.4, 16.42, 16.43, 16.44, 16.45; ZF Servomatic Ltd. 15.22.

Many if not all the companies here have good web pages. You will find a link to them from my site. Thanks again to the listed companies. If I have used any information or mentioned a company name that is not noted here, please accept my apologies and acknowledgments.

Last but by no means least, thank you once again to my family: Vanda, Malcolm and Beth.

1
Development of the automobile electrical system

1.1 A short history

1.1.1 Where did it all begin?

The story of electric power can be traced back to around 600 BC, when the Greek philosopher Thales of Miletus found that amber rubbed with a piece of fur would attract lightweight objects such as feathers. This was due to static electricity. It is thought that, around the same time, a shepherd in what is now Turkey discovered magnetism in lodestones, when he found pieces of them sticking to the iron end of his crook.

William Gilbert, in the sixteenth century, proved that many other substances are 'electric' and that they have two electrical effects. When rubbed with fur, amber acquires 'resinous electricity'; glass, however, when rubbed with silk, acquires 'vitreous electricity'. Electricity repels the same kind and attracts the opposite kind of electricity. Scientists thought that the friction actually created the electricity (their word for charge). They did not realize that an equal amount of opposite electricity remained on the fur or silk.

A German, Otto Von Guerick, invented the first electrical device in 1672. He charged a ball of sulphur with static electricity by holding his hand against it as it rotated on an axle. His experiment was, in fact, well ahead of the theory developed in the 1740s by William Watson, an English physician, and the American statesman Benjamin Franklin, that electricity is in all matter and that it can be transferred by rubbing. Franklin, in order to prove that lightning was a form of electricity, flew a kite during a thunderstorm and produced sparks from a key attached to the string! Some good did come from this dangerous experiment though, as Franklin invented the lightning conductor.

Alessandro Volta, an Italian aristocrat, invented the first battery. He found that by placing a series of glass jars containing salt water, and zinc and copper electrodes connected in the correct order, he could get an electric shock by touching the wires. This was the first wet battery and is indeed the forerunner of the accumulator, which was developed by the French physicist Gaston Planche in 1859. This was a lead-acid battery in which the chemical reaction that produces electricity could be reversed by feeding current back in the opposite direction. No battery or storage cell can supply more than a small amount of power and inventors soon realized that they needed a continuous source of current. Michael Faraday, a Surrey blacksmith's son and an assistant to Sir Humphrey Davy, devised the first electrical generator. In 1831 Faraday made a machine in which a copper disc rotated between the poles of a large magnet. Copper strips provided contacts with the rim of the disc and the axle on which it turned; current flowed when the strips were connected.

William Sturgeon of Warrington, Lancashire, made the first working electric motor in the 1820s. He also made the first working electromagnets and used battery-powered electromagnets in a generator in place of permanent magnets. Several inventors around 1866, including two English electricians – Cromwell Varley and Henry Wilde – produced permanent magnets. Anyos Jedlik, a Hungarian physicist, and the American pioneer electrician, Moses Farmer, also worked in this field. The first really successful generator was the work of a German, Ernst Werner Von Siemens. He produced his generator, which he called a dynamo, in 1867. Today, the term dynamo is applied only to a generator that provides direct current. Generators, which produce alternating current, are called alternators.

The development of motors that could operate from alternating current was the work of an American engineer, Elihu Thomson. Thomson also invented the transformer, which changes the voltage of an electric supply. He demonstrated his invention in 1879 and, 5 years later, three Hungarians, Otto Blathy, Max Deri and Karl Zipernowksy, produced the first commercially practical transformers.

It is not possible to be exact about who conceived particular electrical items in relation to the motor car. Innovations in all areas were thick and fast in the latter half of the nineteenth century.

In the 1860s, Ettiene Lenoir developed the first practical gas engine. This engine used a form of electric ignition employing a coil developed by Ruhmkorff in 1851. In 1866, Karl Benz used a type of magneto that was belt driven. He found this to be unsuitable though, owing to the varying speed of his engine. He solved the problem by using two primary cells to provide an ignition current.

In 1889, Georges Bouton invented contact breakers for a coil ignition system, thus giving positively tuned ignition for the firt time. It is arguable that this is the ancestor of the present

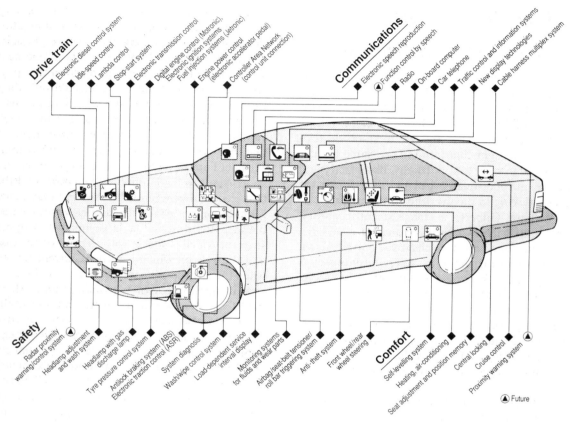

Figure 1.1 Electronics in the car

Figure 1.2 Henry Ford's first car, the Quadricycle

day ignition system. Emile Mors used electric ignition on a low-tension circuit supplied by accumulators that were recharged from a belt-driven dynamo. This was the first successful charging system and can be dated to around 1895.

The now formidable Bosch empire was started in a very small way by Robert Bosch. His most important area of early development was in conjunction with his foreman, Fredrich Simms, when they produced the low-tension magneto at the end of the nineteenth century. Bosch introduced the high-tension magneto to almost universal acceptance in 1902. The 'H' shaped armature of the very earliest magneto is now used as the Bosch trademark on all the company's products.

From this period onwards, the magneto was developed to a very high standard in Europe, while in the USA the coil and battery ignition system took the lead. Charles F. Kettering played a vital role in this area working for the Daytona electrical company (Delco), when he devised the ignition, starting and lighting system for the 1912 Cadillac. Kettering also produced a mercury-type voltage regulator.

The third-brush dynamo, first produced by Dr Hans Leitner and R.H. Lucas, first appeared in about 1905. This gave the driver some control over the charging system. It became known as the constant current charging system. By today's standards this was a very large dynamo and could produce only about 8 A.

Many other techniques were tried over the next decade or so to solve the problem of controlling output on a constantly varying speed dynamo. Some novel control methods were used, some with more success than others. For example, a drive system, which would slip beyond a certain engine speed, was used with limited success, while one of my favourites had a hot wire in the main output line which, as it became red hot, caused current to bypass it and flow through a 'bucking' coil to reduce the dynamo field strength. Many variations of the 'field warp' technique were used. The control of battery charging current for all these constant current systems was poor and often relied on the driver to switch from high to low settings. In fact, one of the early forms of instrumentation was a dashboard hydrometer to check the battery state of charge!

The two-brush dynamo and compensated voltage control unit was used for the first time in the 1930s. This gave far superior control over the charging system and paved the way for the many other electrical systems to come.

In 1936, the much-talked about move to positive earth took place. Lucas played a major part in this change. It was done to allow reduced spark plug firing voltages and hence prolong electrode life. It was also hoped to reduce corrosion between the battery terminals and other contact points around the car.

The 1950s was the era when lighting began to develop towards today's complex arrangements. Flashing indicators were replacing the semaphore arms and the twin filament bulb allowed more suitable headlights to be made. The quartz halogen bulb, however, did not appear until the early 1970s.

Great improvements now started to take place with the fitting of essential items such as heaters, radios and even cigar lighters! Also in the 1960s and 1970s, many more optional extras became available, such as windscreen washers and two-speed wipers. Cadillac introduced full air conditioning and even a time switch for the headlights.

The negative earth system was re-introduced

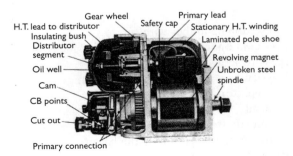

Figure 1.3 Rotating magnet magneto

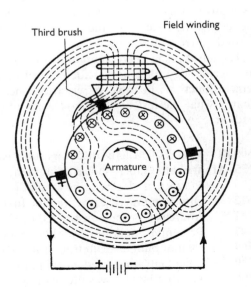

Figure 1.4 Third-brush dynamo

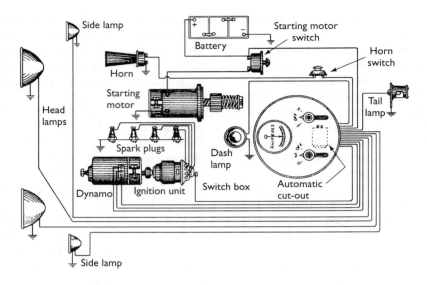

Figure 1.5 A complete circuit diagram

in 1965 with complete acceptance. This did, however, cause some teething problems, particularly with the growing DIY fitment of radios and other accessories. It was also good, of course, for the established auto-electrical trade!

The 1970s also hailed the era of fuel injection and electronic ignition. Instrumentation became far more complex and the dashboard layout was now an important area of design. Heated rear windows that worked were fitted as standard to some vehicles. The alternator, first used in the USA in the 1960s, became the norm by about 1974 in Britain.

The extra power available and the stable supply of the alternator was just what the electronics industry was waiting for and, in the 1980s, the electrical system of the vehicle changed beyond all recognition.

The advances in microcomputing and associated technology have now made control of all vehicle functions possible by electrical means. That is what the rest of this book is about, so read on.

1.1.2 A chronological history

The electrical and electronic systems of the motor vehicle are often the most feared, but at the same time can be the most fascinating aspects of an automobile. The complex circuits and systems now in use have developed in a very interesting way.

For many historical developments it is not possible to be certain exactly who 'invented' a particular component, or indeed when, as developments were taking place in parallel, as well as in series.

It is interesting to speculate on who we could call the founder of the vehicle electrical system. Michael Faraday of course deserves much acclaim, but then of course so does Ettiene Lenoir and so does Robert Bosch and so does Nikolaus Otto and so does...

Perhaps we should go back even further to the ancient Greek philosopher Thales of Miletus who, whilst rubbing amber with fur, discovered static electricity. The Greek word for amber is 'elektron'.

c600 BC Thales of Miletus discovers static electricity by rubbing amber with fur.
c1550 AD William Gilbert showed that many substances contain 'electricity' and that, of the two types of electricity he found different types attract while like types repel.
1672 Otto Von Guerick invented the first electrical device, a rotating ball of sulphur.
1742 Andreas Gordon constructed the first static generator.
1747 Benjamin Franklin flew a kite in a thunderstorm!
1769 Cugnot built a steam tractor in France made mostly from wood.
1780 Luigi Galvani started a chain of events resulting in the invention of the battery.
1800 The first battery was invented by Alessandro Volta.
1801 Trevithick built a steam coach.

1825	Electromagnetism was discovered by William Sturgeon.
1830	Sir Humphery Davy discovered that breaking a circuit causes a spark.
1831	Faraday discovered the principles of induction.
1851	Ruhmkorff produced the first induction coil.
1859	The accumulator was developed by the French physicist Gaston Planche.
1860	Lenoir built an internal-combustion gas engine.
1860	Lenoir developed 'in cylinder' combustion.
1860	Lenoir produced the first spark-plug.
1861	Lenoir produced a type of trembler coil ignition.
1861	Robert Bosch was born in Albeck near Ulm in Germany.
1870	Otto patented the four-stroke engine.
1875	A break spark system was used in the Seigfried Marcus engine.
1876	Otto improved the gas engine.
1879	Hot-tube ignition was developed by Leo Funk.
1885	Benz fitted his petrol engine to a three-wheeled carriage.
1885	The motor car engine was developed by Gottlieb Daimler and Karl Benz.
1886	Daimler fitted his engine to a four-wheeled carriage to produce a four-wheeled motorcar.
1887	The Bosch low-tension magneto was used for stationary gas engines.
1887	Hertz discovered radio waves.
1888	Professor Ayrton built the first experimental electric car.
1889	E. Martin used a mechanical system to show the word 'STOP' on a board at the rear of his car.
1889	Georges Bouton invented contact breakers.
1891	Panhard and Levassor started the present design of cars by putting the engine in the front.
1894	The first successful electric car.
1895	Emile Mors used accumulators that were recharged from a belt-driven dynamo.
1895	Georges Bouton refined the Lenoir trembler coil.
1896	Lanchester introduced epicyclic gearing, which is now used in automatic transmission.
1897	The first radio message was sent by Marconi.
1897	Bosch and Simms developed a low-tension magneto with the 'H' shaped armature, used for motor vehicle ignition.
1899	Jenatzy broke the 100 kph barrier in an electric car.
1899	First speedometer introduced (mechanical).
1899	World speed record 66 mph – in an electric powered vehicle!
1901	The first Mercedes took to the roads.
1901	Lanchester produced a flywheel magneto.
1902	Bosch introduced the high-tension magneto, which was almost universally accepted.
1904	Rigolly broke the 100 mph barrier.
1905	Miller Reese invented the electric horn.
1905	The third-brush dynamo was invented by Dr Hans Leitner and R.H. Lucas.
1906	Rolls-Royce introduced the Silver Ghost.
1908	Ford used an assembly-line production to manufacture the Model T.
1908	Electric lighting appeared, produced by C.A. Vandervell.
1910	The Delco prototype of the electric starter appeared.
1911	Cadillac introduced the electric starter and dynamo lighting.
1912	Bendix invented the method of engaging a starter with the flywheel.
1912	Electric starting and lighting used by Cadillac. This 'Delco' electrical system was developed by Charles F. Kettering.
1913	Ford introduced the moving conveyor belt to the assembly line.
1914	Bosch perfected the sleeve induction magneto.
1914	A buffer spring was added to starters.
1920	Duesenberg began fitting four-wheel hydraulic brakes.
1920	The Japanese made significant improvements to magnet technology.
1921	The first radio set was fitted in a car by the South Wales Wireless Society.
1922	Lancia used a unitary (all-in-one) chassis construction and independent front suspension.

1922	The Austin Seven was produced.
1925	Dr D.E. Watson developed efficient magnets for vehicle use.
1927	Segrave broke the 200 mph barrier in a Sunbeam.
1927	The last Ford model T was produced.
1928	Cadillac introduced the synchromesh gearbox.
1928	The idea for a society of engineers specializing in the auto-electrical trade was born in Huddersfield, Yorkshire, UK.
1929	The Lucas electric horn was introduced.
1930	Battery coil ignition begins to supersede magneto ignition.
1930	Magnet technologies are further improved.
1931	Smiths introduced the electric fuel gauge.
1931	The Vertex magneto was introduced.
1932	The Society of Automotive Electrical Engineers held its first meeting in the Constitutional Club, Hammersmith, London, 21 October at 3.30 pm.
1934	Citroen pioneered front-wheel drive in their 7CV model.
1934	The two-brush dynamo and compensated voltage control unit was first fitted.
1936	An electric speedometer was used that consisted of an AC generator and voltmeter.
1936	Positive earth was introduced to prolong spark-plug life and reduce battery corrosion.
1937	Coloured wires were used for the first time.
1938	Germany produced the Volkswagen Beetle.
1939	Automatic advance was fitted to ignition distributors.
1939	Car radios was banned in Britain for security reasons.
1939	Fuse boxes start to be fitted.
1939	Tachograph recorders were first used in Germany.
1940	The DC speedometer was used, as were a synchronous rotor and trip meter.
1946	Radiomobile company formed.
1947	The transistor was invented.
1948	Jaguar launched the XK120 sports car and Michelin introduced a radial-ply tyre.
1950	Dunlop announced the disc brake.
1951	Buick and Chrysler introduced power steering.
1951	Development of petrol injection by Bosch.

Figure 1.6 Early racing cars required a great deal of skill to perfect the magneto ignition system

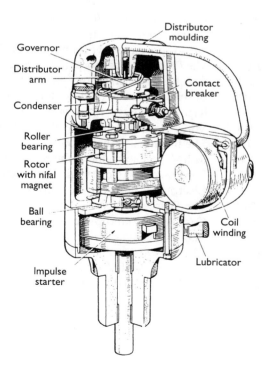

Figure 1.7 Sectional view of the Lucas type 6VRA Magneto

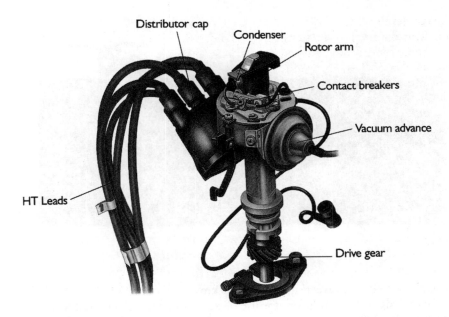

Figure 1.8 Distributor with contact breakers

1951	UK manufacturers start to use 12 V electrical system.
1952	Rover's gas-turbine car set a speed record of 243 kph.
1954	Bosch introduced fuel injection for cars.
1954	Flashing indicators were legalized.
1955	Citroen introduced a car with hydro-pneumatic suspension.
1955	Key starting becomes a standard feature.
1957	Wankel built his first rotary petrol engine.
1957	Asymmetrical headlamps were introduced.
1958	The first integrated circuit was developed.
1959	BMC (now Rover Cars) introduced the Mini.
1960	Alternators started to replace the dynamo.
1963	The electronic flasher unit was developed.
1965	Development work started on electronic control of anti-locking braking system (ABS).
1965	Negative earth system reintroduced.
1966	California brought in legislation regarding air pollution by cars.
1966	In-car record players are not used with great success in Britain due to inferior suspension and poor roads!
1967	The Bosch Jetronic fuel injection system went into production.
1967	Electronic speedometer introduced.
1970	Gabelich drove a rocket-powered car, 'Blue Flame', to a new record speed of 1001.473 kph.
1970	Alternators began to appear in British vehicles as the dynamo began its demise.
1972	Dunlop introduced safety tyres, which seal themselves after a puncture.
1972	Lucas developed head-up instrumentation display.
1974	The first maintenance free breakerless electronic ignition was produced.
1976	Lambda oxygen sensors were produced.
1979	Barrett exceeded the speed of sound in the rocket-engined 'Budweiser Rocket' (1190.377 kph).
1979	Bosch started series production of the Motronic fuel injection system.
1980	The first mass-produced car with four-wheel drive, the Audi Quattro, was available.
1981	BMW introduced the on-board computer.
1981	Production of ABS for commercial vehicles started.
1983	Austin Rover introduced the Maestro, the first car with a talking dashboard.
1983	Richard Noble set an official speed record in the jet-engined 'Thrust 2' of 1019.4 kph.

Figure 1.9 Thrust SSC

1987	The solar-powered 'Sunraycer' travelled 3000 km.
1988	California's emission controls aim for use of zero emission vehicles (ZEVs) by 1998.
1989	The Mitsubishi Gallant was the first mass-produced car with four-wheel steering.
1989	Alternators, approximately the size of early dynamos or even smaller, produced in excess of 100 A.
1990	Fiat of Italy and Peugeot of France launched electric cars.
1990	Fibre-optic systems used in Mercedes vehicles.
1991	The European Parliament voted to adopt stringent control of car emissions.
1991	Gas discharge headlamps were in production.
1992	Japanese companies developed an imaging system that views the road through a camera.
1993	A Japanese electric car reached a speed of 176 kph.
1993	Emission control regulations force even further development of engine management systems.
1994	Head-up vision enhancement systems were developed as part of the Prometheus project.
1995	Greenpeace designed an environmentally friendly car capable of doing 67–78 miles to the gallon (100 km per 3–3.5 litres).
1995	The first edition of *Automobile Electrical and Electronic Systems* was published!
1996	Further legislation on control of emissions.
1997	GM developed a number of its LeSabres for an Automated Highway System.
1998	Thrust SSC broke the sound barrier.
1998	Blue vision headlights started to be used.

Figure 1.10 Modern racing car development has provided many ideas now used in 'normal' cars

Figure 1.11 Jaguar S-Type

1998 Mercedes 'S' class had 40 computers and over 100 motors.
1999 Mobile multimedia became an optional extra.
2000 Second edition of *Automobile Electrical and Electronic Systems* published!
And the story continues with you...

1.2 Where next?

1.2.1 Current developments

Research in motor vehicle technology is an ongoing process. The trend in general seems to be incremental improvements of existing technology. However, electronic control is being introduced in more areas of the vehicle. Some of the areas I expect to be developed further, and hence become more popular, are listed below.

- Electro-hydraulic control of valve lift and timing.
- Total vehicle dynamic control.
- Light-emitting diodes (LEDs) and gas discharge lamps for exterior lighting.
- Electric vehicle traction and batteries.
- Global positioning systems (GPS) systems.
- Integration of personal computing systems into vehicles.

1.2.2 Automobile systems in the next millennium – 'the modern driver'

Just imagine...
The best thing about wire guided car systems is that you can still take control from time to time. I often do some work as my car is taking me on a journey, which is good. Today though is a day off.

'Please enter details of your journey', said the car in its uncannily human voice. The voice can be adjusted, but it's even worse when it sounds like Professor Stephen Hawking. 'I'm going to drive for a change' I told it, but as usual it persisted, 'Would you like me to plan the best route?' 'No', I said, 'not today.' 'All diagnostic routines have been run during the night and no faults found,' it continued. At this point I told it quite succinctly, not to speak again unless in an emergency. After accessing its 'colloquial database', it appeared to understand – and stopped talking.

Today I wanted really to drive. Pulling out of the garage I set off towards my favourite test track. It was the proactive suspension that I wanted to put through its paces. As well as the obvious surface scanning lasers, the new system uses magnetoelastic springs. This system could, in theory, not only change the suspension stiffness on each wheel instantly, it could also change the damping characteristics. We will see.

As usual I tried to feel when the electric motor cut out and the turbine cut in, but as usual, I couldn't. The high performance electro-mechanical torque storage system made sure of that.

Passing other cars on the road reminded me of my first time driving with a joystick instead of a steering wheel, it was weird, too much like a three-dimensional computer simulation. However, now I am used to it, I don't think I could go back.

I was about half way to the test track, according to the guidance system, when the unthinkable happened – the car stopped. 'What's going on' I demanded, and, as the car had interpreted this event as an emergency, it answered, 'An unknown system error has occurred, please wait for further details.' I explained that it should proceed with all haste. Again the 'colloquial database' must have been useful because it said 'Accessing at maximum speed, please be patient.'

Three minutes later the system started up again as if nothing had happened. 'All systems fully functional using first line backups' the car announced, with what could only be described as a little pride in its artificial voice. 'What was wrong?' I asked, which seemed like a reasonable question at the time. 'A comparative run time error occurred in the second parallel processor line due to an incorrect digital signal response from the main sensor area network data bus responsible for critical system monitoring,' the car replied. 'You mean a wire fell off' I said, 'Yes it admitted after consulting its 'concise lexicographical response database'. I think it's about time somebody invented a system that could bypass faults and repair itself, without having to stop the car. Those three minutes could have been important.

At last I reached the test track and switched the car into full sports mode. 'All vehicle control systems adjusted to optimum settings for test track seven,' the car told me. Test track seven is great for putting the car through its paces. It has banked corners, 'S' bends, cobbled surface sections and even a water splash. There were only a few other drivers on the track, so today was going to be the day!

I pulled out on to the track and floored the pedal. The car took off like crazy with the traction control allowing just enough wheel spin to gain maximum possible acceleration. The active steering felt great on the first corner; I could feel it fighting the tendency to oversteer by adjusting the four-wheel steering as well as diverting drive from one wheel to another. Plunging into the water trap at full speed nearly fooled the steering – but not quite. The wipers even switched on just before the water hit the screen. As I accelerated

out of the 'S' bends another car pulled out of a side lane right in front of me – I noticed just in time. I hit the brakes as hard as I could and the ABS stopped me in plenty of time. Off I went again, this time on to the cobbled section, although it didn't feel any different to the rest of the smooth track. I was just about to tell the car to check its magnetoelastic suspension system, when I realized that it must have been working! Just as I was about to finish my first lap, the head-up display flashed 'Automatic Overtake?' in front of me. 'Go for it!' I shouted and the car overtook the one in front like it was standing still. This was a great day for driving.

On the way home, as usual, the car had predicted that I would be going via the local pub and had set a route accordingly. I parked the car (well, it parked itself, really) in the inductive recharge slot, and I went in for a well-earned drink. I couldn't wait to tell my friends what real driving was all about.

Prediction or science fiction?

1.3 Self-assessment

1.3.1 Questions

1. State who invented the spark plug.
2. What significant event occurred in 1800?
3. Make a simple sketch to show the circuit of a magneto.
4. Who did Frederick Simms work for?
5. Explain why positive earth vehicles were introduced.
6. Explain why negative earth vehicles were reintroduced.
7. Which car was first fitted with a starter motor?
8. Charles F. Kettering played a vital role in the early development of the automobile. What was his main contribution and which company did he work for at that time?
9. Describe briefly why legislation has a considerable effect on the development of automotive systems.
10. Pick four significant events from the chronology and describe why they were so important.

1.3.2 Project

Write a short article about driving a car in the year 2020.

2

Electrical and electronic principles

2.1 Safe working practices

2.1.1 Introduction

Safe working practices in relation to electrical and electronic systems are essential, for your safety as well as that of others. You only have to follow two rules to be safe.

- Use your common sense – don't fool about.
- If in doubt – seek help.

The following section lists some particular risks when working with electricity or electrical systems, together with suggestions for reducing them. This is known as risk assessment.

2.1.2 Risk assessment and reduction

Table 2.1 lists some identified risks involved with working on vehicles, in particular the electrical

and electronic systems. The table is by no means exhaustive but serves as a good guide.

2.2 Basic electrical principles

2.2.1 Introduction

To understand electricity properly we must start by finding out what it really is. This means we must think very small (Figure 2.1 shows a representation of an atom). The molecule is the smallest part of matter that can be recognized as that particular matter. Sub-division of the molecule results in atoms, which are the smallest part of matter. An element is a substance that comprises atoms of one kind only.

The atom consists of a central nucleus made up of protons and neutrons. Around this nucleus orbit electrons, like planets around the sun. The

Table 2.1 Risks and risk reduction

Identified risk	Reducing the risk
Electric shock	Ignition HT is the most likely place to suffer a shock, up to 25 000 V is quite normal. Use insulated tools if it is necessary to work on HT circuits with the engine running. Note that high voltages are also present on circuits containing windings, due to back EMF as they are switched off – a few hundred volts is common. Mains supplied power tools and their leads should be in good condition and using an earth leakage trip is highly recommended
Battery acid	Sulphuric acid is corrosive so always use good personal protective equipment (PPE). In this case overalls and, if necessary, rubber gloves. A rubber apron is ideal, as are goggles if working with batteries a lot
Raising or lifting vehicles	Apply brakes and/or chock the wheels when raising a vehicle on a jack or drive-on lift. Only jack under substantial chassis and suspension structures. Use axle stands in case the jack fails
Running engines	Do not wear loose clothing, good overalls are ideal. Keep the keys in your possession when working on an engine to prevent others starting it. Take extra care if working near running drive belts
Exhaust gases	Suitable extraction must be used if the engine is running indoors. Remember, it is not just the carbon monoxide (CO) that might make you ill or even kill you, other exhaust components could cause asthma or even cancer
Moving loads	Only lift what is comfortable for you; ask for help if necessary and/or use lifting equipment. As a general guide, do not lift on your own if it feels too heavy!
Short circuits	Use a jump lead with an in-line fuse to prevent damage due to a short when testing. Disconnect the battery (earth lead off first and back on last) if any danger of a short exists. A very high current can flow from a vehicle battery, it will burn you as well as the vehicle
Fire	Do not smoke when working on a vehicle. Fuel leaks must be attended to immediately. Remember the triangle of fire – Heat/Fuel/Oxygen – don't let the three sides come together
Skin problems	Use a good barrier cream and/or latex gloves. Wash skin and clothes regularly

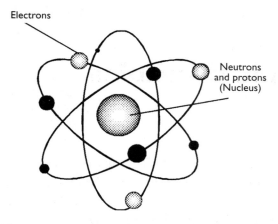

Figure 2.1 The atom

neutron is a very small part of the nucleus. It has equal positive and negative charges and is therefore neutral and has no polarity. The proton is another small part of the nucleus, it is positively charged. The neutron is neutral and the proton is positively charged, which means that the nucleus of the atom is positively charged. The electron is an even smaller part of the atom, and is negatively charged. It orbits the nucleus and is held in orbit by the attraction of the positively charged proton. All electrons are similar no matter what type of atom they come from.

When atoms are in a balanced state, the number of electrons orbiting the nucleus equals the number of protons. The atoms of some materials have electrons that are easily detached from the parent atom and can therefore join an adjacent atom. In so doing these atoms move an electron from the parent atom to another atom (like polarities repel) and so on through material. This is a random movement and the electrons involved are called free electrons

Materials are called conductors if the electrons can move easily. In some materials it is extremely difficult to move electrons from their parent atoms. These materials are called insulators.

2.2.2 Electron flow and conventional flow

If an electrical pressure (electromotive force or voltage) is applied to a conductor, a directional movement of electrons will take place (for example when connecting a battery to a wire). This is because the electrons are attracted to the positive side and repelled from the negative side.

Certain conditions are necessary to cause an electron flow:

- A pressure source, e.g. from a battery or generator.
- A complete conducting path in which the electrons can move (e.g. wires).

An electron flow is termed an electric current. Figure 2.2 shows a simple electric circuit where the battery positive terminal is connected, through a switch and lamp, to the battery negative terminal. With the switch open the chemical energy of the battery will remove electrons from the positive terminal to the negative terminal via the battery. This leaves the positive terminal with fewer electrons and the negative terminal with a surplus of electrons. An electrical pressure therefore exists between the battery terminals.

With the switch closed, the surplus electrons at the negative terminal will flow through the lamp back to the electron-deficient positive terminal. The lamp will light and the chemical energy of the battery will keep the electrons moving in this circuit from negative to positive. This movement from negative to positive is called the electron flow and will continue whilst the battery supplies the pressure – in other words whilst it remains charged.

- Electron flow is from negative to positive.

It was once thought, however, that current flowed from positive to negative and this convention is still followed for most practical purposes. Therefore, although this current flow is not correct, the most important point is that we all follow the same convention.

- Conventional current flow is said to be from positive to negative.

2.2.3 Effects of current flow

When a current flows in a circuit, it can produce only three effects:

- Heat.
- Magnetism.
- Chemical effects.

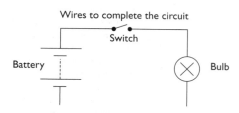

Figure 2.2 A simple electrical circuit

The heating effect is the basis of electrical components such as lights and heater plugs. The magnetic effect is the basis of relays and motors and generators. The chemical effect is the basis for electroplating and battery charging.

In the circuit shown in Figure 2.3 the chemical energy of the battery is first converted to electrical energy, and then into heat energy in the lamp filament.

The three electrical effects are reversible. Heat applied to a thermocouple will cause a small electromotive force and therefore a small current to flow. Practical use of this is mainly in instruments. A coil of wire rotated in the field of a magnet will produce an electromotive force and can cause current to flow. This is the basis of a generator. Chemical action, such as in a battery, produces an electromotive force, which can cause current to flow.

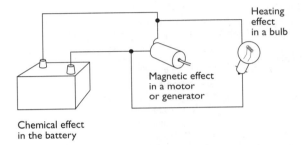

Figure 2.3 A bulb, motor and battery – heat, magnetic and chemical effects

2.2.4 Fundamental quantities

In Figure 2.4, the number of electrons through the lamp every second is described as the rate of flow. The cause of the electron flow is the electrical pressure. The lamp produces an opposition to the rate of flow set up by the electrical pressure. Power is the rate of doing work, or changing energy from one form to another. These quantities as well as several others, are given names as shown in Table 2.2.

If the voltage pressure applied to the circuit was increased but the lamp resistance stayed the same, then the current would also increase. If the voltage was maintained constant but the lamp was changed for one with a higher resistance the current would decrease. Ohm's Law describes this relationship.

Ohm's law states that in a closed circuit 'current is proportional to the voltage and inversely proportional to the resistance'. When 1 volt causes 1 ampere to flow the power used (P) is 1 watt.

Using symbols this means:

Voltage = Current × Resistance
($V = IR$) or ($R = V/I$) or ($I = V/R$)

Power = Voltage × Current
($P = VI$) or ($I = P/V$) or ($V = P/I$)

2.2.5 Describing electrical circuits

Three descriptive terms are useful when discussing electrical circuits.

- *Open circuit*. This means the circuit is broken therefore no current can flow.
- *Short circuit*. This means that a fault has caused a wire to touch another conductor and the current uses this as an easier way to complete the circuit.
- *High resistance*. This means a part of the circuit has developed a high resistance (such as a dirty connection), which will reduce the amount of current that can flow.

2.2.6 Conductors, insulators and semiconductors

All metals are conductors. Silver, copper and aluminium are among the best and are frequently used. Liquids that will conduct an electric current, are called electrolytes. Insulators are generally non-metallic and include rubber, porcelain, glass, plastics, cotton, silk, wax paper and some liquids. Some materials can act as either insulators or conductors depending on conditions. These are called semiconductors and are used to make transistors and diodes.

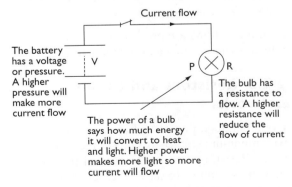

Figure 2.4 An electrical circuit demonstrating links between voltage, current, resistance and power

Table 2.2 Quantities, symbols and units

Name	Definition	Common symbol	Common formula	Unit name	Abbreviation
Electrical charge	One coulomb is the quantity of electricity conveyed by a current of I ampere in I second.	Q	$Q = It$	coulomb	C
Electrical flow or current	The number of electrons having past a fixed point in I second	I	$I = V/R$	ampere	A
Electrical pressure	A pressure of I volt applied to a circuit will produce a current flow of I ampere if the circuit resistance is I ohm	V	$V = IR$	volt	V
Electrical resistance	This is the opposition to current flow in a material or circuit when a voltage is applied across it.	R	$R = V/I$	ohm	Ω
Electrical conductance	Ability of a material to carry an electrical current. One siemens equals I ampere per volt. It was formerly called the mho, or reciprocal ohm.	G	$G = I/R$	siemens	S
Current density	The current per unit area. This is useful for calculating the required conductor cross-sectional areas	J	$J = I/A$ (A = area)		A m^{-2}
Resistivity	A measure of the ability of a material to resist the flow of an electric current. It is numerically equal to the resistance of a sample of unit length and unit cross-sectional area, and its unit is the ohm-meter. A good conductor has a low resistivity ($1.7 \times 10^{-8}\Omega$ m, copper); an insulator has a high resistivity ($10^{15}\Omega$ m, polyethane)	ρ (rho)	$R = \rho L/A$ (L = length A = area)	ohm meter	Ω m
Conductivity	The reciprocal of resistivity	σ (sigma)	$\sigma = I/\rho$	ohm^{-1} meter^{-1}	Ω^{-1} m^{-1}
Electrical power	When a voltage of I volt causes a current of I ampere to flow, the power developed is I watt.	P	$P = IV$ $P = I^2R$ $P = V^2/R$	watt	W
Capacitance	Property of a capacitor that determines how much charge can be stored in it for a given potential difference between its terminals	C	$C = Q/V$ $C = \varepsilon A/d$ (A = plate area, d = distance between, ε = permitivity of dielectric)	farad	F
Inductance	Where a changing current in a circuit builds up a magnetic field which induces an electromotive force either in the same circuit and opposing the current (self-inductance) or in another circuit (mutual inductance).	L	$i = \dfrac{V}{R}\left(1 - e^{-Rt/L}\right)$	henry	H

2.2.7 Factors affecting the resistance of a conductor

In an insulator, a large voltage applied will produce a very small electron movement. In a conductor, a small voltage applied will produce a large electron flow or current. The amount of resistance offered by the conductor is determined by a number of factors.

- Length – the greater the length of a conductor the greater is the resistance.
- Cross-sectional area (CSA) – the larger the cross-sectional area the smaller the resistance.
- The material from which the conductor is made – the resistance offered by a conductor will vary according to the material from which it is made. This is known as the resistivity or specific resistance of the material.
- Temperature – most metals increase in resistance as temperature increases.

Figure 2.5 shows a representation of the factors affecting the resistance of a conductor.

2.2.8 Resistors and circuit networks

Good conductors are used to carry the current with minimum voltage loss due to their low resistance. Resistors are used to control the current flow in a circuit or to set voltage levels. They are made of materials that have a high resistance. Resistors intended to carry low currents are

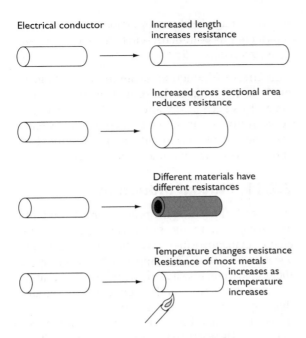

Figure 2.5 Factors affecting electrical resistance

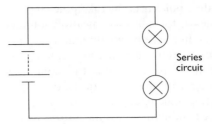

Figure 2.7 Series circuit

often made of carbon. Resistors for high currents are usually wire wound.

Resistors are often shown as part of basic electrical circuits to explain the principles involved. The circuits shown as Figure 2.6 are equivalent. In other words, the circuit just showing resistors is used to represent the other circuit.

When resistors are connected so that there is only one path (Figure 2.7), for the same current to flow through each bulb they are connected in series and the following rules apply.

- Current is the same in all parts of the circuit.

- The applied voltage equals the sum of the volt drops around the circuit.
- Total resistance of the circuit (R_T), equals the sum of the individual resistance values ($R_1 + R_2$ etc).

When resistors or bulbs are connected such that they provide more than one path (Figure 2.8) for the current to flow through and have the same voltage across each component they are connected in parallel and the following rules apply.

- The voltage across all components of a parallel circuit is the same.
- The total current equals the sum of the current flowing in each branch.
- The current splits up depending on each component resistance.
- The total resistance of the circuit (R_T) can be calculated by

$$1/R_\mathrm{T} = 1/R_1 + 1/R_2 \text{ or}$$
$$R_\mathrm{T} = (R_1 \times R_2)/(R_1 + R_2).$$

2.2.9 Magnetism and electromagnetism

Magnetism can be created by a permanent magnet or by an electromagnet (it is one of the three effects of electricity remember). The space around a magnet in which the magnetic effect can be detected is called the magnetic field. The shape of magnetic fields in diagrams is represented by flux lines or lines of force.

Some rules about magnetism:

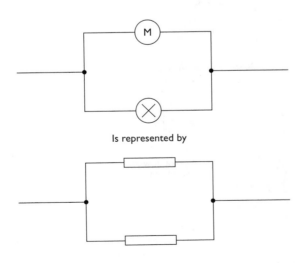

Is represented by

Figure 2.6 An equivalent circuit

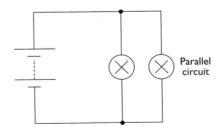

Figure 2.8 Parallel circuit

- Unlike poles attract. Like poles repel.
- Lines of force in the same direction repel sideways, in the opposite direction they attract.
- Current flowing in a conductor will set up a magnetic field around the conductor. The strength of the magnetic field is determined by how much current is flowing.
- If a conductor is wound into a coil or solenoid, the resulting magnetism is the same as a permanent bar magnet.

Electromagnets are used in motors, relays and fuel injectors, to name just a few applications. Force on a current-carrying conductor in a magnetic field is caused because of two magnetic fields interacting. This is the basic principle of how a motor works. Figure 2.9 shows a representation of these magnetic effects.

2.2.10 Electromagnetic induction

Basic laws:

- When a conductor cuts or is cut by magnetism, a voltage is induced in the conductor.
- The direction of the induced voltage depends upon the direction of the magnetic field and the direction in which the field moves relative to the conductor.

- The voltage level is proportional to the rate at which the conductor cuts or is cut by the magnetism.

This effect of induction, meaning that voltage is made in the wire, is the basic principle of how generators such as the alternator on a car work. A generator is a machine that converts mechanical energy into electrical energy. Figure 2.10 shows a wire moving in a magnetic field.

2.2.11 Mutual induction

If two coils (known as the primary and secondary) are wound on to the same iron core then any change in magnetism of one coil will induce a voltage in to the other. This happens when a current to the primary coil is switched on and off. If the number of turns of wire on the secondary coil is more than the primary, a higher voltage can be produced. If the number of turns of wire on the secondary coil is less than the primary a lower voltage is obtained. This is called 'transformer action' and is the principle of the ignition coil. Figure 2.11 shows the principle of mutual induction.

The value of this 'mutually induced' voltage depends on:

- The primary current.

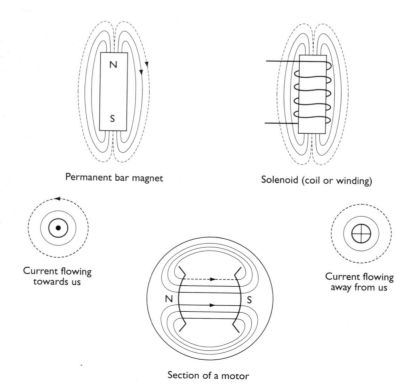

Permanent bar magnet

Solenoid (coil or winding)

Current flowing towards us

Section of a motor

Current flowing away from us

Figure 2.9 Magnetic fields

- The turns ratio between primary and secondary coils.
- The speed at which the magnetism changes.

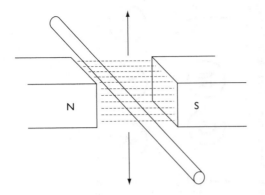

Electricity is induced into the wire
as it moves up or down

Figure 2.10 Induction

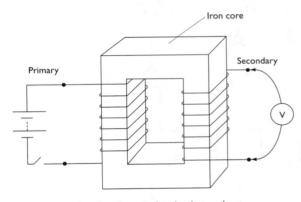

As the switch in the primary is closed and opened, the flow of electricity makes a changing magnetic field around the iron core. The changing magnetism mutually induces electricity into the secondary winding

Figure 2.11 Mutual induction

2.3 Electronic components and circuits

2.3.1 Introduction

This section, describing the principles and applications of various electronic circuits, is not intended to explain their detailed operation. The intention is to describe briefly how the circuits work and, more importantly, how and where they may be utilized in vehicle applications.

The circuits described are examples of those used and many pure electronics books are available for further details. Overall, an understanding of basic electronic principles will help to show how electronic control units work, ranging from a simple interior light delay unit, to the most complicated engine management system.

2.3.2 Components

The main devices described here are often known as discrete components. Figure 2.12 shows the symbols used for constructing the circuits shown later in this section. A simple and brief description follows for many of the components shown.

Resistors are probably the most widely used component in electronic circuits. Two factors must be considered when choosing a suitable resistor, namely the ohms value and the power rating. Resistors are used to limit current flow and provide fixed voltage drops. Most resistors used in electronic circuits are made from small carbon rods, and the size of the rod determines the resistance. Carbon resistors have a negative temperature coefficient (NTC) and this must be considered for some applications. Thin film resistors have more stable temperature properties and are constructed by depositing a layer of carbon onto an insulated former such as glass. The resistance value can be manufactured very accurately by spiral grooves cut into the carbon film. For higher power applications, resistors are usually wire wound. This can, however, introduce inductance into a circuit. Variable forms of most resistors are available in either linear or logarithmic forms. The resistance of a circuit is its opposition to current flow.

A capacitor is a device for storing an electric charge. In its simple form it consists of two plates separated by an insulating material. One plate can have excess electrons compared to the other. On vehicles, its main uses are for reducing arcing across contacts and for radio interference suppression circuits as well as in electronic control units. Capacitors are described as two plates separated by a dielectric. The area of the plates A, the distance between them d, and the permitivity, ε, of the dielectric, determine the value of capacitance. This is modelled by the equation:

$$C = \varepsilon A/d$$

Metal foil sheets insulated by a type of paper are often used to construct capacitors. The sheets are rolled up together inside a tin can. To achieve higher values of capacitance it is necessary to reduce the distance between the plates in

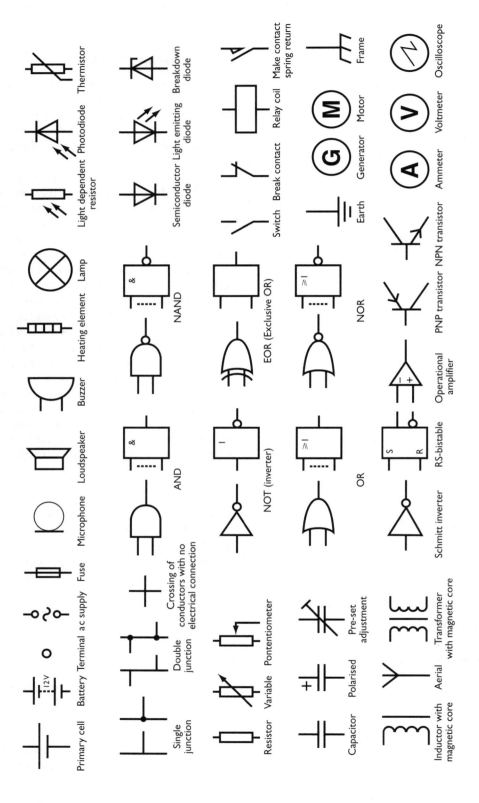

Figure 2.12 Circuit symbols

order to keep the overall size of the device manageable. This is achieved by immersing one plate in an electrolyte to deposit a layer of oxide typically 10^{-4} mm thick, thus ensuring a higher capacitance value. The problem, however, is that this now makes the device polarity conscious and only able to withstand low voltages. Variable capacitors are available that are varied by changing either of the variables given in the previous equation. The unit of capacitance is the farad (F). A circuit has a capacitance of one farad (1 F) when the charge stored is one coulomb and the potential difference is 1 V. Figure 2.13 shows a capacitor charged up from a battery.

Diodes are often described as one-way valves and, for most applications, this is an acceptable description. A diode is a simple PN junction allowing electron flow from the N-type material (negatively biased) to the P-type material (positively biased). The materials are usually constructed from doped silicon. Diodes are not perfect devices and a voltage of about 0.6 V is required to switch the diode on in its forward biased direction. Zener diodes are very similar in operation, with the exception that they are designed to breakdown and conduct in the reverse direction at a pre-determined voltage. They can be thought of as a type of pressure relief valve.

Transistors are the devices that have allowed the development of today's complex and small electronic systems. They replaced the thermal-type valves. The transistor is used as either a solid-state switch or as an amplifier. Transistors are constructed from the same P- and N-type semiconductor materials as the diodes, and can be either made in NPN or PNP format. The three terminals are known as the base, collector and emitter. When the base is supplied with the correct bias the circuit between the collector and emitter will conduct. The base current can be of the order of 200 times less than the emitter current. The ratio of the current flowing through the base compared with the current through the emitter (I_c/I_b), is an indication of the amplification factor of the device and is often given the symbol β.

Another type of transistor is the FET or field effect transistor. This device has higher input impedance than the bipolar type described above. FETs are constructed in their basic form as n-channel or p-channel devices. The three terminals are known as the gate, source and drain. The voltage on the gate terminal controls the conductance of the circuit between the drain and the source.

Inductors are most often used as part of an oscillator or amplifier circuit. In these applications, it is essential for the inductor to be stable and to be of reasonable size. The basic construction of an inductor is a coil of wire wound on a former. It is the magnetic effect of the changes in current flow that gives this device the properties of inductance. Inductance is a difficult property to control, particularly as the inductance value increases due to magnetic coupling with other devices. Enclosing the coil in a can will reduce this, but eddy currents are then induced in the can and this affects the overall inductance value. Iron cores are used to increase the inductance value as this changes the permeability of the core. However, this also allows for adjustable devices by moving the position of the core. This only allows the value to change by a few percent but is useful for tuning a circuit. Inductors, particularly of higher values, are often known as chokes and may be used in DC circuits to smooth the voltage. The value of inductance is the henry (H). A circuit has an inductance of one henry (1 H) when a current, which is changing at one ampere per second, induces an electromotive force of one volt in it.

2.3.3 Integrated circuits

Integrated circuits (ICs) are constructed on a single slice of silicon often known as a substrate. In an IC, Some of the components mentioned previously can be combined to carry out various tasks such as switching, amplifying and logic functions. In fact, the components required for these circuits can be made directly on the slice of silicon. The great advantage of this is not just the size of the ICs but the speed at which they

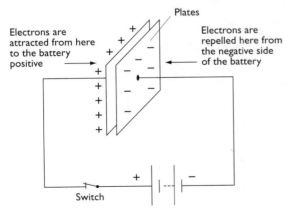

Plates

Electrons are attracted from here to the battery positive

Electrons are repelled here from the negative side of the battery

Switch

When the switch is opened, the plates stay as shown. This is simply called 'charged up'

Figure 2.13 A capacitor charged up

can be made to work due to the short distances between components. Switching speeds in excess of 1 MHz is typical.

There are four main stages in the construction of an IC. The first of these is oxidization by exposing the silicon slice to an oxygen stream at a high temperature. The oxide formed is an excellent insulator. The next process is photo-etching where part of the oxide is removed. The silicon slice is covered in a material called a photoresist which, when exposed to light, becomes hard. It is now possible to imprint the oxidized silicon slice, which is covered with photoresist, by a pattern from a photographic transparency. The slice can now be washed in acid to etch back to the silicon those areas that were not protected by being exposed to light. The next stage is diffusion, where the slice is heated in an atmosphere of an impurity such as boron or phosphorus, which causes the exposed areas to become p- or n-type silicon. The final stage is epitaxy, which is the name given to crystal growth. New layers of silicon can be grown and doped to become n- or p-type as before. It is possible to form resistors in a similar way and small values of capacitance can be achieved. It is not possible to form any useful inductance on a chip. Figure 2.14 shows a representation of the 'packages' that integrated circuits are supplied in for use in electronic circuits.

The range and types of integrated circuits now available are so extensive that a chip is available for almost any application. The integration level of chips has now reached, and in many cases is exceeding, that of VLSI (very large scale integration). This means there can be more than 100 000 active elements on one chip. Development in this area is moving so fast that often the science of electronics is now concerned mostly with choosing the correct combination of chips, and discreet components are only used as final switching or power output stages.

2.3.4 Amplifiers

The simplest form of amplifier involves just one resistor and one transistor, as shown in Figure

2.15. A small change of current on the input terminal will cause a similar change of current through the transistor and an amplified signal will be evident at the output terminal. Note however that the output will be inverted compared with the input. This very simple circuit has many applications when used more as a switch than an amplifier. For example, a very small current flowing to the input can be used to operate, say, a relay winding connected in place of the resistor.

One of the main problems with this type of transistor amplifier is that the gain of a transistor (β) can be variable and non-linear. To overcome this, some type of feedback is used to make a circuit with more appropriate characteristics. Figure 2.16 shows a more practical AC amplifier.

Resistors Rb_1 and Rb_2 set the base voltage of the transistor and, because the base–emitter voltage is constant at 0.6 V, this in turn will set

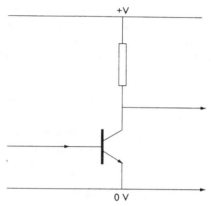

Figure 2.15 Simple amplifier circuit

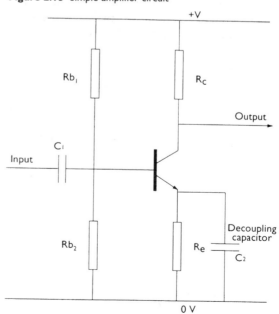

Figure 2.16 Practical AC amplifier circuit

Figure 2.14 Typical integrated circuit package

the emitter voltage. The standing current through the collector and emitter resistors (R_c and R_e) is hence defined and the small signal changes at the input will be reflected in an amplified form at the output, albeit inverted. A reasonable approximation of the voltage gain of this circuit can be calculated as: R_c/R_e

Capacitor C_1 is used to prevent any change in DC bias at the base terminal and C_2 is used to reduce the impedance of the emitter circuit. This ensures that R_e does not affect the output.

For amplification of DC signals, a differential amplifier is often used. This amplifies the voltage difference between two input terminals. The circuit shown in Figure 2.17, known as the long tail pair, is used almost universally for DC amplifiers.

The transistors are chosen such that their characteristics are very similar. For discreet components, they are supplied attached to the same heat sink and, in integrated applications, the method of construction ensures stability. Changes in the input will affect the base–emitter voltage of each transistor in the same way, such that the current flowing through R_e will remain constant. Any change in the temperature, for example, will effect both transistors in the same way and therefore the differential output voltage will remain unchanged. The important property

of the differential amplifier is its ability to amplify the difference between two signals but not the signals themselves.

Integrated circuit differential amplifiers are very common, one of the most common being the 741 op-amp. This type of amplifier has a DC gain in the region of 100 000. Operational amplifiers are used in many applications and, in particular, can be used as signal amplifiers. A major role for this device is also to act as a buffer between a sensor and a load such as a display. The internal circuit of these types of device can be very complicated, but external connections and components can be kept to a minimum. It is not often that a gain of 100 000 is needed so, with simple connections of a few resistors, the characteristics of the op-amp can be changed to suit the application. Two forms of negative feedback are used to achieve an accurate and appropriate gain. These are shown in Figure 2.18 and are

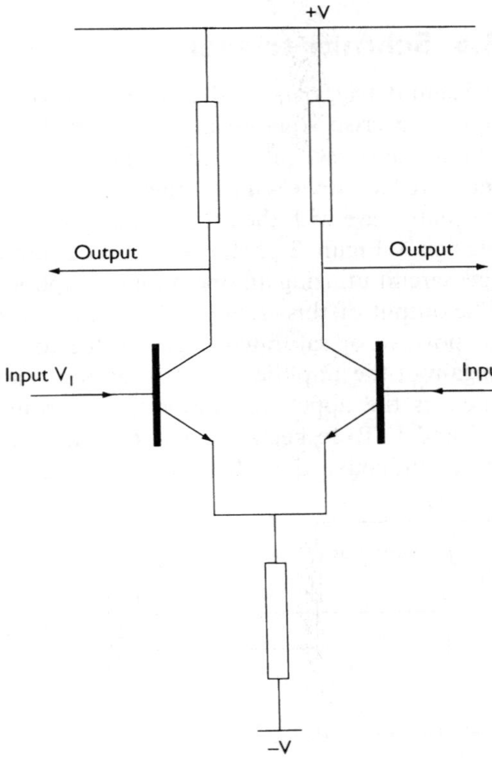

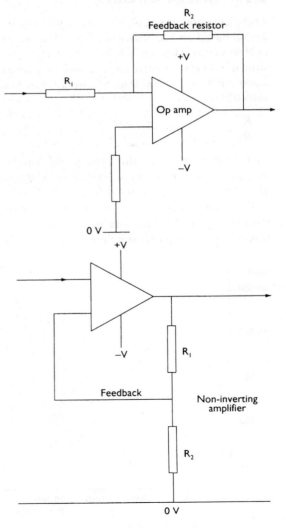

Figure 2.17 DC amplifier, long tail pair

Figure 2.18 Operational amplifier feedback circuits

often referred to as shunt feedback and proportional feedback operational amplifier circuits.

The gain of a shunt feedback configuration is

$$-\frac{R_2}{R_1}$$

The gain with proportional feedback is

$$\frac{R_2}{R_1 + R_2}$$

An important point to note with this type of amplifier is that its gain is dependent on frequency. This, of course, is only relevant when amplifying AC signals. Figure 2.19 shows the frequency response of a 741 amplifier. Op-amps are basic building blocks of many types of circuit, and some of these will be briefly mentioned later in this section.

2.3.5 Bridge Circuits

There are many types of bridge circuits but they are all based on the principle of the Wheatstone bridge, which is shown in Figure 2.20. The meter shown is a very sensitive galvanometer. A simple calculation will show that the meter will read zero when:

$$\frac{R_1}{R_2} = \frac{R_3}{R_4}$$

To use a circuit of this type to measure an unknown resistance very accurately (R_1), R_3 and R_4 are pre-set precision resistors and R_2 is a precision resistance box. The meter reads zero when the reading on the resistance box is equal to the unknown resistor. This simple principle can also

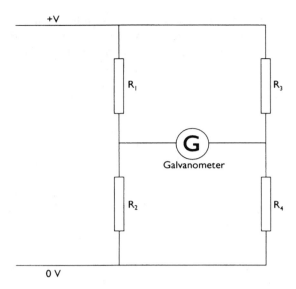

Figure 2.20 Wheatstone bridge

be applied to AC circuits to determine unknown inductance and capacitance.

A bridge and amplifier circuit, which may be typical of a motor vehicle application, is shown in Figure 2.21. In this circuit R_1 has been replaced by a temperature measurement thermistor. The output of the bridge is then amplified with a differential operational amplifier using shunt feedback to set the gain.

2.3.6 Schmitt trigger

The Schmitt trigger is used to change variable signals into crisp square-wave type signals for use in digital or switching circuits. For example, a sine wave fed into a Schmitt trigger will emerge as a square wave with the same frequency as the input signal. Figure 2.22 shows a simple Schmitt trigger circuit utilizing an operational amplifier.

The output of this circuit will be either saturated positive or saturated negative due to the high gain of the amplifier. The trigger points are defined as the upper and lower trigger points (UTP and LTP) respectively. The output signal from an inductive type distributor or a crank

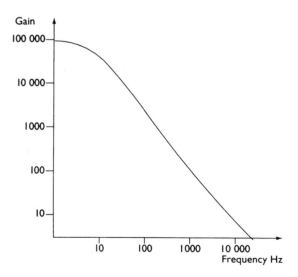

Figure 2.19 Frequency response of a 741 amplifier

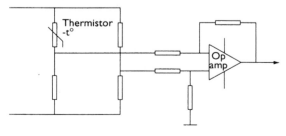

Figure 2.21 Bridge and amplifier circuit

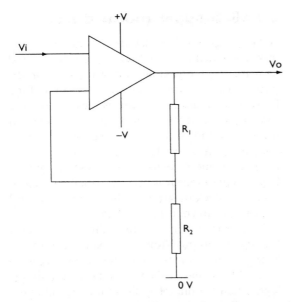

Figure 2.22 Schmitt trigger circuit utilizing an operational amplifier

position sensor on a motor vehicle will need to be passed through a Schmitt trigger. This will ensure that either further processing is easier, or switching is positive. Schmitt triggers can be purchased as integrated circuits in their own right or as part of other ready-made applications.

2.3.7 Timers

In its simplest form, a timer can consist of two components, a resistor and a capacitor. When the capacitor is connected to a supply via the resistor, it is accepted that it will become fully charged in $5CR$ seconds, where R is the resistor value in ohms and C is the capacitor value in farads. The time constant of this circuit is CR, often-denoted τ

The voltage across the capacitor (V_c), can be calculated as follows:

$$V_c = V(1 - e^{-t/CR})$$

where V = supply voltage; t = time in seconds; C = capacitor value in farads; R = resistor value in ohms; e = exponential function.

These two components with suitable values can be made to give almost any time delay, within reason, and to operate or switch off a circuit using a transistor. Figure 2.23. shows an example of a timer circuit using this technique.

2.3.8 Filters

A filter that prevents large particles of contaminates reaching, for example, a fuel injector is an

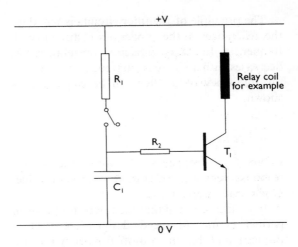

Figure 2.23 Example of a timer circuit

easy concept to grasp. In electronic circuits the basic idea is just the same except the particle size is the frequency of a signal. Electronic filters come in two main types. A low pass filter, which blocks high frequencies, and a high pass filter, which blocks low frequencies. Many variations of these filters are possible to give particular frequency response characteristics, such as band pass or notch filters. Here, just the basic design will be considered. The filters may also be active, in that the circuit will include amplification, or passive, when the circuit does not. Figure 2.24 shows the two main passive filter circuits.

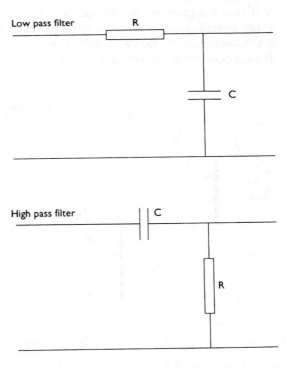

Figure 2.24 Low pass and high pass filter circuits

The principle of the filter circuits is based on the reactance of the capacitors changing with frequency. In fact, capacitive reactance, X_c decreases with an increase in frequency. The roll-off frequency of a filter can be calculated as shown:

$$f = \frac{1}{2\pi RC}$$

where f = frequency at which the circuit response begins to roll off; R = resistor value; C = capacitor value.

It should be noted that the filters are far from perfect (some advanced designs come close though), and that the roll-off frequency is not a clear-cut 'off' but the point at which the circuit response begins to fall.

2.3.9 Darlington pair

A Darlington pair is a simple combination of two transistors that will give a high current gain, of typically several thousand. The transistors are usually mounted on a heat sink and, overall, the device will have three terminals marked as a single transistor – base, collector and emitter. The input impedance of this type of circuit is of the order of 1 MΩ, hence it will not load any previous part of a circuit connected to its input. Figure 2.25 shows two transistors connected as a Darlington pair.

The Darlington pair configuration is used for many switching applications. A common use of a Darlington pair is for the switching of the coil primary current in the ignition circuit.

2.3.10 Stepper motor driver

A later section gives details of how a stepper motor works. In this section it is the circuit used to drive the motor that is considered. For the purpose of this explanation, a driver circuit for a four-phase unipolar motor is described. The function of a stepper motor driver is to convert the digital and 'wattless' (no significant power content) process control signals into signals to operate the motor coils. The process of controlling a stepper motor is best described with reference to a block diagram of the complete control system, as shown in Figure 2.26.

The process control block shown represents the signal output from the main part of an engine management ECU (electronic control unit). The signal is then converted in a simple logic circuit to suitable pulses for controlling the motor. These pulses will then drive the motor via a power stage. Figure 2.27 shows a simplified circuit of a power stage designed to control four motor windings.

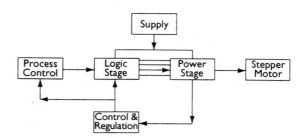

Figure 2.26 Stepper motor control system

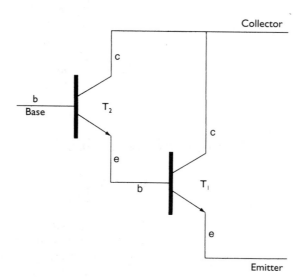

Figure 2.25 Darlington pair

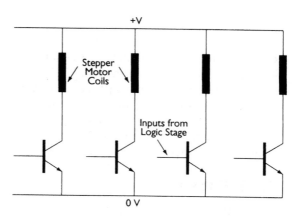

Figure 2.27 Stepper motor driver circuit (power stage)

2.3.11 Digital to analogue conversion

Conversion from digital signals to an analogue signal is a relatively simple process. When an operational amplifier is configured with shunt feedback the input and feedback resistors determine the gain.

$$\text{Gain} = \frac{-R_f}{R_I}$$

If the digital-to-analogue converter circuit is connected as shown in Figure 2.28 then the 'weighting' of each input line can be determined by choosing suitable resistor values. In the case of the four-bit digital signal, as shown, the most significant bit will be amplified with a gain of one. The next bit will be amplified with a gain of 1/2, the next bit 1/4 and, in this case, the least significant bit will be amplified with a gain of 1/8. This circuit is often referred to as an adder. The output signal produced is therefore a voltage proportional to the value of the digital input number.

The main problem with this system is that the accuracy of the output depends on the tolerance of the resistors. Other types of digital-to-analogue converter are available, such as the R2R ladder network, but the principle of operation is similar to the above description.

2.3.12 Analogue to digital conversion

The purpose of this circuit is to convert an analogue signal, such as that received from a temperature thermistor, into a digital signal for use

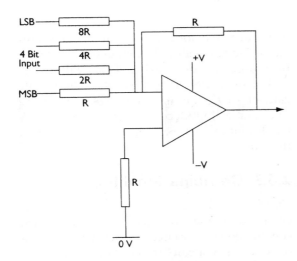

Figure 2.28 Digital-to-analogue converter

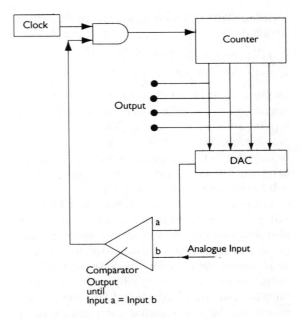

Figure 2.29 Ramp analogue-to-digital converter

by a computer or a logic system. Most systems work by comparing the output of a digital-to-analogue converter (DAC) with the input voltage. Figure 2.29 is a ramp analogue-to-digital converter (ADC). This type is slower than some others but is simple in operation. The output of a binary counter is connected to the input of the DAC, the output of which will be a ramp. This voltage is compared with the input voltage and the counter is stopped when the two are equal. The count value is then a digital representation of the input voltage. The operation of the other digital components in this circuit will be explained in the next section.

ADCs are available in IC form and can work to very high speeds at typical resolutions of one part in 4096 (12-bit word). The speed of operation is critical when converting variable or oscillating input signals. As a rule, the sampling rate must be at least twice the frequency of the input signal.

2.4 Digital electronics

2.4.1 Introduction to digital circuits

With some practical problems, it is possible to express the outcome as a simple yes/no or true/false answer. Let us take a simple example: if the answer to either the first or the second question is 'yes', then switch on the brake warning light, if both answers are 'no' then switch it off.

1. Is the handbrake on?
2. Is the level in the brake fluid reservoir low?

In this case, we need the output of an electrical circuit to be 'on' when *either one or both* of the inputs to the circuit are 'on'. The inputs will be via simple switches on the handbrake and in the brake reservoir. The digital device required to carry out the above task is an OR gate, which will be described in the next section.

Once a problem can be described in logic states then a suitable digital or logic circuit can also determine the answer to the problem. Simple circuits can also be constructed to hold the logic state of their last input – these are, in effect, simple forms of 'memory'. By combining vast quantities of these basic digital building blocks, circuits can be constructed to carry out the most complex tasks in a fraction of a second. Due to integrated circuit technology, it is now possible to create hundreds of thousands if not millions of these basic circuits on one chip. This has given rise to the modern electronic control systems used for vehicle applications as well as all the countless other uses for a computer.

In electronic circuits, true/false values are assigned voltage values. In one system, known as TTL (transistor transistor logic), true or logic '1', is represented by a voltage of 3.5 V and false or logic '0', by 0 V

2.4.2 Logic gates

The symbols and truth tables for the basic logic gates are shown in Figure 2.30. A truth table is used to describe what combination of inputs will produce a particular output.

The AND gate will only produce an output of '1' if both inputs (or all inputs as it can have more than two) are also at logic '1'. Output is '1' when inputs A AND B are '1'.

The OR gate will produce an output when either A OR B (OR both), are '1'. Again more than two inputs can be used.

A NOT gate is a very simple device where the output will always be the opposite logic state from the input. In this case A is NOT B and, of course, this can only be a single input and single output device.

The AND and OR gates can each be combined with the NOT gate to produce the NAND and NOR gates, respectively. These two gates have been found to be the most versatile and are

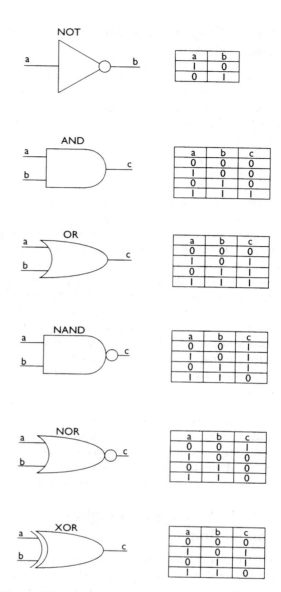

Figure 2.30 Logic gates and truth tables

used extensively for construction of more complicated logic circuits. The output of these two is the inverse of the original AND and OR gates.

The final gate, known as the exclusive OR gate, or XOR, can only be a two-input device. This gate will produce an output only when A OR B is at logic '1' but not when they are both the same.

2.4.3 Combinational logic

Circuits consisting of many logic gates, as described in the previous section, are called combinational logic circuits. They have no memory or counter circuits and can be represented by a simple block diagram with N inputs and Z outputs. The first stage in the design process of creating a combinational logic circuit is to define the required relationship between the inputs and outputs.

Let us consider a situation where we need a circuit to compare two sets of three inputs and, if they are not the same, to provide a single logic '1' output. This is oversimplified, but could be used to compare the actions of a system with twin safety circuits, such as an ABS electronic control unit. The logic circuit could be made to operate a warning light if a discrepancy exists between the two safety circuits. Figure 2.31 shows the block diagram and one suggestion for how this circuit could be constructed.

Referring to the truth tables for basic logic circuits, the XOR gate seemed the most appropriate to carry out the comparison: it will only produce a '0' output when its inputs are the same. The outputs of the three XOR gates are then supplied to a three-input OR gate which, providing all its inputs are '0', will output '0'. If any of its inputs change to '1' the output will change to '1' and the warning light will be illuminated.

Other combinations of gates can be configured to achieve any task. A popular use is to construct an adder circuit to perform addition of two binary numbers. Subtraction is achieved by converting the subtraction to addition, $(4 - 3 = 1$ is the same as $4 + [-3] = 1)$. Adders are also used to multiply and divide numbers, as this is actually repeated addition or repeated subtraction.

2.4.4 Sequential logic

The logic circuits discussed above have been simple combinations of various gates. The output of each system was only determined by the present inputs. Circuits that have the ability to memorize previous inputs or logic states, are known as sequential logic circuits. In these circuits the sequence of past inputs determines the current output. Because sequential circuits store information after the inputs are removed, they are the basic building blocks of computer memories.

Basic memory circuits are called bistables as they have two steady states. They are, however, more often referred to as flip-flops.

There are three main types of flip-flop: an RS memory, a D-type flip-flop and a JK-type flip-flop. The RS memory can be constructed by using two NAND and two NOT gates, as shown in Figure 2.32 next to the actual symbol. If we start with both inputs at '0' and output X is at '1' then as output X goes to the input of the other NAND gate its output will be '0'. If input A is now changed to '1' output X will change to '0', which will in turn cause output Y to go to '1'. The outputs have changed over. If A now reverts to '1' the outputs will remain the same until B goes to '1', causing the outputs to change over again. In this way the circuit remembers which input was last at '1'. If it was A then X is '0' and Y is '1', if it was B then X is '1' and Y is '0'. This is the simplest form of memory circuit. The RS stands for set-reset. The second type of flip-flop is the D-type. It has two inputs labelled CK (for clock) and D; the outputs are labelled Q and $\bar{Q}$. These are often called 'Q' and 'not Q'. The output Q takes on the logic state of D when the clock pulse is applied. The JK-type flip-flop is a combination of the previous two flip-flops. It has two main inputs like the RS type but now labelled J and K and it is controlled by a clock pulse like the D-type. The outputs are again 'Q' and 'not Q'. The circuit remembers the last input to change in the same way as the RS memory

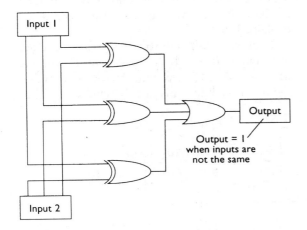

Figure 2.31 Combinational logic to compare inputs

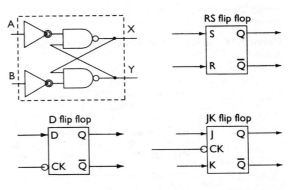

Figure 2.32 D-type and JK-type flip-flop (bistables). A method using NAND gates to make an RS type is also shown

did. The main difference is that the change-over of the outputs will only occur on the clock pulse. The outputs will also change over if both J and K are at logic '1', this was not allowed in the RS type.

2.4.5 Timers and counters

A device often used as a timer is called a 'mono-stable' as it has only one steady state. Accurate and easily controllable timer circuits are made using this device. A capacitor and resistor combination is used to provide the delay. Figure 2.33 shows a monostable timer circuit with the resistor and capacitor attached.

Every time the input goes from 0 to 1 the output Q, will go from 0 to 1 for *t* seconds. The other output Q̄ will do the opposite. Many variations of this type of timer are available. The time delay '*t*' is usually $0.7RC$.

Counters are constructed from a series of bistable devices. A binary counter will count clock pulses at its input. Figure 2.34 shows a four-bit counter constructed from D-type flip-flops. These counters are called 'ripple through' or non-synchronous, because the change of state

ripples through from the least significant bit and the outputs do not change simultaneously. The type of triggering is important for the system to work as a counter. In this case, negative edge triggering is used, which means that the devices change state when the clock pulse changes from '1' to '0'. The counters can be configured to count up or down.

In low-speed applications, 'ripple through' is not a problem but at higher speeds the delay in changing from one number to the next may be critical. To get over this asynchronous problem a synchronous counter can be constructed from JK-type flip-flops, together with some simple combinational logic. Figure 2.35 shows a four-bit synchronous up-counter.

With this arrangement, all outputs change simultaneously because the combinational logic looks at the preceding stages and sets the JK inputs to a '1' if a toggle is required. Counters are also available 'ready made' in a variety of forms including counting to non-binary bases in the up or down mode.

2.4.6 Memory circuits

Electronic circuits constructed using flip-flops as described above are one form of memory. If the flip-flops are connected as shown in Figure 2.36 they form a simple eight-bit word memory. This, however, is usually called a register rather that memory.

Eight bits (binary digits) are often referred to as one byte. Therefore, the register shown has a memory of one byte. When more than one register is used, an address is required to access or store the data in a particular register. Figure 2.37 shows a block diagram of a four-byte memory system. Also shown is an address bus, as each area of this memory is allocated a unique address. A control bus is also needed as explained below.

In order to store information (write), or to get information (read), from the system shown, it is necessary first to select the register containing the required data. This task is achieved by allo-

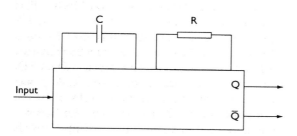

Figure 2.33 Monostable timer circuit with a resistor and capacitor attached

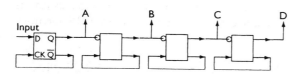

Figure 2.34 Four-bit counter constructed from D-type flip-flops

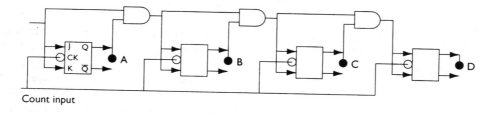

Count input

Figure 2.35 Four-bit synchronous up-counter

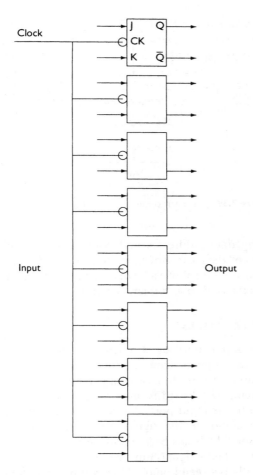

Figure 2.36 Eight-bit register using flip-flops

cating an address to each register. The address bus in this example will only need two lines to select one of four memory locations using an address decoder. The addresses will be binary;

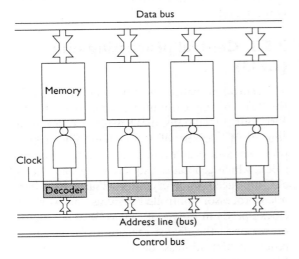

Figure 2.37 Four-byte memory with address lines and decoders

'00', '01', '10' and '11' such that if '11' is on the address bus the simple combinational logic (AND gate), will only operate one register, usually via a pin marked CS or chip select. Once a register has been selected, a signal from the control bus will 'tell' the register whether to read from or write to, the data bus. A clock pulse will ensure all operations are synchronized.

This example may appear to be a complicated way of accessing just four bytes of data. In fact, it is the principle of this technique, that is important, as the same method can be applied to access memory chips containing vast quantities of data. Note that with an address bus of two lines, 4 bytes could be accessed ($2^2=4$). If the number of address lines was increased to eight, then 256 bytes would be available ($2^8=256$). Ten address lines will address one kilobyte of data and so on.

The memory, which has just been described, together with the techniques used to access the data are typical of most computer systems. The type of memory is known as random access memory (RAM). Data can be written to and read from this type of memory but note that the memory is volatile, in other words it will 'forget' all its information when the power is switched off!

Another type of memory that can be 'read from' but not 'written to' is known as read only memory (ROM). This type of memory has data permanently stored and is not lost when power is switched off. There are many types of ROM, which hold permanent data, but one other is worthy of a mention, that is EPROM. This stands for erasable, programmable, read only memory. Its data can be changed with special equipment (some are erased with ultraviolet light), but for all other purposes its memory is permanent. In an engine management electronic control unit (ECU), operating data and a controlling program are stored in ROM, whereas instantaneous data (engine speed, load, temperature etc.) are stored in RAM.

2.4.7 Clock or astable circuits

Control circuits made of logic gates and flip-flops usually require an oscillator circuit to act as a clock. Figure 2.38 shows a very popular device, the 555-timer chip.

The external resistors and capacitor will set the frequency of the output due to the charge time of the capacitor. Comparators inside the chip cause the output to set and reset the

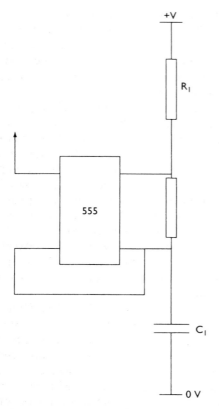

Figure 2.38 A stable circuit using a 555 IC

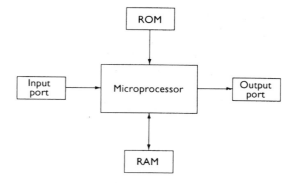

Figure 2.39 Basic microcomputer block diagram

memory (a flip-flop) as the capacitor is charged and discharged alternately to 1/3 and 2/3 of the supply voltage. The output of the chip is in the form of a square wave signal. The chip also has a reset pin to stop or start the output.

2.5 Microprocessor systems

2.5.1 Introduction

The advent of the microprocessor has made it possible for tremendous advances in all areas of electronic control, not least of these in the motor vehicle. Designers have found that the control of vehicle systems – which is now required to meet the customers' needs and the demands of regulations – has made it necessary to use computer control. Figure 2.39 shows a block diagram of a microcomputer containing the four major parts. These are the input and output ports, some form of memory and the CPU or central processing unit (microprocessor). It is likely that some systems will incorporate more memory chips and other specialized components. Three buses car-

rying data, addresses and control signals link each of the parts shown. If all the main elements as introduced above are constructed on one chip, it is referred to as a microcontroller.

2.5.2 Ports

The input port of a microcomputer system receives signals from peripherals or external components. In the case of a personal computer system, a keyboard is one provider of information to the input port. A motor vehicle application could be the signal from a temperature sensor, which has been analogue to digital converted. These signals must be in digital form and usually between 0 and 5 V. A computer system, whether a PC or used on a vehicle, will have several input ports.

The output port is used to send binary signals to external peripherals. A personal computer may require output to a monitor and printer, and a vehicle computer may, for example, output to a circuit that will control the switching of the ignition coil.

2.5.3 Central processing unit (CPU)

The central processing unit or microprocessor is the heart of any computer system. It is able to carry out calculations, make decisions and be in control of the rest of the system. The microprocessor works at a rate controlled by a system clock, which generates a square wave signal usually produced by a crystal oscillator. Modern microprocessor controlled systems can work at clock speeds in excess of 300 MHz. The microprocessor is the device that controls the computer via the address, data and control buses. Many vehicle systems use microcontrollers and these are discussed later in this section.

2.5.4 Memory

The way in which memory actually works was discussed briefly in an earlier section. We will now look at how it is used in a microprocessor controlled system. Memory is the part of the system that stores both the instructions for the microprocessor (the program) and any data that the microprocessor will need to execute the instructions.

It is convenient to think of memory as a series of pigeon-holes, which are each able to store data. Each of the pigeon-holes must have an address, simply to distinguish them from each other and so that the microprocessor will 'know' where a particular piece of information is stored. Information stored in memory, whether it is data or part of the program, is usually stored sequentially. It is worth noting that the microprocessor reads the program instructions from sequential memory addresses and then carries out the required actions. In modern PC systems, memories can be of 128 megabytes or more! Vehicle microprocessor controlled systems do not require as much memory but mobile multimedia systems will.

2.5.5 Buses

A computer system requires three buses to communicate with or control its operations. The three buses are the data bus, address bus and the control bus. Each one of these has a particular function within the system.

The data bus is used to carry information from one part of the computer to another. It is known as a bi-directional bus as information can be carried in any direction. The data bus is generally 4, 8, 16 or 32 bits wide. It is important to note that only one piece of information at a time may be on the data bus. Typically, it is used to carry data from memory or an input port to the microprocessor, or from the microprocessor to an output port. The address bus must first address the data that is accessed.

The address bus starts in the microprocessor and is a unidirectional bus. Each part of a computer system, whether memory or a port, has a unique address in binary format. Each of these locations can be addressed by the microprocessor and the held data placed on the data bus. The address bus, in effect, tells the computer which part of its system is to be used at any one moment.

Finally, the control bus, as the name suggests, allows the microprocessor, in the main, to control the rest of the system. The control bus may have up to 20 lines but has four main control signals. These are read, write, input/output request and memory request. The address bus will indicate which part of the computer system is to operate at any given time and the control bus will indicate how that part should operate. For example, if the microprocessor requires information from a memory location, the address of the particular location is placed on the address bus. The control bus will contain two signals, one memory request and one read signal. This will cause the contents of the memory at one particular address to be placed on the data bus. These data may then be used by the microprocessor to carry out another instruction.

2.5.6 Fetch-execute sequence

A microprocessor operates at very high speed by the system clock. Broadly speaking, the microprocessor has a simple task. It has to fetch an instruction from memory, decode the instruction and then carry out or execute the instruction. This cycle, which is carried out relentlessly (even if the instruction is to do nothing), is known as the fetch–execute sequence. Earlier in this section it was mentioned that most instructions are stored in consecutive memory locations such that the microprocessor, when carrying out the fetch-execute cycle, is accessing one instruction after another from sequential memory locations.

The full sequence of events may be very much as follows.

- The microprocessor places the address of the next memory location on the address bus.
- At the same time a memory read signal is placed on the control bus.
- The data from the addressed memory location are placed on the data bus.
- The data from the data bus are temporarily stored in the microprocessor.
- The instruction is decoded in the microprocessor internal logic circuits.
- The 'execute' phase is now carried out. This can be as simple as adding two numbers inside the microprocessor or it may require data to be output to a port. If the latter is the case, then the address of the port will be placed on the address bus and a control bus 'write' signal is generated.

The fetch and decode phase will take the same time for all instructions, but the execute phase will vary depending on the particular instruction. The actual time taken depends on the complexity of the instructions and the speed of the clock frequency to the microprocessor.

2.5.7 A typical microprocessor

Figure 2.40 shows the architecture of a simplified microprocessor, which contains five registers, a control unit and the arithmetic logic unit (ALU).

The operation code register (OCR) is used to hold the op-code of the instruction currently being executed. The control unit uses the contents of the OCR to determine the actions required.

The temporary address register (TAR) is used to hold the operand of the instruction if it is to be treated as an address. It outputs to the address bus.

The temporary data register (TDR) is used to hold data, which are to be operated on by the ALU, its output is therefore to an input of the ALU.

The ALU carries out additions and logic operations on data held in the TDR and the accumulator.

The accumulator (AC) is a register, which is accessible to the programmer and is used to keep such data as a running total.

The instruction pointer (IP) outputs to the address bus so that its contents can be used to locate instructions in the main memory. It is an incremental register, meaning that its contents can be incremented by one directly by a signal from the control unit.

Execution of instructions in a microprocessor proceeds on a step by step basis, controlled by signals from the control unit via the internal control bus. The control unit issues signals as it receives clock pulses.

The process of instruction execution is as follows:

1. Control unit receives the clock pulse.
2. Control unit sends out control signals.
3. Action is initiated by the appropriate components.
4. Control unit receives the clock pulse.
5. Control unit sends out control signals.
6. Action is initiated by the appropriate components.
   And so on.

A typical sequence of instructions to add a number to the one already in the accumulator is as follows:

1. IP contents placed on the address bus.
2. Main memory is read and contents placed on the data bus.
3. Data on the data bus are copied into OCR.
4. IP contents incremented by one.
5. IP contents placed on the address bus.
6. Main memory is read and contents placed on the data bus.
7. Data on the data bus are copied into TDR.
8. ALU adds TDR and AC and places result on the data bus.
9. Data on the data bus are copied into AC.
10. IP contents incremented by one.

The accumulator now holds the running total. Steps 1 to 4 are the fetch sequence and steps 5 to 10 the execute sequence. If the full fetch–execute sequence above was carried out, say, nine times this would be the equivalent of multiplying the number in the accumulator by 10! This gives an indication as to just how basic the level of operation is within a computer.

Now to take a giant step forwards. It is possible to see how the microprocessor in an engine management ECU can compare a value held in a RAM location with one held in a ROM location. The result of this comparison of, say, instantaneous engine speed in RAM and a pre-programmed figure in ROM, could be to set the ignition timing to another pre-programmed figure.

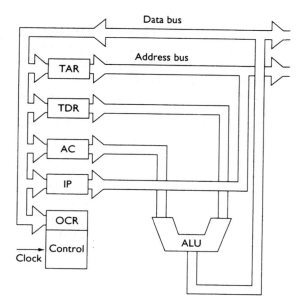

Figure 2.40 Simplified microprocessor with five registers, a control unit and the ALU or arithmetic logic unit

2.5.8 Microcontrollers

As integration technology advanced it became possible to build a complete computer on a single chip. This is known as a microcontroller. The microcontroller must contain a microprocessor, memory (RAM and/or ROM), input ports and output ports. A clock is included in some cases.

A typical family of microcontrollers is the 'Intel' 8051 series. These were first introduced in 1980 but are still a popular choice for designers. A more up-to-date member of this family is the 87C528 microcontroller which has 32K EPROM, 512 bytes of RAM, three (16 bit) timers, four I/O ports and a built in serial interface.

Microcontrollers are available such that a pre-programmed ROM may be included. These are usually made to order and are only supplied to the original customer. Figure 2.41 shows a simplified block diagram of the 8051 microcontroller.

2.5.9 Testing microcontroller systems

If a microcontroller system is to be constructed with the program (set of instructions) permanently held in ROM, considerable testing of the program is required. This is because, once the microcontroller goes into production, tens if not hundreds of thousands of units will be made. A hundred thousand microcontrollers with a hard-wired bug in the program would be a very expensive error!

There are two main ways in which software for a microcontroller can be tested. The first, which is used in the early stages of program development, is by a simulator. A simulator is a program that is executed on a general purpose computer and which simulates the instruction set of the microcontroller. This method does not test the input or output devices.

The most useful aid for testing and debugging

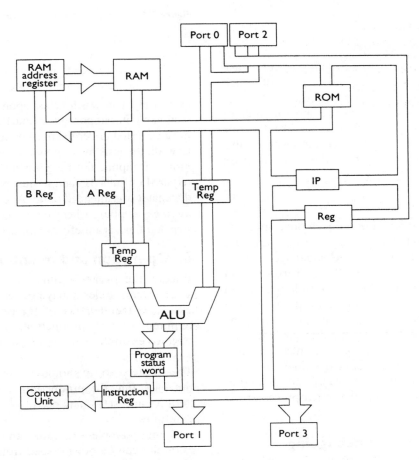

Figure 2.41 Simplified block diagram of the 8051 microcontroller

is an in-circuit emulator. The emulator is fitted in the circuit in place of the microcontroller and is, in turn, connected to a general purpose computer. The microcontroller program can then be tested in conjunction with the rest of the hardware with which it is designed to work. The PC controls the system and allows different procedures to be tested. Changes to the program can easily be made at this stage of the development.

2.5.10 Programming

To produce a program for a computer, whether it is for a PC or a microcontroller-based system is generally a six-stage process.

1. Requirement analysis

This seeks to establish whether in fact a computer-based approach is in fact the best option. It is, in effect, a feasibility study.

2. Task definition

The next step is to produce a concise and unambiguous description of what is to be done. The outcome of this stage is to produce the functional specifications of the program.

3. Program design

The best approach here is to split the overall task into a number of smaller tasks. Each of which can be split again and so on if required. Each of the smaller tasks can then become a module of the final program. A flow chart like the one shown in Figure 2.42 is often the result of this stage, as such charts show the way sub-tasks interrelate.

4. Coding

This is the representation of each program module in a computer language. The programs are often written in a high-level language such as Turbo C, Pascal or even Basic. Turbo C and C++ are popular as they work well in program modules and produce a faster working program than many of the other languages. When the source code has been produced in the high-level language, individual modules are linked and then compiled into machine language – in other words a language consisting of just '1s' and '0s' and in the correct order for the microprocessor to understand.

5. Validation and debugging

Once the coding is completed it must be tested

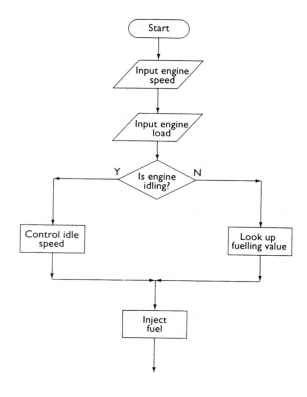

Figure 2.42 Computer program flowchart

extensively. This was touched upon in the previous section but it is important to note that the program must be tested under the most extreme conditions. Overall, the tests must show that, for an extensive range of inputs, the program must produce the required outputs. In fact, it must prove that it can do what it was intended to do! A technique known as single stepping where the program is run one step at a time, is a useful aid for debugging.

6. Operation and maintenance

Finally, the program runs and works but, in some cases, problems may not show up for years and some maintenance of the program may be required for new production; the Millennium bug, for example!

The six steps above should not be seen in isolation, as often the production of a program is iterative and steps may need to be repeated several times.

Some example programs and source code examples can be downloaded from my web site (the URL address is given in the preface).

2.6 Measurement

2.6.1 What is measurement

Measurement is the act of measuring physical quantities to obtain data that are transmitted to recording/display devices and/or to control devices. The term 'instrumentation' is often used in this context to describe the science and technology of the measurement system.

The first task of any measurement system is to translate the physical value to be measured, known as the measurand, into another physical variable, which can be used to operate the display or control device. In the motor vehicle system, the majority of measurands are converted into electrical signals. The sensors that carry out this conversion are often called transducers.

2.6.2 A measurement system

A complete measurement system will vary depending on many factors but many vehicle systems will consist of the following stages.

1. Physical variable.
2. Transduction.
3. Electrical variable.
4. Signal processing.
5. A/D conversion.
6. Signal processing.
7. Display or use by a control device.

Some systems may not require Steps 5 and 6. As an example, consider a temperature measurement system with a digital display. This will help to illustrate the above seven-step process.

1. Engine water temperature.
2. Thermistor.
3. Resistance decreases with temperature increase.
4. Linearization.
5. A/D conversion.
6. Conversion to drive a digital display.
7. Digital read-out as a number or a bar graph.

Figure 2.43 shows a complete measurement system as a block diagram.

2.6.3 Sources of error in measurement

An important question to ask when designing an instrumentation or measurement system is:

What effect will the measurement system have on the variable being measured?

Consider the water temperature measurement example discussed in the previous section. If the transducer is immersed in a liquid, which is at a higher temperature than the surroundings, then the transducer will conduct away some of the heat and lower the temperature of the liquid. This effect is likely to be negligible in this example, but in others, it may not be so small. However, even in this case it is possible that, due to the fitting of the transducer, the water temperature surrounding the sensor will be lower than the rest of the system. This is known as an invasive measurement. A better example may be that if a device is fitted into a petrol pipe to measure flow rate, then it is likely that the device itself will restrict the flow in some way. Returning to the previous example of the temperature transducer it is also possible that the very small current passing through the transducer will have a heating effect.

Errors in a measurement system affect the overall accuracy. Errors are also not just due to invasion of the system. There are many terms associated with performance characteristics of transducers and measurement systems. Some of these terms are considered below.

Accuracy

A descriptive term meaning how close the measured value of a quantity is to its actual value. Accuracy is expressed usually as a maximum error. For example, if a length of about 30 cm is measured with an ordinary wooden ruler then the error may be up to 1 mm too high or too low. This is quoted as an accuracy of +/− 1 mm. This may also be expressed as a percentage which in this case would be 0.33%. An electrical meter is often quoted as the maximum error being a percentage of full-scale deflection. The maximum error or accuracy is contributed to by a number of factors explained below.

Resolution

The 'fineness' with which a measurement can be made. This must be distinguished from accuracy. If a quality steel ruler were made to a very high standard but only had markings or graduations

Figure 2.43 Measrement system block diagram

of one per centimetre it would have a low resolution even though the graduations were very accurate.

Hysteresis

For a given value of the measurand, the output of the system depends on whether the measurand has acquired its value by increasing or decreasing from its previous value. You can prove this next time you weigh yourself on some scales. If you step on gently you will 'weigh less' than if you jump on and the scales overshoot and then settle.

Repeatability

The closeness of agreement of the readings when a number of consecutive measurements are taken of a chosen value during full range traverses of the measurand. If a 5 kg set of weighing scales was increased from zero to 5 kg in 1 kg steps a number of times, then the spread of readings is the repeatability. It is often expressed as a percentage of full scale.

Zero error or zero shift

The displacement of a reading from zero when no reading should be apparent. An analogue electrical test meter, for example, often has some form of adjustment to zero the needle.

Linearity

The response of a transducer is often non-linear (see the response of a thermistor in the next section). Where possible, a transducer is used in its linear region. Non-linearity is usually quoted as a percentage over the range in which the device is designed to work.

Sensitivity or scale factor

A measure of the incremental change in output for a given change in the input quantity. Sensitivity is quoted effectively as the slope of a graph in the linear region. A figure of 0.1 V/°C for example, would indicate that a system would increase its output by 0.1 V for every 1°C increase in temperature of the input.

Response time

The time taken by the output of a system to respond to a change in the input. A system measuring engine oil pressure needs a faster response time than a fuel tank quantity system. Errors in the output will be apparent if the measurement is taken quicker than the response time.

Looking again at the seven steps involved in a measurement system will highlight the potential sources of error.

1. Invasive measurement error.
2. Non-linearity of the transducer.
3. Noise in the transmission path.
4. Errors in amplifiers and other components.
5. Quantization errors when digital conversion takes place.
6. Display driver resolution.
7. Reading error of the final display.

Many good textbooks are available for further study, devoted solely to the subject of measurement and instrumentation. This section is intended to provide the reader with a basic grounding in the subject.

2.7 Sensors and actuators

2.7.1 Thermistors

Thermistors are the most common device used for temperature measurement on a motor vehicle. The principle of measurement is that a change in temperature will cause a change in resistance of the thermistor, and hence an electrical signal proportional to the measurand can be obtained.

Most thermistors in common use are of the negative temperature coefficient (NTC) type. The actual response of the thermistors can vary but typical values for those used in motor vehicles will vary from several kilohms at 0°C to a few hundred ohms at 100°C. The large change in resistance for a small change in temperature makes the thermistor ideal for most vehicles' uses. It can also be easily tested with simple equipment.

Thermistors are constructed of semiconductor materials such as cobalt or nickel oxides. The change in resistance with a change in temperature is due to the electrons being able to break free from the covalent bonds more easily at higher temperatures; this is shown in Figure 2.44(i). A thermistor temperature measuring system can be very sensitive due to large changes in resistance with a relatively small change in temperature. A simple circuit to provide a varying voltage signal proportional to temperature is shown in Figure 2.44(ii). Note the supply must be constant and the current flowing must not significantly heat the thermistor. These could both be sources of error. The

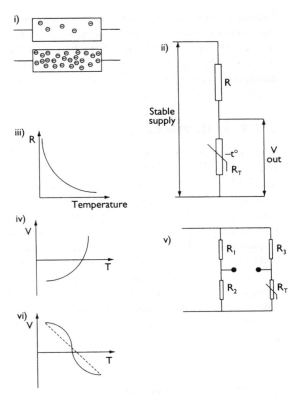

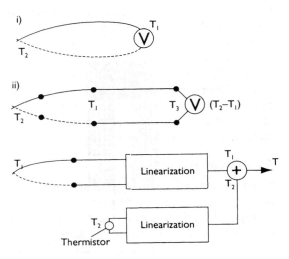

$$R_t = Ae^{(B/T)}$$

where R_t = resistance of the thermistor, T = absolute temperature, B = characteristic temperature of the thermistor (typical value 3000 K), A = constant of the thermistor.

For the bridge configuration as shown V_o is given by:

$$V_o = V_s\left(\frac{R_2}{R_2 + R_1} - \frac{R_1}{R_1 + R_3}\right)$$

By choosing suitable resistor values the output of the bridge will be as shown. This is achieved by substituting the known values of R_t at three temperatures and deciding that, for example, $V_o = 0$ at 0 °C, $V_o = 0.5$ V at 50 °C and $V_o = 1$ V at 100 °C.

2.7.2 Thermocouples

If two different metals are joined together at two junctions, the thermoelectric effect known as the Seebeck effect takes place. If one junction is at a higher temperature than the other junction, then this will be registered on the meter. This is the basis for the sensor known as the thermocouple. Figure 2.45 shows the thermocouple principle and appropriate circuits. Notice that the thermocouple measures a difference in temperature that is $T_1 - T_2$. To make the system of any practical benefit then T_1 must be kept at a known temperature. The lower figure shows a practical circuit in which, if the connections to the meter are at the same temperature, the two voltages produced at these junctions will cancel out. Cold junction compensation circuits can be made to

Figure 2.44 (i) How a thermistor changes resistance; (ii) circuit to provide a varying voltage signal proportional to temperature; (iii) resistance against temperature curve for a thermistor; (iv) non-linearity to compensate partially for the thermistor's non-linearity; (v) bridge circuit to achieve maximum linearity; (vi) final output signal

temperature of a typical thermistor will increase by 1 °C for each 1.3 mW of power dissipated. Figure 2.44(iii) shows the resistance against temperature curve for a thermistor. This highlights the main problem with a thermistor, its non-linear response. Using a suitable bridge circuit, it is possible to produce non-linearity that will partially compensate for the thermistor's non-linearity. This is represented by Figure 2.44(iv). The combination of these two responses is also shown. The optimum linearity is achieved when the mid points of the temperature and the voltage ranges lie on the curve. Figure 2.44(v) shows a bridge circuit for this purpose. It is possible to work out suitable values for R_1, R_2 and R_3. This then gives the more linear output as represented by Figure 2.44(vi). The voltage signal can now be A/D converted if necessary, for further use.

The resistance R_t of a thermistor decreases non-linearly with temperature according to the relationship:

Figure 2.45 Thermocouple principle and circuits

compensate for changes in temperature of T_1. These often involve the use of a thermistor circuit.

Thermocouples are in general used for measuring high temperatures. A thermocouple combination of 70% platinum and 30% rhodium alloy in a junction with 94% platinum and 6% rhodium alloy, is known as a type B thermocouple and has a useful range of 0–1500°C. Vehicle applications are in areas such as exhaust gas and turbo charger temperature measurement.

2.7.3 Inductive sensors

Inductive-type sensors are used mostly for measuring speed and position of a rotating component. They work on the very basic principle of electrical induction (a changing magnetic flux will induce an electromotive force in a winding). Figure 2.46 shows the inductive sensor principle and a typical device used as a crankshaft speed and position sensor.

The output voltage of most inductive-type sensors approximates to a sine wave. The amplitude of this signal depends on the rate of change of flux. This is determined mostly by the original design: by the number of turns, magnet strength and the gap between the sensor and the rotating component. Once in use though, the output voltage increases with the speed of rotation. In the majority of applications, it is the frequency of the signal that is used. The most common way of converting the output of an inductive sensor to a useful signal is to pass it through a Schmitt trigger circuit. This produces constant amplitude but a variable frequency square wave.

In some cases the output of the sensor is used to switch an oscillator on and off or quench the oscillations. A circuit for this is shown in Figure 2.47. The oscillator produces a very high frequency of about 4 MHz and this when switched on and off by the sensor signal and then filtered, produces a square wave. This system has a good resistance to interference.

2.7.4 Hall effect

The Hall effect was first noted by a Dr E.H. Hall: it is a simple principle, as shown in Figure 2.48. If a certain type of crystal is carrying a current in a transverse magnetic field then a voltage will be produced at right angles to the supply current. The magnitude of the voltage is proportional to the supply current and to the magnetic field strength. Figure 2.49 shows part of a Bosch distributor, the principle of which is to 'switch' the magnetic field on and off using a chopper plate. The output of this sensor is almost a square wave with constant amplitude.

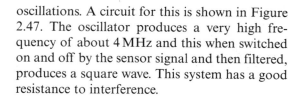

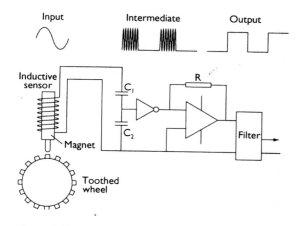

Figure 2.47 Inductive sensor and quenched oscillator circuit

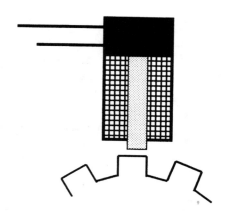

Figure 2.46 Inductive sensor

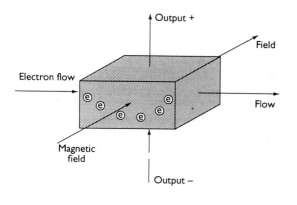

Figure 2.48 Hall effect principle

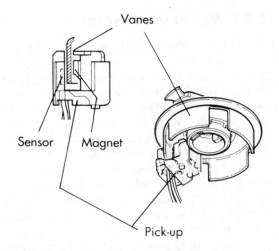

Figure 2.49 Hall effect sensor used in a distributor

The Hall effect can also be used to detect current flowing in a cable. The magnetic field produced around the cable is proportional to the current flowing.

Hall effect sensors are becoming increasingly popular. This is partly due to their reliability but also the fact that they directly produce a constant amplitude square wave in speed measurement applications and a varying DC voltage for either position sensing or current sensing.

2.7.5 Strain gauges

Figure 2.50 shows a simple strain gauge together with a bridge and amplifier circuit used to convert its change in resistance into a voltage signal. The second strain gauge is fitted on the device under test but in a non-strain position to compensate for temperature changes. Quite simply, when a

strain gauge is stretched its resistance will increase, and when it is compressed its resistance decreases. Most strain gauges consist of a thin layer of film that is fixed to a flexible backing sheet, usually paper. This, in turn, is bonded to the part where strain is to be measured.

The sensitivity of a strain gauge is defined by its 'gauge factor'.

$$K = (\Delta R/R)/E$$

where K = gauge factor; R = original resistance; ΔR = change in resistance; E = strain (change in length / original length, $\Delta l/l$).

Most resistance strain gauges have a resistance of about $100\,\Omega$ and a gauge factor of about 2.

Strain gauges are often used indirectly to measure engine manifold pressure. Figure 2.51 shows an arrangement of four strain gauges on a diaphragm forming part of an aneroid chamber used to measure pressure. When changes in

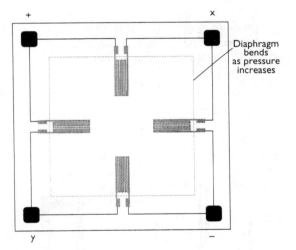

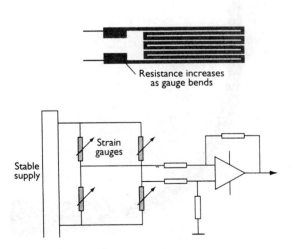

Figure 2.50 Strain gauge and a bridge circuit

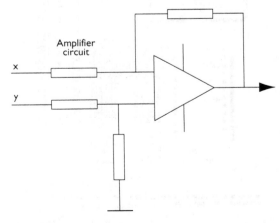

Figure 2.51 Strain gauge pressure sensor, bridge circuit and amplifier

manifold pressure act on the diaphragm the gauges detect the strain. The output of the circuit is via a differential amplifier as shown, which must have a very high input resistance so as not to affect the bridge balance. The actual size of this sensor may be only a few millimetres in diameter. Changes in temperature are compensated for, as all four gauges would be affected in a similar way, thus the bridge balance would remain constant.

2.7.6 Variable capacitance

The value of a capacitor is determined by the surface area of its plates, the distance between the plates and the nature of the dielectric. Sensors can be constructed to take advantage of these properties. Three sensors using the variable capacitance technique are shown in Figure 2.52. These are as follows.

1. Liquid level sensor. The change in liquid level changes the dielectric value.
2. Pressure sensor. Similar to the strain gauge pressure sensor but this time the distance between capacitor plates changes.
3. Position sensor. Detects changes in the area of the plates.

2.7.7 Variable resistance

The two best examples of vehicle applications for variable resistance sensors are the throttle position sensor and the flap-type air flow sensor. Whereas variable capacitance sensors are used to measure small changes, variable resistance sensors generally measure larger changes in position. This is due to a lack of sensitivity inherent in the construction of the resistive track.

The throttle position sensor, as shown in Figure 2.53, is a potentiometer in which, when supplied with a stable voltage (often 5 V) the voltage from the wiper contact will be proportional to the throttle position. In many cases now, the throttle potentiometer is used to indicate the rate of change of throttle position. This information is used when implementing acceleration enrichment or, inversely, over-run fuel cut-off.

The output voltage of a rotary potentiometer can be calculated:

$$V_\text{o} = V_\text{s}\left(\frac{a_\text{i}}{a_\text{e}}\right)$$

where V_o = voltage out; V_s = voltage supply; a_i = angle moved; a_t = total angle possible.

The air flow sensor shown as Figure 2.54 works on the principle of measuring the force exerted on the flap by the air passing through it. A calibrated coil spring exerts a counter force on

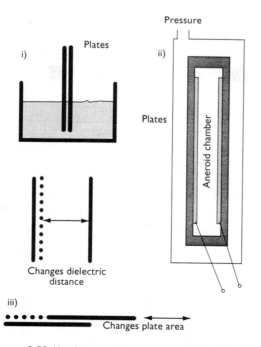

Figure 2.52 Variable capacitance sensors: (i) liquid level; (ii) pressure; (iii) position

Figure 2.53 Throttle potentiometer

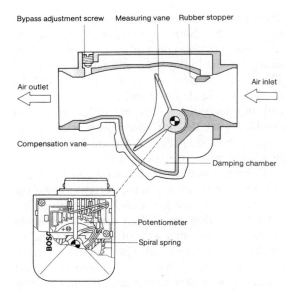

Figure 2.54 Air flow meter (vane type)

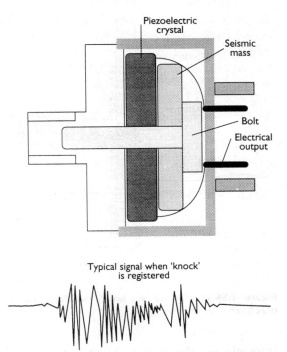

Figure 2.55 Piezoelectric accelerometer or knock sensor

the flap such that the movement of the flap is proportional to the volume of air passing through the sensor. To reduce the fluctuations caused by individual induction strokes a compensation flap is connected to the sensor flap. The fluctuations therefore affect both flaps and are cancelled out. Any damage due to back firing is also minimized due to this design. The resistive material used for the track is a ceramic metal mixture, which is burnt into a ceramic plate at a very high temperature. The slider potentiometer is calibrated such that the output voltage is proportional to the quantity of inducted air.

2.7.8 Accelerometer (knock sensors)

A piezoelectric accelerometer is a seismic mass accelerometer using a piezoelectric crystal to convert the force on the mass due to acceleration into an electrical output signal. The crystal not only acts as the transducer but as the suspension spring for the mass. Figure 2.55 shows a typical accelerometer (or knock sensor) for vehicle use.

The crystal is sandwiched between the body of the sensor and the seismic mass and is kept under compression by the bolt. Acceleration forces acting on the seismic mass cause variations in the amount of crystal compression and hence generate the piezoelectric voltage. The oscillations of the mass are not damped except by the stiffness of the crystal. This means that the sensor will have a very strong resonant fre-

quency but will also be at a very high frequency (in excess of 50 kHz), giving a flat response curve in its working range up to about 15 kHz.

The natural or resonant frequency of a spring mass system is given by:

$$f = \frac{1}{2\pi} \sqrt{\frac{k}{m}}$$

where f = resonant frequency; k = spring constant (very high in this case); m = mass of the seismic mass (very low in this case).

When used as an engine knock sensor, the sensor will also detect other engine vibrations. These are kept to a minimum by only looking for 'knock' a few degrees before and after top dead centre (TDC). Unwanted signals are also filtered out electrically. A charge amplifier is used to detect the signal from this type of sensor. The sensitivity of a vehicle knock sensor is about 20 mV/g (g = 9.81 m/s).

2.7.9 Linear variable differential transformer (LVDT)

This sensor is used for measuring displacement in a straight line (hence linear). Devices are available to measure distances of less than 0.5 mm and over 0.5 m, either side of a central position. Figure 2.56 shows the

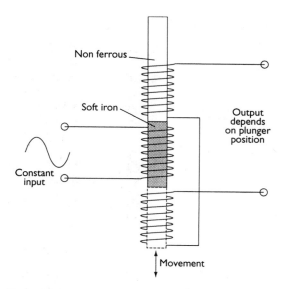

Figure 2.56 Principle of the linear variable differential transducer

principle of the linear variable differential transducer.

The device has a primary winding and two secondary windings. The primary winding is supplied with an AC voltage and AC voltages are induced in the secondary windings by transformer action. The secondary windings are connected in series opposition so that the output of the device is the difference between their outputs. When the ferromagnetic armature is in the central position the output is zero. As the armature now moves one way or the other, the output is increased in one winding and decreased in the other, producing a voltage which, within the working range, is proportional to the displacement. A phase sensitive detector can be used to convert the movement into a DC voltage, often ±5 V. For a device moving ±12 mm this gives a sensitivity of 0.42 V/mm.

LVDTs are used in some manifold pressure sensors where a diaphragm transforms changes in pressure to linear movement.

2.7.10 Hot wire air flow sensor

The distinct advantage of a hot wire air flow sensor is that it measures air mass flow. The basic principle is that, as air passes over a hot wire it tries to cool the wire down. If a circuit is created such as to increase the current through the wire in order to keep the temperature constant, then this current will be proportional to the air flow. A resistor is also incorporated to compensate for temperature variations. The 'hot

wire' is made of platinum, is only a few millimetres long and about 70 μm thick. Because of its small size the time constant of the sensor is very short – in fact in the order of a few milliseconds. This is a great advantage as any pulsations of the air flow will be detected and reacted to in a control unit accordingly. The output of the circuit involved with the hot wire sensor is a voltage across a precision resistor. Figure 2.57 shows a Bosch hot wire air mass sensor.

The resistance of the hot wire and the precision resistor are such that the current to heat the wire varies between 0.5 A and 1.2 A with different air mass flow rates. High resistance resistors are used in the other arm of the bridge and so current flow is very small. The temperature compensating resistor has a resistance of about 500 Ω which must remain constant other than by way of temperature change. A platinum film resistor is used for these reasons. The compensation resistor can cause the system to react to temperature changes within about 3 s.

The output of this device can change if the hot wire becomes dirty. Heating the wire to a very high temperature for 1 s every time the engine is switched off prevents this by burning off any contamination. In some air mass sensors a variable resistor is provided to set the idle mixture.

2.7.11 Thin film air flow sensor

The thin film air flow sensor is similar to the hot wire system. Instead of a hot platinum wire a thin film of nickel is used. The response time of this system is even shorter than the hot wire. Figure 2.58 shows this sensor in more detail.

Figure 2.57 Hot wire mass air flow meter insert type

Figure 2.58 Hot film air mass flow meter

2.7.12 Vortex flow sensor

Figure 2.59 shows the principle of a vortex flow sensor. It has a bluff body, which partially obstructs the flow. Vortices form at the downstream edges of the bluff body at a frequency that is linearly dependent on the flow velocity. Detection of the vortices provides an output signal whose frequency is proportional to flow velocity. Detection of the vortices can be by an ultrasonic transmitter and receiver that will produce a proportional square wave output. The main advantage of this device is the lack of any moving parts, thus eliminating problems with wear.

For a vortex flow sensor to work properly, the flow must be great enough to be turbulent, but not so high as to cause bubbles when measuring fluid flow. As a rough guide, the flow should not exceed 50 m/s.

When used as an engine air flow sensor, this system will produce an output frequency of about 50 Hz at idle speed and in excess of 1 kHz at full load.

2.7.13 Pitot tube

A Pitot tube air flow sensor is a very simple device. It consists of a small tube open to the air flow such that the impact of the air will cause an increase in pressure in the tube compared with the pressure outside the tube. This same system is applied to aircraft to sense air speed when in flight. The two tubes are connected to a differential pressure transducer such as a variable capacitance device. P_1 and P_2 are known as the impact and static pressures, respectively. Figure 2.60 shows a Pitot tube and differential pressure sensor used for air flow sensing.

2.7.14 Turbine fluid flow sensor

Using a turbine to measure fluid flow is an invasive form of measurement. The act of placing a device in the fluid will affect the flow rate. This technique however is still used as, with careful design, the invasion can be kept to a minimum. Figure 2.61 shows a typical turbine flow sensor.

The output of the turbine, rotational speed proportional to flow rate, can be converted to an electrical signal in a number of ways. Often an optical sensor is used as described under the next heading.

2.7.15 Optical sensors

An optical sensor for rotational position is a relatively simple device. The optical rotation sensor and circuit shown in Figure 2.62 consist of a phototransistor as a detector and a light emitting diode light source. If the light is focused to a very narrow beam then the output of the circuit shown will be a square wave with frequency proportional to speed.

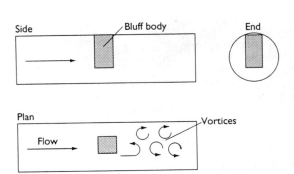

Figure 2.59 Principle of a vortex flow sensor

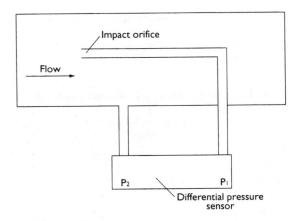

Figure 2.60 Pitot tube and differential pressure sensor for air flow sensing

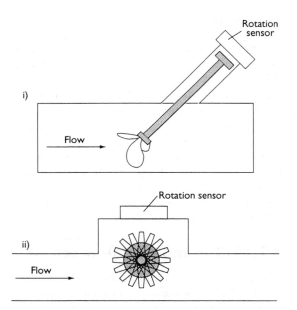

Figure 2.61 Turbine flow sensor

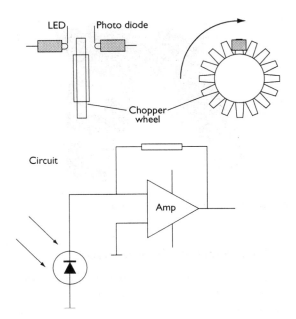

Figure 2.62 Optical sensor

2.7.16 Oxygen sensors

The vehicle application for an oxygen sensor is to provide a closed loop feedback system for engine management control of the air–fuel ratio. The amount of oxygen sensed in the exhaust is directly related to the mixture strength, or air–fuel ratio. The ideal air–fuel ratio of 14.7 : 1 by mass is known as a lambda (λ) value of one. Exhaust gas oxygen (EGO) sensors are placed in the exhaust pipe near to the manifold to ensure adequate heating. The sensors operate reliably at temperatures over 300°C. In some cases, a heating element is incorporated to ensure this temperature is reached quickly. This type of sensor is known as a heated exhaust gas oxygen sensor, or HEGO for short. The heating element (which

consumes about 10 W) does not operate all the time, which ensures that the sensor does not exceed 850°C – the temperature at which damage may occur to the sensor. It is for this reason that the sensors are not often fitted directly in the exhaust manifold. Figure 2.63 shows an exhaust gas oxygen sensor.

The main active component of most types of oxygen sensors is zirconium dioxide (ZrO_2). This ceramic is housed in gas permeable electrodes of platinum. A further ceramic coating is applied to the side of the sensor exposed to the exhaust gas as a protection against residue from the combustion process. The principle of operation is that, at temperatures in excess of 300°C, the zirconium dioxide will conduct the oxygen ions. The

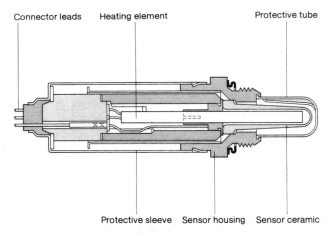

Figure 2.63 Lambda sensor

sensor is designed to be responsive very close to a lambda value of one. As one electrode of the sensor is open to a reference value of atmospheric air, a greater quantity of oxygen ions will be present on this side. Due to electrolytic action these ions permeate the electrode and migrate through the electrolyte (ZrO_2). This builds up a charge rather like a battery. The size of the charge is dependent on the oxygen percentage in the exhaust. A voltage of 400 mV is the normal figure produced at a lambda value of one.

The closely monitored closed loop feedback of a system using lambda sensing allows very accurate control of engine fuelling. Close control of emissions is therefore possible.

2.7.17 Light sensors

A circuit employing a light sensitive resistor is shown in Figure 2.64. The circuit can be configured to switch on or off in response to an increase or decrease in light. Applications are possible for self-dipping headlights, a self-dipping interior mirror, or parking lights that will automatically switch on at dusk.

2.7.18 Thick-film air temperature sensor

The advantage which makes a nickel thick-film thermistor ideal for inlet air temperature sensing is its very short time constant. In other words its resistance varies very quickly with a change in air temperature. Figure 2.65 shows the construction of this device. The response of a thick film sensor is almost linear. It has a sensitivity of about 2 ohms/°C and, as with most metals, it

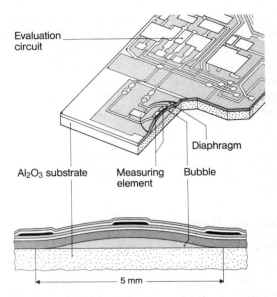

Figure 2.65 Thick-film pressure sensor

has a positive temperature coefficient (PTC) characteristic.

2.7.19 Methanol sensor

In the move towards cleaner exhausts, one idea is to use mixed fuels. Methanol is one potential fuel that can be mixed with petrol. The problem is that petrol has a different stoichiometric air requirement to methanol.

An engine management system can be set for either fuel or a mixture of the fuels. However, the problem with mixing is that the ratio will vary. A special sensor is needed to determine the proportion of methanol, and once fitted this sensor will make it possible to operate the vehicle on any mixture of petrol and methanol.

The methanol sensor (Figure 2.66) is based on the dielectric principle. The measuring cell is a capacitor filled with fuel and the methanol content is calculated from its capacitance. Two

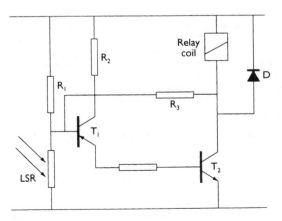

Figure 2.64 Light sensitive resistor circuit

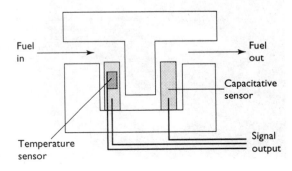

Figure 2.66 Methanol sensor

further measurements are taken – the temperature of the fuel and its conductance. These correction factors ensure cross-sensitivity (a kind of double checking) and the measurement error is therefore very low. The sensor can be fitted to the fuel line so the data it provides to the ECU are current and reliable. The control unit can then adapt the fuelling strategy to the fuel mix currently in use. Some further development is taking place but this sensor looks set to play a major part in allowing the use of alternative fuels in the near future.

2.7.20 Rain sensor

Rain sensors are used to switch on wipers automatically. Most work on the principle of reflected light. The device is fitted inside the windscreen and light from an LED is reflected back from the outer surface of the glass. The amount of light reflected changes if the screen is wet, even with a few drops of rain. Figure 2.67 shows a typical sensor.

2.7.21 Dynamic vehicle position sensors

These sensors are used for systems such as active suspension, stability control and general systems where the movement of the vehicle is involved. Most involve the basic principle of an accelerometer; that is, a ball hanging on a string or a seismic mass acting on a sensor.

2.7.22 Sensors: summary

The above brief look at various sensors hardly scratches the surface of the number of types, and the range of sensors available for specific tasks. The subject of instrumentation is now a science in its own right. The overall intention of this section has been to highlight some of the problems and solutions to the measurement of variables associated with vehicle technology.

Sensors used by motor vehicle systems are following a trend towards greater integration of processing power in the actual sensor. Four techniques are considered, starting with the conventional system. Figure 2.68 shows each level of sensor integration in a block diagram form.

Conventional

Analogue sensor in which the signal is transmitted to the ECU via a simple wire circuit. This technique is very susceptible to interference.

Integration level I

Analogue signal processing is now added to the sensor, this improves the resistance to interference.

Integration level 2

At the second level of integration, analogue to-digital conversion is also included in the sensor. This signal is made bus compatible (CAN for example) and hence becomes interference proof.

Integration level 3

The final level of integration is to include 'intelligence' in the form of a microcomputer as part of the sensor. The digital output will be interference proof. This level of integration will also allow built in monitoring and diagnostic ability. These types of sensor are very expensive at the time of writing but the price is falling and will continue to do so as more use is made of the 'intelligent sensor'.

Figure 2.67 Rain sensor

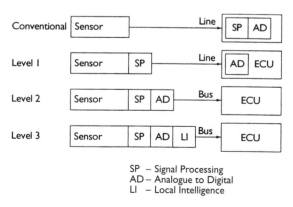

Figure 2.68 Block diagram of four types of sensors and their differing aspects

2.7.23 Actuators: introduction

There are many ways of providing control over variables in and around the vehicle. 'Actuators' is a general term used here to describe a control mechanism. When controlled electrically actuators will work either by the thermal or magnetic effect. In this section, the term actuator will be used to mean a device that converts electrical signals into mechanical movement. This section is not written with the intention of describing all available types of actuator. Its intention is to describe some of the principles and techniques used in controlling a wide range of vehicle systems.

2.7.24 Solenoid actuators

The basic operation of solenoid actuators is very simple. The term 'solenoid' means: 'many coils of wire wound onto a hollow tube'. However, the term is often misused, but has become so entrenched that terms like 'starter solenoid' – when really it is a starter relay – are in common use.

A good example of a solenoid actuator is a fuel injector. Figure 2.69 shows a typical

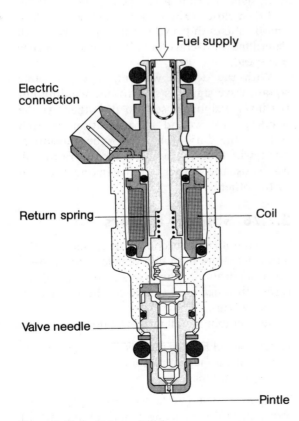

Figure 2.69 Fuel injector (MK I)

example. When the windings are energized the armature is attracted due to magnetism and compresses the spring. In the case of a fuel injector, the movement is restricted to about 0.1 mm. The period that an injector remains open is very small – under various operating conditions, between 1.5 and 10 ms is typical. The time it takes an injector to open and close is also critical for accurate fuel metering. Further details about injection systems are discussed in Chapters 9 and 10.

The reaction time for a solenoid-operated device, such as a fuel injector, depends very much on the inductance of the winding. Figure 2.70 shows a graph of solenoid-operated actuator variables.

A suitable formula to show the relationship between some of the variables is as follows:

$$i = \frac{V}{R}(1 - e^{-Rt/L})$$

where i = instantaneous current in the winding, V = supply voltage, R = total circuit resistance, L = inductance of the injector winding, t = time current has been flowing, e = base of natural logs.

The resistance of commonly used injectors is about 16 Ω. Some systems use ballast resistors in series with the fuel injectors. This allows lower inductance and resistance operating windings to be used, thus speeding up reaction time. Other types of solenoid actuators, for example door lock actuators, have less critical reaction times. However, the basic principle remains the same.

2.7.25 Motorized actuators

Permanent magnet electric motors are used in many applications and are very versatile. The output of a motor is, of course, rotation, but this

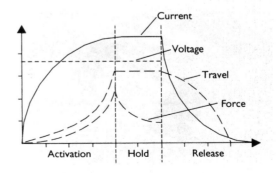

Figure 2.70 Solenoid-operated actuator variables

can be used in many ways. If the motor drives a rotating 'nut' through which a plunger is fitted, and on which there is a screw thread, the rotary action can easily be converted to linear movement. In most vehicle applications the output of the motor has to be geared down, this is to reduce speed and increase torque. Permanent magnet motors are almost universally used now in place of older and less practical motors with field windings. Some typical examples of where these motors are used are:

- windscreen wipers
- windscreen washers
- headlight lift
- electric windows
- electric sun roof
- electric aerial operation
- seat adjustment
- mirror adjustment
- headlight washers
- headlight wipers
- fuel pumps
- ventilation fans.

One disadvantage of simple motor actuators is that no direct feedback of position is possible. This is not required in many applications; however, in some cases, such as seat adjustment when a 'memory' of the position may be needed, a variable resistor type sensor can be fitted to provide feedback. A typical motor actuator is shown in Figure 2.71.

A rotary idle actuator is shown in Figure 2.72. This device is used to control idle speed by controlling air bypass. There are two basic types in common use. These are single winding types, which have two terminals, and double winding types, which have three terminals. Under ECU control, the motor is caused to open and close a shutter, thus controlling air bypass. These actuators only rotate about 90° to open and close the valve. As these are permanent magnet motors,

Figure 2.71 Seat adjustment motor

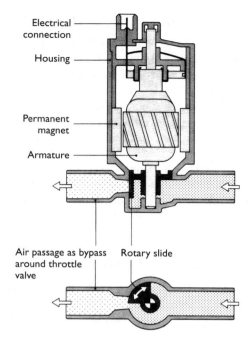

Figure 2.72 Rotary idle actuator

the term 'single or double windings' refers to the armature.

The single winding type is fed with a square wave signal causing it to open against a spring and then close again, under spring tension. The on/off ratio or duty cycle of the square wave will determine the average valve open time and hence idle speed.

With the double winding type the same square wave signal is sent to one winding but the inverse signal is sent to the other. As the windings are wound in opposition to each other if the duty cycle is 50% then no movement will take place. Altering the ratio will now cause the shutter to move in one direction or the other.

2.7.26 Stepper motors

Stepper motors are becoming increasingly popular as actuators in motor vehicles and in many other applications. This is mainly because of the ease with which they can be controlled by electronic systems.

Stepper motors fall into three distinct groups:

1. variable reluctance motors
2. permanent magnet motors
3. hybrid motors.

Figure 2.73 shows the basic principle of variable reluctance, permanent magnet and hybrid

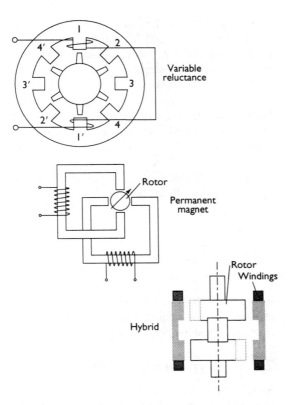

Figure 2.73 Basic principle of variable reluctance, permanent magnet and hybrid stepper motors

stepper motors. The operation of each is described briefly but note that the underlying principle is the same for each type.

Variable reluctance motors rely on the physical principle of maximum flux. A number of windings are set in a circle on a toothed stator. The rotor also has teeth and is made of a permeable material. Note in this example that the rotor has two teeth less than the stator. When current is supplied to a pair of windings of one phase, the rotor will line up with its teeth positioned such as to achieve maximum flux. It is now simply a matter of energizing the windings in a suitable order to move the rotor. For example, if phase four is energized, the motor will 'step' once in a clockwise direction. If phase two is energized the step would be anti-clockwise.

These motors do not have a very high operating torque and have no torque in the non-excited state. They can, however, operate at relatively high frequencies. The step angles are usually 15°, 7.5°, 1.8° or 0.45°.

Permanent magnet stepper motors have a much higher starting torque and also have a holding torque when not energized. The rotor is

now, in effect, a permanent magnet. In a *variable reluctance motor* the direction of current in the windings does not change; however, it is the change in direction of current that causes the *permanent magnet motor* to step. Permanent magnet stepper motors have step angles of 45°, 18°, 15° or 7.5°. Because of their better torque and holding properties, permanent magnet motors are becoming increasingly popular. For this reason, this type of motor will be explained in greater detail.

The hybrid stepper motor as shown in Figure 2.74 is, as the name suggests, a combination of the previous two motors. These motors were developed to try and combine the high speed operation and good resolution of the variable reluctance type with the better torque properties of the permanent magnet motor. A pair of toothed wheels is positioned on either side of the

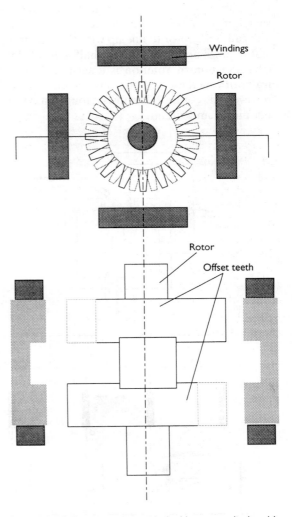

Figure 2.74 Stepper motor with double stators displaced by one pole pitch

magnet. The teeth on the 'North' and 'South' wheels are offset such as to take advantage of the variable reluctance principle but without losing all the torque benefits. Step angles of these motors are very small: 1.8°, 0.75° or 0.36°.

All of the above-mentioned types of motor have been, and are being, used in various vehicle applications. These applications range from idle speed air bypass and carburettor choke control to speedometer display drivers.

Let us look now in more detail at the operation and construction of the permanent magnet stepper motor. The most basic design for this type of motor comprises two double stators displaced by one pole pitch. The rotor is often made of barium-ferrite in the form of a sintered annular magnet. As the windings shown in Figure 2.75 are energized first in one direction then the other, the motor will rotate in 90° steps. The step angle is simply 360° divided by the number of stator poles. Half steps can be achieved by switching off a winding before it is reversed. This will cause the rotor to line up with the remaining stator poles and implement a half step of 45°. The direction of rotation is determined by the order in which the windings are switched on, off or reversed. Figure 2.75 shows a four-phase stepper motor and circuit.

Impulse sequence graphs for two phase stepper motors are shown in Figure 2.76. The first graph is for full steps, and the second graph for implementing half steps.

The main advantage of a stepper motor is that feedback of position is not required. This is because the motor can be indexed to a known starting point and then a calculated number of steps will move the motor to any suitable position.

The calculations often required for stepper applications are listed below:

$$\alpha = 360/z$$
$$z = 360/\alpha$$
$$f_z = (nz)/60$$
$$n = (fz \times 60)/z$$
$$w = (fz \times 2\pi)/z$$

where α = step angle, n = revolutions per minute, w = angular velocity, f_z = step frequency, z = steps per revolution.

2.7.27 Synchronous motors

Synchronous motors are used when a drive is required that must be time synchronized. They always rotate at a constant speed, which is determined by the system frequency and the number of pole pairs in the motor.

$$n = (f \times 60)/p$$

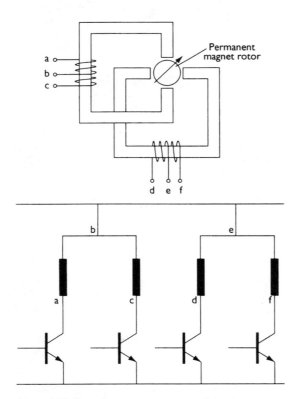

Figure 2.75 Four-phase stepper motor and circuit

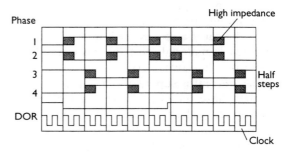

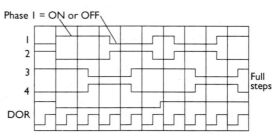

Figure 2.76 Impulse sequence graphs for two-phase stepper motors: the first graph is for half steps, the second for implementing full steps

where n = rpm; f = frequency; p = number of pole pairs.

Figure 2.77 shows a reversing synchronous motor and its circuit together with the speed torque characteristic. This shows a constant speed and a break off at maximum torque. Maximum torque is determined by supply voltage.

2.7.28 Thermal actuators

An example of a thermal actuator is the movement of a traditional-type fuel or temperature gauge needle (see Chapter 13). A further example is an auxiliary air device used on many earlier fuel injection systems. The principle of this device is shown in Figure 2.78. When current is supplied to the terminals, a heating element operates and causes a bimetallic strip to bend, which closes a simple valve.

The main advantage of this type of actuator, apart from its simplicity, is that if placed in a suitable position its reaction time will vary with the temperature of its surroundings. This is ideal for applications such as fast idle on cold starting control, where once the engine is hot no action is required from the actuator.

2.8 New developments

Development in electronics, particularly digital electronics, is so rapid that it is difficult to keep up. I have tried to provide a basic background in this

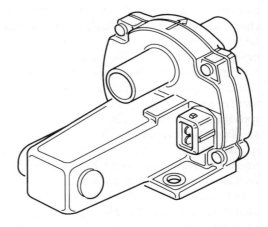

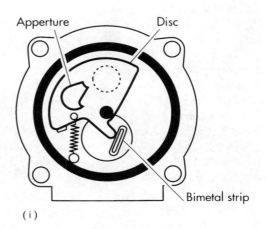

(i)

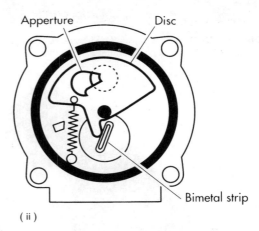

(ii)

Figure 2.78 Diagram showing operation of extra air valve (electrical): (i) bypass channel closed; (ii) bypass channel partially open

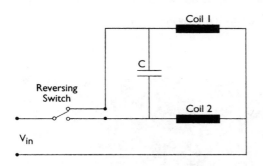

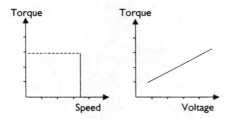

Figure 2.77 Reversing synchronous motor and circuit and its speed torque characteristic

chapter, because this is timeless. More systems are becoming 'computer' based, and it is these digital aspects that are developing. The trend is towards greater integration and communication between systems. This allows for built-in fault diagnostics as well as monitoring of system performance to ensure compliance with legislation (particularly relating to emissions). The move towards greater on-board diagnostics (OBD) will continue.

A number of new sensors are becoming available. Hall effect sensors are being used in place of inductive sensors for applications such as engine speed and wheel speed. The two main advantages are that measurement of lower (or even zero) speed is possible and the voltage output of the sensors is independent of speed. Figure 2.79 shows a Hall effect sensor used to sense wheel speed.

An interesting sensor used to monitor oil quality is now available. The type shown in Figure 2.80 from the Kavlico Corporation works by

monitoring changes in the dielectric constant of the oil. The dielectric constant increases as antioxidant additives in the oil deplete. The value also rapidly increases if coolant contaminates the oil. The sensor output increases as the dielectric constant increases.

Lucas Varity has an oil condition sensor that utilizes high frequency shear waves to measure the viscosity of the oil. The sensor is shown in Figure 2.81, and it will enable vehicle operators to optimize engine oil change times.

Knock sensing on petrol/gasoline engine vehicles has been used since the mid 1980s to improve performance, reduce emissions and improve economy. These sensors give a good 'flat' response over the 2–20 kHz range. The diesel knock sensor shown in Figure 2.82 works between 7 and 20 kHz. With suitable control

Figure 2.79 Hall effect sensor in SSI package with dressed cable for ABS wheel-speed applications

Figure 2.81 Oil condition sensor

Figure 2.80 Oil quality sensor

Figure 2.82 Diesel knock sensor

electronics, the engine can be run near the detonation border line (DBL). This improves economy, performance and emissions.

One development in actuator technology is the rotary electric exhaust gas recirculation (EEGR) valve for use in diesel engine applications (Lucas Varity). This device is shown in Figure 2.83. The main claims for this valve are its self-cleaning action, accurate gas flow control and its reaction speed.

An low *g* accelerometer is available from Texas Instruments. The sensor shown in Figure 2.84 can be constructed to operate from 0.4 to 10 *g*. This sensor is used for ride control, antilock brakes (ABS) and safety restraint systems (SRS).

Figure 2.83 Rotary electric exhaust gas recirculation valve

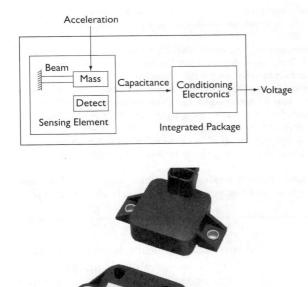

Figure 2.84 Capacitive low *g* acceleration sensor concept

2.9 Diagnostics – electronics, sensors and actuators

2.9.1 Introduction

Individual electronic components can be tested in a number of ways but a digital multimeter is normally the favourite option. Table 2.3 suggests some methods of testing components removed from the circuit.

2.9.2 Testing sensors

Testing sensors to diagnose faults is usually a matter of measuring their output signals. In some cases the sensor will produce this signal on its own (an inductive sensor for example). In other cases, it will be necessary to supply the correct voltage to the device to make it work (a Hall sensor for example). In this case, it is normal to check that the vehicle circuit is supplying the voltage before proceeding to test the sensor. Table 2.4 lists some common sensors together with suggested test methods (correct voltage supply is assumed).

2.9.3 Testing actuators

Testing actuators is simple, as many are operated by windings. The resistance can be measured with an ohmmeter. Injectors, for example, often have a resistance of about 16 Ω. A good tip is that where an actuator has more than one winding (a stepper motor for example), the resistance of each should be about the same. Even if the expected value is not known, it is likely that if the windings all read the same then the device is in order.

With some actuators, it is possible to power them up from the vehicle battery. A fuel injector should click for example, and a rotary air bypass device should rotate about half a turn. Be careful with this method as some actuators could be damaged.

Remember: if in doubt – seek advice!

Table 2.3 Electronic component testing

Component	Test method
Resistor	Measure the resistance value with an ohmmeter and compare this with the value written or colour coded on the component.
Capacitor	A capacitor can be difficult to test without specialist equipment but try this: charge the capacitor up to 12 V and connect it to a digital voltmeter. As most digital meters have an internal resistance of about 10 MΩ calculate the expected discharge time ($T = 5CR$) and see if the device complies. A capacitor from a contact breaker ignition system should take about 5 seconds to discharge in this way.
Inductor	An inductor is a coil of wire so a resistance check is the best method to test for continuity.
Diode	Many multimeters have a diode test function. If so, the device should read open circuit in one direction, and about 0.4–0.6 V in the other direction. This is its switch-on voltage. If no meter is available with this function then wire the diode to a battery via a small bulb, it should light with the diode one way and not the other.
LED	Most LEDs can be tested by connecting them to a 1.5 V battery. Note the polarity though, the longest leg or the flat side of the case is negative.
Transistor (bipolar)	Some multimeters even have transistor testing connections but, if not available, the transistor can be connected into a simple circuit as in Figure 2.85 and voltage tests carried out as shown. This also illustrates a method of testing electronic circuits in general. It is fair to point out that, without specific data, it is difficult for the non-specialist to test unfamiliar circuit boards. It is always worth checking for obvious breaks and dry joints though.
Digital components	A logic probe can be used. This is a device with a very high internal resistance so it does not affect the circuit under test. Two different coloured lights are used, one glows for a 'logic 1' and the other for 'logic 0'. Specific data are required in most cases but basic tests can be carried out.

Table 2.4 Testing sensors

Sensor	Test method
Inductive (reluctance)	A simple resistance test is good. Values vary from about 800 to 1200 Ω. The 'sine wave' output can be viewed on a 'scope' or measured with an AC voltmeter.
Hall effect	The square wave output can be seen on a scope or the voltage output measured with a DC voltmeter. This varies between 0 and 8 V for a Hall sensor used in a distributor depending on whether the chip is magnetized or not.
Thermistor	Most thermistors have a negative temperature coefficient (NTC). This means the resistance falls as temperature rises. A resistance check with an ohmmeter should give readings broadly as follows: 0 °C = 4500 Ω, 20 °C = 1200 Ω and 100 °C = 200 Ω.
Flap air flow	The main part of this sensor is a variable resistor. If the supply is left connected then check the output on a DC voltmeter. The voltage should change *smoothly* from about 0 to the supply voltage (often 5 V).
Hot wire air flow	This sensor includes some electronic circuits to condition the signal from the hot wire. The normal supply is either 5 or 12 V. The output should change between about 0 and 5 V as the air flow changes.
Throttle potentiometer	This sensor is a variable resistor. If the supply is left connected then check the output on a DC voltmeter. The voltage should change *smoothly* from about 0 to the supply voltage (often 5 V). If no supply then check the resistance, again it should change smoothly.
Oxygen (lambda)	The lambda sensor produces its own voltage, like a battery. This can be measured with the sensor connected to the system. The voltage output should vary smoothly between 0.2 and 0.8 V as the mixture is controlled by the ECU
Pressure	The normal supply to an externally mounted manifold absolute pressure (MAP) sensor is 5 V. The output should change between about 0 and 5 V as the manifold pressure changes – as a rough guide, 2.5 V at idle speed

2.10 Self-assessment

2.10.1 Questions

1. Describe briefly the difference between 'electron flow' and 'conventional flow'.
2. Sketch the symbols for 10 basic electronic components.
3. Explain what is meant by the 'frequency response' of an operational amplifier.
4. Draw the circuit to show how a resistor and capacitor can be connected to work as a timer. Include values and calculate the time constant for your circuit.
5. State four sources of error in a measurement system.

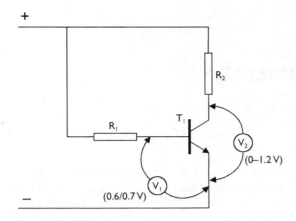

+

R₂

R₁　T₁

V₂
(0–1.2 V)

–　(0.6/0.7 V)　V₁

Figure 2.85 Transistor test

6. Describe how a knock sensor operates to produce a signal.
7. Make a sketch to show how a rotary idle speed actuator works and describe how it can vary idle speed when only able to take up a closed or open position.
8. Draw a graph showing the output signal of a Hall sensor used in an ignition distributor.
9. Describe the operation of a permanent magnet stepper motor and state three merits of this actuator.
10. Outline the six-stage process generally required to produce a program for a computer.

2.10.2 Project

Discuss the developments of sensors and actuators. Consider the reasons for these developments and use examples. Why is the integration of electronics within sensors an issue? Produce a specification sheet for an anti-lock brake system wheel speed sensor, detailing what it must be able to do.

3

Tools and test equipment

3.1 Basic equipment

3.1.1 Introduction

Diagnostic techniques are very much linked to the use of test equipment. In other words you must be able to interpret the results of tests. In most cases this involves comparing the result of a test to the reading given in a data book or other source of information. By way of an introduction, Table 3.1 lists some of the basic words and descriptions relating to tools and equipment. Figure 3.1 shows a selection of basic tools.

3.1.2 Basic hand tools

You cannot learn to use tools from a book, it is clearly a very practical skill. However, you can follow the recommendations made here and, of course, by the manufacturers. Even the range of basic hand tools is now quite daunting and very expensive. One thing to highlight, as an example, is the number of different types of screwdriver ends, as shown in Figure 3.2. These are worthy of mention because often using the wrong driver and damaging the screw head causes a lot of trouble. And of course, as well as all these different types they are all available in many different sizes!

It is worth repeating the general advice and instructions for the use of hand tools.

- Only use a tool for its intended purpose.
- Always use the correct size tool for the job you are doing.
- Pull a wrench rather than push it whenever possible.
- Do not use a file or similar, without a handle.
- Keep all tools clean and replace them in a suitable box or cabinet.
- Do not use a screwdriver as a pry bar.

Table 3.1 Tools and test equipment

Hand tools	Spanners and hammers and screwdrivers and all the other basic bits!
Special tools	A collective term for items not held as part of a normal tool kit. Or items required for just one specific job.
Test equipment	In general, this means measuring equipment. Most tests involve measuring something and comparing the result of that measurement with data. The devices can range from a simple ruler to an engine analyser.
Dedicated test equipment	Some equipment will only test one specific type of system. The large manufacturers supply equipment dedicated to their vehicles. For example, a diagnostic device which plugs in to a certain type of fuel injection ECU.
Accuracy	Careful and exact, free from mistakes or errors and adhering closely to a standard.
Calibration	Checking the accuracy of a measuring instrument.
Serial port	A connection to an electronic control unit, a diagnostic tester or computer for example. 'Serial' means the information is passed in a 'digital' string, like pushing black and white balls through a pipe in a certain order.
Code reader or scanner	This device reads the 'black and white balls' mentioned above or the on–off electrical signals, and converts them into a language we can understand.
Combined diagnostic and information system	Usually now PC-based, these systems can be used to carry out tests on vehicle systems and they also contain an electronic workshop manual. Test sequences guided by the computer can also be carried out.
Oscilloscope	The main part of the 'scope' is the display, which is like a TV or computer screen. A 'scope' is a voltmeter but instead of readings in numbers it shows the voltage levels by a trace or mark on the screen. The marks on the screen can move and change very fast allowing one to see the way voltages change.

Figure 3.1 A selection of basic tools

▬	Flat Tip
■	Scrulox®
✚	Phillips®
✱	Torx®
⬡	Hex
⬮	Clutch 'G'
✤	Pozidriv®
⋏	Tri-Wing®
▬	Clutch 'A'
✿	Tamper Resistant Torx®

Figure 3.2 Many types of 'driver' shapes are now in use

- Always follow manufacturers recommendations (you cannot remember everything!).
- Look after your tools and they will look after you.

3.1.3 Accuracy of test equipment

Accuracy can mean a number of slightly different things.

- Careful and exact.
- Free from mistakes or errors; precise.
- Adhering closely to a standard.

Consider measuring a length of wire with a steel rule. How accurately could you measure it? To the nearest 0.5 mm? This raises a number of issues: first, you could make an error reading the ruler. Secondly, why do we need to know the length of a bit of wire to the nearest 0.5 mm? Thirdly, the ruler may have stretched or expanded, and so does not give the correct reading.

The first and second of these issues can be dispensed with by knowing how to read the test equipment correctly and also knowing the appropriate level of accuracy required. A micrometer for a plug gap? A ruler for valve clearances? I think you get the idea. The accuracy of the equipment itself is another issue.

Accuracy is a term meaning how close the measured value of something is to its actual value. For example, if a length of about 30 cm is measured with an ordinary wooden ruler, then the error may be up to 1 mm too high or too low. This is quoted as an accuracy of ±1 mm. This may also be given as a percentage, which in this case would be 0.33%.

The resolution, or in other words the 'fineness', with which a measurement can be made, is related to accuracy. If a steel ruler was made to a very high standard but only had markings of one graduation per centimetre it would have a very low resolution even though the graduations were very accurate. In other words the equipment is accurate but your reading will not be.

To ensure instruments are, and remain, accurate there are just two simple guidelines.

- Look after the equipment; a micrometer thrown on the floor will not be accurate.
- Ensure instruments are calibrated regularly – this means being checked against known good equipment.

Table 3.2 (see page 58) is a summary of the steps to ensure a measurement is accurate.

3.2 Multimeters

3.2.1 Basic test meters

An essential tool for working on vehicle electrical and electronic systems is a good digital multimeter. Digital meters are most suitable for accuracy of reading as well as their available

Table 3.2 Ensuring a measurement in accurate

Step	Example
Decide on the level of accuracy required.	Do we need to know that the battery voltage is 12.6 V or 12.635 V?
Choose the correct instrument for the job.	A micrometer to measure the thickness of a shim.
Ensure the instrument has been looked after and calibrated when necessary.	Most instruments will go out of adjustment after a time. You should arrange for adjustment at regular intervals. Most tool suppliers will offer the service or, in some cases, you can compare older equipment to new stock.
Study the instructions for the instrument in use and take the reading with care. Ask yourself if the reading is about what you expected.	Is the piston diameter 70.75 mm or 170.75 mm?
Make a note if you are taking several readings.	Don't take a chance, write it down!

facilities. The list of functions given in Table 3.3, which is broadly in order, starting from essential to desirable, should be considered.

A way of determining the quality of a meter as well as by the facilities provided, is to consider the following:

- accuracy,
- loading effect of the meter,
- protection circuits.

The loading effect is a consideration for any form of measurement. The question to ask is: 'Does the instrument change the conditions, so making the reading incorrect?' With a multimeter this relates to the internal resistance of the meter. It is recommended that the internal resistance of a meter should be a minimum of 10 MΩ. This not only ensures greater accuracy but also prevents the meter damaging sensitive circuits.

Consider Figure 3.3, which shows two equal resistors connected in series across a 12 V supply. It is clear that the voltage across each resistor should be 6 V. However, the internal resistance of the meter will affect the circuit conditions and change the voltage reading. If the resistor values were 100 kΩ the effect of meter internal resistance would be as follows.

Meter resistance 1 MΩ

Parallel combined value of 1 MΩ and 100 kΩ = 91 kΩ.

The voltage drop in the circuit across this would be:

$$91/(100 + 91) \times 12 = 5.71 \text{ V}$$

This is an error of about 5%

Meter resistance 10 MΩ

Parallel combined value of 10 MΩ and 100 KΩ = 99 KΩ.

Table 3.3 The range and accuracy for various functions

Function	Range	Accuracy
DC Voltage	500 V	0.3%
DC Current	10 A	1.0%
Resistance	0–10 MΩ	0.5%
AC Voltage	500 V	2.5%
AC Current	10 A	2.5%
Dwell	3,4,5,6,8 cylinders	2.0%
RPM	10,000 rpm	0.2%
Duty cycle	% on/off	0.2%/kHz
Frequency	over 100 kHz	0.01%
Temperature	> 900°C	0.3% +3°C
High current clamp	1000 A (DC)	Depends on conditions
Pressure	3 bar	10.0% of standard scale

The voltage drop in the circuit across this would be:

$$99/(100 + 99) \times 12 = 5.97 \text{ V}$$

This is an error of about 0.5%

Note that this 'invasive measurement' error is in addition to the basic accuracy of the meter.

Protection circuits are worth a mention as many motor vehicle voltage readings are prone to high voltage transient spikes, which can damage low quality equipment. A fused current

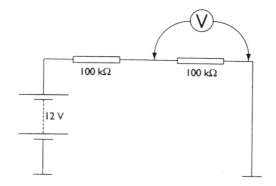

Figure 3.3 Two equal resistors connected in series across a 12 V supply with a meter for test purposes

range is also to be recommended. Figure 3.4 shows a basic block diagram of a digital voltmeter. Note how this closely represents any digital instrumentation system.

Figure 3.5 shows a digital multimeter in use. A final note relating to using a digital meter is that only two further skills are required – firstly where to put the probes, and secondly what the reading you get actually means!

3.3 Specialist equipment

3.3.1 Oscilloscopes

Two types of oscilloscope are available, these are either analogue or digital. Figure 3.6 shows the basic operation of an analogue oscilloscope. Heating a wire creates a source of electrons, which are then accelerated by suitable voltages and focused into a beam. This beam is directed towards a fluorescent screen where it causes light to be given off. This is the basic cathode ray tube. The plates shown in Figure 3.6 are known as X and Y plates as they make the electron beam draw a 'graph' of a voltage signal. The X plates are supplied with a saw-tooth signal, which

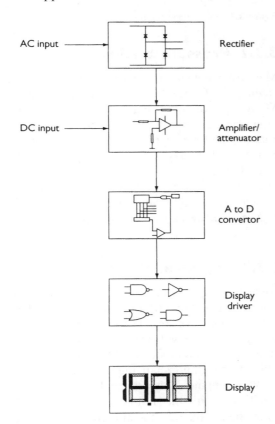

Figure 3.4 Digital voltmeter

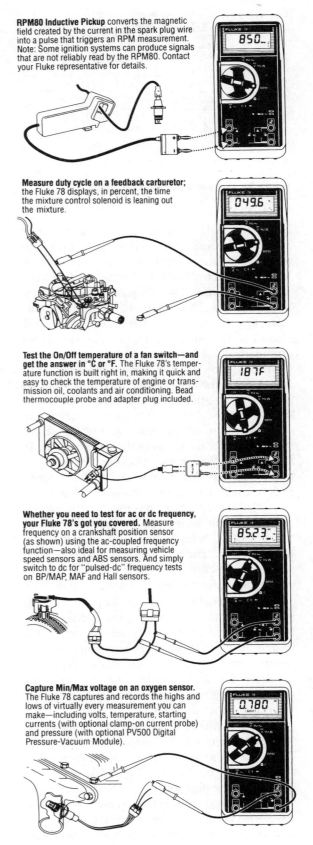

RPM80 Inductive Pickup converts the magnetic field created by the current in the spark plug wire into a pulse that triggers an RPM measurement. Note: Some ignition systems can produce signals that are not reliably read by the RPM80. Contact your Fluke representative for details.

Measure duty cycle on a feedback carburetor; the Fluke 78 displays, in percent, the time the mixture control solenoid is leaning out the mixture.

Test the On/Off temperature of a fan switch—and get the answer in °C or °F. The Fluke 78's temperature function is built right in, making it quick and easy to check the temperature of engine or transmission oil, coolants and air conditioning. Bead thermocouple probe and adapter plug included.

Whether you need to test for ac or dc frequency, your Fluke 78's got you covered. Measure frequency on a crankshaft position sensor (as shown) using the ac-coupled frequency function—also ideal for measuring vehicle speed sensors and ABS sensors. And simply switch to dc for "pulsed-dc" frequency tests on BP/MAP, MAF and Hall sensors.

Capture Min/Max voltage on an oxygen sensor. The Fluke 78 captures and records the highs and lows of virtually every measurement you can make—including volts, temperature, starting currents (with optional clamp-on current probe) and pressure (with optional PV500 Digital Pressure-Vacuum Module).

Figure 3.5 A digital multimeter in use

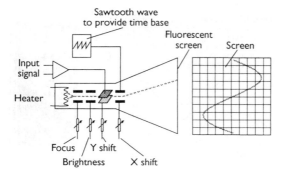

Figure 3.6 Principle of analogue oscilloscope

causes the beam to move across the screen from left to right and then to 'fly back' and start again. The beam moves because the electron beam is attracted towards whichever plate has a positive potential. The Y plates can now be used to show voltage variations of the signal under test. The frequency of the saw-tooth signal, known as the time base, can be adjusted either automatically as is the case with many analysers, or manually on a stand-alone oscilloscope. The signal from the item under test can either be amplified or attenuated (reduced), much like changing the scale on a voltmeter. The trigger, in other words when the trace across the screen starts, can be caused internally or externally. In the case of the engine analyser, triggering is often external – each time an individual spark fires or each time number one spark plug fires.

A digital oscilloscope has much the same end result as the analogue type but the signal can be thought of as being plotted rather than drawn on the screen. The test signal is A/D converted and the time base is a simple timer or counter circuit. Because the signal is plotted digitally on a screen from data in memory, the picture can be saved, frozen or even printed. The speed of data conversion and the sampling rate as well as the resolution of the screen are very important to ensure accurate results. This technique is becoming the norm, as including scales and notes or superimposing two or more traces for comparison can enhance the display.

A very useful piece of equipment becoming very popular is the 'Scopemeter' (Figure 3.7). This is a hand-held digital oscilloscope, that allows data to be stored and transferred to a PC for further investigation. The Scopemeter can be used for a large number of vehicle tests. The waveforms used as examples in this chapter were 'captured' using a Scopemeter. This type of test equipment is highly recommended.

Figure 3.7 Bosch 'scopemeter'

3.3.2 Pressure testing

Measuring the fuel pressure in a fuel injection engine is of great value when fault-finding. Many types of pressure gauge are available and often come as part of a kit consisting of various adapters and connections. The principle of the gauges is that they contain a very small tube wound in a spiral. As fuel under pressure is forced into a spiral tube, the tube unwinds causing the needle to move over a graduated scale. Figure 3.8 shows a selection of pressure testing equipment.

3.3.3 Engine analysers

Some form of engine analyser has become an almost essential tool for fault-finding in modern vehicle engine systems. The latest machines are now generally based around a personal computer. This allows more facilities, which can be added to by simply changing the software.

Whilst engine analysers are designed to work specifically with the motor vehicle, it is worth remembering that the machine consists basically of three parts.

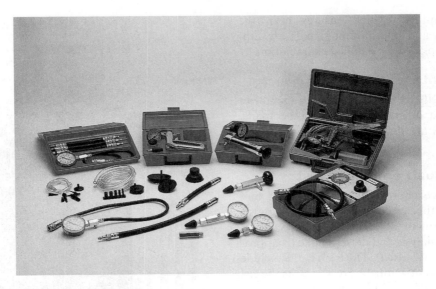

Figure 3.8 Pressure testing equipment

- Multimeter.
- Gas analyser.
- Oscilloscope.

This is not intended to imply that other tests available, such as cylinder balance, are less valid, but to show that the analyser is not magic, it is just able to present results of electrical tests in a convenient way to allow diagnosis of faults. The key component of any engine analyser is the oscilloscope facility, which allows the user to 'see' the signal under test.

The following is a description of the facilities available on a typical engine analyser. It is a new concept in garage equipment design, based on a personal computer and specially engineered for workshop use, enabling a flexibility of use far exceeding the ability of the machines previously available.

Software is used to give the machine its 'personality' as an engine analyser, system tester, wheel aligner (or any of the other uses made of personal computers). Either an infrared handset or a standard 'qwerty' keyboard controls the machine. The information is displayed on a super VGA monitor giving high resolution colour graphics. Output can be sent to a standard printer when a hard copy is required for the customer.

Many external measurement modules and software application programmes are available. The modules are connected to the host computer by high speed RS422 or RS232 serial communication links. Application software and DOS are loaded onto a hard disk. Vehicle spe-

cific data can also be stored on disk to allow fast easy access to information but also to allow a guided test procedure.

The modern trend with engine analysers seems to be to allow both guided test procedures with pass/fail recommendations for the less skilled technician, and freedom to test any electrical device using the facilities available in any reasonable way. This is more appropriate for the highly skilled technician. Some of the routines available on modern engine analysers are listed below.

Tune-up

A full prompted sequence that assesses each component in turn with results and diagnosis displayed at the end of each component test. Stored data allow pass/fail diagnosis by automatically comparing results of tests with data on the disk. Printouts can be taken to show work completed.

Symptom analysis

This allows direct access to specific tests relating to reported driveability problems.

Waveforms

A comprehensive range of digitized waveforms can be displayed with colour highlights. The display can be frozen or recalled to look for intermittent faults. A standard laboratory scope mode is available to allow examination of EFI or ABS traces for example. Printouts can be made from any display. An interesting feature is 'transient capture', which ensures even the fastest spikes and intermittent signals are captured and displayed for detailed examination.

Adjustments

Selecting specific components from a menu can enable simple quick adjustment to be made. Live readings are displayed appropriate to the selection.

MOT (annual) emissions test

Full MOT procedure tests are integrated and displayed on the screen with pass/fail diagnosis to the department of transport specifications, for both gas analysis and diesel smoke (if appropriate options are fitted). The test results include engine rpm and oil temperature as well as the gas readings. These can all be printed for garage or customer use.

The connections to the vehicle for standard use are much the same for most equipment manufacturers. These are listed in Table 3.4.

Figure 3.9 shows a 'Sun' digital oscilloscope engine analyser. This test equipment has many features as listed previously, and others such as the following.

- High technology test screens. Vacuum waveform, cylinder time balance bar graph, power balance waveform and dual trace laboratory scope waveform.
- Scanner interface. This allows the technician to observe all related information at the same time.
- Expanded memory. This feature allows many screens to be saved at once, then recalled at a later time for evaluation and reference.

The tests are user controlled whereas some machines have pre-programmed sequences. Some of the screens available are given in Table 3.5.

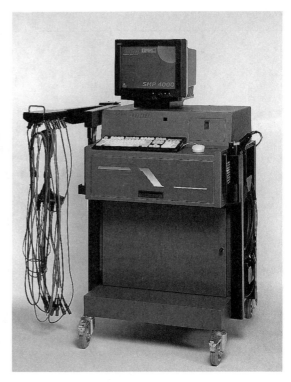

Figure 3.9 Engine analyser

3.3.4 Exhaust gas measurement

It has now become standard to measure four of the main exhaust gases, namely:

- carbon monoxide (CO),
- carbon dioxide (CO_2),
- hydrocarbons (HC),
- oxygen (O_2).

The emission test module is often self contained with its own display but can be linked to the main analyser display. Often, the lambda value and the air–fuel ratio are displayed in addition to the four gasses. The Greek symbol lambda (λ) is used to represent the ideal air–fuel ratio (AFR) of $14.7:1$ by mass. In other words, just

Table 3.4 Connections

Connection	Purpose or example of use
Battery positive	Battery and charging voltages
Battery negative	A common earth connection
Coil positive	To check supply voltage to coil
Coil negative (adaptors are available for DIS)	To look at dwell, rev/min and primary waveforms
Coil HT lead clamp (adaptors are available for DIS)	Secondary waveforms
Number one cylinder plug lead clamp	Timing light and sequence of waveforms
Battery cable amp clamp	Charging and starting current
Oil temperature probe (dip stick hole)	Oil temperature
Vacuum connection	Engine load

Table 3.5 Screens on a digital oscilloscope

Primary	Secondary	Diagnostic	Cylinder test
Primary waveform	Secondary waveform	Voltage waveform	Vacuum waveform
Primary parade waveform	Secondary parade waveform	Lab scope waveform	Power balance waveform
Dwell bar graph	kV histogram	Fuel injector waveform	Cylinder time balance bar graph
Duty cycle/dwell bar graph	kV bar graph	Alternator waveform	Cylinder shorting even/odd bar graph
Duty cycle/voltage bar graph	Burn time bar graph		Cranking amps bar graph

the right amount of air to burn up all the fuel. Typical gas, lambda and AFR readings are given in Table 3.6 for a closed loop lambda control system, before (or without a catalytic connecter) and after the catalytic converter. These are for a modern engine in excellent condition (and are examples only – always check current data).

The composition of exhaust gas is now a critical measurement and hence a certain degree of accuracy is required. To this end the infrared measurement technique has become the most suitable for CO, CO_2 and HC. Each individual gas absorbs infrared radiation at a specific rate. Oxygen is measured by electro-chemical means in much the same way as the on-vehicle lambda sensor.

CO is measured as shown in Figure 3.10. The emitter is heated to about $700\,^{\circ}C$ which, by using a suitable reflector, produces a beam of infrared light. This beam is passed via a chopper disk, through a measuring cell to a receiver chamber. This sealed chamber contains a gas with a defined content of CO (in this case). This gas absorbs some of the CO specific radiation and its temperature increases. This causes expansion and therefore a gas flow from chamber 1 to chamber 2. This flow is detected by a flow sensor, which produces an AC output signal. This is converted and calibrated to a zero CO reading. The AC signal is produced due to the action of the chopper disk. If the chopper disk was not used then the flow from chamber 1 to chamber 2 would only take place when the machine was switched on or off.

If the gas to be measured is now pumped through the measuring cell, some of the infrared radiation will be absorbed before it reaches the receiver chamber. This varies the heating effect on the CO specific gas and hence the measured flow between chambers 1 and 2 will change. The flow meter will produce a change in its AC signal, which is converted, and then output to a suitable display. A similar technique is used for the measurement of CO_2 and HC. At present it is not possible to measure nitrogen oxides (NOx) without the most sophisticated laboratory equipment. Research is being carried out in this area.

Good four-gas emission analysers often have the following features:

- A stand-alone unit is not dependent on other equipment.
- Graphical screens simultaneously display up to four values as graphs, and the display order is user selectable. Select from HC, CO, CO_2, O_2 and rev/min for graphical display.

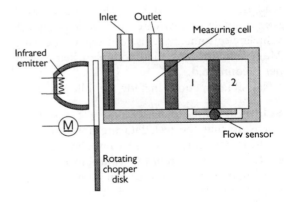

Figure 3.10 Carbon monoxide measurement technique

Table 3.6 Gas, lambda and AFR readings

Reading:	CO(%)	HC(ppm)	CO_2(%)	O_2(%)	Lambda (λ)	AFR
Before catalyst	0.6	120	14.7	0.7	1.0	14.7
After catalyst	0.2	12	15.3	0.1	1.0	14.7

- The user can create personalized letterheads for screen printouts.
- The non-dispersive infrared (NDIR) method of detection (each individual gas absorbs infrared light at a specific rate) is used.
- Display screens may be frozen or stored in memory for future retrieval.
- Recalibrate at the touch of a button (if the calibration gas and a regulator are used).
- Display exhaust gas concentrations in real time numerics or create live exhaust gas data graphs in selectable ranges.
- Calculate and display lambda (λ) (the ideal air–fuel ratio in about 14.7 : 1).
- Display engine rev/min in numeric or graphical form and display oil temperature along with current time and date.
- Display engine diagnostic data from a scanner.
- Operate from mains supply or a 12 V battery.

Accurate measurement of exhaust gas is not only required for MOT tests but is essential to ensure an engine is correctly tuned. Table 3.6 lists typical values measured from a typical exhaust. Note the toxic emissions are small, but nonetheless dangerous.

3.3.5 Serial port communications – the scanner

Serial communication is an area that is continuing to grow. A special interface is required to read data. This standard is designed to work with a single or two-wire port, which connects vehicle electronic systems to a diagnostic plug. Many functions are then possible when a scanner is connected.

Possible functions include the following.

- Identification of ECU and the system to ensure that the test data are appropriate to the system currently under investigation.
- Read-out of current live values from sensors so that spurious figures can be easily recognized. Information – such as engine speed, temperature air flow and so on – can be displayed and checked against test data.
- System function simulation allows actuators to be tested by moving them and watching for a suitable response.
- Programming of system changes. Basic idle CO or changes in basic timing can be programmed into the system.

At present, several standards exist, which means several different types of serial readers are needed, or at best several different adapters and program modules. A new standard, called On-Board Diagnostics II (OBD II), has been developed by the Society of Automotive Engineers (SAE). In the USA, all new vehicles must conform to this standard. This means that just one scan tool will work with all new vehicles. A similar standard, known as EOBD, has also recently been adopted in Europe.

A company called GenRad produces scanners to meet these standards. Figure 3.11 shows an example. This scanner allows the technician to perform all the necessary operations, such as fault code reading, via a single common connector. The portable hand-held tool has a large graphics display allowing clear instructions and data. Context-sensitive help is available to eliminate the need to refer back to manuals to look up fault code definitions. It has a memory, so data can be reused even after disconnecting the tool from the power supply. This scanner will even connect to the Controller Area Network (CAN) systems.

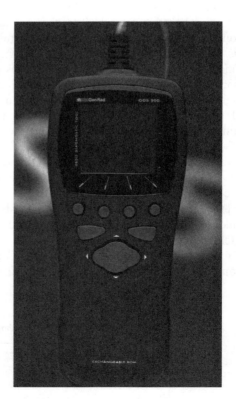

Figure 3.11 OBD II scanner

3.4 Dedicated equipment

3.4.1 Introduction

As the electronic complexity of the modern vehicle continues to increase, developments in suitable test equipment must follow. The term 'dedicated' implies test equipment used for only one specific system. Figure 3.12 is a representation of one type of dedicated test equipment. A special plug and socket is used to 'break in' to the ECU wiring, whilst in many cases still allowing the vehicle system to function normally.

Readings can be taken between various points and compared with set values, thus allowing diagnosis. Ford have used a system such as this for many years, known simply as a breakout box. A multimeter takes the readings between predetermined test points on the box which are connected to the ECU wiring.

A further development of this system is a digitally controlled tester that will run very quickly through a series of tests and display the results. These can be compared with stored data allowing a pass/fail output.

Many electronic systems now have ECUs that contain self-diagnosis circuits. This is represented in Figure 3.13. Activating the blink code output can access the information held in the ECU memory. This is done in some cases by connecting two wires and then switching on the ignition. A further refinement is to read the information via a serial link, which requires suitable test equipment.

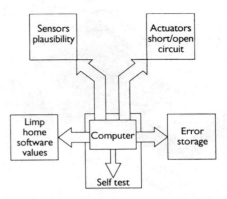

Figure 3.13 Many electronic systems now have ECUs containing self-diagnosis circuits

3.4.2 Serial port communications

A special interface of the type that is stipulated by ISO 9141 is required to read data. This standard is designed to work with a single- or two-wire port allowing many vehicle electronic systems to be connected to a central diagnostic plug. The sequence of events to extract data from the ECU is as follows.

- Test unit transmits a code word.
- ECU responds by transmitting a baud rate recognition word.
- Test unit adopts the appropriate setting.
- ECU transmits fault codes.

The test unit converts these to suitable output text. Further functions are possible, including the following.

- Identification of the ECU and system to ensure that the test data are appropriate to the system currently under investigation.
- Read out of current live values from sensors. Spurious figures can be easily recognized. Information such as engine speed, temperature, airflow and so on, can be displayed and checked against test data.
- System function stimulation to allow actuators to be tested by moving them and watching for a suitable response.
- Programming of system changes such as basic idle CO or changes in basic timing can be programmed into the system.

3.4.3 Laser 2000 electronic systems tester

The Lucas Laser 2000 tester (Figure 3.14) is designed to find faults on vehicles with electronic

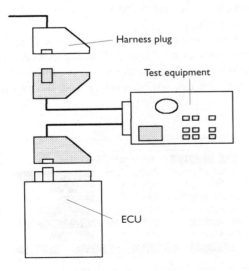

Figure 3.12 One type of dedicated test equipment where a special plug and socket is used to 'break in' to the ECU wiring

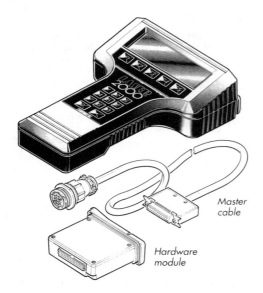

Figure 3.14 Lucas Laser 2000 Tester

systems. These may include fuelling, anti-lock braking traction control etc. Fault finding on such systems can be difficult and slow due to the complexity of the electronics.

Some ECUs have on-board diagnostics (OBD), which means that the ECU monitors its inputs and outputs and, if any are incorrect, stores a fault code. A warning lamp will also be illuminated to alert the driver. These codes can be read out and a test procedure followed.

More advanced ECUs have a data link (serial communications as described above), so fault codes can be displayed using a tester, as a number or with a text description. Other information, such as operating values, can also be passed to the tester. Commands can also be sent from the tester to the ECU, for example to operate a solenoid.

Depending upon the diagnostic features of the ECU, the Laser 2000 system can provide all these options. Furthermore, data may be logged whilst driving the vehicle, fault codes can be erased and the results of up to four tests may be stored within the Laser 2000 system. The output may be printed if required.

The Laser 2000 system is configured for different systems by a hardware module, a software module, a master cable and an adaptor cable. The hardware module is to allow future upgrades, one module covers most current systems. The software contains the test routines for a system or range of systems. Adaptor cables are required to connect to each vehicle diagnostic connector. The Laser 2000 tester is designed to

be user friendly – it is menu driven by function keys relating to the menu on the screen.

One of the most interesting features, besides those mentioned above, is the Laser 2000's snapshot mode. In this mode, as well as displaying live data, the tester can record values over a period. This is very useful for identifying intermittent faults as the recorded data can be replayed slowly, the results being displayed in numerical as well as graphical format. Figure 3.15 shows a typical snapshot screen on the Laser 2000 system. Gradual changes are to be expected but sudden changes in, say, air flow could indicate a fault in the air flow sensor or wiring. Snapshot data may also be printed in tabular form via the RS232 interface.

3.5 On-board diagnostics

Figure 3.16 shows the Bosch Motronic M5 with the On-Board Diagnostics II (OBD II) system. On-board diagnostics are becoming essential for the longer term operation of a system, such as producing a clean exhaust. In the USA, a very comprehensive diagnosis of all the components in the system that affect the exhaust is now required. It can be expected that, in due course, a similar requirement will be made within the EC. Any fault detected must be indicated to the driver by a warning light.

Digital electronics allow both sensors and actuators to be monitored. This is done by allocating values to all operating states of the sensors and actuators. If a deviation from these figures is detected, this is stored in the memory and can be output in the workshop to assist with fault-finding.

Monitoring of the ignition system is very important as misfiring not only produces more emissions of hydrocarbons, but the unburnt fuel enters the catalytic converter and burns there.

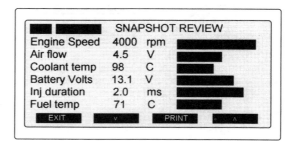

Figure 3.15 Snapshot of the Laser 2000 Screen

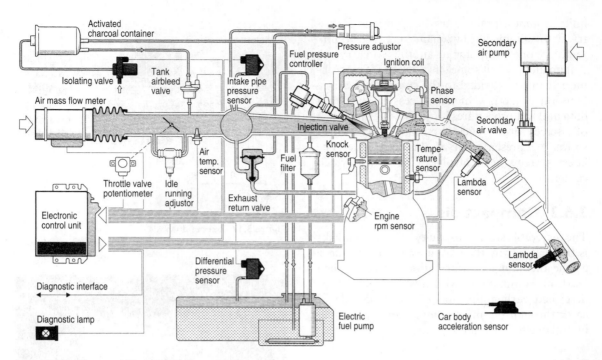

Figure 3.16 Motronic M5 with OBD II

This can cause higher than normal temperatures and may damage the catalytic converter. An accurate crankshaft speed sensor is used to monitor ignition and combustion in the cylinders. Misfiring alters the torque of the crankshaft for an instant, which causes irregular rotation. This allows a misfire to be recognized instantly.

A number of further sensors are required for the OBD II functions. Another lambda sensor, after the catalytic converter, monitors its operation. An intake pressure sensor and a valve are needed to control the activated charcoal filter to reduce and monitor evaporative emissions from the fuel tank. A differential pressure sensor also monitors the fuel tank permeability. As well as the driver's fault lamp, a considerable increase in the electronics is required in the control unit in order to operate this system. A better built-in monitoring system, it is thought, will have a greater effect in reducing vehicle emissions than tighter annual testing.

The diagnostic socket used by systems conforming to OBD II standards should have the following pin configuration:

1. Manufacturer's discretion.
2. Bus + Line, SAE J1850.
3. Manufacturer's discretion.
4. Chassis ground.
5. Signal ground.
6. Manufacturer's discretion.
7. K Line, ISO 9141.
8. Manufacturer's discretion.
9. Manufacturer's discretion.
10. Bus – Line, SAE J1850.
11. Manufacturer's discretion.
12. Manufacturer's discretion.
13. Manufacturer's discretion.
14. Manufacturer's discretion.
15. L line, ISO 9141.
16. Vehicle battery positive.

With future standards and goals set it should be beneficial for vehicle manufacturers to begin implementation of at least the common connector in the near future. Many diagnostic system manufacturers would welcome this move. If lack of standardization continues it will become counter-productive for all concerned.

3.6 New developments in testing

3.6.1 Networking

The next development in diagnosis and testing is likely to be increased networking of vehicles, via adaptors, to computers locally, and then via modems and the telephone lines. It is already

quite common practice in the computer industry, with suitable hardware and software, to link
remotely one computer to another to carry out
diagnostics and, in some cases repairs. This technique can be extended to the computerized systems on modern vehicles. Access to the latest
data and test procedures is available at the 'touch
of a screen' or the 'click of a mouse'. Figure 3.17
shows a representation of this technique. The
latest systems even involve a hand-held video
camera.

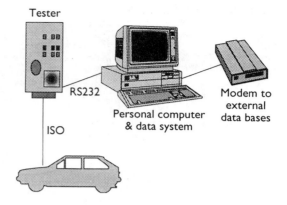

Figure 3.17 Remote diagnosis

3.6.2 Compact disks

The incredible storage capacity of compact disks
is the reason why they are used more and more
as the medium for information storage. When
used in conjunction with test equipment as
described earlier, the operator will be able to
work through the most complex of faults, with
interactive help from the computer.

3.6.3 Integrated diagnostic and measurement systems

The information in the following sections is from
a company known as GenRad. Other companies
produce systems and some vehicle manufacturers have developed similar equipment. GenRad
is a leader in this area, hence I have chosen this
company's products to illustrate the current state
of diagnostic and similar systems.

Figure 3.18 GenRad diagnostic system

3.6.4 GenRad Diagnostic System (GDS)

This system as shown in Figure 3.18 is a fully
featured measurement system, integrated with
guided diagnostics for fast and accurate fault
diagnosis on complex electrical and electronic
systems. Hand-held and portable, the unit is
much more than a fault code reader. Capable of
capturing intermittent faults as well as multiple
and interrelated faults, the diagnostic system,
known as the GDS, can save significant amounts
of service time.

The system is highly user friendly; communication with the technician takes place via an
easy-to-read liquid crystal display (LCD) with a
rugged touchscreen. The display has also been
designed for enhanced visibility in a variety of
lighting conditions. The GDS is ergonomically
designed and features an impact resistant case
with integral carrying handle for easy portabil-
ity. It also features a stand to ease positioning

and colour coded connector inserts for probe
identification.

The GDS has 1 Mbyte of SRAM as standard
with optional extra memory giving a total capacity of up to 4 Mbytes for extended diagnostic
routines. The GDS's versatility means that the
unit also forms the basis of a comprehensive
engine monitoring and analysis system, including ignition, cranking, charging and fuel injection diagnostics by the addition of software and
transducers.

An optional extra to the GDS is a spread
spectrum wireless link, allowing communication between more than one unit and other PC
systems. Measurement parameters and data
can be communicated, allowing remote analysis by an operator where access is limited.
Applications can also be loaded to the GDS
via an RF wireless link, precluding the need for
an operator to return to a base station to
download software.

General specification

Display	LCD Graphics 320 × 240 pixels
Touchscreen	Analogue, 256 × 256 resolution
Memory	1, 2 or 4 Mbyte RAM options
Power sources	Internal 1.8 Ah NiCd rechargeable battery or vehicle battery. Base station with automatic 3 hour fast charge to internal battery.
Battery life	Typically 60 minutes with display and stimulus active (4 V into 1000 Ω).
Approx. weight	2.5 kg
Approx. size	230 mm × 310 mm × 70 mm
Operating temperature range	0°C to +50°C
Storage temperature range	20°C to +60°C
Ingress protection	IP54

Capabilities

- Dual channel AC and DC auto-ranging bipolar voltage measurement.
- Dual channel sampling oscilloscope.
- Single channel positive and negative peak voltage detect.
- Auto-ranging resistance measurement.
- Dual channel timing (period, pulse width, duty cycle and frequency).
- Triggered voltage measurement.
- Waveform generator/sensor stimulation/resistance simulation.
- Current.
- Pressure.
- Temperature.

Communications

- CAN
- ISO 9141
- J1850 VPW
- J1850 PWM
- KWP2000

Accessories

- Automatic breakout box option.
- Base Station with CD-ROM drive, unit docking, charge, date download, RS232 or radio modem interface.

3.6.5 Multi Protocol Adaptor (MPA)

GenRad's Multi Protocol Adaptor (MPA), shown in Figure 3.19, provides an interface between the serial communications on a vehicle and the host diagnostic unit. The unit provides a communications path between the vehicle and

Figure 3.19 Multi protocol adaptor

the host unit across the selected serial protocol. The host diagnostic unit is also able to download executable code or applications to the MPA, allowing the host computer to be disconnected or freed up to perform other tasks. This increases the diagnostic unit's flexibility and efficiency.

The stand-alone 'flight recorder' capability of the MPA is instrumental in the detection of specific real time events, such as intermittent faults, which are notoriously difficult to trace. This is invaluable for a technician in improving accuracy and efficiency. The MPA also provides a vehicle flash programming output, which may be used to provide vehicle ECU programming 'in-situ' via the vehicle diagnostic connector.

The MPA can be customized to meet specific requirements, for example, to allow a vehicle manufacturer to meet environmental protection legislation.

Also available as an option to the MPA is a Controller Area Network (CAN) active cable that provides a trouble-free connection directly onto the vehicle's CAN bus. Using an active CAN cable, the technician is able to connect off-board test and diagnostic equipment with long cables to the vehicle's CAN bus via the OBD II (J1962) connector.

General specification

Host computer interface	RS 485/RS 232 (opto-isolated)
Vehicle serial communications interface	ISO 9141 – 2, J1850, CAN (vers2B)
Vehicle FLASH programming supply	12 V to 19 V @ 100 mA
Auxiliary and ignition relay control	2 × 2 amp open drain outputs
RAM (non-volatile to 24 hours)	128 kbytes to 4 Mbytes
BOOT ROM	512 kbytes (in-circuit reprogrammable)
Analogue measurements	10 channels, 8 bits
Power switching to hose	Vbatt @ 2 amp (non-isolated but switchable under from the host computer)
Vehicle connection	Rugged captive J1962 cable (2 metres)
Host connection	6 m cable, connector to suit host type

3.6.6 The Electronic Service Bay

The Electronic Service Bay (ESB) is an open system providing electronic information exchange and guided diagnostics for the service technician, serving all aspects of the dealership service operations. With increasingly sophisticated vehicles being brought to market and with the huge variety of vehicle variants throughout the world, the service technician is often faced with a vast amount of information from which specific details need to be retrieved.

In addition, vehicle manufacturers, as information providers, are subject to pressures to disseminate more and more data throughout their networks, including external directives such as new legislation. In short, there has been an explosion of information in the Service Bay – and at a time when quality of service has itself become a major differentiator. For a correct and timely repair the service technician must be able to access accurate information that is up to date and vehicle specific.

To address this, GenRad's Electronic Service Bay brings relevant information from several sources together either to run on a standard PC or to run on a GDS3000. Using a GDS3000 as the hardware platform gives the added and powerful advantage of bringing all the sourced information together on one single portable unit for the service technician. It is a rugged unit, making it ideal for workshop or roadside use. Information is presented in a consistent, convenient way and it is designed to be user-friendly, so that computing expertise is not essential for the operator.

The information available in the Electronic Service Bay is also valuable to other areas of the Dealership – especially the parts counter. Components of GenRad's Electronic Service Bay can also be used on a standard office computer – so the information is available to anyone who may need it. The information stored in the ESB is displayed in Data Viewers. The following Data Viewers are included.

Service manuals

Information, which is currently delivered in paper manuals, is now presented as an electronic book. Electronic books are much more flexible than their paper equivalents – and can easily be read in random order (rather than sequentially). The ability to navigate freely around a book is very appropriate to a reference manual.

Diagnostic sequences

A large proportion of existing service manuals is devoted to fault-finding sequences. In the Electronic Service Bay these sequences are presented as step-by-step instructions – hence eliminating the need for large sections of existing manuals. These automated sequences are able to take measurements and read values from the vehicle without the need for the technician to use a variety of special tools. In addition, Automatic Diagnostics are often able to use techniques that would be difficult for a service technician even if he or she had the special tools. As an example, the diagnostic sequence may include oscilloscope functionality with automatic interpretation of the results.

Parts catalogues

Apart from ordering parts, assembly and compatibility information can also be made available to the service technician by displaying parts catalogues on the same screen as other information.

Service bulletins

Even with electronic publishing and delivery there are always last minute changes caused by unexpected vehicle and component problems. If a technician is to make effective use of Service Bulletins, it is important that they are presented at the right time. For this reason Service Bulletins are connected with other elements of the Electronic Service Bay so that they appear alongside the other information.

Although the Electronic Service Bay contains all these separate items, it is much more than 'putting them all on one computer.' The real need is to make all this information available at the right time. To make this work, the Electronic Service Bay observes the following principles.

- All the Data Viewers work in the same way. This means standard buttons and navigation mechanisms, so once one viewer has been mastered, the others are immediately familiar.
- There is never any need to enter the same information twice. Once the system knows about a vehicle, then all the Data Viewers can use this knowledge.
- The Data Viewers are simple to operate. They require only a touchscreen for all operations. Keyboards are avoided.

GenRad's ESB runs on a desktop or portable PC with Windows 95 or Windows NT (minimum specifications 486 DX2 66 MHz, 16 MB RAM,

VGA display) or with the GDS3000 which allows integration of analogue measurement and vehicle communications.

3.7 Diagnostic procedures

3.7.1 Introduction

Finding the problem when complex automotive systems go wrong – is easy. Well, it is easy if you have the necessary knowledge. This knowledge is in two parts:

1. an understanding of the system in which the problem exists,
2. the ability to apply a logical diagnostic routine.

It is also important to be clear about two definitions:

Symptom(s) – what the user of the system (vehicle or whatever) notices,
Fault – the error in the system that causes the symptom(s).

The knowledge requirement and use of diagnostic skills can be illustrated with a very simple example in the next section.

3.7.2 The 'theory' of diagnostics

One theory of diagnostics can be illustrated by the following example.

After connecting a hosepipe to the tap and turning on the tap, no water comes out of the end. Your knowledge of this system tells you that water should come out providing the tap is on, because the pressure from a tap pushes water through the pipe, and so on. This is where diagnostic skills become essential. The following stages are now required.

1. Confirm that no water is coming out by looking down the end of the pipe!
2. Does water come out of the other taps, or did it come out of this tap before you connected the hose?
3. Consider what this information tells you, for example, the hose must be blocked or kinked.
4. Walk the length of the pipe looking for a kink.
5. Straighten out the hose.
6. Check that water now comes out and that no other problems have been created.

The procedure just followed made the hose work but it is also guaranteed to find a fault in any system. It is easy to see how it works in connection with a hosepipe, but I'm sure anybody could have found that fault! The skill is to be able to apply the same logical routine to more complex situations. The routine can be summarized by the following six steps.

1. Verify the fault.
2. Collect further information.
3. Analyse the evidence.
4. Carry out further tests to locate the fault.
5. Fix the fault.
6. Check this and other associated systems for correct operation.

Steps 3 and 4 will form a loop until the fault is located. Remember that with a logical process you will not only ensure you do find the fault, you will also save time and effort.

3.7.3 Waveforms

In this section I will first explain the principle of using an oscilloscope for displaying waveforms and then examine a selection of actual waveforms. You will find that both the words 'waveform' and 'patterns' are used in books and workshop manuals – they mean the same thing.

When you look at a waveform on a screen you must remember that the height of the scale represents voltage and the width represents time. Both of these axes can have their scales changed. They are called axes because the 'scope' is drawing a graph of the voltage at the test points over a period of time. The time scale can vary from a few μs to several seconds. The voltage scale can vary from a few mV to several kV. For most test measurements only two connections are needed, just like a voltmeter. The time scale will operate at intervals pre-set by the user. It is also possible to connect a 'trigger' wire so that, for example, the time scale starts moving across the screen each time the ignition coil fires. This keeps the display in time with the speed of the engine. When you use a full engine analyser, all the necessary connections are made as listed in a previous section. Figure 3.20 shows an example waveform.

For each of the following waveforms I have noted what is being measured, the time and voltage settings, and the main points to examine for correct operation. All the waveforms shown are from a correctly operating vehicle. The skill you will learn by practice is to note when your own measurements vary from those shown here.

- Inductive pulse generator output (Figure 3.21).

- Hall effect pulse generator output (Figure 3.22).
- Primary circuit pattern (Figure 3.23).
- Secondary circuit pattern – one cylinder (Figure 3.24).
- Secondary circuit pattern – four cylinders called parade (Figure 3.25).
- Alternator ripple voltage (Figure 3.26).
- Injector waveform (Figure 3.27).

- Injector waveform with current limiting (Figure 3.28).
- Air flow meter output (Figure 3.29).
- Lambda sensor voltage (Figure 3.30).
- Full load switch operation (Figure 3.31).
- ABS wheel speed sensor output signal (Figure 3.32).
- Vehicle speed sensor (Figure 3.33).

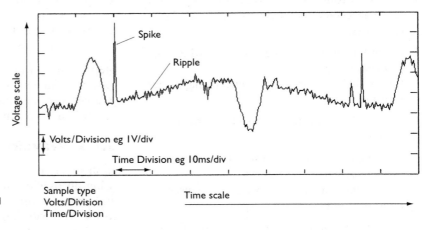

Figure 3.20 How to 'read' an oscilloscope trace (a random signal is shown)

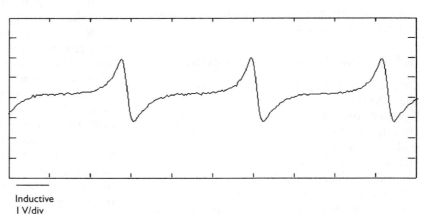

Figure 3.21 Inductive pulse generator output

Inductive
1 V/div
10 ms/div

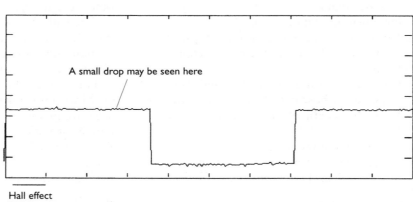

Hall effect
3 V/div
5 ms/div

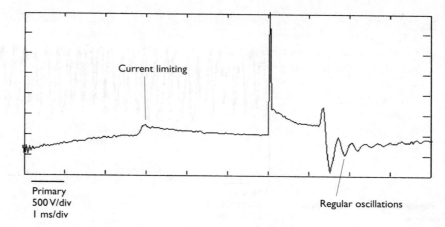

Figure 3.23 Primary circuit pattern

Primary
500 V/div
1 ms/div

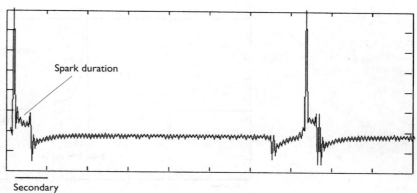

Figure 3.24 Secondary circuit pattern – one cylinder

Secondary
2 kV/div
5 ms/div

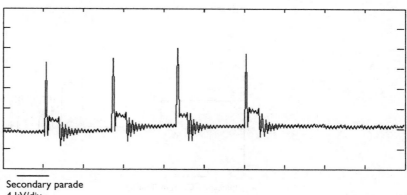

Figure 3.25 Secondary circuit pattern – four cylinders

Secondary parade
4 kV/div
1 ms/div

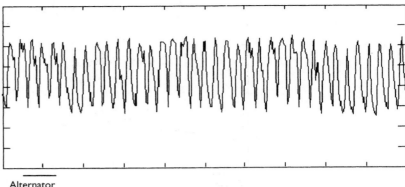

Figure 3.26 Alternator ripple
voltage

Alternator
200 mV/div dc
200 ms/div

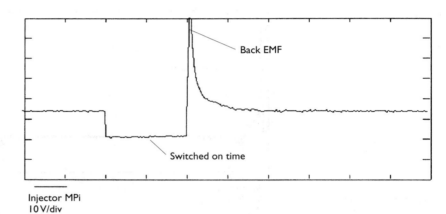

Figure 3.27 Injector waveform

Injector MPi
10 V/div
2 ms/div

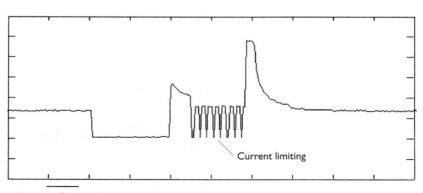

Figure 3.28 Injector waveform
with current limiting

Injector - Current limiting
10 V/div
500 us/div

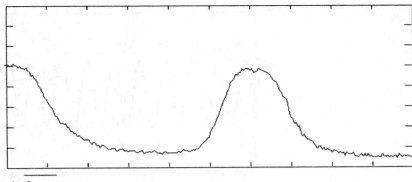

Figure 3.29 Air flow meter output

Air flow meter
1 V/div
100 ms\div

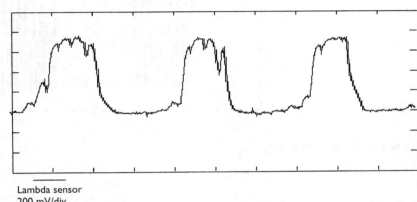

Figure 3.30 Lambda sensor voltage

Lambda sensor
200 mV/div
200 ms/div

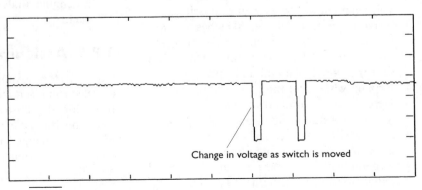

Change in voltage as switch is moved

Figure 3.31 Full load switch operation

Full load switch
5 V/div dc
10 ms/div

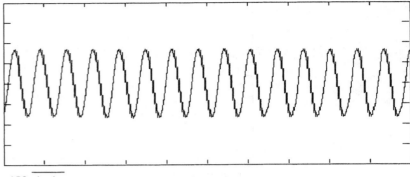

Figure 3.32 ABS wheel speed sensor output signal

ABS wheel sensor
1 V/div
50 ms/div

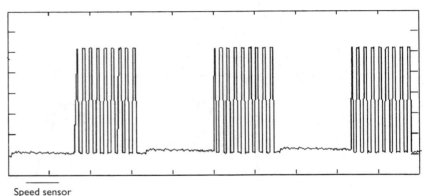

Speed sensor
1 V/div dc
5 ms/div

Figure 3.33 Venicle speed sensor

3.8 Self-assessment

3.8.1 Questions

1. State five essential characteristics of an electrical test multimeter.
2. Describe why the internal resistance of a voltmeter should be as high as possible.
3. Make two clearly labelled sketches to show the waveforms on an oscilloscope when testing the output from an ignition Hall effect sensor at low speed and high speed.
4. Explain what is meant by a 'serial port'.
5. Describe briefly four advantages for the technician, of a standardized diagnostic plug.
6. State why a 'code reader' or 'scanner' is an important piece of test equipment.
7. Explain what is meant by an 'integrated diagnostic and measurement system.

8. Using the six-stage diagnostic procedure discussed in this chapter, write out an example relating to testing a charging system.
9. Describe the meaning of accuracy in relation to test equipment.
10. List the main test connections required for an engine analyser and state the purpose of each.

3.8.2 Assignment

Consider the advantages and disadvantages of an 'electronic service bay'. Discuss the implications for the customer and for the repairer. Examine how the service and repair environment has changed over the last 20 years and comment on what may be the situation in the next 20.

4

Electrical systems and circuits

4.1 The systems approach

4.1.1 What is a system?

System is a word used to describe a collection of related components, which interact as a whole. A motorway system, the education system or computer systems, are three varied examples. A large system is often made up of many smaller systems, which in turn can each be made up of smaller systems and so on. Figure 4.1 shows how this can be represented in a visual form.

One further definition of a system: 'A group of devices serving a common purpose'.

Using the systems approach helps to split extremely complex technical entities into more manageable parts. It is important to note however, that the links between the smaller parts and the boundaries around them are also very important. System boundaries will also overlap in many cases.

The modern motor vehicle is a very complex system and, in itself, forms just a small part of a larger transport system. It is the ability for the motor vehicle to be split into systems on many levels that aids both in its design and construction. In particular, the systems approach helps to understand how something works and, furthermore, how to go about repairing it when it doesn't!

4.1.2 Vehicle systems

Splitting the vehicle into systems is not an easy task because it can be done in many different ways. From the viewpoint of this book a split between mechanical systems and electrical systems would seem a good start. This division though, can cause as many problems as it solves. For example, in which half do we put anti-lock brakes, mechanical or electrical? The answer is of course both! However, even with this problem it still makes it easier to be able to consider just one area of the vehicle and not have to try to comprehend the whole. Most of the chapters in this book are major sub-systems of the vehicle and, indeed, the sub-headings are further divisions. Figure 4.2 shows a simplified vehicle system block diagram.

Once a complex set of interacting parts, such as a motor vehicle, has been systemized, the function or performance of each part can be examined in more detail. Functional analysis determines what each part of the system should do and, in turn, can determine how each part actually works. It is again important to stress that the links and interactions between various sub-systems are a very important consideration. An example of this would be how the power requirements of the vehicle lighting system will have an effect on the charging system operation.

To analyse a system further, whatever way it has been subdivided from the whole, consideration should be given to the inputs and the outputs of the system.

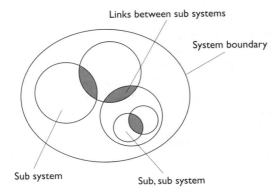

Figure 4.1 A system can be made up of smaller systems

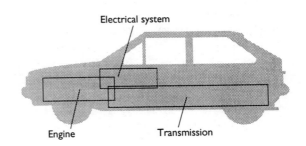

Figure 4.2 Simplified vehicle system block diagram

Many of the complex electronic systems on a vehicle lend themselves to this form of analysis. Considering the electronic control unit (ECU) of the system as the control element, and looking at its inputs and outputs, is the recommended approach.

4.1.3 Open loop systems

An open loop system is designed to give the required output whenever a given input is applied. A good example of an open loop vehicle system would be the headlights. With the given input of the switch being operated, the output required is that the headlights will be illuminated. This can be taken further by saying that an input is also required from the battery and a further input of, say, the dip switch. The feature that determines that a system is open loop, is that no feedback is required for the system to operate. Figure 4.3 shows this example in block diagram form.

4.1.4 Closed loop systems

A closed loop system is identified by a feedback loop. It can be described as a system where there is a possibility of applying corrective measures if the output is not quite what is desired. A good example of this in a vehicle is an automatic temperature control system. The interior temperature of the vehicle is determined by the output from the heater, which is switched on or off in response to a signal from a temperature sensor inside the cabin. The feedback loop is due to the fact that the output from the system, i.e., temperature, is also an input to the system. This is represented in Figure 4.4.

The feedback loop in any closed loop system can be in many forms. The driver of a car with a conventional heating system can form a feedback loop by turning the heater down when he

Figure 4.3 Open loop system

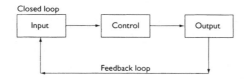

Figure 4.4 Closed loop system

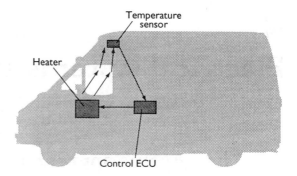

Figure 4.5 Closed loop heating system

or she is too hot and turning it back up when cold. The feedback to a voltage regulator in an alternator is an electrical signal using a simple wire.

4.1.5 Summary

Many complex vehicle systems are represented in this book as block diagrams. In this way several inputs can be shown supplying information to an ECU which, in turn, controls the system outputs. Figure 4.5 shows the cabin temperature control system using a block diagram.

4.2 Electrical wiring, terminals and switching

4.2.1 Cables

Cables used for motor vehicle applications are almost always copper strands insulated with PVC. Copper, beside its very low resistivity of about $1.7 \times 10^{-8}\,\Omega$ m, has ideal properties such as ductility and malleability. This makes it the natural choice for most electrical conductors. PVC as the insulation is again ideal, as it not only has very high resistance, the order of $10^{15}\,\Omega$ m, but is also very resistant to petrol, oil, water and other contaminants.

The choice of cable size depends on the current drawn by the consumer. The larger the cable used then the smaller the volt drop in the circuit, but the cable will be heavier. This means a trade-off must be sought between the allowable volt drop and maximum cable size. Table 4.1 lists some typical maximum volt drops in a circuit.

In general, the supply to a component must not be less than 90% of the system supply. If a

Table 4.1 Typical maximum volt drops

Circuit (12 V)	Load	Cable drop (V)	Maximum drop (V) incl. connections
Lighting circuit	<15 W	0.1	0.6
Lighting circuit	>15 W	0.3	0.6
Charging circuit	Nominal	0.5	0.5
Starter circuit	Maximum at 20°C	0.5	0.5
Starter solenoid	Pull-in	1.5	1.9
Other circuits	Nominal	0.5	1.5

vehicle is using a 24 V supply, the figures in Table 4.1 should be doubled.

Volt drop in a cable can be calculated as follows:

Calculate the current $\quad I = P/V_s$

Volt drop $\qquad\qquad V_d = I\rho l/A$

where: I = current in amps, P = power rating of component in watts, V_s = system supply in volts, V_d = volt drop in volts, ρ = resistivity of copper in Ω m, l = length of the cable in m, A = cross-sectional area in m^2.

A transposition of this formula will allow the required cable cross-section to be calculated

$$A = I\rho l/Vd$$

where I = maximum current in amps, and V_d = maximum allowable volt drop in volts.

Cable is available in stock sizes and Table 4.2 lists some typical sizes and uses. The current rating is assuming that the cable length is not excessive and that operating temperature is within normal limits. Cables normally consist of multiple strands to provide greater flexibility.

4.2.2 Colour codes and terminal designations

As seems to be the case for any standardization, a number of colour code and terminal designation systems are in operation. For reference purposes I will just make mention of three. First, the British Standard system (BS AU 7a: 1983). This system uses 12 colours to determine the main purpose of the cable and tracer colours to further define its use. The main colour uses and some further examples are given in Table 4.3.

A European system used by a number of manufacturers is based broadly on Table 4.4. Please note that there is no correlation between the 'Euro' system and the British standard colour codes. In particular, note the use of the colour brown in each system. After some practice with the use of colour code systems the job of the technician is made a lot easier when fault-finding an electrical circuit.

A system now in use almost universally is the terminal designation system in accordance with DIN 72 552. This system is to enable easy and correct connections to be made on the vehicle,

Table 4.2 Cables and their applications

Cable strand diameter (mm)	Cross-sectional area (mm^2)	Continuous current rating (A)	Example applications
9/0.30	0.6	5.75	Sidelights etc
14/0.25	0.7	6.0	Clock, radio
14/0.30	1.0	8.75	Ignition
28/0.30	2.0	17.5	Headlights, HRW
44/0.30	3.1	27.5	
65/0.30	4.6	35.0	Main supply
84/0.30	5.9	45.0	
97/0.30	6.9	50.0	Charging wires
120/0.30	8.5	60.0	
37/0.90	23.5	350.0	Starter supply
to		to	
61/0.90	39.0	700.0	

Table 4.3 British Standard colour codes for cables

Colour	Symbol	Destination/use
Brown	N	Main battery feed
Blue	U	Headlight switch to dip switch
Blue/White	U/W	Headlight main beam
Blue/Red	U/R	Headlight dip beam
Red	R	Sidelight main feed
Red/Black	R/B	Left-hand sidelights and number plate
Red/Orange	R/O	Right-hand sidelights
Purple	P	Constant fused supply
Green	G	Ignition controlled fused supply
Green/Red	G/R	Left-hand side indicators
Green/White	G/W	Right-hand side indicators
Light Green	LG	Instruments
White	W	Ignition to ballast resistor
White/Black	W/B	Coil negative
Yellow	Y	Overdrive and fuel injection
Black	B	All earth connections
Slate	S	Electric windows
Orange	O	Wiper circuits (fused)
Pink/White	K/W	Ballast resistor wire
Green/Brown	G/N	Reverse
Green/Purple	G/P	Stop lights
Blue/Yellow	U/Y	Rear fog light

Table 4.4 European colour codes for cables

Colour	Symbol	Destination/use
Red	Rt	Main battery feed
White/Black	Ws/Sw	Headlight switch to dip switch
White	Ws	Headlight main beam
Yellow	Ge	Headlight dip beam
Grey	Gr	Sidelight main feed
Grey/Black	Gr/Sw	Left-hand sidelights
Grey/Red	Gr/Rt	Right-hand sidelights
Black/Yellow	Sw/Ge	Fuel injection
Black/Green	Sw/Gn	Ignition controlled supply
Black/White/Green	Sw/Ws/Gn	Indicator switch
Black/White	Sw/Ws	Left-hand side indicators
Black/Green	Sw/Gn	Right-hand side indicators
Light Green	LGn	Coil negative
Brown	Br	Earth
Brown/White	Br/Ws	Earth connections
Pink/White	KW	Ballast resistor wire
Black	Sw	Reverse
Black/Red	Sw/Rt	Stop lights
Green/Black	Gn/Sw	Rear fog light

particularly in after-sales repairs. It is important, however, to note that the designations are not to identify individual wires but are to define the terminals of a device. Listed in Table 4.5 are some of the most popular numbers.

Ford motor company now uses a circuit numbering and wire identification system. This is in use worldwide and is known as Function, System-Connection (FSC). The system was devel-

Table 4.5 Popular terminal designation numbers

1	Ignition coil negative
4	Ignition coil high tension
15	Switched positive (ignition switch output)
30	Input from battery positive
31	Earth connection
49	Input to flasher unit
49a	Output from flasher unit
50	Starter control (solenoid terminal)
53	Wiper motor input
54	Stop lamps
55	Fog lamps
56	Headlamps
56a	Main beam
56b	Dip beam
58L	Left-hand sidelights
58R	Right-hand sidelights
61	Charge warning light
85	Relay winding out
86	Relay winding input
87	Relay contact input (change over relay)
87a	Relay contact output (break)
87b	Relay contact output (make)
L	Left-hand side indicators
R	Right-hand side indicators
C	Indicator warning light (vehicle)

oped to assist in vehicle development and production processes. However, it is also very useful to help the technician with fault-finding. Many of the function codes are based on the DIN system. Note that earth wires are now black!

The system works as follows (see Tables 4.6 and 4.7):

31S-AC3A ‖ 1.5 BK/RD

Function
  31 = ground/earth
  S = additionally switched circuit

System
  AC = headlamp levelling

Connection
  3 = switch connection
  A = branch

Size
  1.5 = 1.5 mm^2

Colour
  BK = Black (determined by function 31)
  RD = Red stripe

As a final point to this section, it must be noted that the colour codes and terminal designations given are for illustration only. Further reference should be made for specific details to manufacturer's information.

Table 4.6 New Ford colour codes table

Code	Colour
BK	Black
BN	Brown
BU	Blue
GN	Green
GY	Grey
LG	Light-Green
OG	Orange
PK	Pink
RD	Red
SR	Silver
VT	Violet
WH	White
YE	Yellow

Table 4.7 System codes table

Letter	Main system	Examples
D	Distribution systems	DE = earth
A	Actuated systems	AK = wiper/washer
B	Basic systems	BA = charging BB = starting
C	Control systems	CE = power steering
G	Gauge systems	GA = level/pressure/temperature
H	Heated systems	HC = heated seats
L	Lighting systems	LE = headlights
M	Miscellaneous systems	MA = air bags
P	Powertrain control systems	PA = engine control
W	Indicator systems ('indications' not turn signals)	WC = bulb failure
X	Temporary for future features	XS = too much!

4.2.3 Harness design

The vehicle wiring harness has developed over the years from a loom containing just a few wires, to the looms used at present on top range vehicles containing well over 1000 separate wires. Modern vehicles tend to have wiring harnesses constructed in a number of ways. The most popular is still for the bundle of cables to be spirally wrapped in non-adhesive PVC tape. The tape is non-adhesive so as to allow the bundle of wires to retain some flexibility, as shown in Figure 4.6.

Another technique often used is to place the cables side by side and plastic weld them to a backing strip as shown in Figure 4.7. This method allows the loom to be run in narrow areas, for example behind the trim on the inner sill or under carpets.

A third way of grouping cables, as shown in Figure 4.8 is to place them inside PVC tubes. This has the advantage of being harder wearing and, if suitable sealing is arranged, can also be waterproof.

When deciding on the layout of a wiring loom within the vehicle, many issues must be considered. Some of there are as follows.

1. Cable runs must be as short as possible.
2. The loom must be protected against physical damage.
3. The number of connections should be kept to a minimum.
4. Modular design may be appropriate.

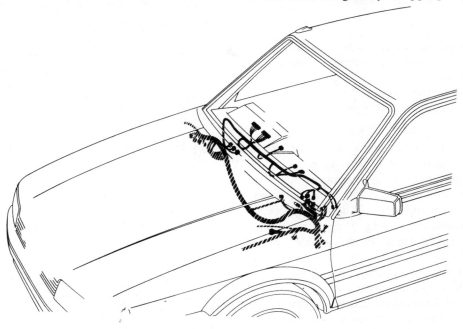

Figure 4.6 PVC wound harness

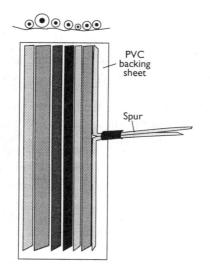

Figure 4.7 Cables side by side and plastic welded to a backing strip

Figure 4.8 PVC tube and tape harness

5. Accident damage areas to be considered.
6. Production line techniques should be considered.
7. Access must be possible to main components and sub-assemblies for repair purposes.

From the above list – which is by no means definitive – it can be seen that, as with most design problems, some of the main issues for consideration are at odds with each other. The more connections involved in a wiring loom, then the more areas for potential faults to develop. However, having a large multiplug assembly, which connects the entire engine wiring to the rest of the loom, can have considerable advantages. During production, the engine and all its ancillaries can be fitted as a complete unit if supplied ready wired, and in the after-sales repair market, engine replacement and repairs are easier to carry out.

Because wiring looms are now so large, it is often necessary to split them into more manageable sub-assemblies. This will involve more connection points. The main advantage of this is that individual sections of the loom can be replaced if damaged.

Keeping cable runs as short as possible will not only reduce volt drop problems but will allow thinner wire to be used, thus reducing the weight of the harness, which can now be quite considerable.

The overall layout of a loom on a vehicle will broadly follow one of two patterns; that is, an 'E' shape or an 'H' shape (Figure 4.9). The 'H' is the more common layout. It is becoming the norm to have one or two main junction points as part of the vehicle wiring with these points often being part of the fuse box and relay plate.

Figure 4.10 shows a more realistic representation of the harness layout. This figure also serves to show the level of complexity and number of connection points involved. It is the aim of multiplexed systems (discussed later) to reduce these problems and provide extra 'communication' and diagnostic facilities.

4.2.4 Printed circuits

The printed circuit is used almost universally on the rear of the instrument pack and other similar places. This allows these components to be supplied as complete units and also reduces the amount and complexity of the wiring in what are usually cramped areas.

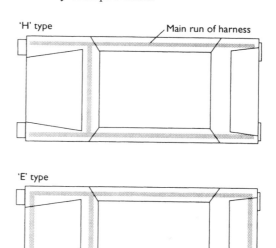

Figure 4.9 'H' and 'E' wiring layouts

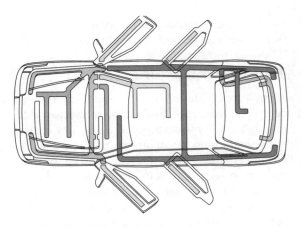

Figure 4.10 Typical wiring harness layout

The printed circuits are constructed using a thin copper layer that is bonded to a plastic sheet – on both sides in some cases. The required circuit is then printed on to the copper using a material similar to wax. The unwanted copper can then be etched away with an acid wash. A further layer of thin plastic sheet can insulate the copper strips if required.

Figure 4.11 shows a picture of a typical printed circuit from an instrument panel and gives some indication as to how many wires would be required to do the same job. Connection to the main harness is by one or more multiplugs.

4.2.5 Fuses and circuit breakers

Some form of circuit protection is required to protect the electrical wiring of a vehicle and also to protect the electrical and electronic compo-

nents. It is now common practice to protect almost all electrical circuits with a fuse. The simple definition of a fuse is that it is a deliberate weak link in the circuit. If an overload of current occurs then the fuse will melt and disconnect the circuit before any serious damage is caused. Automobile fuses are available in three types, glass cartridge, ceramic and blade type. The blade type is the most popular choice owing to its simple construction and reliability against premature failure due to vibration. Figure 4.12 shows different types of fuse.

Fuses are rated with a continuous and peak current value. The continuous value is the current that the fuse will carry without risk of failure, whereas the peak value is the current that the fuse will carry for a short time without failing. The peak value of a fuse is usually double the continuous value. Using a lighting circuit as an example, when the lights are first switched on a very high surge of current will flow due to the low (cold) resistance of the bulb filaments. When the filament resistance increases with temperature, the current will reduce, thus illustrating the need for a fuse to be able to carry a higher current for a short time.

To calculate the required value for a fuse, the maximum possible continuous current should be worked out. It is then usual to choose the next highest rated fuse available. Blade fuses are available in a number of continuous rated values as listed in Table 4.8 together with their colour code.

The chosen value of a fuse as calculated above must protect the consumer as well as the wiring. A good example of this is a fuse in a

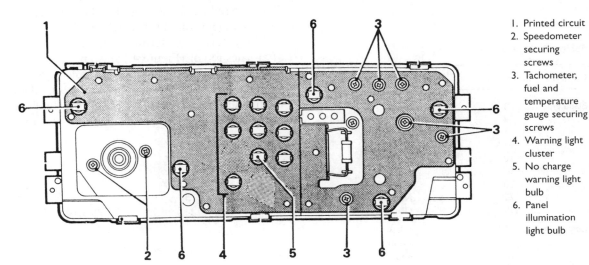

1. Printed circuit
2. Speedometer securing screws
3. Tachometer, fuel and temperature gauge securing screws
4. Warning light cluster
5. No charge warning light bulb
6. Panel illumination light bulb

Figure 4.11 Four-gauge instrument pack

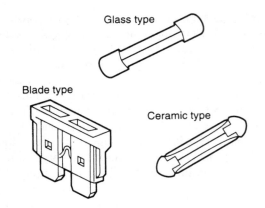

Figure 4.12 Different types of fuse

wiper motor circuit. If a value were used that is much too high, it would probably still protect against a severe short circuit. However, if the wiper blades froze to the screen, a large value fuse would not necessarily protect the motor from overload.

It is now common practice to use fusible links in the main output feeds from the battery as protection against major short circuits in the event of an accident or error in wiring connections. These links are simply heavy duty fuses and are rated in values such as 50, 100 or 150 A.

Occasionally, circuit breakers are used in place of fuses, this being more common on heavy vehicles. A circuit breaker has the same rating and function as a fuse but with the advantage that it can be reset. The disadvantage is the much higher cost. Circuit breakers use a bimetallic strip which, when subjected to excessive current, will bend and open a set of con-

tacts. A latch mechanism prevents the contacts from closing again until a reset button is pressed.

4.2.6 Terminations

Many types of terminals are available and have developed from early bullet-type connectors into the high quality waterproof systems now in use. A popular choice for many years was the spade terminal. This is still a standard choice for connection to relays for example, but is now losing ground to the smaller blade or round terminals as shown in Figure 4.13. Circular multipin connectors are used in many cases, the pins varying in size from 1 mm to 5 mm. With any type of multipin connector, provision must always be made to prevent incorrect connection.

Protection against corrosion of the actual connector is provided in a number of ways. Earlier methods included applying suitable grease to the pins to repel water. It is now more usual to use rubber seals to protect the terminals, although a small amount of contact lubricant can still be used.

Many multiway connectors employ some kind of latch to prevent individual pins working loose, and also the complete plug and socket assembly is often latched. Figure 4.14 shows several types of connector.

For high quality electrical connections, the contact resistance of a terminal must be kept to a minimum. This is achieved by ensuring a tight join with a large surface area in contact, and by using a precious metal coating often containing silver. It is worth noting that many connections are only designed to be removed a limited number of times before deterioration in effectiveness. This is to reduce the cost of manufacture but can cause problems on older vehicles.

Many forms of terminal are available for after-sales repair (Figure 4.15), some with more success than others. A good example is sealed terminals, which in some cases are specified by the manufacturers for repair purposes. These are pre-insulated polyamide terminations that pro-

Table 4.8 Blade fuses

Continuous current (A)		Colour
Blade type	3	Violet
	4	Pink
	5	Clear/Beige
	7.5	Brown
	10	Red
	15	Blue
	20	Yellow
	25	Neutral/White
	30	Green
Ceramic type	5	Yellow
	8	White
	16	Red
	25	Blue

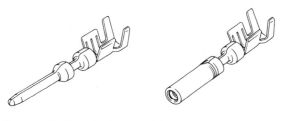

Figure 4.13 'Round' crimp terminals

Figure 4.14 Terminals and connector blocks

vide a tough, environment resistant connection for most wire sizes used on motor vehicles. They simultaneously insulate, seal and protect the joint from abrasion and mechanical abuse. The stripped wire is inserted into the metallic barrel and crimped in the usual way. The tubing is then heated and adhesive flows under pressure from the tubing, filling any voids and providing an

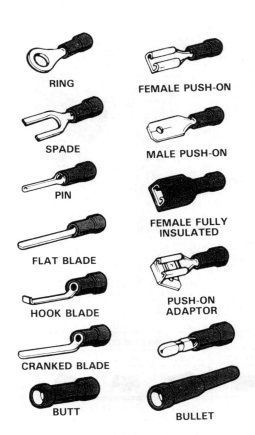

Figure 4.15 Crimp terminals for repair work

excellent seal with the cable. The seal prevents the ingress of water and other fluids, preventing electrolytic action. The connection is also resistant to temperature changes.

4.2.7 Switches

Developments in ergonomics and styling have made the simple switch into quite a complex issue. The method of operation of the switch must meet various criteria. The grouping of switches to minimize driver fatigue and distraction, access to a switch in an emergency and hazards from switch projections under impact conditions are just some of the problems facing the designer. It has now become the norm for the main function switches to be operated by levers mounted on the steering column. These functions usually include; lights, dip, flash, horn, washers and wipers. Other control switches are mounted within easy reach of the driver on or near the instrument fascia panel. As well as all the design constraints already mentioned, the reliability of the switch is important. Studies have shown that, for example, a headlamp dip switch may be operated in the region of 22 000 times during 80 000 km (50 000) miles of vehicle use (about 4 years). This places great mechanical and electrical stress on the switch.

A simple definition of a switch is 'a device for breaking and making the conducting path for the current in a circuit'. This means that the switch can be considered in two parts; the contacts, which perform the electrical connection, and the mechanical arrangement, which moves the contacts. There are many forms of operating mechanisms, all of which make and break the contacts. Figure 4.16 shows just one common method of sliding contacts.

The characteristics the contacts require are simple:

1. Resistance to mechanical and electrical wear.
2. Low contact resistance.
3. No build up of surface films.
4. Low cost.

Materials often used for switch contacts include copper, phosphor bronze, brass, beryllium copper and in some cases silver or silver alloys. Gold is used for contacts in very special applications. The current that a switch will have to carry is the major consideration as arc erosion of the contacts is the largest problem. Silver is one of the best materials for switch contacts and one way of getting around the obvious problem of cost is to

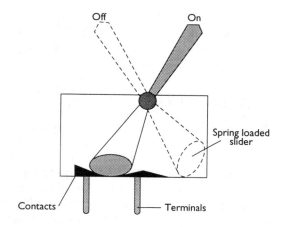

Figure 4.16 Switch with sliding contacts

have only the contact tips made from silver, by resistance welding the silver to, for example, brass connections. It is common practice now to use switches to operate a relay that in turn will operate the main part of the circuit. This allows far greater freedom in the design of the switch due to very low current, but it may be necessary to suppress the inductive arc caused by the relay winding. It must also not be forgotten that the relay is also a switch, but as relays are not con-

strained by design issues the very fast and positive switching action allows higher currents to be controlled.

The electrical life of a switch is dependent on its frequency of operation, the on–off ratio of operation, the nature of the load, arc suppression and other circuit details, the amount of actuator travel used, ambient temperature and humidity and vibration levels, to name just a few factors.

The range of size and types of switches used on the motor vehicle is vast, from the contacts in the starter solenoid, to the contacts in a sunroof micro switch. Figure 4.17 shows just one type of motor vehicle switch together with its specifications.

Some of the terms used to describe switch operation are listed below.

Free position Position of the actuator when no force is applied.
Pretravel Movement of the actuator between the free and operating position.
Operating position Position the actuator takes when contact changeover takes place.
Release position Actuator position when the mechanism resets.

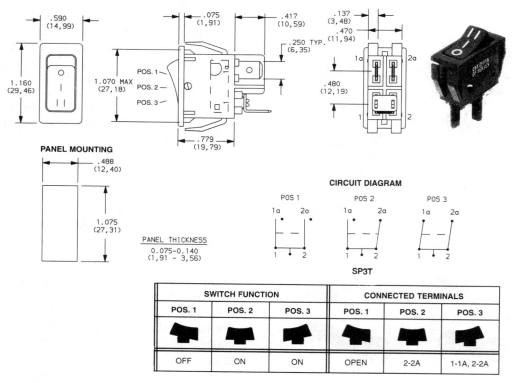

Figure 4.17 Single-pole triple-throw rocker switch

Overtravel Movement of the actuator beyond the operating position.

Total travel Sum of pretravel and overtravel.

Actuating force Force required to move the actuator from the free to the operating position.

Release force Force required to allow the mechanism to reset.

The number of contacts, the number of poles and the type of throw are the further points to be considered in this section. Specific vehicle current consumers require specific switching actions. Figure 4.18 shows the circuit symbols for a selection of switches and switching actions. Relays are also available with contacts and switching action similar to those shown.

So far, all the switches mentioned have been manually operated. Switches are also available, however, that can operate due to temperature, pressure and inertia, to name just three. These three examples are shown in Figure 4.19. The temperature switch shown is typical of those used to operate radiator cooling fans and it operates by a bimetal strip which bends due to tempera-

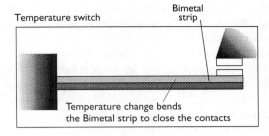

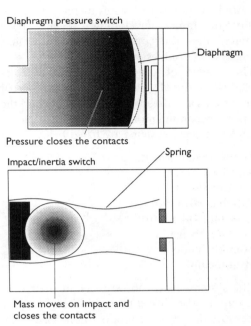

Figure 4.19 Temperature, pressure and inertia switches

ture and causes a set of contacts to close. The pressure switch shown could be used to monitor over-pressure in an air conditioning system and simply operates by pressure on a diaphragm which, at a pre-determined pressure, will overcome spring tension and close (or open) a set of contacts. Finally, the inertia switch is often used to switch off the supply to a fuel injection pump in the event of an impact to the vehicle.

4.3 Multiplexed wiring systems

4.3.1 Limits of the conventional wiring system

The complexity of modern wiring systems has been increasing steadily over the last 25 years or so and, in recent years, has increased dramatically. It has now reached a point where the size and weight of the wiring

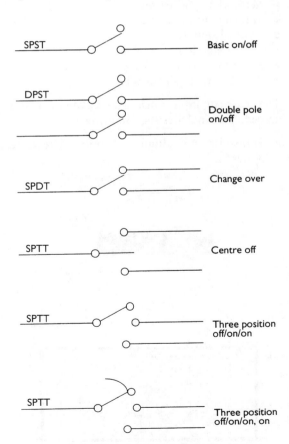

Figure 4.18 Circuit symbols for a selection of switches and switching actions

harness is a major problem. The number of separate wires required on a top-of-the-range vehicle can be in the region of 1500! The wiring loom required to control all functions in or from the driver's door can require up to 50 wires, the systems in the dashboard area alone can use over 100 wires and connections. This is clearly becoming a problem as, apart from the obvious issues of size and weight, the number of connections and the number of wires increase the possibility of faults developing. It has been estimated that the complexity of the vehicle wiring system doubles every 10 years.

The number of systems controlled by electronics is continually increasing. A number of these systems are already in common use and the others are becoming more widely adopted. Some examples of these systems are listed below:

● Engine management.
● Anti-lock brakes.
● Traction control.
● Variable valve timing.
● Transmission control.
● Active suspension.
● Communications.
● Multimedia.

All the systems listed above work in their own right but are also linked to each other. Many of the sensors that provide inputs to one electronic control unit are common to all or some of the others. One solution to this is to use one computer to control all systems. This, however, would be very expensive to produce in small numbers. A second solution is to use a common data bus. This would allow communication between modules and would make the information from the various vehicle sensors available to all sensors.

Taking this idea a stage further, if data could be transmitted along one wire and made available to all parts of the vehicle, then the vehicle, wiring could be reduced to just three wires. These wires would be a mains supply, an earth connection and a signal wire. The idea of using just one line for many signals is not new and has been in use in areas such as telecommunications for many years. Various signals can be 'multiplexed' on to one wire in two main ways – frequency division and time division multiplexing. Frequency division is similar to the way radio signals are transmitted. It is oversimplifying a complex subject, but a form of time division multiplexing is generally used for transmission of digital signals.

A ring main or multiplexed wiring system is represented in Figure 4.20. This shows that the data bus and the power supply cables must 'visit' all areas of the vehicle electrical system. To illustrate the operation of this system, consider the events involved in switching the sidelights on and off. First, in response to the driver pressing the light switch, a unique signal is placed on the data bus. This signal is only recognized by special receivers built as part of each light unit assembly, and these in turn will make a connection between the power ring main and the lights. The events are similar to turn off the lights, except that the code placed on the data bus will be different and will be recognized only by the appropriate receivers as an off code.

4.3.2 Multiplex data bus

In order to transmit different data on one line, a number of criteria must be carefully defined and agreed. This is known as the communications protocol. Some of the variables that must be defined are as follows:

● Method of addressing.
● Transmission sequence.
● Control signals.
● Error detection.
● Error treatment.
● Speed or rate of transmission.

The physical layer must also be defined and agreed. This includes the following:

● Transmission medium, e.g. copper wire, fibre optics etc.

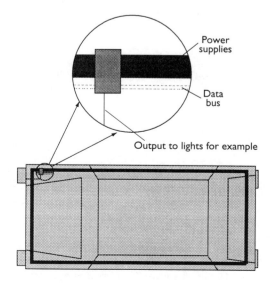

Figure 4.20 Multiplexed 'ring main' wiring system

- Type of transmission coding, e.g. analogue or digital.
- Type of signals, e.g. voltage, current or frequency etc.

The circuit to meet these criteria is known as the bus interface and will often take the form of a single integrated circuit. This IC will, in some cases, have extra circuitry in the form of memory for example. It may, however, be appropriate for this chip to be as cheap as possible due to the large numbers required on a vehicle. As is general with any protocol system, it is hoped that one only will be used. This, however, is not always the case.

4.3.3 Bosch CAN (Controller Area Networks)

Bosch has developed the protocol known as 'CAN' or Controller Area Network. This system is claimed to meet practically all requirements with a very small chip surface (easy to manufacture, therefore cheaper). CAN is suitable for transmitting data in the area of drive line components, chassis components and mobile communications. It is a compact system, which will make it practical for use in many areas. Two variations on the physical layer are available that suit different transmission rates. One is for data transmission of between 100K and 1M baud (bits per second), to be used for rapid control devices. The other will transmit between 10K and 100K baud as a low-speed bus for simple switching and control operations.

CAN modules are manufactured by a number of semiconductor firms such as Intel and Motorola. A range of modules is available in either Voll-CAN for fast buses and basic-CAN for lower data rates. These are available in a stand-alone format or integrated into various microprocessors. All modules have the same CAN protocol. It is expected that this protocol will become standardized by the International Standards Organization (ISO).

Many sensors and actuators are not yet 'busable' and, although prototype vehicles have been produced, the conventional wiring cannot completely be replaced. The electronic interface units must be placed near, or ideally integrated into, sensors and actuators. Particularly in the case of engine type sensors and actuators due to heat and vibration, this will require further development to ensure reliability and low price. Figure 4.21 shows the CAN bus system on a vehicle.

Significant use is now made of the data bus to allow ECUs to communicate. Figure 4.22 shows an example from a Volvo.

4.3.4 CAN signal format

The CAN message signal consists of a sequence of binary digits (bits). A voltage (or light in fibre optics) being present indicates the value '1' while none present indicates '0'. The actual message can vary between 44 and 108 bits in length. This is made up of a start bit, name, control bits, the data itself, a cyclic redundancy check (CRC) for error detection, a confirmation signal and finally a number of stop bits (Figure 4.23).

The name portion of the signal identifies the message destination and also its priority. As the

━━━ CAN bus for bodywork electrics and electronics

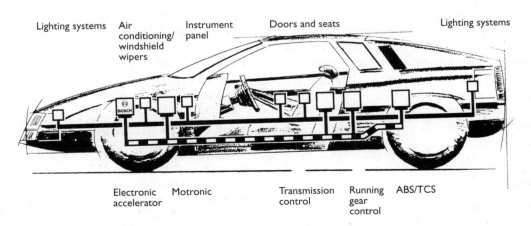

Lighting systems Air conditioning/ windshield wipers Instrument panel Doors and seats Lighting systems

Electronic accelerator Motronic Transmission control Running gear control ABS/TCS

Figure 4.21 Bosch CAN (Controller Area Network) data bus

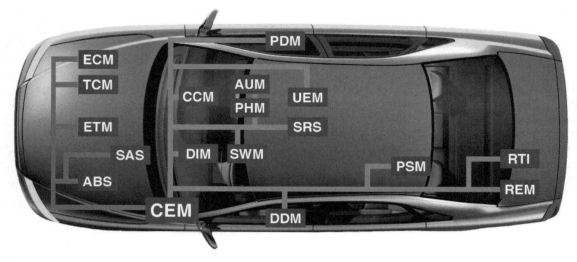

Figure 4.22 ECUs can communicate

transmitter puts a message on the bus it also reads the name back from the bus. If the name is not the same as the one it sent then another transmitter must be in operation that has a higher priority. If this is the case it will stop transmission of its own message. This is very important in the case of motor vehicle data transmission.

Errors in a message are recognized by the cyclic redundancy check. This is achieved by assembling all the numbers in a message into a complex algorithm and this number is also sent. The receiver uses the same algorithm and checks that the two numbers tally. If an error is recognized the message on the bus is destroyed. This is recognized by the transmitter, which then sends the message again. This technique, when combined with additional tests, makes it possible for no faulty messages to be transmitted without being discovered.

The fact that each station in effect monitors its own output, interrupts disturbed transmissions and acknowledges correct transmissions, means that faulty stations can be recognized and

uncoupled (electronically) from the bus. This will prevent other transmissions being disturbed incorrectly.

All messages are sent to all units and each unit makes the decision whether the message should be acted upon or not. This means that further systems can be added to the bus at any time and can make use of data on the bus without affecting any of the other systems.

Interference protection is required in some cases. Bus lines, which consist of copper wires, act as transmitting and receiving antennae in a vehicle. Suitable protective circuits can be used at lower frequencies and the bus can therefore be designed in the form of an unscreened two-wire line. These measures can only be used to a limited extent and screening is recommended. The use of optical fibres would completely solve the radiated interference problem. However, the coupling of transmitters and receivers as well as connections and junctions has, up until now, either not been reliable enough or too expensive. These problems are currently being examined and it is expected that the problem will be solved in the near future. Figure 4.24 shows a method of bus connection for a wire data bus.

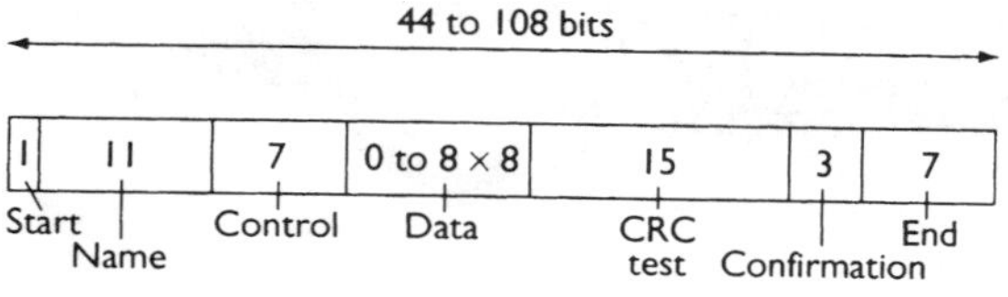

4.3.5 Local intelligence

A decision has to be made on a vehicle as to where the 'intelligence' will be located. The first solution is to use a local module, which will drive the whole of a particular sub-system. It is connected by means of conventional wires. This solves the problem of the number of wires run-

I	II	7	0 to 8 × 8	15	3	7

← 44 to 108 bits →

Start — Name — Control — Data — CRC test — Confirmation — End

Figure 4.23 A CAN 'word' is made up of a start bit, name, control bits, the data itself, a cyclic redundancy check (CRC) for error detection, a confirmation signal and stop bits

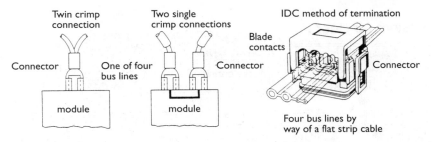

Figure 4.24 Data bus connections

ning from the vehicle body to the door but still involves a lot of wiring and connectors in the door. This can reduce reliability.

A second solution is to use intelligent actuators. This system involves the control electronics or intelligence being integrated into the actuators. In other words, the operating element accommodates the electronic functions that are necessary to code instructions and relay them. The actuators with their built-in control electronics perform the operating functions, such as adjusting the position of the mirror or opening and closing the windows.

The intelligence integrated into all the basic components in the form of a microprocessor with a basic CAN interface enables detailed self-diagnosis. A complete check at the end of the assembly line on the vehicle manufacturer's premises and rapid fault diagnosis in the workshop are both possible thanks to this intelligence. Figure 4.25 represents the two methods.

Much development is taking place on intelligent actuators and sensors, as this method appears to be the best choice for the future. Figure 4.26 shows a representation of a complete multiplexed subsystem.

4.3.6 Summary of CAN

CAN is a shared broadcast bus that runs at speeds up to 1 Mbit/s. It is based around sending messages (or *frames*), which are of variable length, between 0 and 8 bytes. Each frame has an *identifier*, which must be unique (i.e. two nodes on the same bus must not send frames with the same identifier). The interface between the CAN bus and the CPU is usually called the *CAN controller*.

The CAN protocol comes in two versions: CAN 1.0 and CAN 2.0. CAN 2.0 is backwards compatible with CAN 1.0, and most new controllers are CAN 2.0. There are two parts to the CAN 2.0 standard: part A and part B. With CAN 1.0 and CAN 2.0A, identifiers must be 11-

bits long. With CAN 2.0B identifiers can be 11-bits (a 'standard' identifier) or 29-bits (an 'extended' identifier). To comply with CAN 2.0 a controller must be either 2.0 part B passive, or 2.0 part B active. If it is passive, then it must ignore extended frames (CAN 1.0 controllers will generate error frames when they see frames with 29-bit identifiers). If it is active then it must allow extended frames to be received and transmitted. There are some compatibility rules for sending and receiving the two types of frames.

- The architecture of controllers is not covered by the CAN standard, so there is a variation in how they are used. There are, though, two general approaches: *BasicCAN* and *FullCAN* (not to be confused with CAN 1.0 and 2.0, or standard identifiers and extended identifiers); they differ in the buffering of messages.

Solution 1: local module

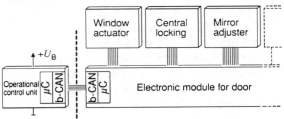

Solution 2: intelligent actuator

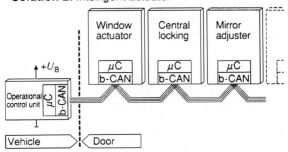

Figure 4.25 A complete check at the end of the assembly line on the vehicle manufacturer's premises and rapid fault diagnosis in the workshop are both possible thanks to local intelligence

Subsystem for doors

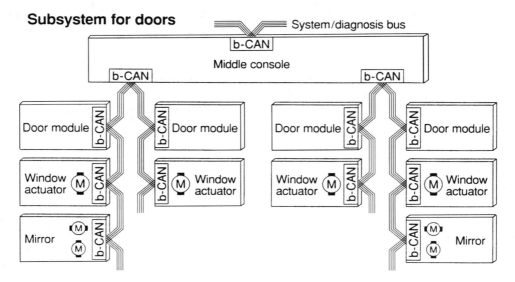

Figure 4.26 Complete multiplexed sub-system

- In a BasicCAN controller the architecture is similar to a UART, except that complete frames are sent instead of characters: there is (typically) a single transmit buffer, and a double-buffered receive buffer. The CPU puts a frame in the transmit buffer, and takes an interrupt when the frame is sent; the CPU receives a frame in the receive buffer, takes an interrupt and empties the buffer (before a subsequent frame is received). The CPU must manage the transmission and reception, and handle the storage of the frames.
- In a FullCAN controller the frames are stored in the controller. A limited number of frames can be dealt with (typically 16); because there can be many more frames on the network, each buffer is tagged with the identifier of the frame mapped to the buffer. The CPU can update a frame in the buffer and mark it for transmission; buffers can be

examined to see if a frame with a matching identifier has been received. Figure 4.27 represents the dual data bus system where a high-speed data bus is used for key engine and chassis systems, and a low-speed bus for other systems.

4.4 Circuit diagrams and symbols

4.4.1 Symbols

The selection of symbols given in Chapter 3, is intended as a guide to some of those in use. Some manufacturers use their own variation but a standard is developing. The idea of a symbol is to represent a component in a very simple but easily recognizable form. The symbol for a motor or for a small electronic unit deliberately

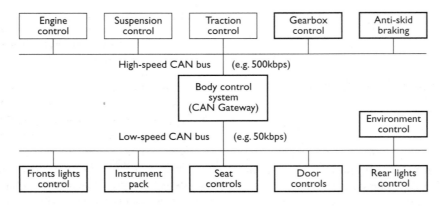

Figure 4.27 The vehicle dual data bus system

easily recognizable form. The symbol for a motor or for a small electronic unit deliberately leaves out internal circuitry in order to concentrate on the interconnections between the various devices.

Examples of how these symbols are used are given in the next three sections, which show three distinct types of wiring diagrams. Due to the complexity of modern wiring systems it is now common practice to show just part of the whole system on one sheet. For example, lights on one page, auxiliary circuits on the next, and so on.

4.4.2 Conventional circuit diagrams

The conventional type of diagram shows the electrical connections of a circuit but makes no attempt to show the various parts in any particular order or position. Figure 4.28 shows an example of this type of diagram.

4.4.3 Layout or wiring diagrams

A layout circuit diagram makes an attempt to show the main electrical components in a position similar to those on the actual vehicle. Owing to the complex circuits and the number of individual wires, some manufacturers now use two diagrams – one to show electrical connections and the other to show the actual layout of the wiring harness and components. Citroen, amongst others, has started to use this system. An example of this is reproduced in Figure 4.29.

4.4.4 Terminal diagrams

A terminal diagram shows only the connections of the devices and not any of the wiring. The terminal of each device, which can be represented pictorially, is marked with a code. This code indicates the device terminal designation, the destination device code and its terminal designation and, in some cases, the wire colour code. Figure 4.30 shows an example of this technique.

4.4.5 Current flow diagrams

Current flow diagrams are now very popular. The idea is that the page is laid out such as to show current flow from the top to the bottom. These diagrams often have two supply lines at the top of the page marked 30 (main battery positive supply) and 15 (ignition controlled supply). At the bottom of the page is a line marked 31 (earth or chassis connection). Figure 4.31 is a representation of this technique.

4.5 Case study

4.5.1 The smart electrical system of the future – Volvo S80

Although a car is not primarily experienced through its electrical system, the revolutionary new electrical system in the Volvo S80 has a natural position alongside engines, transmissions and chassis performance. To give an idea of what has happened to the electrical system in cars over the years, the first Volvo back in 1927 had four fuses, protecting a mere 30 m of electrical cable.

Seventy years later, the Volvo of 1997 had 54 fuses for 1200 m of cables and a host of functions that were totally unknown in 1927. For example, the total computer power in the car is more than 6 Mb. By tradition, each function had its own system and each system had one supplier. The capacity of the electrical system was measured in terms of the sum of the number of components. However, this could not continue because the need for a radical change was pressing.

A new system that could handle everything was needed. All the components had to be able to communicate and 'understand' one another's language as well as being integrated in one system. The Volvo S80 not only has a new electrical system – many cars have advanced electrical systems – but it uses the multiplex system for communications.

The electrical system is designed as a communication network of 18 computers with central control units and no fewer than 24 modules for most electrical functions. These modules function like computers and control the electrical functions in the car whenever necessary. Figure 4.22 earlier in this chapter shows the links between these systems.

Multiplex technology involves only two cables. One of them is able to carry all the signals in the system at the same time. The other is the electrical cable, which carries the necessary power. These cables run around the entire car and are known as the *databus*. The information travels in digital packages. All the small network modules are able to recognize 'their' signal for action and do as they are told.

When the signal 'open left front window'

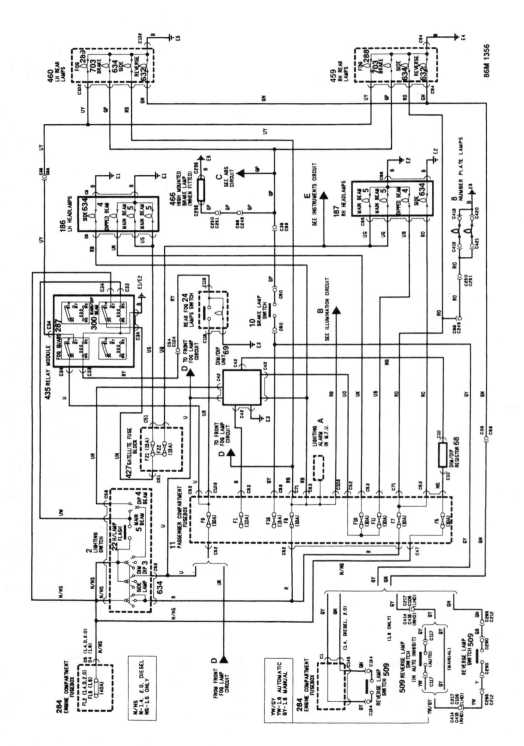

Figure 4.28 Conventional circuit diagram

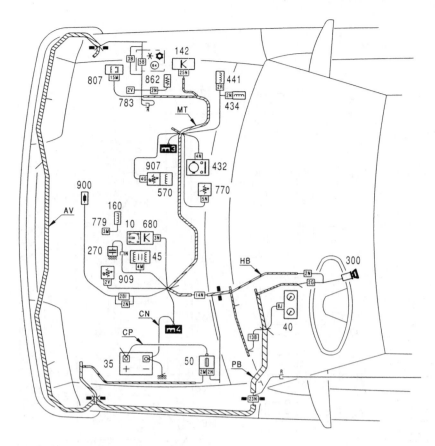

Figure 4.29 Layout diagram

arrives, for example, only one module (in the front door) reacts to it, receives it and transmits an 'order' to the electric motor to lower the window. Signals are able continuously to alert and activate the different modules as a result of the capacity of the system, which also operates at two speeds depending on the function. The engine and transmission management uses a high-speed databus, whereas all the other functions use a slightly slower data bus.

The benefits of the multiplex system are considerable:

- fewer cables and connections in the car;
- improved reliability;
- communication between all the components;
- software adaptations;
- easier and improved opportunities for the retro-installation of electrical functions.

The system also has the benefit of self-diagnosis for all functions, including engine management, making the OBD (on-board diagnostics) unit even more important than before. Diagnosis is easier, as is servicing. Any information about a fault or malfunction is passed on to the driver by indicator lamps and a message display in the instrument cluster. All the cables in the system are fitted in well-protected cable ducts. The multiplex system in each car is programmed according to model specifications and fitted options.

4.6 Electromagnetic compatibility (EMC)

4.6.1 Definitions

EMC – Electromagnetic compatibility
The ability of a device or system to function without error in its intended electromagnetic environment.

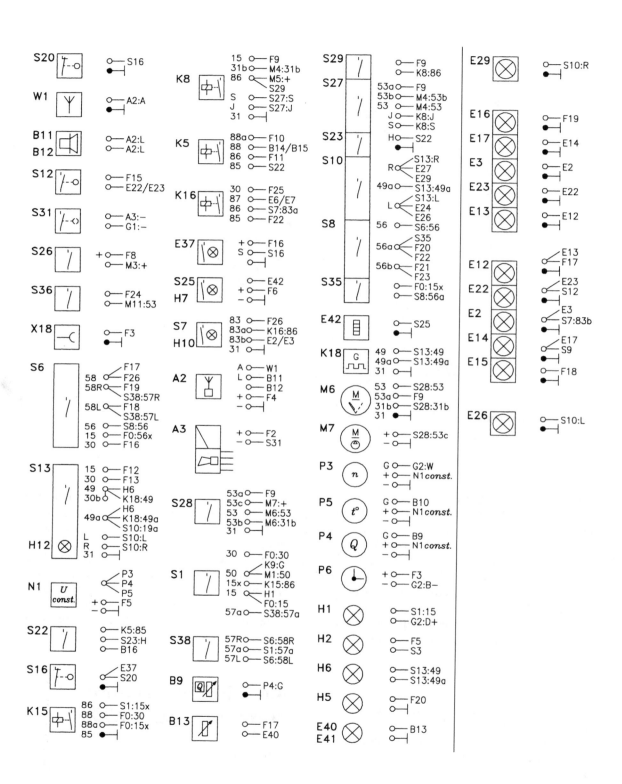

Figure 4.30 Terminal diagram

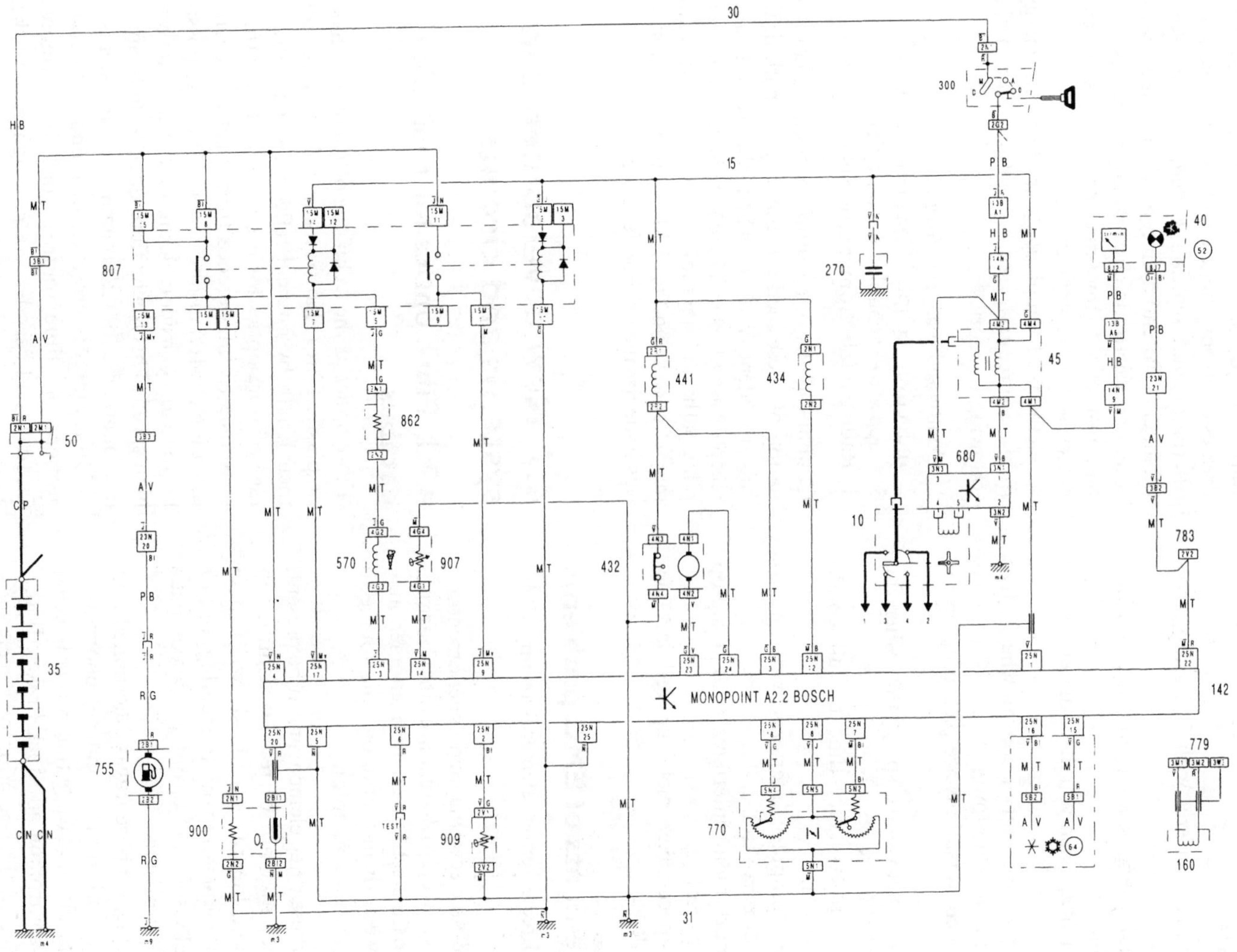

Figure 4.31 Current flow diagram

EMI – Electromagnetic interference

Electromagnetic emissions from a device or system that interfere with the normal operation of another device or system.

4.6.2 Examples of EMC problems

- A computer interferes with FM radio reception.
- A car radio buzzes when you drive under a power line.
- A car misfires when you drive under a power line.
- A helicopter goes out of control when it flies too close to a radio tower.
- CB radio conversations are picked up on the stereo.
- The screen on a video display jitters when fluorescent lights are on.
- The clock resets every time the air conditioner kicks in.
- A laptop computer interferes with an aircraft's rudder control!
- The airport radar interferes with a laptop computer display.
- A heart pacemaker picks up cellular telephone calls!

4.6.3 Elements of EMC problems

There are three essential elements to any EMC problem:

1. Source of an electromagnetic phenomenon.
2. Receptor (or victim) that cannot function properly due to the electromagnetic phenomenon.
3. Path between them that allows the source to interfere with the receptor.

Each of these three elements must be present, although they may not be readily identified in every situation. Identifying at least two of these elements and eliminating (or attenuating) one of them generally solves electromagnetic compatibility (EMC) problems.

For example, suppose it was determined that radiated emissions from a mobile telephone were inducing currents on a cable that was connected to an ECU controlling anti-lock brakes. If this adversely affected the operation of the circuit a possible coupling path could be identified.

Shielding, filtering, or re-routing of the cable may be the answer. If necessary, filtering or redesigning the circuit would be further possible methods of attenuating the coupling path to the point where the problem is non-existent.

Potential sources of electromagnetic compatibility problems include radio transmitters, power lines, electronic circuits, lightning, lamp dimmers, electric motors, arc welders, solar flares and just about anything that utilizes or creates electromagnetic energy. On a vehicle, the alternator and ignition system are the worst offenders. Potential receptors include radio receivers, electronic circuits, appliances, people, and just about anything that utilizes or can detect electromagnetic energy.

Methods of coupling electromagnetic energy from a source to a receptor fall into one of the following categories:

1. Conducted (electric current).
2. Inductively coupled (magnetic field).
3. Capacitively coupled (electric field).
4. Radiated (electromagnetic field).

Coupling paths often utilize a complex combination of these methods making the path difficult to identify even when the source and receptor are known. There may be multiple coupling paths and steps taken to attenuate one path may enhance another. EMC therefore is a serious issue for the vehicle designer.

4.7 New developments in systems and circuits

4.7.1 Fibre optics for multiplex databus

'Fibre optics' is the technique of using thin glass or plastic fibres that transmit light throughout their length by internal reflections. The advantage of fibre optics for use as a databus is their resistance to interference from electromagnetic radiation (EMR) or interference. It is also possible to send a considerable amount of data at very high speed. This is why fibre-optic technology is in common use for telecommunication systems.

Disadvantages, however, are found in the connection of fibre optics, and furthermore, encoders and decoders to 'put' signals onto the databus are more complex than when a normal wire is used. Figure 4.32 shows some current techniques for connecting fibre-optic cables.

4.7.2 The need for multiplexing

As an example of how the need for multiplexed systems is increasing, look at Figure 4.33. This figure shows the block diagram for an intelligent lighting system, but note how many sensor inputs are required. Much of this data would already be available on a databus.

This issue is one of the main reason for the development of multiplexed systems.

4.8 Self-assessment

4.8.1 Questions

1. Make a list of 10 desirable properties of a wiring terminal/connection.
2. Explain why EMC is such an important issue for automotive electronic system designers.
3. Describe why it is an advantage to consider vehicle systems as consisting of inputs, control and outputs.
4. Calculate the ideal copper cable size required for a fuel pump circuit. The pump draws 8 A from a 12 V battery. The maximum allowable volt drop is 0.5 V.
5. Explain what 'contact resistance' of a switch means.
6. State why a fuse has a continuous and a peak rating.
7. Describe the operation of a vehicle using the CAN system.
8. Explain the term 'error checking' in relation to a multiplexed wiring system.
9. State four types of wiring diagrams and list two advantages and two disadvantages for each.
10. Describe briefly the way in which a wiring colour code or a wiring numbering system can assist the technician when diagnosing electrical faults.

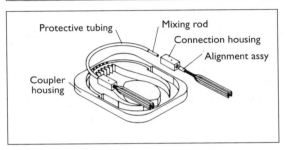

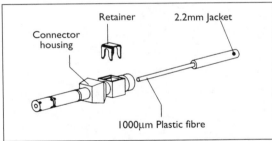

Figure 4.32 Fibre-optic connectors

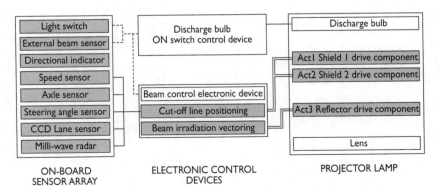

Figure 4.33 Block diagram of control system for low beam lamps

4.8.2 Project

Prepare two papers, the first outlining the benefits of using standard wiring looms and associated techniques and the second outlining the benefits of using a multiplexed system. After completion of the two papers make a judgement on which technique is preferable for future use.

Make sure you support your judgement with reasons!

5
Batteries

5.1 Vehicle batteries

5.1.1 Requirements of the vehicle battery

The vehicle battery is used as a source of energy in the vehicle when the engine, and hence the alternator, is not running. The battery has a number of requirements, which are listed below broadly in order of importance.

- To provide power storage and be able to supply it quickly enough to operate the vehicle starter motor.
- To allow the use of parking lights for a reasonable time.
- To allow operation of accessories when the engine is not running.
- To act as a swamp to damp out fluctuations of system voltage.
- To allow dynamic memory and alarm systems to remain active when the vehicle is left for a period of time.

The first two of the above list are arguably the most important and form a major part of the criteria used to determine the most suitable battery for a given application. The lead-acid battery, in various similar forms, has to date proved to be the most suitable choice for vehicle use. This is particularly so when the cost of the battery is taken into account.

The final requirement of the vehicle battery is that it must be able to carry out all the above listed functions over a wide temperature range. This can be in the region of −30 to +70°C. This is intended to cover very cold starting conditions as well as potentially high under-bonnet temperatures.

5.1.2 Choosing the correct battery

The correct battery depends, in the main, on just two conditions.

1. The ability to power the starter to enable minimum starting speed under very cold conditions.
2. The expected use of the battery for running accessories when the engine is not running.

The first of these two criteria is usually the deciding factor. Figure 5.1 shows a graph comparing the power required by the starter and the power available from the battery, plotted against temperature. The point at which the lines cross is the cold start limit of the system (see also the chapter on starting systems). European standards generally use the figure of −18°C as the cold start limit and a battery to meet this requirement is selected.

Research has shown that under 'normal' cold operating conditions in the UK, most vehicle batteries are on average only 80% charged. Many manufacturers choose a battery for a vehicle that will supply the required cold cranking current when in the 80% charged condition at −7°C.

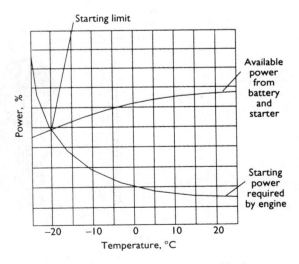

Figure 5.1 Comparison of the power required by the starter and the power available from the battery plotted against temperature

5.1.3 Positioning the vehicle battery

Several basic points should be considered when choosing the location for the vehicle battery:

- Weight distribution of vehicle components.
- Proximity to the starter to reduce cable length.
- Accessibility.
- Protection against contamination.
- Ambient temperature.
- Vibration protection.

As usual, these issues will vary with the type of vehicle, intended use, average operating temperature and so on. Extreme temperature conditions may require either a battery heater or a cooling fan. The potential build-up of gases from the battery may also be a consideration.

5.2 Lead-acid batteries

5.2.1 Construction

Even after well over 100 years of development and much promising research into other techniques of energy storage, the lead-acid battery is still the best choice for motor vehicle use. This is particularly so when cost and energy density are taken into account.

Incremental changes over the years have made the sealed and maintenance-free battery now in common use very reliable and long lasting. This may not always appear to be the case to some end-users, but note that quality is often related to the price the customer pays. Many bottom-of-the-range cheap batteries, with a 12 month guarantee, will last for 13 months!

The basic construction of a nominal 12 V lead-acid battery consists of six cells connected in series. Each cell, producing about 2 V, is housed in an individual compartment within a polypropylene, or similar, case. Figure 5.2 shows a cut-away battery showing the main component parts. The active material is held in grids or baskets to form the positive and negative plates. Separators made from a microporous plastic insulate these plates from each other.

The grids, connecting strips and the battery posts are made from a lead alloy. For many years this was lead antimony (PbSb) but this has now been largely replaced by lead calcium (PbCa). The newer materials cause less gassing of the electrolyte when the battery is fully charged. This has been one of the main reasons why sealed batteries became feasible, as water loss is considerably reduced.

However, even modern batteries described as sealed do still have a small vent to stop the pressure build-up due to the very small amount of

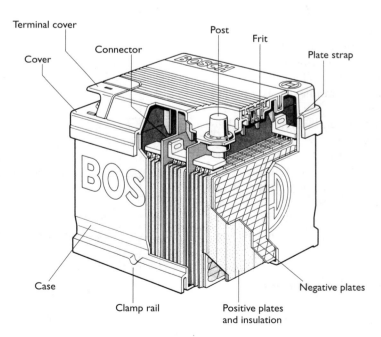

Figure 5.2 Lead-acid battery

gassing. A further requirement of sealed batteries is accurate control of charging voltage.

5.2.2 Battery rating

In simple terms, the characteristics or rating of a particular battery are determined by how much current it can produce and how long it can sustain this current.

The rate at which a battery can produce current is determined by the speed of the chemical reaction. This in turn is determined by a number of factors:

- Surface area of the plates.
- Temperature.
- Electrolyte strength.
- Current demanded.

The actual current supplied therefore determines the overall capacity of a battery. The rating of a battery has to specify the current output and the time.

Ampere hour capacity

This is now seldom used but describes how much current the battery is able to supply for either 10 or 20 hours. The 20-hour figure is the most common. For example, a battery quoted as being 44 Ah (ampere-hour) will be able, if fully charged, to supply 2.2 A for 20 hours before being completely discharged (cell voltage above 1.75 V).

Reserve capacity

A system used now on all new batteries is reserve capacity. This is quoted as a time in minutes for which the battery will supply 25 A at 25°C to a final voltage of 1.75 V per cell. This is used to give an indication of how long the battery could run the car if the charging system was not working. Typically, a 44 Ah battery will have a reserve capacity of about 60 minutes.

Cold cranking amps

Batteries are given a rating to indicate performance at high current output and at low temperature. A typical value of 170 A means that the battery will supply this current for one minute at a temperature of −18°C, at which point the cell voltage will fall to 1.4 V (BS – British Standards).

Note that the overall output of a battery is much greater when spread over a longer time. As mentioned above, this is because the chemical reaction can only work at a certain speed. Figure 5.3

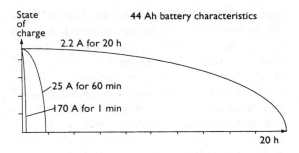

Figure 5.3 Battery discharge characteristics compared

shows the above three discharge characteristics and how they can be compared.

The cold cranking amps (CCA) capacity rating methods do vary to some extent; British standards, DIN standards and SAE standards are the three main examples.

Standard	Time
BS	60 seconds
DIN	30 seconds
SAE	30 seconds

In summary, the capacity of a battery is the amount of electrical energy that can be obtained from it. It is usually given in ampere-hours (Ah), reserve capacity (RC) and cold cranking amps (CCA).

- A 40 Ah battery means it should give 2 A for 20 hours.
- The reserve capacity indicates the time in minutes for which the battery will supply 25 A at 25°C.
- Cold cranking current indicates the maximum battery current at −18°C (0°F) for a set time (standards vary).

A battery for normal light vehicle use may be rated as follows: 44 Ah, 60 RC and 170 A CCA (BS). A 'heavy duty' battery will have the same Ah rating as its 'standard duty' counterpart, but it will have a higher CCA and RC.

5.3 Maintenance and charging

5.3.1 Maintenance

By far the majority of batteries now available are classed as 'maintenance free'. This implies that little attention is required during the life of the battery. Earlier batteries and some heavier types do, however, still require the electrolyte level to be checked and topped up periodically. Battery posts are still a little prone to corrosion and hence the usual service of cleaning with hot water if appropriate and the application of petroleum jelly or proprietary terminal grease is still recommended. Ensuring that the battery case and, in particular, the top remains clean, will help to reduce the rate of self-discharge.

The state of charge of a battery is still very important and, in general, it is not advisable to allow the state of charge to fall below 70% for long periods as the sulphate on the plates can harden, making recharging difficult. If a battery is to be stored for a long period (more than a few weeks), then it must be recharged every so often to prevent it from becoming sulphated. Recommendations vary but a recharge every six weeks is a reasonable suggestion.

5.3.2 Charging the lead-acid battery

The recharging recommendations of battery manufacturers vary slightly. The following methods, however, are reasonably compatible and should not cause any problems. The recharging process must 'put back' the same ampere-hour capacity as was used on discharge plus a bit more to allow for losses. It is therefore clear that the main question about charging is not how much, but at what rate.

The old recommendation was that the battery should be charged at a tenth of its ampere-hour capacity for about 10 hours or less. This is assuming that the ampere-hour capacity is quoted at the 20 hour rate, as a tenth of this figure will make allowance for the charge factor. This figure is still valid, but as ampere-hour capacity is not always used nowadays, a different method of deciding the rate is necessary. One way is to set a rate at 1/16 of the reserve capacity, again for up to 10 hours. The final suggestion is to set a charge rate at 1/40 of the cold start performance figure, also for up to 10 hours. Clearly, if a battery is already half charged, half the time is required to recharge to full capacity.

The above suggested charge rates are to be recommended as the best way to prolong battery life. They do all, however, imply a constant current charging source. A constant voltage charging system is often the best way to charge a battery. This implies that the charger, an alternator on a car for example, is held at a constant level and the state of charge in the battery will determine how much current will flow. This is often the fastest way to recharge a flat battery. The two ways of charging are represented in Figure 5.4. This shows the relationship between charging voltage and the charging current. If a constant voltage of less than 14.4 V is used then it is not possible to cause excessive gassing and this method is particularly appropriate for sealed batteries.

Boost charging is a popular technique often applied in many workshops. It is not recommended as the best method but, if correctly administered and not repeated too often, is suitable for most batteries. The key to fast or boost charging is that the battery temperature should not exceed 43°C. With sealed batteries it is particularly important not to let the battery create excessive gas in order to prevent the build-up of pressure. A rate of about five times the 'normal' charge setting will bring the battery to 78–80% of its full capacity within approximately one hour. Table 5.1 summarizes the charging techniques for a lead-acid battery. Figure 5.5 shows a typical battery charger.

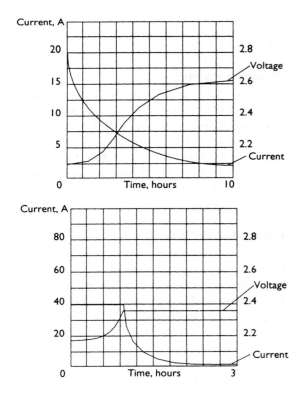

Figure 5.4 Two ways of charging a battery showing the relationship between charging voltage and charging current

Table 5.1 Charging techniques for a lead-acid battery

Charging method	Notes
Constant voltage	Will recharge any battery in 7 hours or less without any risk of overcharging (14.4 V maximum).
Constant current	Ideal charge rate can be estimated as: 1/10 of Ah capacity, 1/16 of reserve capacity or 1/40 of cold start current (charge time of 10–12 hours or pro rata original state).
Boost charging	At no more than five times the ideal rate, a battery can be brought up to about 70% of charge in about one hour.

Figure 5.5 Battery charger

5.4 Diagnosing lead-acid battery faults

5.4.1 Servicing batteries

In use, a battery requires very little attention other than the following when necessary:

- Clean corrosion from terminals using hot water.
- Terminals should be smeared with petroleum jelly or Vaseline, *not* ordinary grease.
- Battery tops should be clean and dry.
- If not sealed, cells should be topped up with distilled water 3 mm above the plates.
- The battery should be securely clamped in position.

5.4.2 Battery faults

Any electrical device can suffer from two main faults; these are either open circuit or short cir-cuit. A battery is no exception but can also suffer from other problems, such as low charge or low capacity. Often a problem – apparently with the vehicle battery – can be traced to another part of the vehicle such as the charging system. Table 5.2 lists all of the common problems encountered with lead-acid batteries, together with typical causes.

Repairing modern batteries is not possible. Most of the problems listed will require the battery to be replaced. In the case of sulphation it is sometimes possible to bring the battery back to life with a very long low current charge. A fortieth of the ampere-hour capacity or about a 1/200 of the cold start performance, for about 50 hours, is an appropriate rate.

Table 5.2 Common problems with lead-acid batteries and their likely causes

Symptom or fault	Likely causes
Low state of charge	Charging system fault
	Unwanted drain on battery
	Electrolyte diluted
	Incorrect battery for application
Low capacity	Low state of charge
	Corroded terminals
	Impurities in the electrolyte
	Sulphated
	Old age – active material fallen from the plates
Excessive gassing and temperature	Overcharging
	Positioned too near exhaust component
Short circuit cell	Damaged plates and insulators
	Build-up of active material in sediment trap
Open circuit cell	Broken connecting strap
	Excessive sulphation
	Very low electrolyte
Service life shorter than expected	Excessive temperature
	Battery has too low a capacity
	Vibration excessive
	Contaminated electrolyte
	Long periods of not being used
	Overcharging

5.4.3 Testing batteries

For testing the state of charge of a non-sealed type of battery, a hydrometer can be used, as shown in Figure 5.6. The hydrometer comprises a syringe that draws electrolyte from a cell, and a float that will float at a particular depth in the electrolyte according to its density. The density or specific gravity is then read from the graduated scale on the float. A fully charged cell should show 1.280, 1.200 when half charged and 1.130 if discharged.

Most vehicles are now fitted with maintenance-free batteries and a hydrometer cannot be used to find the state of charge. This can only be determined from the voltage of the battery, as given in Table 5.3. An accurate voltmeter is required for this test.

A heavy-duty (HD) discharge tester as shown

Table 5.3 State of charge of a battery

Battery volts at 20°C	State of Charge
12.0 V	Discharged (20% or less)
12.3 V	Half charged (50%)
12.7 V	Charged (100%)

in Figure 5.7 is an instrument consisting of a low-value resistor and a voltmeter connected to a pair of heavy test prods. The test prods are firmly pressed on to the battery terminals. The voltmeter reads the voltage of the battery on heavy discharge of 200–300 A.

Assuming a battery to be in a fully charged condition, a serviceable battery should read about 10 V for a period of about 10 s. A sharply falling battery voltage to below 3 V indicates an

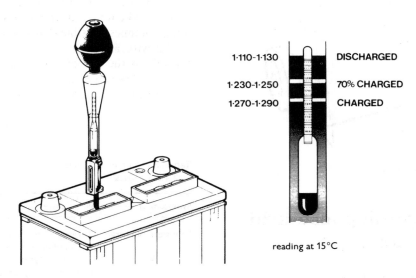

1·110-1·130	DISCHARGED
1·230-1·250	70% CHARGED
1·270-1·290	CHARGED

reading at 15°C

Figure 5.6 Hydrometer test of a battery

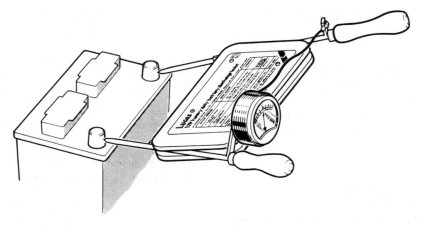

Figure 5.7 Heavy duty discharge test

unserviceable cell. Note also if any cells are gassing, as this indicates a short circuit. A zero or extremely low reading can indicate an open circuit cell. When using the HD tester, the following precautions must be observed:

- Blow gently across the top of the battery to remove flammable gases.
- The test prods must be positively and firmly pressed into the lead terminals of the battery to minimize sparking.
- It should not be used while a battery is on charge.

5.4.4 Safety

The following points must be observed when working with batteries:

- Good ventilation.
- Protective clothing.
- Supply of water available (running water preferable).
- First aid equipment available, including eyewash.
- No smoking or naked lights permitted.

5.5 Advanced battery technology

5.5.1 Electrochemistry

Electrochemistry is a very complex and wide-ranging science. This section is intended only to scratch the surface by introducing important terms and concepts. These will be helpful with the understanding of vehicle battery operation.

The branch of electrochemistry of interest here is the study of galvanic cells and electrolysis. When an electric current is passed through an electrolyte it causes certain chemical reactions and a migration of material. Some chemical reactions, when carried out under certain conditions will produce electrical energy at the expense of the free energy in the system.

The reactions of most interest are those that are reversible, in other words they can convert electrical energy into chemical energy and vice versa. Some of the terms associated with electrochemistry can be confusing. The following is a selection of terms and names with a brief explanation of each.

> **Anion**: The negative charged ion that travels to the positive terminal during electrolysis.
> **Anode**: Positive electrode of a cell.

> **Catalyst**: A substance that significantly increases the speed of a chemical reaction without appearing to take part in it.
> **Cation**: The positively charged ion that travels to the negative terminal during electrolysis.
> **Cathode**: The negative electrode of a cell.
> **Diffusion**: The self-induced mixing of liquids or gases.
> **Dissociation**: The molecules or atoms in a solution decomposing into positive and negative ions. For example, sulphuric acid (H_2SO_4) dissociates into H^{++}, H^{++} (two positive ions or cations, which are attracted to the cathode), and SO_4^{--} (negative ions or anions, which are attracted to the anode).
> **Electrode**: Plates of a battery or an electrolysis bath suspended in the electrolyte.
> **Electrolysis**: Conduction of electricity between two electrodes immersed in a solution containing ions (electrolyte), which causes chemical changes at the electrodes.
> **Electrolyte**: An ion-conducting liquid covering both electrodes.
> **Ion**: A positively or negatively charged atomic or molecular particle.
> **Secondary galvanic cell:** A cell containing electrodes and an electrolyte, which will convert electrical energy into chemical energy when being charged, and the reverse during discharge.

5.5.2 Electrolytic conduction

Electricity flows through conductors in one of two ways. The first is by electron movement, as is the case with most metals. The other type of flow is by ionic movement, which may be charged atoms or molecules. For electricity to flow through an electrolyte, ion flow is required.

To explain electrolytic conduction, which is current flow through a liquid, sulphuric acid (H_2SO_4) is the best electrolyte example to choose. When in an aqueous solution (mixed with water), sulphuric acid dissociates into H^{++}, H^{++} and SO_4^{--}, which are positive and negative ions. The positive charges are attracted to the negative electrode and the negative charges are attracted to the positive electrode. This movement is known as ion flow or ion drift.

5.5.3 Ohm's Law and electrolytic resistance

The resistance of any substance depends on the following variables:

- Nature of the material.
- Temperature.
- Length.
- Cross-sectional area.

This is true for an electrolyte as well as solid conductors. Length and cross-sectional area have straightforward effects on the resistance of a sample, be it a solid or a liquid. Unlike most metals however, which have a positive temperature coefficient, electrolytes are generally the opposite and have a negative temperature coefficient.

The nature of the material or its conductance (the reciprocal of resistance) is again different between solids and liquids. Different substances have different values of resistivity, but with electrolytes the concentration is also important.

5.5.4 Electrochemical action of the lead-acid battery

A fully charged lead-acid battery consists of lead peroxide (PbO_2) as the positive plates, spongy lead (Pb) as the negative plates and diluted sulphuric acid (H_2SO_4) + (H_2O). The dilution of the electrolyte is at a relative density of 1.28. The lead is known as the active material and, in its two forms, has different valencies. This means a different number of electrons exists in the outer shell of the pure lead than when present as a compound with oxygen. The lead peroxide has, in fact, a valency of +iv (four electrons missing).

As discussed earlier in this chapter, when sulphuric acid is in an aqueous solution (mixed with water), it dissociates into charged ions H^{++}, H^{++} and SO_4^{--}. From the 'outside', the polarity of the electrolyte appears to be neutral as these charges cancel out. The splitting of the electrolyte into these parts is the reason that a charging or discharging current can flow through the liquid.

The voltage of a cell is created due to the ions (charged particles) being forced into the solution from the electrodes by the solution pressure. Lead will give up two positively charged atoms, which have given up two electrons, into the liquid. As a result of giving up two positively charged particles, the electrode will now have an excess of electrons and hence will take on a negative polarity with respect to the electrolyte. If a further electrode is immersed into the electrolyte, different potentials will develop at the two electrodes and therefore a potential difference will exist between the two. A lead-acid battery has a nominal potential difference of 2 V. The electrical pressure now present between the plates results in equilibrium within the electrolyte. This is because the negative charges on one plate exert an attraction on the positive ions that have entered the solution. This attraction has the same magnitude as the solution pressure and hence equilibrium is maintained.

When an external circuit is connected to the cell, the solution pressure and attraction force are disrupted. This allows additional charged particles to be passed into and through the electrolyte. This will only happen, however, if the external voltage pressure is greater that the electrical tension within the cell. In simple terms this is known as the charging voltage.

When a lead-acid cell is undergoing charging or discharging, certain chemical changes take place. These can be considered as two reactions, one at the positive plate and one at the negative plate. The electrode reaction at the positive plate is a combination of equations (a) and (b).

(a) $PbO_2 + 4H^+ + 2e^- \rightarrow Pb^{++} + 2H_2O$

The lead peroxide combines with the dissociated hydrogen and tends to become lead and water.

(b) $Pb^{++} + SO_4^{--} \rightarrow PbSO_4$

The lead now tends to combine with the sulphate from the electrolyte to become lead sulphate. This gives the overall reaction at the positive pole as:

(c)(a + b) $PbO_2 + 4H^+ + SO_4^{--} + 2e^- \rightarrow$ $PbSO_4 + 2H_2O$

There is a production of water (a) and a deposition of lead sulphate (b) together with a consumption of sulphuric acid.

The electrode reaction at the negative plate is:

(d) $Pb \rightarrow Pb^{++} + 2e^-$

The neutral lead loses two negative electrons to the solution, and becomes positively charged.

(e) $Pb^{++} SO_4^{--} \rightarrow PbSO_4$

This then tends to attract the negatively charged sulphate from the solution and the pole becomes lead sulphate. The overall reaction at the negative pole is therefore:

(f)(d + e) $Pb + SO_4^{--} \rightarrow PbSO_4 + 2e^-$

This reaction leads to a consumption of sulphuric acid and the production of water as the battery is discharged.

The reverse of the above process is when the

battery is being charged. The process is the reverse of that described above. The reactions involved in the charging process are listed below.

The charging reaction at the negative electrode:

(g) $PbSO_4 + 2e^- + 2H^+ \rightarrow Pb + H_2SO_4$

The electrons from the external circuit ($2e^-$) combine with the hydrogen ions in the solution ($2H^+$) and then the sulphate to form sulphuric acid as the plate tends to become lead. The reaction at the positive pole is:

(h) $PbSO_4 - 2e^- + 2H_2O \rightarrow PbO_2 + H_2SO_4 + 2H^+$

The electrons given off to the external circuit ($2e^-$), release hydrogen ions into the solution ($2H^+$). This allows the positive plate to tend towards lead peroxide, and the concentration of sulphuric acid in the electrolyte to increase.

The net two-way chemical reaction is the sum of the above electrode processes:

(i) (c + f or g + h)
$PbO_2 + 2H_2SO_4 + Pb \leftrightarrow 2PbSO_4 + 2H_2O$

This two-way or reversible chemical reaction (charged on the left and discharged on the right), describes the full process of the charge and discharge cycle of the lead-acid cell.

The other reaction of interest in a battery is that of gassing after it has reached the fully charged condition. This occurs because once the plates of the battery have become 'pure' lead and lead peroxide, the external electrical supply will cause the water in the electrolyte to decompose. This gassing voltage for a lead-acid battery is about 2.4 V. This gassing causes hydrogen and oxygen to be given off resulting in loss of water (H_2O), and an equally undesirable increase in electrolyte acid density.

The reaction, as before, can be considered for each pole of the battery in turn.

At the positive plate:

(j) $2H_2O - 4e^- \rightarrow O_2 + 4H^+$

At the negative plate:

(k) $4H^+ + 4e^- \rightarrow 2H_2$

The sum of these two equations gives the overall result of the reaction;

(l) (j + k) $2H_2O \rightarrow O_2 + 2H_2$

It is acceptable for gassing to occur for a short time to ensure all the lead sulphate has been converted to either lead or lead peroxide. It is the material of the grids inside a battery that con-tribute to the gassing. With sealed batteries this is a greater problem but has been overcome to a large extent by using lead-calcium for the grid material in place of the more traditional lead-antimony.

The voltage of a cell and hence the whole battery is largely determined by the concentration of the acid in the electrolyte. The temperature also has a marked effect. This figure can be calculated from the mean electrical tension of the plates and the concentration of ions in solution. Table 5.4 lists the results of these calculations at 27°C. As a rule of thumb, the cell voltage is about 0.84 plus the value of the relative density.

Table 5.4 Factors affecting the voltage of a battery

Acid density	Cell voltage	Battery voltage	% charge
1.28	2.12 V	12.7 V	100
1.24	2.08 V	12.5 V	70
1.20	2.04 V	12.3 V	50
1.15	1.99 V	12.0 V	20
1.12	1.96 V	11.8 V	0

It is accepted that the terminal voltage of a lead-acid cell must not be allowed to fall below 1.8 V as, apart from the electrolyte tending to become very close to pure water, the lead sulphate crystals grow markedly making it very difficult to recharge the battery.

5.5.5 Characteristics

The following headings are the characteristics of a battery that determine its operation and condition.

Internal resistance

Any source of electrical energy can be represented by the diagram shown in Figure 5.8. This shows a perfect voltage source in series with a resistor. This is used to represent the reason why the terminal voltage of a battery drops when a load is placed across it. As an open circuit, no current flows through the internal resistance and hence no voltage is dropped. When a current is drawn from the source a voltage drop across the internal resistance will occur. The actual value can be calculated as follows.

Connect a voltmeter across the battery and note the open circuit voltage, for example 12.7 V. Connect an external load to the battery, and measure the current, say 50 A. Note again the on-load terminal voltage of the battery, for example 12.2 V.

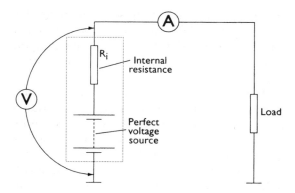

Figure 5.8 Equivalent circuit of an electrical supply showing a perfect voltage source in series with a resistor

A calculation will determine the internal resistance:

$$R_i = (U - V)/I$$

where U = open circuit voltage, V = on-load voltage, I = current, R_i = internal resistance.

For this example the result of the calculation is 0.01 Ω.

Temperature and state of charge affect the internal resistance of a battery. The internal resistance can also be used as an indicator of battery condition – the lower the figure, the better the condition.

Efficiency

The efficiency of a battery can be calculated in two ways, either as the ampere-hour efficiency or the power efficiency.

$$\text{Ah efficiency} = (\text{Ah discharging}/\text{Ah charging}) \times 100\%$$

At the 20 hour rate this can be as much as 90%. This is often quoted as the reciprocal of the efficiency figure; in this example about 1.1, which is known as the charge factor.

$$\text{Energy efficiency} = (P_d \times t_d)/(P_c \times t_c) \times 100\%$$

where P_d = discharge power, t_d = discharge time, P_c = charging power, t_c = charging time.

A typical result of this calculation is about 75%. This figure is lower than the Ah efficiency as it takes into account the higher voltage required to force the charge into the battery.

Self-discharge

All batteries suffer from self-discharge, which means that even without an external circuit the state of charge is reduced. The rate of discharge is of the order of 0.2–1% of the Ah capacity per day. This increases with temperature and the age of the battery. It is caused by two factors. First, the chemical process inside the battery changes due to the material of the grids forming short circuit voltaic couples between the antimony and the active material. Using calcium as the mechanical improver for the lead grids reduces this. Impurities in the electrolyte, in particular trace metals such as iron, can also add to self-discharge.

Second, a leakage current across the top of the battery, particularly if it is in a poor state of cleanliness, also contributes to the self-discharge. The fumes from the acid together with particles of dirt can form a conducting film. This problem is much reduced with sealed batteries.

5.6 Developments in electrical storage

5.6.1 Lead-acid battery developments

Lead-acid batteries have not changed much from the very early designs (invented by Gaston Plante in 1859). Incremental changes and, in particular, the development of accurate charging system control has allowed the use of sealed and maintenance-free batteries. Figure 5.9 shows a typical modern battery.

The other main developments have been to design batteries for particular purposes. This is particularly appropriate for uses such as supplementary batteries in a caravan or as power supplies for lawn mowers and other traction uses. These batteries are designed to allow deep discharge and, in the case of caravan batteries, may also have vent tubes fitted to allow gases to be vented outside. Some batteries are designed to withstand severe vibration for use on plant-type vehicles.

The processes in lead-acid batteries are very similar, even with variations in design. However, batteries using a gel in place of liquid electrolyte are worth a mention. These batteries have many advantages in that they do not leak and are more resistant to poor handling.

The one main problem with using a gel electrolyte is that the speed of the chemical reaction is reduced. Whilst this is not a problem for some types of supply, the current required by a vehicle

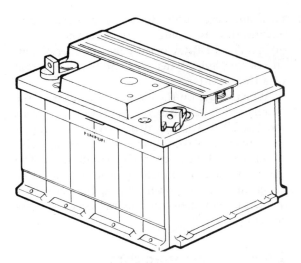

Figure 5.9 Modern vehicle battery

starter is very high for a short duration. The cold cranking amps (CCA) capacity of this type of battery is therefore often lower than the equivalent-sized conventional battery.

The solid-gel type electrolyte used in some types of these batteries is thixotropic. This means that, due to a high viscosity, the gel will remain immobile even if the battery is inverted. A further advantage of a solid gel electrolyte is that a network of porous paths is formed through the electrolyte. If the battery is over-charged, the oxygen emitted at the positive plate will travel to the negative plate, where it combines with the lead and sulphuric acid to form lead sulphate and water:

$$O_2 + 2Pb = 2PbO$$
$$PbO + H_2SO_4 = PbSO_4 + H_2O$$

This reforming of the water means the battery is truly maintenance free. The recharging procedure is very similar to the more conventional batteries.

To date, gel-type batteries have not proved successful for normal motor vehicle use, but are an appropriate choice for specialist performance vehicles that are started from an external power source. Ordinary vehicle batteries using a gel electrolyte appeared on the market some years ago accompanied by great claims of reliability and long life. However, these batteries did not become very popular. This could have been because the cranking current output was not high enough due to the speed of the chemical reaction.

An interesting development in 'normal' lead-acid batteries is the use of lead-antimony (PbSb) for the positive plate grids and lead-calcium (PbCa) for the negative plate grids. This results in a significant reduction in water loss and an increase in service life. The plates are sealed in microporous pocket-type separators, on each side of which are glass-fibre reinforcing mats. The pocket separators collect all the sludge and hence help to keep the electrolyte in good condition.

5.6.2 Alkaline batteries

Lead-acid batteries traditionally required a considerable amount of servicing to keep them in good condition, although this is not now the case with the advent of sealed and maintenance-free batteries. However, when a battery is required to withstand a high rate of charge and discharge on a regular basis, or is left in a state of disuse for long periods, the lead-acid cell is not ideal. Alkaline cells on the other hand require minimum maintenance and are far better able to withstand electrical abuse such as heavy discharge and over-charging.

The disadvantages of alkaline batteries are that they are more bulky, have lower energy efficiency and are more expensive than a lead-acid equivalent. When the lifetime of the battery and servicing requirements are considered, the extra initial cost is worth it for some applications. Bus and coach companies and some large goods-vehicle operators have used alkaline batteries.

Alkaline batteries used for vehicle applications are generally the nickel-cadmium type, as the other main variety (nickel-iron) is less suited to vehicle use. The main components of the nickel-cadmium – or Nicad – cell for vehicle use are as follows:

- positive plate – nickel hydrate (NiOOH);
- negative plate – cadmium (Cd);
- electrolyte – potassium hydroxide (KOH) and water (H_2O).

The process of charging involves the oxygen moving from the negative plate to the positive plate, and the reverse when discharging. When fully charged, the negative plate becomes pure cadmium and the positive plate becomes nickel hydrate. A chemical equation to represent this reaction is given next but note that this is simplifying a more complex reaction.

$$2NiOOH + Cd + 2H_2O + KOH \leftrightarrow$$
$$2Ni(OH)_2 + CdO_2 + KOH$$

The $2H_2O$ is actually given off as hydrogen (H) and oxygen (O_2) as gassing takes place all the time during charge. It is this use of water by the cells that indicates they are operating, as will have been noted from the equation. The electrolyte does not change during the reaction. This means that a relative density reading will not indicate the state of charge. These batteries do not suffer from over-charging because once the cadmium oxide has changed to cadmium, no further reaction can take place.

The cell voltage of a fully charged cell is 1.4 V but this falls rapidly to 1.3 V as soon as discharge starts. The cell is discharged at a cell voltage of 1.1 V. Figure 5.10 shows a simplified representation of a Nicad battery cell.

Ni-MH or nickel-metal-hydride batteries show some promise for electric vehicle use.

5.6.3 The ZEBRA battery

The Zero Emissions Battery Research Activity (ZEBRA) has adopted a sodium-nickel-chloride battery for use in its electric vehicle programme. This battery functions on an electrochemical principle. The base materials are nickel and sodium chloride. When the battery is charged, nickel chloride is produced on one side of a ceramic electrolyte and sodium is produced on the other. Under discharge, the electrodes change back to the base materials. Each cell of the battery has a voltage of 2.58 V.

The battery operates at an internal temperature of 270–350°C, requiring a heat-insulated

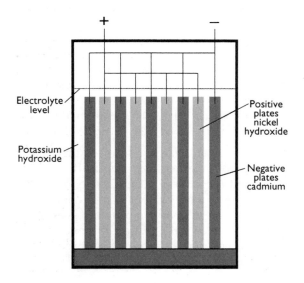

Figure 5.10 Simplified representation of a Nicad alkaline battery cell

enclosure. The whole unit is 'vacuum packed' to ensure that the outer surface never exceeds 30°C. The ZEBRA battery has an energy density of 90 Wh/kg, which is more than twice that of a lead-acid type.

When in use on the electric vehicle (EV), the battery pack consists of 448 individual cells rated at 289 V. The energy density is 81 Wh/kg; it has a mass of 370 kg (over 1/4 of the total vehicle mass) and measures $993{\times}793{\times}280\,mm^3$. The battery pack can be recharged in just one hour using an external power source. It is currently in use/development on the Mercedes A-class vehicle.

5.6.4 Ultra-capacitors

Ultra-capacitors are very high capacity but (relatively) low size capacitors. This is achieved by employing several distinct electrode materials prepared using special processes. Some state-of-the-art ultra-capacitors are based on high surface area, ruthenium dioxide (RuO_2) and carbon electrodes. Ruthenium is extremely expensive and available only in very limited amounts.

Electrochemical capacitors are used for high-power applications such as cellular electronics, power conditioning, industrial lasers, medical equipment, and power electronics in conventional, electric and hybrid vehicles. In conventional vehicles, ultra-capacitors could be used to reduce the need for large alternators for meeting intermittent high peak power demands related to power steering and braking. Ultra-capacitors recover braking energy dissipated as heat and can be used to reduce losses in electric power steering.

One system in use on a hybrid bus uses 30 ultra-capacitors to store 1600 kJ of electrical energy (20 farads at 400 V). The capacitor bank has a mass of 950 kg. Use of this technology allows recovery of energy, such as when braking, that would otherwise have been lost. The capacitors can be charged in a very short space of time. The energy in the capacitors can also be used very quickly, such as for rapid acceleration.

5.6.5 Fuel cells

The energy of oxidation of conventional fuels, which is usually manifested as heat, may be converted directly into electricity in a fuel cell. All oxidations involve a transfer of electrons between the fuel and oxidant, and this is employed in a fuel cell to convert the energy directly into electricity. All battery cells involve an oxide reduction at the positive pole and an oxidation at the negative during some part of their chemical process. To achieve the separation of these reactions in a fuel cell, an anode, a cathode and electrolyte are required. The electrolyte is fed directly with the fuel.

It has been found that a fuel of hydrogen when combined with oxygen proves to be a most efficient design. Fuel cells are very reliable and silent in operation, but at present are very expensive to construct. Figure 5.11 shows a simplified representation of a fuel cell.

Operation of one type of fuel cell is such that as hydrogen is passed over an electrode (the anode) of porous nickel, which is coated with a catalyst, the hydrogen diffuses into the electrolyte. This causes electrons to be stripped off the hydrogen atoms. These electrons then pass through the external circuit. Negatively charged hydrogen anions (OH^-) are formed at the electrode over which oxygen is passed such that it also diffuses into the solution. These anions move through the electrolyte to the anode. The electrolyte, which is used, is a solution of potassium hydroxide (KOH). Water is formed as the by-product of a reaction involving the hydrogen ions, electrons and oxygen atoms. If the heat generated by the fuel cell is used, an efficiency of over 80% is possible, together with a very good energy density figure. A single fuel cell unit is often referred to as a 'stack'.

The working temperature of these cells varies but about 200 °C is typical. High pressure is also used and this can be of the order of 30 bar. It is the pressures and storage of hydrogen that are the main problems to be overcome before the fuel cell will be a realistic alternative to other forms of storage for the mass market. The next section, however, explains one way around the 'hydrogen' problem.

Fuel cells in use on 'urban transport' vehicles typically use $20 \times 10\,kW$ stacks (200 kW) operating at 650 V.

5.6.6 Fuel cell developments

Some vehicle manufacturers have moved fuel cell technology nearer to production reality with an on-board system for generating hydrogen from methanol. Daimler-Benz, now working with the Canadian company Ballard Power Systems, claimed the system as a 'world first'. The research vehicle is called NECAR (New Electric Car). It is based on the Mercedes-Benz's 'A-class' model (Figure 5.12). In the system, a reformer converts the methanol into hydrogen by water vapour reformation. The hydrogen gas is then supplied to fuel cells to react with atmospheric oxygen, which in turn produces electric energy.

The great attraction of methanol is that it can easily fit into the existing gasoline/diesel infrastructure of filling stations and does not need highly specialized equipment or handling. It is easy to store on board the vehicle, unlike hydrogen which needs heavy and costly tanks. At the time of writing the NECAR (Figure 5.13) has a range of about 400 km on a 40 litre methanol tank. Consideration is also being given to multifuel hydrogen sourcing.

The methanol reformer technology used has benefited from developments that have allowed the system to become smaller and more efficient compared with earlier efforts. The result is a 470 mm high unit located in the rear of the A-class, in which the reformer directly injects hydrogen into the fuel cells. Hydrogen production occurs at a temperature of some 280 °C. Methanol and water vaporize to yield hydrogen (H), carbon dioxide (CO_2), and carbon monoxide (CO). After catalytic oxidation of the CO, the hydrogen gas is fed to the negative pole of the fuel cell where a special plastic foil, coated with a platinum catalyst and sandwiched between two electrodes, is located. The conversion of the hydrogen into positively charged protons and negatively charged electrons begins with the arrival of oxygen at the positive pole.

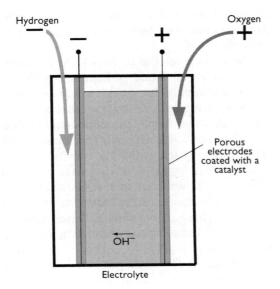

Figure 5.11 Representation of a fuel cell

Figure 5.12 Mercedes-Benz A-class

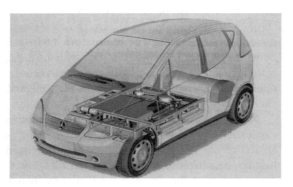

Figure 5.13 NECAR – Fuel cells in a Mercedes A-class

Table 5.5 The potential energy density of several battery types

Battery type	Cell voltage	Energy density
Lead-acid	2 V	30 Wh/kg
Nickel-iron/cadmium	1.22 V	45 Wh/kg
Nickel-metal-hydride	1.2 V	50–80 Wh/kg
Sodium-sulphur	2–2.5 V	90–100 Wh/kg
Sodium-nickel-chloride	2.58 V	90–100 Wh/kg
Lithium	3.5 V	100 Wh/kg
H_2/O_2 Fuel cell	~30 V	500 Wh/kg

The foil is only permeable to protons; therefore, a voltage builds up across the fuel cell.

5.6.7 Sodium sulphur battery

Much research is underway to improve on current battery technology in order to provide a greater energy density for electric vehicles. (Electric traction will be discussed further in a later chapter.) A potential major step forwards however is the sodium sulphur battery, which has now reached production stage. Table 5.5 compares the potential energy density of several types of battery. Wh/kg means watt hours per kilogram or the power it will supply, for how long per kilogram.

Sodium-sulphur batteries have recently reached the production stage and, in common with the other types listed, have much potential; however,

all types have specific drawbacks. For example, storing and carrying hydrogen is one problem of fuel cells.

The sodium-sulphur or NaS battery consists of a cathode of liquid sodium into which is placed a current collector. This is a solid electrode of β-alumina. A metal can that is in contact with the anode (a sulphur electrode) surrounds the whole assembly. The major problem with this system is that the running temperature needs to be 300–350°C. A heater rated at a few hundred watts forms part of the charging circuit. This maintains the battery temperature when the vehicle is not running. Battery temperature is maintained when in use due to I^2R losses in the battery.

Each cell of this battery is very small, using only about 15 g of sodium. This is a safety feature because, if the cell is damaged, the sulphur on the outside will cause the potentially dangerous sodium to be converted into polysulphides – which are comparatively harmless. Small cells also have the advantage that they can be distributed around the car. The capacity of each cell is about 10 Ah. These cells fail in an open circuit condition and hence this must be taken into account, as the whole string of cells used to create the required voltage would be rendered inoperative. The output voltage of each cell is about 2 V. Figure 5.14 shows a representation of a sodium-sulphur battery cell.

A problem still to be overcome is the casing material, which is prone to fail due to the very corrosive nature of the sodium. At present, an expensive chromized coating is used.

This type of battery, supplying an electric motor, is becoming a competitor to the internal combustion engine. The whole service and charging infrastructure needs to develop but looks promising. It is estimated that the cost of running an electric vehicle will be as little as 15% of the petrol version, which leaves room to absorb the extra cost of production.

5.6.8 The Swing battery

Some potential developments in battery technology are major steps in the right direction but many new methods involve high temperatures. One major aim of battery research is to develop a high performance battery, that works at a normal operating temperature. One new idea is called the 'Swing battery'. Figure 5.15 shows the chemical process of this battery.

The Swing concept batteries use lithium ions. These batteries have a carbon anode and a cathode made of transition metal oxides. Lithium ions are in constant movement between these very thin electrodes in a non-aqueous electrolyte. The next step planned by the company is to use a solid polymer electrolyte, based on polyethylene oxide instead of the liquid electrolyte.

The Swing process takes place at normal temperatures and gives a very high average cell voltage of 3.5 V, compared with cell voltages of approximately 1.2 V for nickel-cadmium and about 2.1 V for lead-acid or sodium-sulphur batteries. Tests simulating conditions in electric vehicles have demonstrated specific energies of about 100 Wh/kg and 200 Wh/l.

The complexity of the electrical storage system increases with higher operating temperatures, an increased numbers of cells and with the presence of agitated or recycled electrolytes. To ensure reliable and safe operation, higher and higher demands will be made on the battery management system. This will clearly introduce more cost to the vehicle system as a whole. Consideration must be given not only to specific energy storage but also to system complexity and safety.

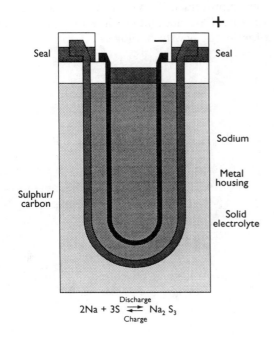

Figure 5.14 Sodium sulphur battery

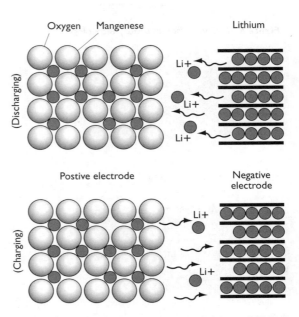

Figure 5.15 Chemical process of the 'swing' battery (3.5 V/cell at room temperature)

Figure 5.16 is a comparison of batteries considering energy density and safety factors.

The high temperature systems have, however, proved their viability for use in vehicles. They have already passed a series of abuse tests and other systems are in preparation. A sodium-sulphur battery when fully charged, which is rated at 20 kWh, contains about 10 kg of liquid sodium. Given 100 000 vehicles, 1000 tonnes of liquid sodium will be in use. These quantities have to be encapsulated in two hermetically sealed containers. The Swing concept is still new but offers a potentially safe system for use in the future.

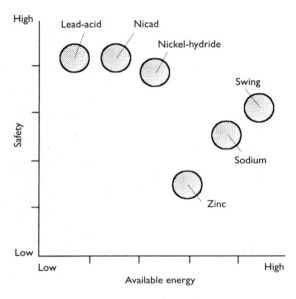

Figure 5.16 Comparison of battery technologies

5.7 Self-assessment

5.7.1 Questions

1. Describe what a 'lead-acid' battery means.
2. State the three ways in which a battery is generally rated.
3. Make a clearly labelled sketch to show how a 12 V battery is constructed.
4. Explain why a battery is rated or described in different ways.
5. List six considerations when deciding where a vehicle battery should be positioned.
6. Describe how to measure the internal resistance of a battery.
7. Make a table showing three ways of testing the state of charge of a lead-acid battery together with the results.
8. Describe the two methods of recharging a battery.
9. State how the ideal charge rate for a lead-acid battery can be determined.
10. Explain why the 'energy density' of a battery is important.

5.7.2 Assignment

Carry out research into the history of the vehicle battery and make notes of significant events. Read further about 'new' types of battery and suggest some of their advantages and disadvantages. What are the main limiting factors to battery improvements? Why is the infrastructure for battery 'service and repair' important for the adoption of new technologies?

6
Charging systems

6.1 Requirements of the charging system

6.1.1 Introduction

The 'current' demands made by modern vehicles are considerable. The charging system must be able to meet these demands under all operating conditions and still 'fast charge' the battery.

The main component of the charging system is the alternator and on most modern vehicles – with the exception of its associated wiring – this is the only component in the charging system. Figure 6.1 shows an alternator in common use. The alternator generates AC but must produce DC at its output terminal as only DC can be used to charge the battery and run electronic circuits. The output of the alternator must be a constant voltage regardless of engine speed and current load.

To summarize, the charging system must meet the following criteria (when the engine is running).

- Supply the current demands made by all loads.
- Supply whatever charge current the battery demands.

- Operate at idle speed.
- Supply constant voltage under all conditions.
- Have an efficient power-to-weight ratio.
- Be reliable, quiet, and have resistance to contamination.
- Require low maintenance.
- Provide an indication of correct operation.

6.1.2 Vehicle electrical loads

The loads placed on an alternator can be considered as falling under three separate headings: continuous, prolonged and intermittent. The charging system of a modern vehicle has to cope with high demands under many varied conditions. To give some indication as to the output that may be required, consider the power used by each individual component and add this total to the power required to charge the battery. Table 6.1 lists the typical power requirements of various vehicle systems. The current draw (to the nearest 0.5 A) at 14 and 28 V (nominal; alternator output voltages for 12 and 24 V systems) is also given for comparison.

Figure 6.2 shows how the demands on the alternator have increased over the years, together with a prediction of the future.

Not shown in Table 6.1 are consumers, such

Figure 6.1 A14 VI alternator

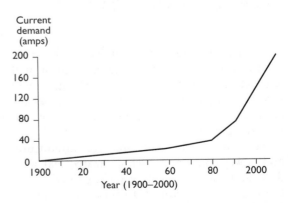

Figure 6.2 How the demands on the alternator have changed

Table 6.1 Typical power requirements of some common vehicle electrical components

Continuous loads	Power	Current at 14 V	28 V
Ignition	30 W	2.0	1.0
Fuel injection	70 W	5.0	2.5
Fuel pump	70 W	5.0	2.5
Instruments	10 W	1.0	0.5
Total	180 W	13.0	6.5

Prolonged loads	Power	Current at 14 V	28 V
Side and tail lights	30 W	2.0	1.0
Number plate lights	10 W	1.0	0.5
Headlights main beam	200 W	15.0	7.0
Headlights dip beam	160 W	12.0	6.0
Dashboard lights	25 W	2.0	1.0
Radio / Cassette / CD	15 W	1.0	0.5
Total (Av. main & dip)	260 W	19.5	9.5

Intermittent loads	Power	Current at 14 V	28 V
Heater	50 W	3.5	2.0
Indicators	50 W	3.5	2.0
Brake lights	40 W	3.0	1.5
Front wipers	80 W	6.0	3.0
Rear wipers	50 W	3.5	2.0
Electric windows	150 W	11.0	5.5
Radiator cooling fan	150 W	11.0	5.5
Heater blower motor	80 W	6.0	3.0
Heated rear window	120 W	9.0	4.5
Interior lights	10 W	1.0	0.5
Horns	40 W	3.0	1.5
Rear fog lights	40 W	3.0	1.5
Reverse lights	40 W	3.0	1.5
Auxiliary lamps	110 W	8.0	4.0
Cigarette lighter	100 W	7.0	3.5
Headlight wash wipe	100 W	7.0	3.5
Seat movement	150 W	11.0	5.5
Seat heater	200 W	14.0	7.0
Sun-roof motor	150 W	11.0	5.5
Electric mirrors	10 W	1.0	0.5
Total	1.7 kW	125.5	63.5

The average consumption of the intermittent loads is estimated using a factor of 0.1 (0.1 × 1.7 kW = 170 W).

as electrically pre-heated catalytic converters, electrical power assisted steering and heated windscreens, to list just three. Changes will therefore continue to take place in the vehicle electrical system and the charging system will have to keep up!

The intermittent loads are used infrequently and power consumers such as heated rear windows and seat heaters are generally fitted with a timer relay. The factor of 0.1 is therefore applied to the total intermittent power requirement, for the purpose of further calculations. This assumes the vehicle will be used under *normal* driving conditions.

The consumer demand on the alternator is the sum of the constant loads, the prolonged loads and the intermittent loads (with the factor applied). In this example:

$$180 + 260 + 170 = 610 \, \text{W} \ (43 \, \text{A at } 14 \, \text{V})$$

The demands placed on the charging system therefore are extensive. This load is in addition to the current required to recharge the battery. Further sections in this chapter discuss how these demands are met.

6.2 Charging system principles

6.2.1 Basic principles

Figure 6.3 shows a representation the vehicle charging system as three blocks, the alternator, battery and vehicle loads. When the alternator voltage is less than the battery (engine slow or not running for example), the direction of current flow is from the battery to the vehicle loads. The alternator diodes prevent current flowing into the alternator. When the alternator output is greater than the battery voltage, current will flow from the alternator to the vehicle loads and the battery.

From this simple example it is clear that the alternator output voltage must be greater than the battery voltage at all times when the engine is running. The actual voltage used is critical and depends on a number of factors.

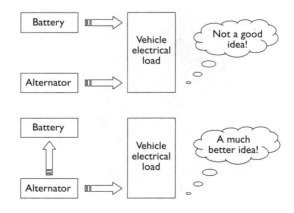

Figure 6.3 Vehicle charging system

6.2.2 Charging voltages

The main consideration for the charging voltage is the battery terminal voltage when fully charged. If the charging system voltage is set to this value then there can be no risk of over-charging the battery. This is known as the constant voltage charging technique. The chapter on batteries discusses this issue in greater detail. The figure of $14.2 \pm 0.2\,V$ is the accepted charging voltage for a $12\,V$ system. Commercial vehicles generally employ two batteries in series at a nominal voltage of $24\,V$, the accepted charge voltage would therefore be doubled. These voltages are used as the standard input for all vehicle loads. For the purpose of clarity the text will just consider a $12\,V$ system.

The other areas for consideration when determining the charging voltage are any expected voltage drops in the charging circuit wiring and the operating temperature of the system and battery. The voltage drops must be kept to a minimum, but it is important to note that the terminal voltage of the alternator may be slightly above that supplied to the battery.

6.3 Alternators and charging circuits

6.3.1 Generation of electricity

Figure 6.4 shows the basic principle of a three-phase alternator together with a representation of its output. Electromagnetic induction in caused by a rotating magnet inside a stationary loop or loops of wire. In a practical alternator, the rotating magnet is an electromagnet that is supplied via two slip rings.

Figure 6.5 shows the most common design, which is known as a claw pole rotor. Each end of the rotor will become a north or a south pole and hence each claw will be alternately north and south. It is common practice, due to reasons of efficiency, to use claw pole rotors with 12 or 16 poles.

The stationary loops of wire are known as the stator and consist of three separate phases, each with a number of windings. The windings are mechanically spaced on a laminated core (to reduce eddy currents), and must be matched to the number of poles on the rotor. Figure 6.6 shows a typical example.

The three-phase windings of the stator can be connected in two ways, known as star or delta windings – as shown in Figure 6.7. The current and voltage output characteristics are different for star- and delta-wound stators.

Star connection can be thought of as a type of series connection of the phases and, to this end, the output voltage across any two phases will be the vector sum of the phase voltages. Current output will be the same as the phase current. Star-wound stators therefore produce a

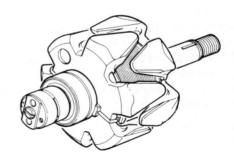

Figure 6.5 Rotor

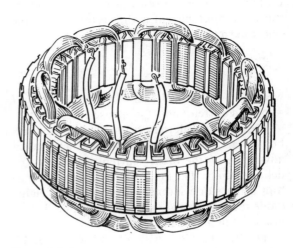

Figure 6.6 Stator

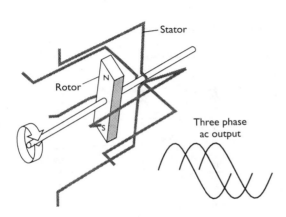

Figure 6.4 Principle of a three-phase alternator

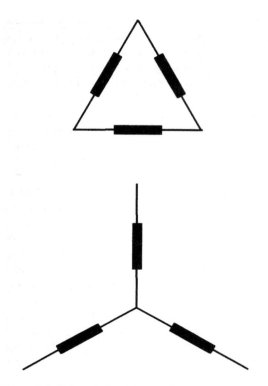

Figure 6.7 Delta and star stator windings

higher voltage, whereas delta-wound stators produce a higher current.

The voltage and current in three-phase stators can be calculated as follows.

Star-wound stators can be thought of as a type of series circuit.

$$V = V_p\sqrt{3}$$
$$I = I_p$$

A delta connection can similarly be thought of as a type of parallel circuit. This means that the output voltage is the same as the phase voltage but the output current is the vector sum of the phase currents.

$$V = V_p$$
$$I = I_p\sqrt{3}$$

where V = output voltage; V_p = phase voltage; I = output current; and I_p = phase current.

Most vehicle alternators use the star windings but some heavy-duty machines have taken advantage of the higher current output of the delta windings. The majority of modern alternators using star windings incorporate an eight-diode rectifier so as to maximize output. This is discussed in a later section.

The frequency of an alternator output can be calculated. This is particularly important if an

AC tapping from the stator is used to run a vehicle rev-counter:

$$f = \frac{pn}{60}$$

where f = frequency in Hz; n = alternator speed in rev/min; and p = number of pole pairs (a 12 claw rotor has 6 pole pairs).

An alternator when the engine is at idle, will have a speed of about 2000 rev/min, which, with a 12 claw rotor will produce a frequency of $6 \times 2000/60 = 200$ Hz.

A terminal provided on many alternators for this output is often marked W. The output is half-wave rectified and is used, in particular, on diesel engines to drive a rev-counter. It is also used on some petrol engine applications to drive an electric choke.

6.3.2 Rectification of AC to DC

In order for the output of the alternator to charge the battery and run other vehicle components it must be converted from alternating current (AC) to direct current (DC). The component most suitable for this task is the silicon diode. If single-phase AC is passed through a diode, its output is half-wave rectified as shown in Figure 6.8. In this example, the diode will only allow the positive half cycles to be conducted towards the positive of the battery. The negative cycles are blocked.

Figure 6.9 shows a four-diode bridge rectifier

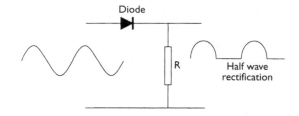

Figure 6.8 Half-wave rectification

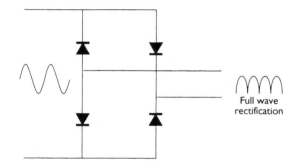

Figure 6.9 Full-wave bridge rectifier (single phase)

to full-wave rectify single phase AC. A diode is often considered to be a one-way valve for electricity. While this is a good analogy it is important to remember that while a good quality diode will block reverse flow up to a pressure of about 400 V, it will still require a small voltage pressure of about 0.6 V to conduct in the forward direction.

In order to full-wave rectify the output of a three-phase machine, six diodes are required. These are connected in the form of a bridge, as shown in Figure 6.10. The 'bridge' consists of three positive diodes and three negative diodes. The output produced by this configuration is shown compared with the three-phase signals.

A further three positive diodes are often included in a rectifier pack. These are usually smaller than the main diodes and are only used to supply a small current back to the field windings in the rotor. The extra diodes are known as the auxiliary, field or excitation diodes. Figure 6.11 shows the layout of a nine-diode rectifier.

Owing to the considerable currents flowing through the main diodes, some form of heat sink is required to prevent thermal damage. In some cases diodes are connected in parallel to carry higher currents without damage. Diodes in the rectifier pack also serve to prevent reverse current flow from the battery to the alternator. This also allows alternators to be run in parallel without balancing, as equalizing current cannot flow from one to the other. Figure 6.12 shows examples of some common rectifier packs.

When a star-wound stator is used, the addition of the voltages at the neutral point of the star is, in theory, 0 V. In practice, however, due to slight inaccuracies in the construction of the stator and rotor, a potential develops at this point. This potential (voltage) is known as the third harmonic and is shown in Figure 6.13. Its frequency is three times the fundamental frequency of the phase windings. By employing two extra diodes, one positive and one negative connected to the star point, the energy can be collected. This can increase the power output of an alternator by up to 15%.

Figure 6.14 shows the full circuit of an alternator using an eight-diode main rectifier and

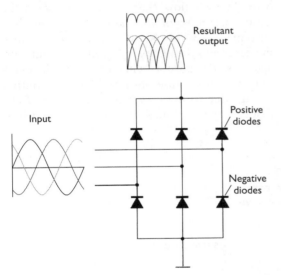

Figure 6.10 Three-phase bridge rectifier

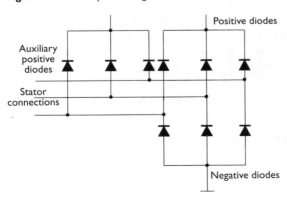

Figure 6.11 Nine-diode rectifier

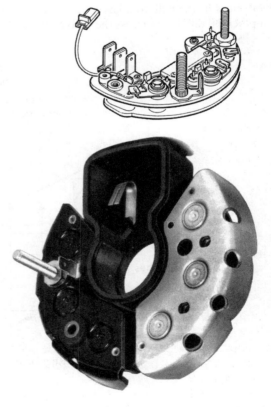

Figure 6.12 Rectifier packs in common use

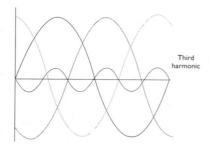

Figure 6.13 The third harmonic

three field diodes. The voltage regulator, which forms the starting point for the next section, is also shown in this diagram. The warning light in an alternator circuit, in addition to its function of warning of charging faults, also acts to supply the initial excitation to the field windings. An alternator will not always self-excite as the residual magnetism in the fields is not usually enough to produce a voltage that will overcome the 0.6 or 0.7 V needed to forward bias the rectifier diodes. A typical wattage for the warning light bulb is 2 W. Many manufacturers also connect a resistor in parallel with the bulb to assist in excitation and allow operation if the bulb blows. The charge warning light bulb is extinguished when the alternator produces an output from the field diodes as this causes both sides of the bulb to take on the same voltage (a potential difference across the bulb of 0 V).

6.3.3 Regulation of output voltage

To prevent the vehicle battery from being over-charged the regulated system voltage should be kept below the gassing voltage of the lead-acid battery. A figure of 14.2 ± 0.2 V is used for all 12 V charging systems. Accurate voltage control is vital with the ever-increasing use of electronic systems. It has also enabled the wider use of sealed batteries, as the possibility of over-charging is minimal. Figure 6.15 shows two common voltage regulators. Voltage regulation is a difficult task on a vehicle alternator because of the constantly changing engine speed and loads on the alternator. The output of an alternator without regulation would rise linearly in proportion with engine speed. Alternator output is also proportional to magnetic field strength and this, in turn, is proportional to the field current. It is the task of the regulator to control this field current in response to alternator output voltage. Figure 6.16 shows a flow chart which represents the action of the regulator, showing how the field current is switched off as output voltage increases and then back on again as output voltage falls. The abrupt switching of the field current does not cause abrupt changes in output voltage due to the very high inductance of the field (rotor) windings. In addition, the whole

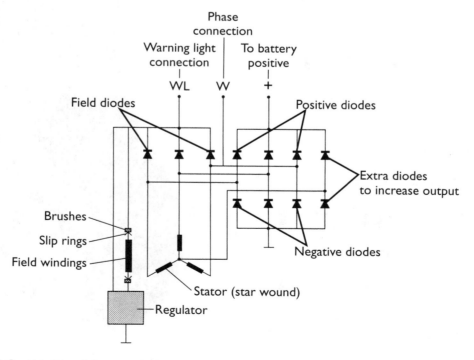

Figure 6.14 Complete internal alternator circuit

Charging systems **123**

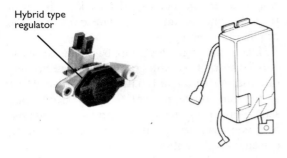

Figure 6.15 Voltage regulators

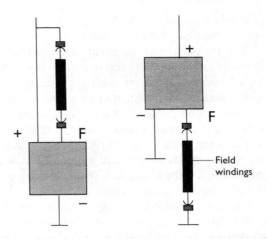

Figure 6.17 How the voltage regulator is incorporated in the field circuit

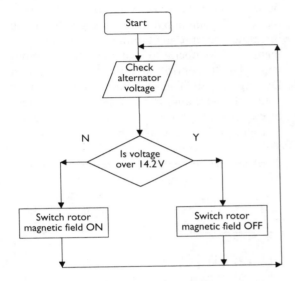

Figure 6.16 Action of the voltage regulator

switching process only takes a few milliseconds. Many regulators also incorporate some temperature compensation to allow a higher charge rate in colder conditions and to reduce the rate in hot conditions.

When working with regulator circuits, care must be taken to note 'where' the field circuit is interrupted. For example, some alternator circuits supply a constant feed to the field windings from the excitation diodes and the regulator switches the earth side. In other systems, one side of the field windings is constantly earthed and the regulator switches the supply side. Figure 6.17 shows these two methods.

Alternators do not require any extra form of current regulation. This is because if the output voltage is regulated the voltage supplied to the field windings cannot exceed the pre-set level. This in turn will only allow a certain current to flow due to the resistance of the windings and

hence a limit is set for the field strength. This will then limit the maximum current the alternator can produce.

Regulators can be mechanical or electronic, and the latter are now almost universal on modern cars. The mechanical type uses a winding connected across the output of the alternator. The magnetism produced in this winding is proportional to the output voltage. A set of normally closed contacts is attached to an armature, which is held in position by a spring. The supply to the field windings is via these contacts. When the output voltage rises beyond a pre-set level, say 14 V, the magnetism in the regulator winding will overcome spring tension and open the contacts. This switches off the field current and causes the alternator output to fall. As the output falls below a pre-set level, the spring will close the regulator contacts again and so the process continues. Figure 6.18. shows a simplified circuit of a mechanical regulator. This

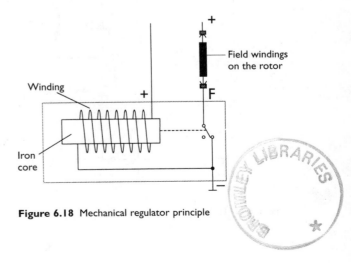

Figure 6.18 Mechanical regulator principle

BROMLEY LIBRARIES

principle has not changed from the very early voltage control of dynamo output.

The problem with mechanical regulators is the wear on the contacts and other moving parts. This has been overcome with the use of electronic regulators which, due to more accurate tolerances and much faster switching, are far superior, producing a more stable output. Due to the compactness and vibration resistance of electronic regulators they are now fitted almost universally on the alternator, reducing the number of connecting cables required.

The key to electronic voltage regulation is the Zener diode. As discussed in Chapter 3, this diode can be constructed to break down and conduct in the reverse direction at a precise level. This is used as the sensing element in an electronic regulator. Figure 6.19 shows a simplified electronic voltage regulator.

This regulator operates as follows. When the alternator first increases in speed the output will be below the pre-set level. Under these circumstances transistor T_2 will be switched on by a feed to its base via resistor R_3. This allows full field current to flow, thus increasing voltage output. When the pre-set voltage is reached, the Zener diode will conduct. Resistors R_1 and R_2 are a simple series circuit to set the voltage appropriate to the value of the diode when the supply is, say, 14.2 V. Once Z_D conducts, transistor T_1 will switch on and pull the base of T_2 down to ground. This switches T_2 off and so the field current is interrupted, causing output voltage to fall. This will cause Z_D to stop conducting, T_1 will switch off, allowing T_2 to switch back on and so the cycle will continue. The conventional diode, D_1, absorbs the back EMF from the field windings and so prevents damage to the other components.

Electronic regulators can be made to sense either the battery voltage, the machine voltage (alternator), or a combination of the two. Most systems in use at present tend to be machine sensed as this offers some protection against over-voltage in the event of the alternator being driven with the battery disconnected.

Figure 6.20 shows the circuit of a hybrid integrated circuit (IC) voltage regulator. The hybrid system involves the connection of discrete components on a ceramic plate using film techniques. The main part of the regulator is an integrated circuit containing the sensing elements and temperature compensation components. The IC controls an output stage such as a Darlington pair. This technique produces a very compact device and, because of the low number of components and connections, is very reliable.

Figure 6.21 is a graph showing how the IC

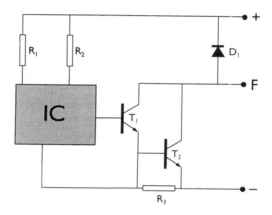

Figure 6.20 Hybrid IC regulator circuit

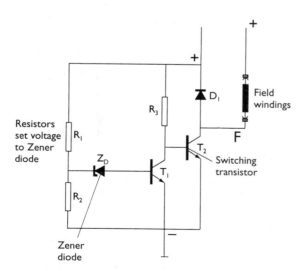

Figure 6.19 Electronic voltage regulator

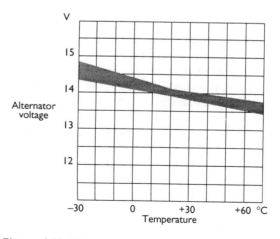

Figure 6.21 How the regulator response changes with temperature

regulator response changes with temperature. This change is important to ensure correct charging under 'summer' and 'winter' conditions. When a battery is cold, the electrolyte resistance increases. This means a higher voltage is necessary to cause the correct recharging current.

Over-voltage protection is required in some applications in order to prevent damage to electronic components. When an alternator is connected to a vehicle battery system, the voltage, even in the event of regulator failure, will not often exceed about 20 V due to the low resistance and swamping effect of the battery. If an alternator is run with the battery disconnected (which is not recommended), a heavy duty Zener diode connected across the output of the WL/field diodes will offer some protection as, if the system voltage exceeds its breakdown figure, it will conduct and cause the system voltage to be kept within reasonable limits.

6.3.4 Charging circuits

For many applications, the charging circuit is one of the simplest on the vehicle. The main output is connected to the battery via a suitably sized cable (or in some cases two cables to increase reliability and flexibility), and the warning light is connected to an ignition supply on one side and to the alternator terminal at the other. A wire may also be connected to the phase terminal if it is utilized. Figure 6.22 shows two typical

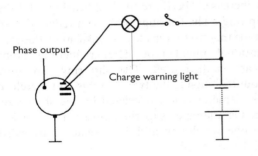

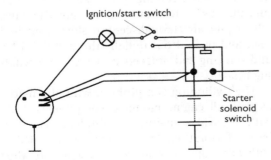

Figure 6.22 Example charging circuits

wiring circuits. Note that the output of the alternator is often connected to the starter main supply simply for convenience of wiring. If the wires are kept as short as possible this will reduce voltage drop in the circuit. The voltage drop across the main supply wire when the alternator is producing full output current, should be less than 0.5 V.

Some systems have an extra wire from the alternator to 'sense' battery voltage directly. An ignition feed may also be found and this is often used to ensure instant excitement of the field windings. A number of vehicles link a wire from the engine management ECU to the alternator. This is used to send a signal to increase engine idle speed if the battery is low on charge.

6.4 Case studies

6.4.1 An alternator in common use

Figure 6.23 shows the Lucas model A127 alternator used in large numbers by several vehicle manufacturers. The basic data relating to this machine are listed below.

- 12 V negative earth.
- Regulated voltage 14.0–14.4 V.
- Machine sensed.
- Maximum output when hot, 65 A (earth-return).
- Maximum speed 16 500 rev/min.
- Temperature range −40 to +105 °C.
- European plug and stud termination (7 mm).

This alternator has a frame diameter of 127 mm, a 15 mm drive shaft and weighs about 4 kg. It is a star-wound machine.

6.4.2 Bosch compact alternator

The Bosch compact alternator is becoming very popular with a number of European manufacturers and others. Figure 6.24 shows a cut-away picture of this machine. The key points are as follows:

- 20–70% more power than conventional units.
- 15–35% better power-to-weight ratio.
- Maximum speed up to 20 000 rev/min.
- Twin interior cooling fans.
- Precision construction for reduced noise.
- Versions available: 70, 90 and up to 170 A.

The compact alternator follows the well-known claw pole design. Particular enhancements have

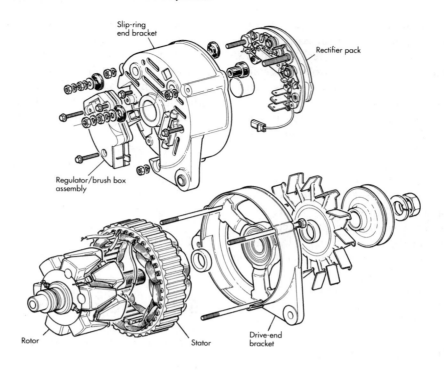

Figure 6.23 Lucas A127 alternator

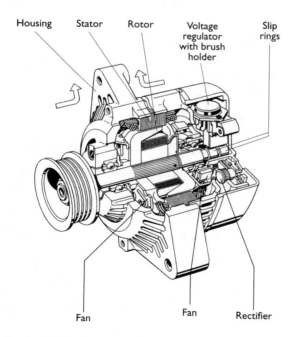

Figure 6.24 Bosch compact alternator

that reduces the voltage drop across the main power transistor from 1.2 V to 0.6 V. This allows a greater field current to flow, which again will improve efficiency.

The top speed of an alternator is critical as it determines the pulley ratio between the engine and alternator. The main components affected by increased speed are the ball bearings and the slip rings. The bearings have been replaced with a type that uses a plastic cage instead of the conventional metal type. Higher melting point grease is also used. The slip rings are now mounted outside the two bearings and therefore the diameter is not restricted by the shaft size. Smaller diameter slip rings give a much lower peripheral velocity, and thus greater shaft speed can be tolerated.

Increased output results in increased temperature so a better cooling system was needed. The machine uses twin internal asymmetric fans, which pull air through central slots front and rear, and push it out radially through the drive and slip ring end brackets over the stator winding heads.

High vibration is a problem with alternators as with all engine mounted components. Cars with four-valve engines can produce very high levels of vibration. The alternator is designed to withstand up to 80 g. New designs are thus required for the mounting brackets.

been made to the magnetic circuit of the rotor and stator. This was achieved by means of modern 'field calculation' programmes. The optimization reduces the iron losses and hence increases efficiency.

A new monolithic circuit regulator is used

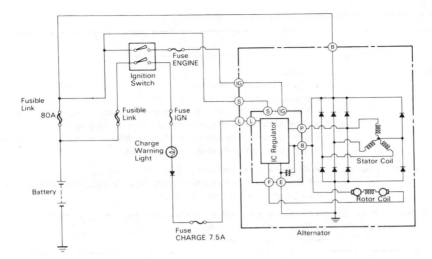

Figure 6.25 Japanese alternator circuit

6.4.3 Japanese alternator

Figure 6.25 shows the internal and external circuit of a typical alternator used on a number of Japanese vehicles. It is an eight-diode machine and uses an integrated circuit regulator. Four electrical connections are made to the alternator; the main output wire (B), an ignition feed (IG), a battery sensing wire (S) and the warning light (L).

Figure 6.26 shows all the main components of the alternator. Internal cooling fans are used to draw air through the slots in the end brackets. The diameter of the slip rings is only about 14 mm. This keeps the surface speed (m/s) to a minimum, allowing greater rotor speeds (rev/min).

The IC regulator ensures consistent output voltage with built-in temperature compensation.

Figure 6.26 Alternator components

The ignition-controlled feed is used to ensure that the machine charges fully at low engine speed. Because of this ignition voltage supply to the fields, the cut-in speed is low.

6.5 Diagnosing charging system faults

6.5.1 Introduction

As with all systems, the six stages of fault-finding should be followed.

1. Verify the fault.
2. Collect further information.
3. Evaluate the evidence.
4. Carry out further tests in a logical sequence.
5. Rectify the problem.
6. Check all systems.

The procedure outlined in the next section is related primarily to stage 4 of the process. Table 6.2 lists some common symptoms of a charging system malfunction together with suggestions for the possible fault.

6.5.2 Testing procedure

After connecting a voltmeter across the battery and an ammeter in series with the alternator output wire(s), as shown in Figure 6.27, the process of checking the charging system operation is as follows.

Table 6.2 Common symptoms and faults of a charging system malfunction

Symptom	Possible fault
Battery loses charge	• Defective battery. • Slipping alternator drive belt. • Battery terminals loose or corroded. • Alternator internal fault (diode open circuit, brushes worn or regulator fault etc). • Open circuit in alternator wiring, either main supply, ignition or sensing wires if fitted. • Short circuit component causing battery drain even when all switches are off. • High resistance in the main charging circuit.
Charge warning light stays on when engine is running	• Slipping or broken alternator drive belt. • Alternator internal fault (diode open circuit, brushes worn or regulator fault etc). • Loose or broken wiring/connections.
Charge warning light does not come on at any time	• Alternator internal fault (brushes worn open circuit or regulator fault etc). • Blown warning light bulb. • Open circuit in warning light circuit.

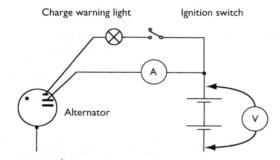

Figure 6.27 Alternator testing

1. Hand and eye checks (drive belt and other obvious faults) – belt at correct tension, all connections clean and tight.
2. Check battery (see Chapter 5) – must be 70% charged.
3. Measure supply voltages to alternator – battery volts.
4. Maximum output current (discharge battery slightly by leaving lights on for a few minutes, leave lights on and start engine) – ammeter should read within about 10% of rated maximum output.
5. Regulated voltage (ammeter reading 10 A or less) – $14.2 \pm 0.2\,V$.
6. Circuit volt drop – 0.5 V maximum.

If the alternator is found to be defective then a quality replacement unit is the normal recommendation. Figure 6.28 explains the procedure used by Bosch to ensure quality exchange units. Repairs are possible but only if the general state of the alternator is good.

6.6 Advanced charging system technology

6.6.1 Charging system – problems and solutions

The charging system of a vehicle has to cope under many varied conditions. An earlier section gave some indication as to the power output that may be required. Looking at two of the operating conditions that may be encountered makes the task of producing the required output even more difficult.

The first scenario is the traffic jam, on a cold night, in the rain! This can involve long periods when the engine is just idling, but use of nearly all electrical devices is still required. The second scenario is that the car has been parked in the open on a frosty night. The engine is started, seat heaters, heated rear window and blower fan are switched on whilst a few minutes are spent scraping the screen and windows. All the lights and wipers are now switched on and a journey of half an hour through busy traffic follows. The seat heaters and heated rear window can generally be assumed to switch off automatically after about 15 minutes.

Tests and simulations have been carried out using the above examples as well as many others. At the end of the first scenario the battery state of charge will be about 35% less than its original level; in the second case the state of charge will be about 10% less. These situations are worst case scenarios, but nonetheless possible. If the situations were repeated without other journeys

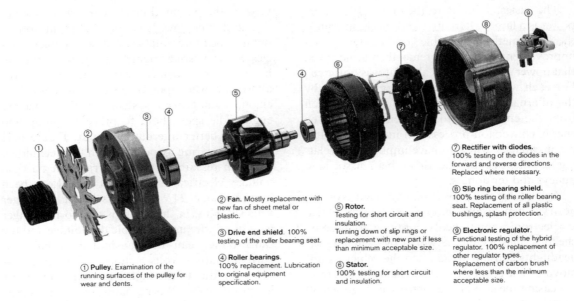

② Fan. Mostly replacement with new fan of sheet metal or plastic.

③ Drive end shield. 100% testing of the roller bearing seat.

④ Roller bearings. 100% replacement. Lubrication to original equipment specification.

① Pulley. Examination of the running surfaces of the pulley for wear and dents.

⑤ Rotor. Testing for short circuit and insulation. Turning down of slip rings or replacement with new part if less than minimum acceptable size.

⑥ Stator. 100% testing for short circuit and insulation.

⑦ Rectifier with diodes. 100% testing of the diodes in the forward and reverse directions. Replaced where necessary.

⑧ Slip ring bearing shield. 100% testing of the roller bearing seat. Replacement of all plastic bushings, splash protection.

⑨ Electronic regulator. Functional testing of the hybrid regulator. 100% replacement of other regulator types. Replacement of carbon brush where less than the minimum acceptable size.

Figure 6.28 Alternator overhaul procedure (Bosch)

in between, then the battery would soon be incapable of starting the engine. Combining this with the ever-increasing power demands on the vehicle alternator makes this problem difficult to solve. It is also becoming even more important to ensure the battery remains fully charged, as ECUs with volatile memories and alarm systems make a small but significant drain on the battery when the vehicle is parked.

A number of solutions are available to try and ensure the battery will remain in a state near to full charge at all times. A larger capacity battery could be used to 'swamp' variations in electrical use and operating conditions. Some limit, however, has to be set due to the physical size of the battery. Five options for changes to the power supply system are represented graphically in Figure 6.29 and are listed below.

- Fitting a more powerful alternator.
- Power management system.
- Two-stage alternator drive mechanism or increased alternator speed.
- Increased engine idle speed.
- Dual voltage systems.

The five possible options listed above have some things in their favour and some against, not least of which are the technical and economic factors. For the manufacturers, I would predict that a combination of a more powerful alternator, which can be run at a higher speed, together with a higher or dual voltage system, would be the way forward. This is likely to be the most cost effective and technically feasible solution. Each of the suggestions is now discussed in more detail.

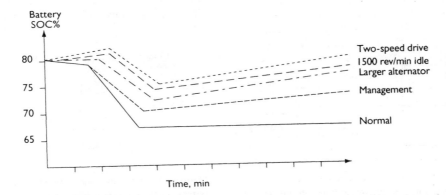

Figure 6.29 Graphical representation comparing various charging techniques when applied to a vehicle used for winter commuting

The easiest solution to the demand for more power is a larger alternator, and this is, in reality, the only method available as an after-market improvement. It must be remembered, however, that power supplied by an alternator is not 'free'. For each watt of electrical power produced by the alternator, between 1.5 and 2 W are taken from the engine due to the inefficiency of the energy conversion process. An increase in alternator capacity will also have implications relating to the size of the drive belt, associated pulleys and tensioners.

An intelligent power management system, however, may become more financially attractive as electronic components continue to become cheaper. This technique works by switching off headlights and fog lights when the vehicle is not moving. The cost of this system may be less than increasing the size of the alternator. Figure 6.30 shows the operating principle of this system. A speed sensor signal is used via an electronic processing circuit to trigger a number of relays. The relays can be used to interrupt the chosen lighting circuits. An override switch is provided, for use in exceptional conditions.

A two-speed drive technique which uses a ratio of 5:1 for engine speeds under 1200 rev/min and usually about 2.5:1 at higher speeds shows some promise but adds more complications to the drive system. Due to improvements in design, however, modern alternators are now being produced that are capable of running at speeds up to 20 000 rev/min. If the maximum engine speed is considered to be about 6000 rev/min, a pulley ratio of about 3.3:1 can be used. This will allow the alternator to run as fast as 2300 rev/min, even with a low engine idle speed of 700 rev/min. The two-speed drive is only at the prototype stage at present.

Increased idle speed may not be practical in view of the potential increase in fuel consumption and emissions. It is nonetheless an option, but may be more suitable for diesel-engined vehicles. Some existing engine management systems, however, are provided with a signal from the alternator when power demand is high. The engine management system can then increase engine idle speed both to prevent stalling and ensure a better alternator output. Figure 6.31 shows the wiring associated with this technique.

Much research is being carried out on dual voltage electrical systems. It has long been known that a 24 V system is better for larger vehicles. This, in the main, is due to the longer lengths of wire used. Double the voltage and the same power can be transmitted at half the current (watts = volts × amps). This causes less volt drop due to the higher resistance in longer lengths of cable. Wiring harnesses used on passenger cars are becoming increasingly heavy and unmanageable. If a higher supply voltage was used, the cross-section of individual cables could be halved with little or no effect. Because heavy vehicle electrics have been 24 V for a long time, most components (bulbs etc) are already available if a change in strategy by the vehicle manufacturers takes place. Under discussion is a −12, 0, +12 V technique using three bus bars or rails. High power loads can be connected between −12 and +12 (24 V), and loads which must be supplied by 12 V can be balanced between the −12, 0 and 0, +12 voltage supply rails. A representation of this is shown in Figure 6.32. Note, however, that running some bulbs (such as for high power headlights) can be a problem because the filament has to be very thin. Some commercial (24 V) vehicles actually use a 12 V supply to the headlights for this reason.

6.6.2 Charge balance calculation

The charge balance or energy balance of a charging system is used to ensure that the

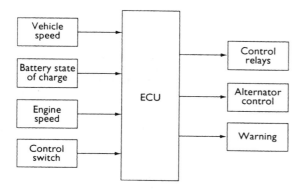

Figure 6.30 Operating principles of a power management system

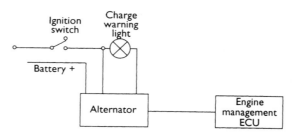

Figure 6.31 Alternator wiring to allow engine management system to sense current demand and control engine idle speed to prevent stalling

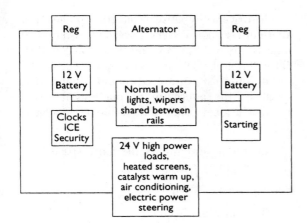

Figure 6.32 Dual rail power supply technique

alternator can cope with all the demands placed on it and still charge the battery. The following steps help to indicate the size of alternator required or to check if the one fitted to a vehicle is suitable.

As a worked example, the figures from Table 6.1 will be used. The calculations relate to a passenger car with a 12 V electrical system. A number of steps are involved.

1. Add the power used by all the continuous and prolonged loads.
2. Total continuous and prolonged power (P_1) = 440 W.
3. Calculate the current at 14 V ($I = W/V$) = 31.5 A.
4. Determine the intermittent power (factored by 0.1) (P_2) = 170 W.
5. Total power ($P_1 + P_2$) = 610 W.
6. Total current = 610/14 = 44 A.

Electrical component manufacturers provide tables to recommend the required alternator, calculated from the total power demand and the battery size. However, as a guide for 12 V passenger cars, the rated output should be about 1.5 times the total current demand (in this example 44 × 1.5 = 66 A). Manufacturers produce machines of standard sizes, which in this case would probably mean an alternator rated at 70 A. In the case of vehicles with larger batteries and starters, such as for diesel-powered engines and commercial vehicles, a larger output alternator may be required.

The final check is to ensure that the alternator output at idle is large enough to supply all continuous and prolonged loads (P_1) and still charge the battery. Again the factor of 1.5 can be

applied. In this example the alternator should be able to supply (31.5 × 1.5) = 47 A, at engine idle. On normal systems this relates to an alternator speed of about 2000 rev/min (or less). This can be checked against the characteristic curve of the alternator.

6.6.3 Alternator characteristics

Alternator manufacturers supply 'characteristic curves' for their alternators. These show the properties of the alternator under different conditions. The curves are plotted as output current (at stabilized voltage), against alternator rev/min and input power against input rev/min. Figure 6.33 shows a typical alternator characteristic curve.

It is common to mark the following points on the graph.

- Cut in speed.
- Idle speed range.
- Speed at which 2/3 of rated output is reached.
- Rated output speed.
- Maximum speed.
- Idle current output range.
- Current 2/3 of rated output.
- Rated output.
- Maximum output.

The graphs are plotted under specific conditions such as regulated output voltage and constant temperature (27°C is often used). The graph is often used when working out what size alternator will be required for a specific application.

The power curve is used to calculate the type of drive belt needed to transmit the power or torque to the alternator. As an aside, the power curve and the current curve can be used together to calculate the efficiency of the alternator. At

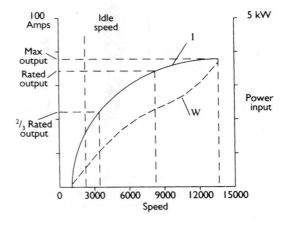

Figure 6.33 Typical alternator characteristic curve

any particular speed when producing maximum output for that speed, the efficiency of any machine is calculated from:

Efficiency = Power out/Power in

In this case, the efficiency at 8000 rev/min is

(Power out = 14 V × 58 A = 812 W)
812 W / 4000 W = 0.203 or about 20%

Efficiency at 2/3 maximum output

(Power out = 14 V × 38 A = 532 W)
532/2000 = 0.266 or about 27%

These figures help to illustrate how much power is lost in the generation process. The inefficiency is mainly due to iron losses, copper losses, windage (air friction) and mechanical friction. The energy is lost as heat.

6.6.4 Mechanical and external considerations

Most light vehicle alternators are mounted in similar ways. This usually involves a pivoted mounting on the side of the engine with an adjuster on the top or bottom to set drive belt tension. It is now common practice to use 'multi-V' belts driving directly from the engine crankshaft pulley. This type of belt will transmit greater torque and can be worked on smaller diameter pulleys or with tighter corners than the more traditional 'V' belt. Figure 6.34 is an extract from information regarding the mounting and drive belt fitting for a typical alternator.

The drive ratio between the crank pulley and alternator pulley is very important. A typical ratio is about 2.5 : 1. In simple terms, the alternator should be driven as fast as possible at idle speed, but must not exceed the maximum rated speed of the alternator at maximum engine speed. The ideal ratio can therefore be calculated as follows:

Maximum ratio = max alternator speed / max engine speed.

For example:

15000 rev/min / 6000 rev/min = 2.5 : 1.

During the design stage the alternator will often have to be placed in a position determined by the space available in the engine compartment. However, where possible the following points should be considered:

- Adequate cooling.
- Suitable protection from contamination.
- Access for adjustment and servicing.
- Minimal vibration if possible.
- Recommended belt tension.

6.7 New developments in charging systems

6.7.1 General developments

Alternators are being produced capable of ever greater outputs in order to supply the constantly

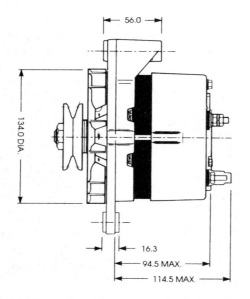
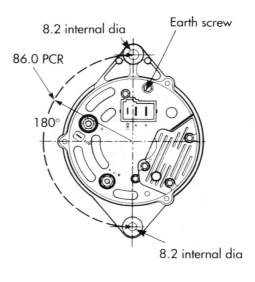

Figure 6.34 Extract from information supplied by Lucas Automotive Ltd. relating to the Plus Pac alternator

increasing demands placed on them by manufacturers. The main problem to solve is that of producing high output at lower engine speeds. A solution to this is a variable drive ratio, but this is fraught with mechanical problems. The current solution is tending towards alternators capable of much higher maximum speeds, which allows a greater drive ratio and hence greater speed at lower engine rev/min.

The main design of alternators does not appear to be changing radically; however, the incremental improvements have allowed far more efficient machines to be produced.

6.7.2 Water-cooled alternators

Valeo have an interesting technique involving running the engine coolant through the alternator. A 120–190 A output range is available. Compared with conventional air-cooled alternators the performance of these new machines has been enhanced more particularly in the following areas:

- Improved efficiency (10–25%).
- Increased output at engine idle speed.
- Noise reduction (10–12 dB due to fan elimination).
- Resistance to corrosion (machine is enclosed).
- Resistance to high ambient temperature (>130°C).

Additional heating elements can be integrated into the alternator to form a system that donates an additional 2–3 kW to the coolant, enabling faster engine warm up after a cold start. This contributes to reduced pollution and increased driver comfort.

Valeo have also developed an alternator with a 'self-start' regulator. This can be thought of as an independent power centre because the warning light and other wires (not the main feed!) can be eliminated. This saves manufacturing costs and also ensures that output is maintained at idle speed.

6.7.3 Integrated alternator

A very interesting development is the Integrated Starter Alternator Damper (ISAD) system. Refer to Section 7.7 for more details.

6.8 Self-assessment

6.8.1 Questions

1. State the ideal charging voltage for a 12 V (nominal) battery.
2. Describe the operation of an alternator with reference to a rotating 'permanent magnet'.
3. Make a clearly labelled sketch to show a typical external alternator circuit.
4. Explain how and why the output voltage of an alternator is regulated.
5. Describe the differences between a star-wound and a delta-wound stator.
6. Explain why connecting two extra diodes to the centre of a star-wound stator can increase the output of an alternator.
7. Draw a typical characteristic curve for an alternator. Label each part with an explanation of its purpose.
8. Describe briefly how a rectifier works.
9. Explain the difference between a battery-sensed and a machine-sensed alternator.
10. List five charging system faults and the associated symptoms.

6.6.1 Assignment

Investigate and test the operation of a charging system on a vehicle. Produce a report in the standard format (as set out in *Advanced Automotive Fault Diagnosis*, Tom Denton (2000), Arnold).

Make recommendations on how the system could be improved.

Figure 6.35 A Bosch water-cooled alternator

7

Starting systems

7.1 Requirements of the starting system

7.1.1 Engine starting requirements

An internal combustion engine requires the following criteria in order to start and continue running.

- Combustible mixture.
- Compression stroke.
- A form of ignition.
- The minimum starting speed (about 100 rev/min).

In order to produce the first three of these, the minimum starting speed must be achieved. This is where the electric starter comes in. The ability to reach this minimum speed is again dependent on a number of factors.

- Rated voltage of the starting system.
- Lowest possible temperature at which it must still be possible to start the engine. This is known as the starting limit temperature.
- Engine cranking resistance. In other words the torque required to crank the engine at its starting limit temperature (including the initial stalled torque).
- Battery characteristics.
- Voltage drop between the battery and the starter.
- Starter-to-ring gear ratio.
- Characteristics of the starter.
- Minimum cranking speed of the engine at the starting limit temperature.

It is not possible to view the starter as an isolated component within the vehicle electrical system, as Figure 7.1 shows. The battery in particular is of prime importance.

Another particularly important consideration in relation to engine starting requirements is the starting limit temperature. Figure 7.2 shows

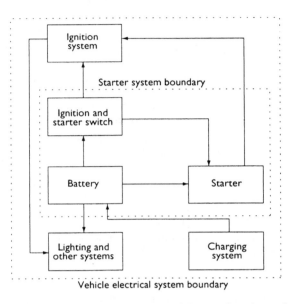

Figure 7.1 Starting system as part of the complete electrical system

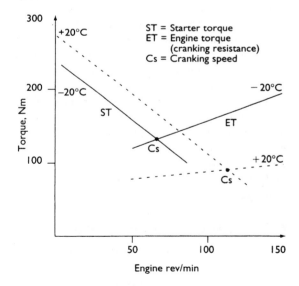

Figure 7.2 Starter torque and engine cranking torque

how, as temperature decreases, starter torque also decreases and the torque required to crank the engine to its minimum speed increases.

Typical starting limit temperatures are $-18°C$

to −25°C for passenger cars and −15°C to −20°C for trucks and buses. Figures from starter manufacturers are normally quoted at both +20°C and −20°C.

7.1.2 Starting system design

The starting system of any vehicle must meet a number of criteria in excess of the eight listed above.

- Long service life and maintenance free.
- Continuous readiness to operate.
- Robust, such as to withstand starting forces, vibration, corrosion and temperature cycles.
- Lowest possible size and weight.

Figure 7.3 shows the starting system general layout. It is important to determine the minimum cranking speed for the particular engine. This varies considerably with the design and type of engine. Some typical values are given in Table 7.1 for a temperature of −20°C.

The rated voltage of the system for passenger cars is, almost without exception, 12 V. Trucks and buses are generally 24 V as this allows the use of half the current that would be required with a 12 V system to produce the same power. It will also considerably reduce the voltage drop in

Table 7.1 Typical minimum cranking speeds

Engine	Minimum cranking speed
Reciprocating spark ignition	60–90 rev/min
Rotary spark ignition	150–180 rev/min
Diesel with glow plugs	60–140 rev/min
Diesel without glow plugs	100–200 rev/min

the wiring, as the length of wires used on commercial vehicles is often greater than passenger cars.

The rated output of a starter motor can be determined on a test bench. A battery of maximum capacity for the starter, which has a 20% drop in capacity at −20°C, is connected to the starter by a cable with a resistance of 1 mΩ. These criteria will ensure the starter is able to operate even under the most adverse conditions. The actual output of the starter can now be measured under typical operating conditions. The rated power of the motor corresponds to the power drawn from the battery less copper losses (due to the resistance of the circuit), iron losses (due to eddy currents being induced in the iron parts of the motor) and friction losses.

Figure 7.4 shows an equivalent circuit for a starter and battery. This indicates how the starter output is very much determined by line resistance and battery internal resistance. The lower the total resistance, the higher the output from the starter.

There are two other considerations when designing a starting system. The location of the starter on the engine is usually pre-determined, but the position of the battery must be considered. Other constraints may determine this, but if the battery is closer to the starter the cables will be shorter. A longer run will mean cables with a greater cross-section are needed to ensure a low resistance. Depending on the intended use of the vehicle, special sealing arrangements on the starter may be necessary to prevent the ingress of contaminants. Starters are available designed with this in mind. This may be appropriate for off-road vehicles.

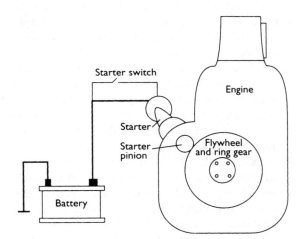

Figure 7.3 Starter system general layout

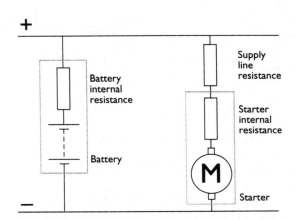

Figure 7.4 Equivalent circuit for a starter system

7.1.3 Choosing a starter motor

As a guide, the starter motor must meet all the criteria previously discussed. Referring back to Figure 7.2 (the data showing engine cranking torque compared with minimum cranking speed) will determine the torque required from the starter.

Manufacturers of starter motors provide data in the form of characteristic curves. These are discussed in more detail in the next section. The data will show the torque, speed, power and current consumption of the starter at $+20°C$ and $-20°C$. The power rating of the motor is quoted as the maximum output at $-20°C$ using the recommended battery.

Figure 7.5 shows how the required power output of the starter relates to the engine size.

As a very general guide the stalled (locked) starter torque required per litre of engine capacity at the starting limit temperature is as shown in Table 7.2

A greater torque is required for engines with a lower number of cylinders due to the greater piston displacement per cylinder. This will determine the peak torque values. The other main factor is compression ratio.

To illustrate the link between torque and power, we can assume that, under the worst conditions $(-20°C)$, a four-cylinder 2 litre engine requires 480 Nm to overcome static friction and 160 Nm to maintain the minimum cranking speed of 100 rev/min. With a starter pinion-to-ring gear ratio of $10:1$, the motor must there-

Table 7.2 Torque required for various engine sizes

Engine cylinders	Torque per litre
2	12.5 Nm
4	8.0 Nm
6	6.5 Nm
8	6.0 Nm
12	5.5 Nm

fore, be able to produce a maximum stalled torque of 48 Nm and a driving torque of 16 Nm. This is working on the assumption that stalled torque is generally three to four times the cranking torque.

Torque is converted to power as follows:

$$P = T\omega$$

where P = power; T = torque and ω = angular velocity.

$$\omega = \frac{2\pi n}{60}$$

where n = rev/min.

In this example, the power developed at 1000 rev/min with a torque of 16 Nm (at the starter) is about 1680 W. Referring back to Figure 7.5, the ideal choice would appear to be the starter marked (e).

The recommended battery would be 55 Ah and 255 A cold start performance.

7.2 Starter motors and circuits

7.2.1 Starting system circuits

In comparison with most other circuits on the modern vehicle, the starter circuit is very simple. The problem to be overcome, however, is that of volt drop in the main supply wires. The starter is usually operated by a spring-loaded key switch, and the same switch also controls the ignition and accessories. The supply from the key switch, via a relay in many cases, causes the starter solenoid to operate, and this in turn, by a set of contacts, controls the heavy current. In some cases an extra terminal on the starter solenoid provides an output when cranking, which is usually used to bypass a dropping resistor on the ignition or fuel pump circuits. The basic circuit for the starting system is shown in Figure 7.6.

The problem of volt drop in the main supply circuit is due to the high current required by the

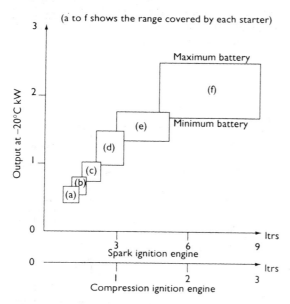

Figure 7.5 Power output of the starter compared with engine size

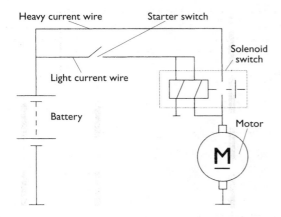

Figure 7.6 Basic starting circuit

starter, particularly under adverse starting conditions such as very low temperatures.

A typical cranking current for a light vehicle engine is of the order of 150 A, but this may peak in excess of 500 A to provide the initial stalled torque. It is generally accepted that a maximum volt drop of only 0.5 V should be allowed between the battery and the starter when operating. An Ohm's law calculation indicates that the maximum allowed circuit resistance is 2.5 mΩ when using a 12 V supply. This is a worst case situation and lower resistance values are used in most applications. The choice of suitable conductors is therefore very important.

7.2.2 Principle of operation

The simple definition of any motor is a machine to convert electrical energy into mechanical energy. The starter motor is no exception. When current flows through a conductor placed in a magnetic field, a force is created acting on the conductor relative to the field. The magnitude of this force is proportional to the field strength, the length of the conductor in the field and the current flowing in the conductor.

In any DC motor, the single conductor is of no practical use and so the conductor is shaped into a loop or many loops to form the armature. A many-segment commutator allows contact via brushes to the supply current.

The force on the conductor is created due to the interaction of the main magnetic field and the field created around the conductor. In a light vehicle starter motor, the main field was traditionally created by heavy duty series windings wound around soft iron pole shoes. Due to improvements in magnet technology, permanent magnet fields allowing a smaller and lighter construction are replacing wire-wound fields. The strength of the magnetic field created around the conductors in the armature is determined by the value of the current flowing. The principle of a DC motor is shown in Figure 7.7.

Most starter designs use a four-pole four-brush system. Using four field poles concentrates the magnetic field in four areas as shown in Figure 7.8. The magnetism is created in one of three ways, permanent magnets, series field windings or series–parallel field windings.

Figure 7.9 shows the circuits of the two methods where field windings are used. The series–parallel fields can be constructed with a lower resistance, thereby increasing the current and hence torque of the motor. Four brushes are

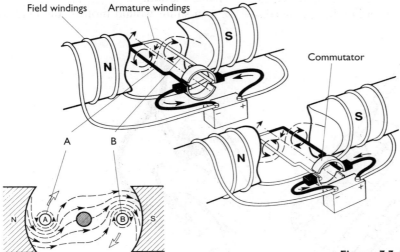

Figure 7.7 Interaction of two magnetic fields results in rotation when a commutator is used to reverse the supply each half turn

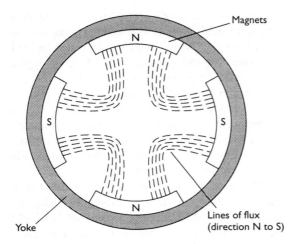

Figure 7.8 Four-pole magnetic field

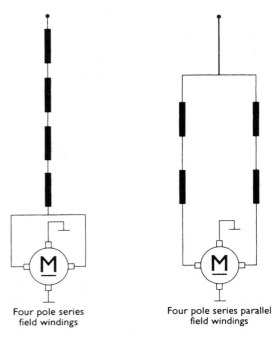

Four pole series
field windings

Four pole series parallel
field windings

Figure 7.9 Starter internal circuits

used to carry the heavy current. The brushes are made of a mixture of copper and carbon, as is the case for most motor or generator brushes. Starter brushes have a higher copper content to minimize electrical losses. Figure 7.10 shows some typical field coils with brushes attached. The field windings on the right are known as wave wound.

The armature consists of a segmented copper commutator and heavy duty copper windings. The windings on a motor armature can, broadly speaking, be wound in two ways. These are known as lap winding and wave winding. Figure 7.11 shows the difference between these two methods. Starter motors tend to use wave winding as this technique gives the most appropriate torque and speed characteristic for a four-pole system.

A starter must also have some method of engaging with, and release from, the vehicle's flywheel ring gear. In the case of light vehicle starters, this is achieved either by an inertia-

type engagement or a pre-engagement method. These are both discussed further in subsequent sections.

7.2.3 DC motor characteristics

It is possible to design a motor with characteristics that are most suitable for a particular task. For a comparison between the main types of DC motor, the speed–torque characteristics are shown in Figure 7.12. The four main types of motor are referred to as shunt wound, series wound, compound wound and permanent magnet excitation.

In shunt wound motors, the field winding is

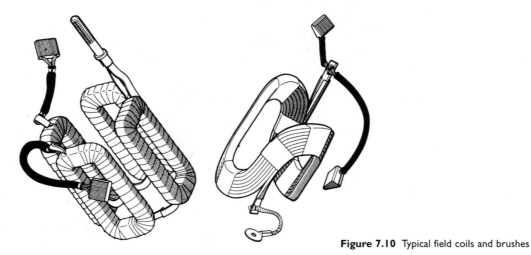

Figure 7.10 Typical field coils and brushes

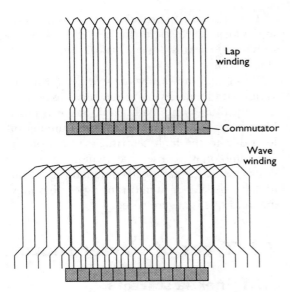

Figure 7.11 Typical lap and wave wound armature circuits

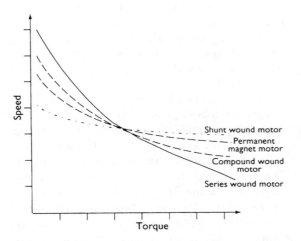

Figure 7.12 Speed and torque characteristics of DC motors

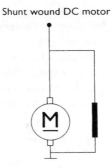

Figure 7.13 Shunt wound motor (parallel wound)

Figure 7.14 Series wound motor

connected in parallel with the armature as shown in Figure 7.13. Due to the constant excitation of the fields, the speed of this motor remains constant, virtually independent of torque.

Series wound motors have the field and armature connected in series. Because of this method of connection, the armature current passes through the fields making it necessary for the field windings to consist usually of only a few turns of heavy wire. When this motor starts under load the high initial current, due to low resistance and no back EMF, generates a very strong magnetic field and therefore high initial torque. This characteristic makes the series wound motor ideal as a starter motor. Figure 7.14 shows the circuit of a series wound motor.

The compound wound motor, as shown in Figure 7.15, is a combination of shunt and series wound motors. Depending on how the field windings are connected, the characteristics can vary. The usual variation is where the shunt winding is connected, which is either across the armature or across the armature and series winding. Large starter motors are often compound wound and can be operated in two stages. The first stage involves the shunt winding being connected in series with the armature. This unusual connection allows for low meshing torque due to the resistance of the shunt winding. When the pinion of the starter is fully in mesh with the ring gear, a set of contacts causes the main supply to be passed through the series winding and armature giving full torque. The shunt winding will now be connected in parallel and will act in such a way as to limit the maximum speed of the motor.

Permanent magnet motors are smaller and simpler compared with the other three discussed. Field excitation, as the name suggests, is by permanent magnet. This excitation will remain constant under all operating conditions. Figure 7.16 shows the accepted representation for this type of motor.

The characteristics of this type of motor are broadly similar to the shunt wound motors.

However, when one of these types is used as a starter motor, the drop in battery voltage tends to cause the motor to behave in a similar way to a series wound machine. In some cases though, the higher speed and lower torque characteristic are enhanced by using an intermediate transmission gearbox inside the starter motor.

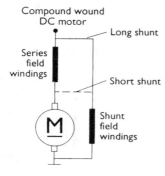

Figure 7.15 Compound wound motor

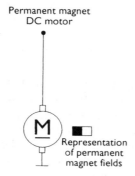

Figure 7.16 Permanent magnet motor

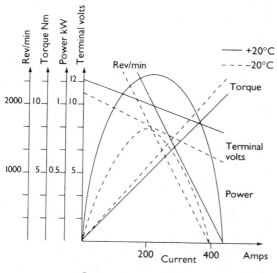

Curves of a 12 V 0.9 kW starter using the maximum size battery of 55 Ah, 255 A

Figure 7.17 Starter motor characteristic curves

Information on particular starters is provided in the form of characteristic curves. Figure 7.17 shows the details for a typical light vehicle starter motor.

This graph shows how the speed of the motor varies with load. Owing to the very high speeds developed under no load conditions, it is possible to damage this type of motor. Running off load due to the high centrifugal forces on the armature may cause the windings to be destroyed. Note that the maximum power of this motor is developed at mid-range speed but maximum torque is at zero speed.

7.3 Types of starter motor

7.3.1 Inertia starters

In all standard motor vehicle applications it is necessary to connect the starter to the engine ring gear only during the starting phase. If the connection remained permanent, the excessive speed at which the starter would be driven by the engine would destroy the motor almost immediately.

The inertia type of starter motor has been the technique used for over 80 years, but is now becoming redundant. The starter shown in Figure 7.18 is the Lucas M35J type. It is a four-pole, four-brush machine and was used on small to medium-sized petrol engined vehicles. It is capable of producing 9.6 Nm with a current draw of 350 A. The M35J uses a face-type commutator and axially aligned brush gear. The fields are wave wound and are earthed to the starter yoke.

The starter engages with the flywheel ring gear by means of a small pinion. The toothed pinion and a sleeve splined on to the armature shaft are threaded such that when the starter is operated, via a remote relay, the armature will

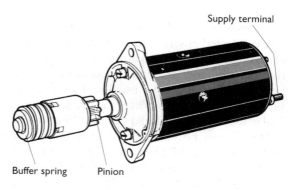

Figure 7.18 Inertia type starter

cause the sleeve to rotate inside the pinion. The pinion remains still due to its inertia and, because of the screwed sleeve rotating inside it, the pinion is moved to mesh with the ring gear.

When the engine fires and runs under its own power, the pinion is driven faster than the armature shaft. This causes the pinion to be screwed back along the sleeve and out of engagement with the flywheel. The main spring acts as a buffer when the pinion first takes up the driving torque and also acts as a buffer when the engine throws the pinion back out of mesh.

One of the main problems with this type of starter was the aggressive nature of the engagement. This tended to cause the pinion and ring gear to wear prematurely. In some applications the pinion tended to fall out of mesh when cranking due to the engine almost, but not quite, running. The pinion was also prone to seizure often due to contamination by dust from the clutch. This was often compounded by application of oil to the pinion mechanism, which tended to attract even more dust and thus prevent engagement.

The pre-engaged starter motor has largely overcome these problems.

7.3.2 Pre-engaged starters

Pre-engaged starters are fitted to the majority of vehicles in use today. They provide a positive engagement with the ring gear, as full power is not applied until the pinion is fully in mesh. They prevent premature ejection as the pinion is held into mesh by the action of a solenoid. A one-way clutch is incorporated into the pinion to prevent the starter motor being driven by the engine. One example of a pre-engaged starter in common use is shown in Figure 7.19, the Bosch EF starter.

Figure 7.20 shows the circuit associated with operating this type of pre-engaged starter. The basic operation of the pre-engaged starter is as follows. When the key switch is operated, a supply is made to terminal 50 on the solenoid. This causes two windings to be energized, the hold-on winding and the pull-in winding. Note that the pull-in winding is of very low resistance and hence a high current flows. This winding is connected in series with the motor circuit and the current flowing will allow the motor to rotate slowly to facilitate engagement. At the same time, the magnetism created in the solenoid attracts the plunger and, via an operating lever, pushes the pinion into mesh with the flywheel ring gear. When the pinion is fully in mesh the plunger, at the end of its travel, causes a heavy-duty set of copper contacts to close. These contacts now supply full battery power to the main circuit of the starter motor. When the main contacts are closed, the pull-in winding is effectively switched off due to equal voltage supply on both ends. The hold-on winding holds the plunger in position as long as the solenoid is supplied from the key switch.

When the engine starts and the key is released, the main supply is removed and the plunger and pinion return to their rest positions under spring tension. A lost motion spring located on the plunger ensures that the main

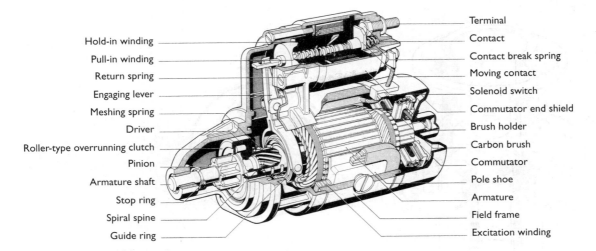

Hold-in winding	Terminal
Pull-in winding	Contact
Return spring	Contact break spring
Engaging lever	Moving contact
Meshing spring	Solenoid switch
Driver	Commutator end shield
Roller-type overrunning clutch	Brush holder
Pinion	Carbon brush
Armature shaft	Commutator
Stop ring	Pole shoe
Spiral spine	Armature
Guide ring	Field frame
	Excitation winding

Figure 7.19 Pre-engaged starter

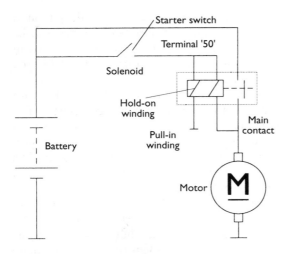

Figure 7.20 Starter circuit

contacts open before the pinion is retracted from mesh.

During engagement, if the teeth of the pinion hit the teeth of the flywheel (tooth to tooth abutment), the main contacts are allowed to close due to the engagement spring being compressed. This allows the motor to rotate under power and the pinion will slip into mesh.

Figure 7.21 shows a sectioned view of a one-way clutch assembly. The torque developed by the starter is passed through the clutch to the ring gear. The purpose of this free-wheeling device is to prevent the starter being driven at an excessively high speed if the pinion is held in mesh after the engine has started. The clutch consists of a driving and driven member with several rollers between the two. The rollers are spring loaded and either wedge-lock the two members together by being compressed

against the springs, or free-wheel in the opposite direction.

Many variations of the pre-engaged starter are in common use, but all work on similar lines to the above description. The wound field type of motor has now largely been replaced by the permanent magnet version.

7.3.3 Permanent magnet starters

Permanent magnet starters began to appear on production vehicles in the late 1980s. The two main advantages of these motors, compared with conventional types, are less weight and smaller size. This makes the permanent magnet starter a popular choice by vehicle manufacturers as, due to the lower lines of today's cars, less space is now available for engine electrical systems. The reduction in weight provides a contribution towards reducing fuel consumption.

The standard permanent magnet starters currently available are suitable for use on spark ignition engines up to about 2 litre capacity. They are rated in the region of 1 kW. A typical example is the Lucas Model M78R/M80R shown in Figure 7.22.

The principle of operation is similar in most respects to the conventional pre-engaged starter motor. The main difference being the replacement of field windings and pole shoes with high quality permanent magnets. The reduction in weight is in the region of 15% and the diameter of the yoke can be reduced by a similar factor.

Permanent magnets provide constant excitation and it would be reasonable to expect the speed and torque characteristic to be constant.

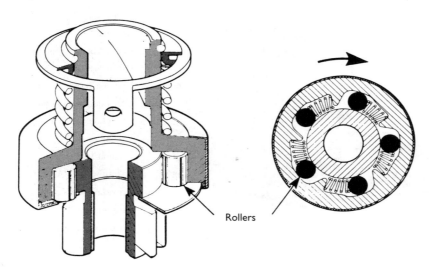

Figure 7.21 One-way roller clutch drive pinion

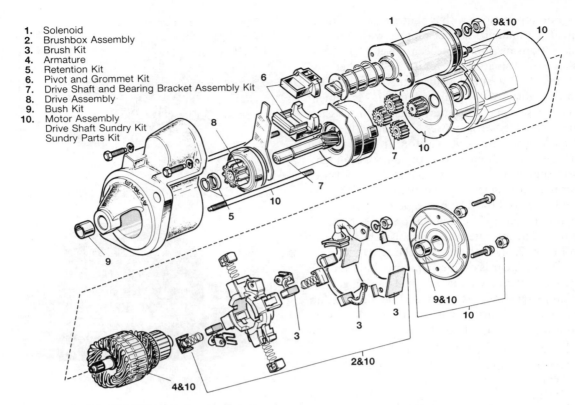

1. Solenoid
2. Brushbox Assembly
3. Brush Kit
4. Armature
5. Retention Kit
6. Pivot and Grommet Kit
7. Drive Shaft and Bearing Bracket Assembly Kit
8. Drive Assembly
9. Bush Kit
10. Motor Assembly
    Drive Shaft Sundry Kit
    Sundry Parts Kit

Figure 7.22 Lucas M78R/M80R starter

However, due to the fall in battery voltage under load and the low resistance of the armature windings, the characteristic is comparable to series wound motors. In some cases, flux concentrating pieces or interpoles are used between the main magnets. Due to the warping effect of the magnetic field, this tends to make the characteristic curve very similar to that of the series motor.

Development by some manufacturers has also taken place in the construction of the brushes. A copper and graphite mix is used but the brushes are made in two parts allowing a higher copper content in the power zone and a higher graphite content in the commutation zone. This results in increased service life and a reduction in voltage drop, giving improved starter power. Figure 7.23 shows three modern permanent magnet (PM) starters.

▲ D7 E starter motor. ▶

D6 RA starter motor.

Figure 7.23 Modern permanent magnet starters (Valeo)

For applications with a higher power requirement, permanent magnet motors with intermediate transmission have been developed. These allow the armature to rotate at a higher and more efficient speed whilst still providing the torque, due to the gear reduction. Permanent magnet starters with intermediate transmission are available with power outputs of about 1.7 kW and are suitable for spark ignition engines up to about 3 litres, or compression ignition engines up to about 1.6 litres. This form of permanent magnet motor can give a weight saving of up to 40%. The principle of operation is again similar to the conventional pre-engaged starter. The intermediate transmission, as shown in Figure 7.24, is of the epicyclic type.

The sun gear is on the armature shaft and the planet carrier drives the pinion. The ring gear or annulus remains stationary and also acts as an intermediate bearing. This arrangement of gears gives a reduction ratio of about 5 : 1. This can be calculated by the formula:

$$\text{Ratio} = \frac{AS}{S}$$

where A = number of teeth on the annulus, and S = number of teeth on the sun gear.

The annulus gear in some types is constructed from a high grade polyamide compound with mineral additives to improve strength and wear resistance. The sun and planet gears are conventional steel. This combination of materials gives a quieter and more efficient operation. Figure 7.25 shows a PM starter with intermediate

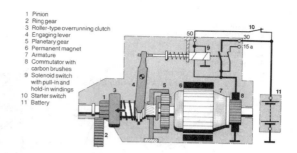

1 Pinion
2 Ring gear
3 Roller-type overrunning clutch
4 Engaging lever
5 Planetary gear
6 Permanent magnet
7 Armature
8 Commutator with carbon brushes
9 Solenoid switch with pull-in and hold-in windings
10 Starter switch
11 Battery

1 Drive end shield, 2 Pinion, 3 Solenoid switch, 4 Terminal, 5 Commutator end shield, 6 Brush plate with carbon brushes, 7 Commutator, 8 Armature, 9 Permanent magnet, 10 Field frame, 11 Planetary gear (intermediate transmission), 12 Engaging lever, 13 Pinion-engaging drive.

Figure 7.25 Pre-engaged starter and details (Bosch)

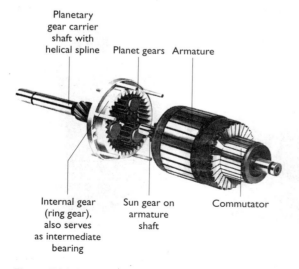

Planetary gear carrier shaft with helical spline   Planet gears   Armature

Internal gear (ring gear), also serves as intermediate bearing   Sun gear on armature shaft   Commutator

Figure 7.24 Starter motor intermediate transmission

transmission, together with its circuit and operating mechanism.

7.3.4 Heavy vehicle starters

The subject area of this book is primarily the electrical equipment on cars. This short section is included for interest, hence further reference should be made to other sources for greater detail about heavy vehicle starters.

The types of starter that are available for heavy duty applications are as many and varied as the applications they serve. In general, higher voltages are used, which may be up to 110 V in specialist cases, and two starters may even be running in parallel for very high power and torque requirements.

Large road vehicles are normally 24 V and employ a wide range of starters. In some cases the design is simply a large and heavy duty version of the pre-engaged type discussed earlier. The Delco-Remy 42-MT starter shown in Figure 7.26 is a good example of this type. This starter may also be fitted with a thermal cut-out to prevent overheating damage due to excessive cranking. Rated at 8.5 kW, it is capable of producing over 80 Nm torque at 1000 rev/min.

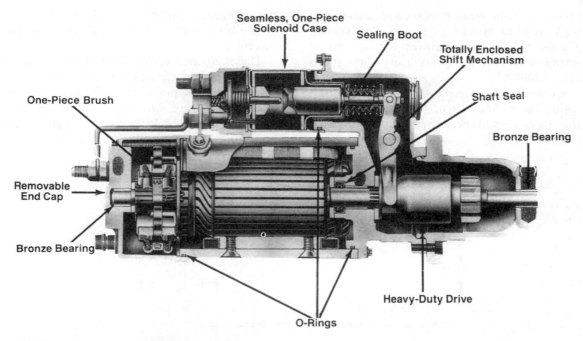

Figure 7.26 Delco-Remy 42-MT starter

Other methods of engaging the pinion include sliding the whole armature or pushing the pinion with a rod through a hollow armature. This type uses a solenoid to push the pinion into mesh via a rod through the centre of the armature.

Sliding-armature-type starters work by positioning the field windings forwards from the main armature body, such that the armature is attracted forwards when power is applied. A trip lever mechanism will then only allow full power when the armature has caused the pinion to mesh.

7.3.5 Starter installation

Starters are generally mounted in a horizontal position next to the engine crank-case with the drive pinion in a position for meshing with the flywheel or drive plate ring gear.

The starter can be secured in two ways: either by flange or cradle mounting. Flange mounting is the most popular technique used on small and medium-sized vehicles and, in some cases, it will incorporate a further support bracket at the rear of the starter to reduce the effect of vibration. Larger vehicle starters are often cradle mounted but again also use the flange mounting method, usually fixed with at least three large bolts. In both cases the starters must have some kind of pilot, often a ring machined on the drive end bracket, to ensure correct positioning with respect to the ring gear. This will ensure correct gear backlash and a suitable out of mesh clearance. Figure 7.27 shows the flange mountings method used for most light vehicle starter motors.

Clearly the main load on the vehicle battery is the starter and this is reflected in the size of supply cable required. Any cable carrying a current will experience power loss known as I^2R loss. In order to reduce this power loss, the current or the resistance must be reduced. In the case of the starter the high current is the only way of delivering the high torque. This is the reason for using heavy conductors to the starter to ensure low resistance, thus reducing the volt drop and

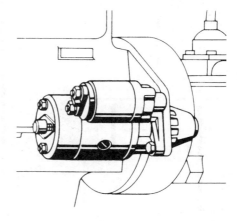

Figure 7.27 Flange mounting is used for most light vehicle starter motors

power loss. The maximum allowed volt drop is 0.5 V on a 12 V system and 1 V on a 24 V system. The short circuit (initial) current for a typical car starter is 500 A and for very heavy applications can be 3000 A.

Control of the starter system is normally by a spring-loaded key switch. This switch will control the current to the starter solenoid, in many cases via a relay. On vehicles with automatic transmission, an inhibitor switch to prevent the engine being started in gear will also interrupt this circuit.

Diesel engined vehicles may have a connection between the starter circuit and a circuit to control the glow plugs. This may also incorporate a timer relay. On some vehicles the glow plugs are activated by a switch position just before the start position.

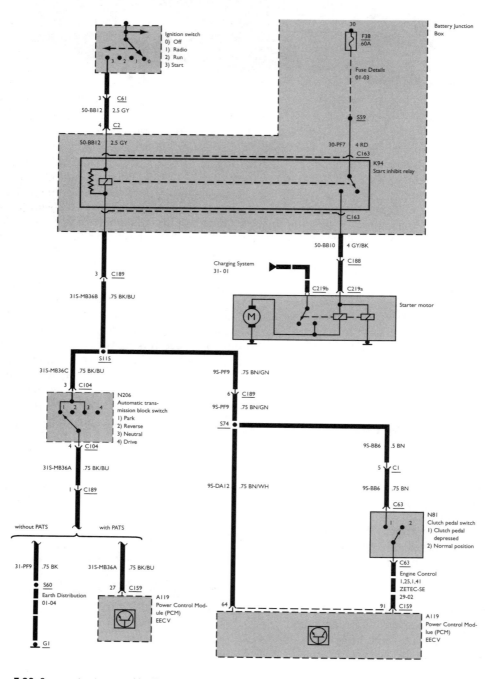

Figure 7.28 Starter circuit as used by Ford

7.3.6 Summary

The overall principle of starting a vehicle engine with an electric motor has changed little in over 80 years. Of course, the motors have become far more reliable and longer lasting. It is interesting to note that, assuming average mileage, the modern starter is used about 2000 times a year in city traffic! This level of reliability has been achieved by many years of research and development.

7.4 Case studies

7.4.1 Ford

The circuit shown is Figure 7.28 is from a vehicle fitted with manual or automatic transmission. The inhibitor circuits will only allow the starter to operate when the automatic transmission is in 'park' or 'neutral'. Similarly for the manual version, the starter will only operate if the clutch pedal is depressed.

The starter relay coil is supplied with the positive connection by the key switch. The earth path is connected through the appropriate inhibitor switch. To prevent starter operation when the engine is running the power control module (EEC V) controls the final earth path of the relay.

A resistor fitted across the relay coil reduces back EMF. The starter in current use is a standard pre-engaged, permanent magnet motor.

7.4.2 Toyota

The starter shown in Figure 7.29 has been in use for several years but is included because of its unusual design. The drive pinion incorporates the normal clutch assembly but is offset from the armature. Drive to the pinion is via a gear set with a ratio of about 3:1. The idle gear means the pinion rotates in the same direction as the armature.

Ball bearings are used on each end of the armature and pinion. The idler gear incorporates a roller bearing. The solenoid acts on the spring and steel ball to move the pinion into mesh. The electrical operation of the machine is

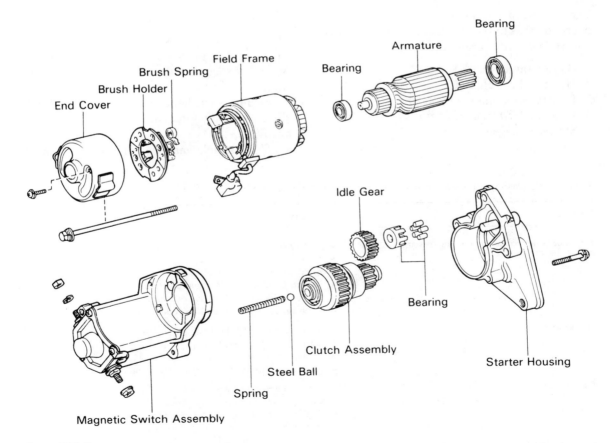

Figure 7.29 Toyota starter motor components

standard. It has four brushes and four field poles.

7.5 Diagnosing starting system faults

7.5.1 Introduction

As with all systems, the six stages of fault-finding should be followed.

1. Verify the fault.
2. Collect further information.
3. Evaluate the evidence.
4. Carry out further tests in a logical sequence.
5. Rectify the problem.
6. Check all systems.

The procedure outlined in the next section is related primarily to stage 4 of the process. Table 7.3 lists some common symptoms of a charging system malfunction together with suggestions for the possible fault.

7.5.2 Circuit testing procedure

The process of checking a 12 V starting system operation is as follows (tests 3 to 8 are all carried out while trying to crank the engine).

1. Battery (at least 70%).
2. Hand and eye checks.
3. Battery volts (minimum 10 V).
4. Solenoid lead (same as battery).
5. Starter voltage (no more that 0.5 V less than battery).
6. Insulated line volt drop (maximum 0.25 V).
7. Solenoid contacts volt drop (almost 0 V).
8. Earth line volt drop (maximum 0.25 V).

The idea of these tests is to see if the circuit is supplying all the available voltage to the starter. If it is, then the starter is at fault, if not then the circuit is at fault.

If the starter is found to be defective then a replacement unit is the normal recommendation. Figure 7.30 explains the procedure used by Bosch to ensure quality exchange units. Repairs are possible but only if the general state of the motor is good.

7.6 Advanced starting system technology

7.6.1 Speed, torque and power

To understand the forces acting on a starter motor let us first consider a single conducting wire in a magnetic field. The force on a single conductor in a magnetic field can be calculated by the formula:

$$F = BIl$$

where F = force in N, l = length of conductor in the field in m, B = magnetic field strength in Wb/m^2, I = current flowing in the conductor in amps.

Fleming's left-hand rule will serve to give the direction of the force (the conductor is at 90° to the field).

This formula may be further developed to

Table 7.3 Common symptoms of a charging system malfunction and possible faults

Symptom	Possible fault
Engine does not rotate when trying to start	• Battery connection loose or corroded. • Battery discharged or faulty. • Broken, loose or disconnected wiring in the starter or circuit. • Defective starter switch or automatic gearbox inhibitor switch. • Starter pinion or flywheel ring gear loose. • Earth strap broken, loose or corroded.
Starter noisy	• Starter pinion or flywheel ring gear loose. • Starter mounting bolts loose. • Starter worn (bearings etc). • Discharged battery (starter may jump in and out).
Starter turns engine slowly	• Discharged battery (slow rotation). • Battery terminals loose or corroded. • Earth strap or starter supply loose or disconnected. • High resistance in supply or earth circuit. • Internal starter fault.

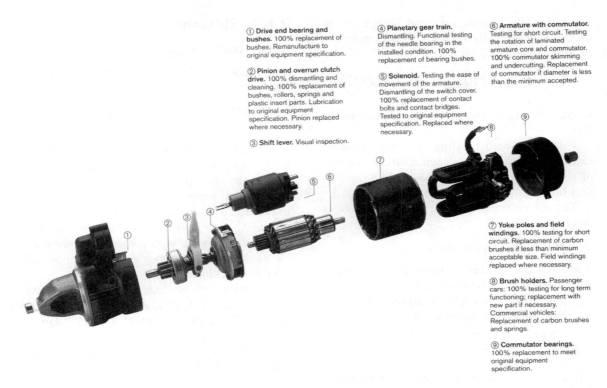

① **Drive end bearing and bushes.** 100% replacement of bushes. Remanufacture to original equipment specification.

② **Pinion and overrun clutch drive.** 100% dismantling and cleaning. 100% replacement of bushes, rollers, springs and plastic insert parts. Lubrication to original equipment specification. Pinion replaced where necessary.

③ **Shift lever.** Visual inspection.

④ **Planetary gear train.** Dismantling. Functional testing of the needle bearing in the installed condition. 100% replacement of bearing bushes.

⑤ **Solenoid.** Testing the ease of movement of the armature. Dismantling of the switch cover. 100% replacement of contact bolts and contact bridges. Tested to original equipment specification. Replaced where necessary.

⑥ **Armature with commutator.** Testing for short circuit. Testing the rotation of laminated armature core and commutator. 100% commutator skimming and undercutting. Replacement of commutator if diameter is less than the minimum accepted.

⑦ **Yoke poles and field windings.** 100% testing for short circuit. Replacement of carbon brushes if less than minimum acceptable size. Field windings replaced where necessary.

⑧ **Brush holders.** Passenger cars: 100% testing for long term functioning; replacement with new part if necessary. Commercial vehicles: Replacement of carbon brushes and springs.

⑨ **Commutator bearings.** 100% replacement to meet original equipment specification.

Figure 7.30 Quality starter overhaul procedure

calculate the stalled torque of a motor with a number of armature windings as follows:

$$T = BIlrZ$$

where T = torque in Nm, r = armature radius in m, and Z = number of active armature conductors.

This will only produce a result for stalled or lock torque because, when a motor is running, a back EMF is produced in the armature windings. This opposes the applied voltage and hence reduces the current flowing in the armature winding. In the case of a series wound starter motor, this will also reduce the field strength B. The armature current in a motor is given by the equation:

$$I = \frac{V - e}{R}$$

where I = armature current in amps, V = applied voltage in volts, R = resistance of the armature in ohms, e = total back EMF in volts.

From the above is should be noted that, at the instant of applying a voltage to the terminals of a motor, the armature current will be at a maximum since the back EMF is zero. As soon as the speed increases so will the back EMF and hence the armature current will decrease. This is why a starter motor produces 'maximum torque at zero rev/min'.

For any DC machine the back EMF is given by:

$$e = \frac{2p\phi nZ}{c}$$

where e = back EMF in volts, p = number of pairs of poles, ϕ = flux per pole in webers, n = speed in revs/second, Z = number of armature conductors, $c = 2p$ for lap-wound and 2 for a wave-wound machine.

The formula can be re-written for calculating motor speed:

$$n = \frac{ce}{2p\phi Z}$$

If the constants are removed from this formula it clearly shows the relationship between field flux, speed and back EMF,

$$n \propto \frac{e}{\phi}$$

To consider the magnetic flux (ϕ) it is necessary to differentiate between permanent magnet starters and those using excitation via windings. Permanent magnetism remains reasonably constant. The construction and design of the magnet determine its strength. Flux density can be calculated as follows:

$$B = \frac{\phi}{A} \quad \text{(units: T (tesla) or Wb/m}^2\text{)}$$

where A = area of the pole perpendicular to the flux.

Pole shoes with windings are more complicated as the flux density depends on the material of the pole shoe as well as the coil and the current flowing.

The magneto-motive force (MMF) of a coil is determined thus:

MMF = NI Ampere turns

where N = the number of turns on the coil and I = the current flowing in the coil.

Magnetic field strength H requires the active length of the coil to be included:

$$H = \frac{NI}{l}$$

where l = active length of the coil, H = magnetic field strength.

In order to convert this to flux density B, the permeability of the pole shoe must be included:

$$B = H\mu_0\mu_r$$

where μ_0 = permeability of free space (4×10^{-7} Henry/metre), and μ_r = relative permeability of the core to free space

To calculate power consumed is a simple task using the formula:

$$P = T\omega$$

where P = power in watts, T = torque in Nm, and ω = angular velocity in rad/s.

Here is a simple example of the use of this formula. An engine requires a minimum cranking speed of 100 rev/min and the required torque to achieve this is 9.6 Nm.

At a 10:1 ring gear to pinion ratio this will require a 1000 rev/min starter speed (n). To convert this to rad/s:

$$\omega = \frac{2\pi n}{60}$$

This works out to 105 rad/s.

$$P = T\omega$$

$$9.6 \times 105 = 1000 \text{ W or 1 kW}$$

7.6.2 Efficiency

Efficiency = Power out/Power in ($\times$ 100%)

The efficiency of most starter motors is of the order of 60%.

1 kW/60% = 1.67 kW (the required input power)

The main losses, which cause this, are iron losses, copper losses and mechanical losses. Iron losses are due to hysteresis loss caused by changes in magnetic flux, and also due to induced eddy currents in the iron parts of the motor. Copper losses are caused by the resistance of the windings; sometimes called I^2R losses. Mechanical losses include friction and windage (air) losses.

Using the previous example of a 1 kW starter it can be seen that, at an efficiency of 60%, this motor will require a supply of about 1.7 kW.

From a nominal 12 V supply and allowing for battery volt drop, a current of the order of 170 A will be required to achieve the necessary power.

7.7 New developments in starting systems

7.7.1 General

Many incremental developments have taken place to improve the starter motor and its circuit. Two of the more recent were the introduction of permanent magnet motors and the use of epicyclic gear reduction to increase torque.

Starters are now available with an expected life of 50 000 reliable starts. This corresponds to over half a million kilometres on the road.

7.7.2 High voltage systems

The first cars equipped with high-voltage electrical systems are likely to appear in the year 2000. The first electrical system with three times the present 12 V standard will probably appear in a Mercedes-Benz or BMW. By 2005, higher voltage systems could begin to be seen on cheaper cars. This change could force a complete redesign of every electrical component in the car.

The starter-generator, which combines a starter motor and a high-output generator into a single unit, will become more viable. It will be lighter and more efficient than current separate alternators and starters. It will fit on the crankshaft between the engine and transmission.

Many engineers favour a dual-voltage system that includes both 14 V and 42 V circuits. Traditional low-voltage devices such as electronics could run off the 14 V system, while the 42 V network would power the newer systems. The dual systems would be complicated, but the advantages could be considerable. Some manufacturers want the industry to set a 42 V standard.

7.7.3 Integrated starters

A device called a 'dynastart' was used on a number of vehicles from the 1930s through to the 1960s. This device was a combination of the starter and a dynamo. The device, directly mounted on the crankshaft, was a compromise and hence not very efficient.

The method is now known as an Integrated Starter Alternator Damper (ISAD). It consists of an electric motor, which functions as a control element between the engine and the transmission, and can also be used to start the engine and deliver electrical power to the batteries and the rest of the vehicle systems. The electric motor replaces the mass of the flywheel.

The motor transfers the drive from the engine and is also able to act as a damper/vibration absorber unit. The damping effect is achieved by a rotation capacitor. A change in relative speed between the rotor and the engine due to the vibration, causes one pole of the capacitor to be charged. The effect of this is to take the energy from the vibration.

Using ISAD to start the engine is virtually noiseless, and cranking speeds of 700 rev/min are possible. Even at –25°C it is still possible to crank at about 400 rev/min. A good feature of this is that a stop/start function is possible as an economy and emissions improvement technique. Because of the high speed cranking, the engine will fire up in about 0.1–0.5 seconds.

The motor can also be used to aid with acceleration of the vehicle. This feature could be used to allow a smaller engine to be used or to enhance the performance of a standard engine.

When used in alternator mode, the ISAD can produce up to 2 kW at idle speed. It can supply power at different voltages as both AC and DC. Through the application of intelligent control electronics, the ISAD can be up to 80% efficient.

Citroen have used the ISAD system in a Xsara model prototype. The car can produce 150 Nm for up to 30 seconds, which is significantly more than the 135 Nm peak torque of the 1580 cc, 65 kW fuel injected version. Figure 7.31 shows a cutaway of the ISAD on the Xsara engine. Citroen call the system 'Dynalto'. A 220 V outlet is even provided inside the car to power domestic electrical appliances!

7.7.4 Electronic starter control

'Valeo' have developed an electronic switch that can be fitted to its entire range of starters. Starter control will be supported by an ECU. The electronic starter incorporates a static relay on a circuit board integrated into the solenoid switch. This will prevent cranking when the engine is running.

'Smart' features can be added to improve comfort, safety and service life.

- Starter torque can be evaluated in real time to tell the precise instant of engine start. The starter can be simultaneously shut off to reduce wear and noise generated by the freewheel phase.
- Thermal protection of the starter components allows optimization of the components to save weight and to give short circuit protection.
- Electrical protection also reduces damage from misuse or system failure.
- Modulating the solenoid current allows redesign of the mechanical parts allowing a softer operation and weight reduction.

It will even be possible to retrofit this system to existing systems.

Figure 7.31 Dynalto system

7.8 Self-assessment

7.8.1 Questions

1. State four advantages of a pre-engaged starter when compared with an inertia type.
2. Describe the operation of the pull-in and hold-on windings in a pre-engaged starter solenoid.
3. Make a clearly labelled sketch of the engagement mechanism of a pre-engaged starter.
4. Explain what is meant by 'voltage drop' in a starter circuit and why it should be kept to a minimum.
5. Describe the engagement and disengagement of an inertia starter.
6. State two advantages and two disadvantages of a permanent magnet starter.
7. Calculate the gear ratio of an epicyclic gear set as used in a starter. The annulus has 40 teeth and the sun gear has 16 teeth.
8. Describe the operation of a roller-type one-way clutch.
9. Make a sketch to show the speed torque characteristics of a series, shunt and compound motor.
10. Describe the difference between a lap- and a wave-wound armature.

7.8.2 Assignment

A starter motor has to convert a very large amount of energy in a very short time. Motors rated at several kW are in common use. The overall efficiency of the motor is low. For example, at cranking speed:

> Input power to a motor ($W = VI$)
> (about 2000 W)

Output power from the motor can be calculated:

$$P = \frac{2\pi n T}{60}$$

(about 1100 W)

where $V = 10\,\text{V}$ (terminal voltage), $I = 200\,\text{A}$ (current), $n = 1500\,\text{rev/min}$, $T = 7\,\text{Nm}$ (torque), therefore, efficiency (P_{out}/P_{in}) 1100/2000 = 55%

A large saving in battery power would be possible if this efficiency were increased. Discuss how to improve the efficiency of the starting system. Would it be cost effective?

8
Ignition systems

8.1 Ignition fundamentals

8.1.1 Functional requirements

The fundamental purpose of the ignition system is to supply a spark inside the cylinder, near the end of the compression stroke, to ignite the compressed charge of air–fuel vapour.

For a spark to jump across an air gap of 0.6 mm under normal atmospheric conditions (1 bar), a voltage of 2–3 kV is required. For a spark to jump across a similar gap in an engine cylinder, having a compression ratio of 8 : 1, approximately 8 kV is required. For higher compression ratios and weaker mixtures, a voltage up to 20 kV may be necessary. The ignition system has to transform the normal battery voltage of 12 V to approximately 8–20 kV and, in addition, has to deliver this high voltage to the right cylinder, at the right time. Some ignition systems will supply up to 40 kV to the spark plugs.

Conventional ignition is the forerunner of the more advanced systems controlled by electronics. It is worth mentioning at this stage that the fundamental operation of most ignition systems is very similar. One winding of a coil is switched on and off causing a high voltage to be induced in a second winding. A coil-ignition system is composed of various components and sub-assemblies, the actual design and construction of which depend mainly on the engine with which the system is to be used.

When considering the design of an ignition system many factors must be taken into account, the most important of these being:

- Combustion chamber design.
- Air–fuel ratio.
- Engine speed range.
- Engine load.
- Engine combustion temperature.
- Intended use.
- Emission regulations.

8.1.2 Types of ignition system

The basic choice for types of ignition system can be classified as shown in Table 8.1.

8.1.3 Generation of high tension

If two coils (known as the primary and secondary) are wound on to the same iron core then any change in magnetism of one coil will induce a voltage into the other. This happens when a current is switched on and off to the primary coil. If the number of turns of wire on the secondary coil is more than the primary, a higher voltage can be produced. This is called *transformer action* and is the principle of the ignition coil.

The value of this 'mutually induced' voltage depends upon:

- The primary current.
- The turns ratio between the primary and secondary coils.
- The speed at which the magnetism changes.

Table 8.1 Types of ignition system

Type	Conventional	Electronic	Programmed	Distributorless
Trigger	Mechanical	Electronic	Electronic	Electronic
Advance	Mechanical	Mechanical	Electronic	Electronic
Voltage source	Inductive	Inductive	Inductive	Inductive
Distribution	Mechanical	Mechanical	Mechanical	Electronic

Figure 8.1 shows a typical ignition coil in section. The two windings are wound on a laminated iron core to concentrate the magnetism. Some coils are oil filled to assist with cooling.

8.1.4 Advance angle (timing)

For optimum efficiency the ignition advance angle should be such as to cause the maximum combustion pressure to occur about 10° after top dead centre (TDC). The ideal ignition timing is dependent on two main factors, engine speed and engine load. An increase in engine speed requires the ignition timing to be advanced. The cylinder charge, of air–fuel mixture, requires a certain time to burn (normally about 2 ms). At higher engine speeds the time taken for the piston to travel the same distance reduces. Advancing the time of the spark ensures full burning is achieved.

A change in timing due to engine load is also required as the weaker mixture used on low load conditions burns at a slower rate. In this situation, further ignition advance is necessary. Greater load on the engine requires a richer mixture, which burns more rapidly. In this case some retardation of timing is necessary. Overall, under any condition of engine speed and load an ideal advance angle is required to ensure maximum pressure is achieved in the cylinder just after top dead centre. The ideal advance angle may be further refined by engine temperature and any risk of detonation.

Spark advance is achieved in a number of ways. The simplest of these being the mechanical system comprising a centrifugal advance mechanism and a vacuum (load sensitive) control unit. Manifold vacuum is almost inversely proportional to the engine load. I prefer to consider manifold pressure, albeit less than atmospheric pressure, as the manifold absolute pressure (MAP) is proportional to engine load. Digital ignition systems may adjust the timing in relation to the temperature as well as speed and load. The values of all ignition timing functions are combined either mechanically or electronically in order to determine the ideal ignition point.

The energy storage takes place in the ignition coil. The energy is stored in the form of a magnetic field. To ensure the coil is charged before the ignition point a dwell period is required. Ignition timing is at the end of the dwell period.

8.1.5 Fuel consumption and exhaust emissions

The ignition timing has a significant effect on fuel consumption, torque, drivability and exhaust emissions. The three most important pollutants are hydrocarbons (HC), carbon monoxide (CO) and nitrogen oxides (NOx).

The HC emissions increase as timing is advanced. NOx emissions also increase with advanced timing due to the higher combustion temperature. CO changes very little with timing and is mostly dependent on the air–fuel ratio.

As is the case with most alterations of this type, a change in timing to improve exhaust emissions will increase fuel consumption. With the leaner mixtures now prevalent, a larger advance is required to compensate for the slower burning rate. This will provide lower consumption and high torque but the mixture must be controlled accurately to provide the best compromise with regard to the emission problem. Figure 8.2 shows the effect of timing changes on emissions, performance and consumption.

8.1.6 Conventional ignition components

Spark plug

Seals electrodes for the spark to jump across in the cylinder. Must withstand very high voltages, pressures and temperatures.

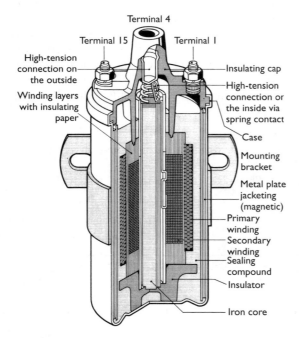

Figure 8.1 Typical ignition coil

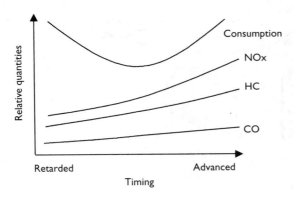

Figure 8.2 Effect of changes in ignition timing at a fixed engine speed

Ignition coil

Stores energy in the form of magnetism and delivers it to the distributor via the HT lead. Consists of primary and secondary windings.

Ignition switch

Provides driver control of the ignition system and is usually also used to cause the starter to crank.

Ballast resistor

Shorted out during the starting phase to cause a more powerful spark. Also contributes towards improving the spark at higher speeds.

Contact breakers (breaker points)

Switches the primary ignition circuit on and off to charge and discharge the coil.

Capacitor (condenser)

Suppresses most of the arcing as the contact breakers open. This allows for a more rapid break of primary current and hence a more rapid collapse of coil magnetism, which produces a higher voltage output.

HT Distributor

Directs the spark from the coil to each cylinder in a pre-set sequence.

Centrifugal advance

Changes the ignition timing with engine speed. As speed increases the timing is advanced.

Vacuum advance

Changes timing depending on engine load. On conventional systems the vacuum advance is most important during cruise conditions.

Figure 8.3 shows some conventional and electronic ignition components.

The circuit of a contact breaker ignition system is shown in Figure 8.4.

8.1.7 Plug leads (HT)

HT, or high tension (which is just an old fashioned way of saying high voltage) components and systems, must meet or exceed stringent ignition product requirements, such as:

- Insulation to withstand 40 000 V systems.
- Temperatures from −40°C to +260°C (−40°F to 500°F).
- Radio frequency interference suppression.
- 160 000 km (100 000 mile) product life.
- Resistance to ozone, corona, and fluids.
- 10-year durability.

Delphi produces a variety of cable types that meet the increased energy needs of leaner-burning engines without emitting electromagnetic interference (EMI). The cable products offer metallic and non-metallic cores, including composite, high-temperature resistive and wire-wound inductive cores. Conductor construction includes copper, stainless steel, Delcore, CHT, and wire-wound. Jacketing materials include organic and inorganic compounds, such as CPE, EPDM and silicone. Figure 8.5 shows the construction of these leads. Table 8.2 summarizes some of the materials used for different temperature ranges.

8.2 Electronic ignition

8.2.1 Introduction

Electronic ignition is now fitted to almost all spark ignition vehicles. This is because the conventional mechanical system has some major disadvantages.

- Mechanical problems with the contact breakers, not the least of which is the limited lifetime.
- Current flow in the primary circuit is limited to about 4 A or damage will occur to the contacts – or at least the lifetime will be seriously reduced.
- Legislation requires stringent emission limits, which means the ignition timing must stay in tune for a long period of time.
- Weaker mixtures require more energy from the spark to ensure successful ignition, even at very high engine speed.

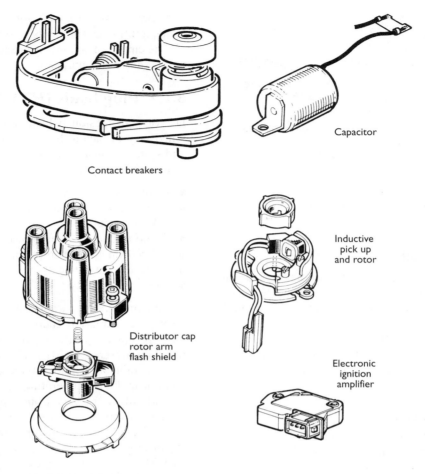

Figure 8.3 Conventional and electronic ignition components

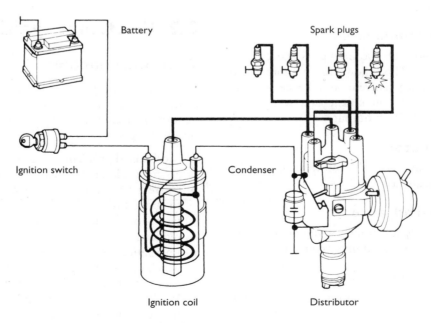

Figure 8.4 Contact breaker ignition system

Resistive ignition cable

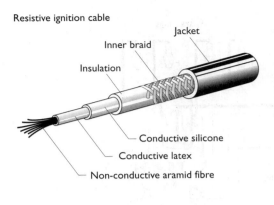

Wire-wound ignition cable

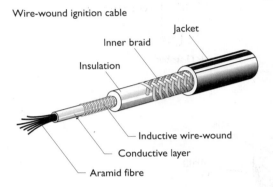

Figure 8.5 Ignition plug leads

These problems can be overcome by using a power transistor to carry out the switching function and a pulse generator to provide the timing signal. Very early forms of electronic ignition used the existing contact breakers as the signal provider. This was a step in the right direction but did not overcome all the mechanical limitations, such as contact bounce and timing slip. Most (all?) systems nowadays are constant energy, ensuring high performance ignition even at high engine speed. Figure 8.6 shows the circuit of a standard electronic ignition system.

8.2.2 Constant dwell systems

The term 'dwell' when applied to ignition is a measure of the time during which the ignition coil is charging, in other words when the primary current is flowing. The dwell in conventional systems was simply the time during which the contact breakers were closed. This is now often expressed as a percentage of one charge–discharge cycle. Constant dwell electronic ignition systems have now been replaced almost without exception by constant energy systems discussed in the next section.

An old but good example of a constant dwell system is the Lucas OPUS (oscillating pick-up system) ignition. Figure 8.7 shows the pulse generator assembly with a built-in amplifier. The timing rotor is in the form of a plastic drum with a ferrite rod for each cylinder embedded around its edge. This rotor is mounted on the shaft of the distributor. The pick-up is mounted on the base plate and comprises an 'E'-shaped ferrite core with primary and secondary windings enclosed in a plastic case. Three wires are connected from the pick-up to the amplifier module.

The amplifier module contains an oscillator used to energize the primary pick-up winding, a smoothing circuit and the power switching stage. The mode of operation of this system is that the oscillator supplies a 470 kHz AC signal to the pick-up primary winding. When none of the ferrite rods are in proximity to the pick-up the power transistor allows primary ignition to flow. As the distributor rotates and a ferrite rod passes the pick-up, the magnetic linkage allows an output from the pick-up secondary winding. Via the smoothing stage and the power stage of the module, the ignition coil will now switch off, producing the spark.

Whilst this was a very good system in its time, constant dwell still meant that at very high engine speeds, the time available to charge the coil could only produce a lower power spark. Note that as engine speed increases, the dwell angle or dwell percentage remains the same but the actual time is reduced.

Table 8.2 Materials used for various ignition components for different temperatures

Ignition component	Operating temperature (continuous)		
	110°C	175°C	232°C
Terminals	Zinc plated	Phosphor bronze or stainless steel	Stainless steel
Boot material	EPDM or silicone	Silicone	High-temperature silicone
Jacket	CPE	Silicone	Silicone
Insulation	EPDM	EPDM	Silicone
Conductor	Delcore copper or stainless steel	Delcore or CHT	CHT or wire-wound core

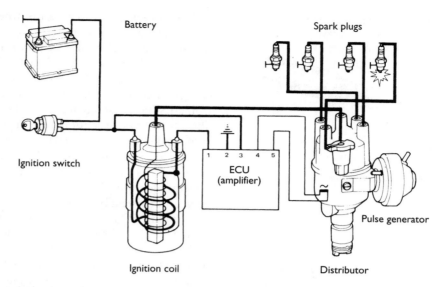

Figure 8.6 Electronic ignition system

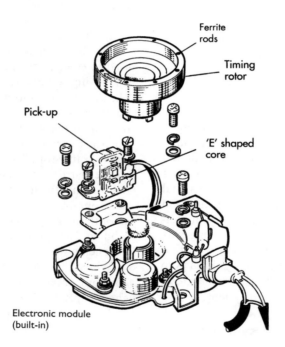

Figure 8.7 OPUS ignition system

8.2.3 Constant energy systems

In order for a constant energy electronic ignition system to operate, the dwell must increase with engine speed. This will only be of benefit, however, if the ignition coil can be charged up to its full capacity, in a very short time (the time available for maximum dwell at the highest expected engine speed). To this end, constant energy coils are very low resistance and low inductance. Typical resistance values are less

than 1 W (often 0.5 W). Constant energy means that, within limits, the energy available to the spark plug remains constant under all operating conditions.

An energy value of about 0.3 mJ is all that is required to ignite a static stoichiometric mixture. In the case of lean or rich mixtures together with high turbulence, energy values in the region of 3–4 mJ are necessary. This has made constant energy ignition essential on all of today's vehicles in order to meet the expected emission and

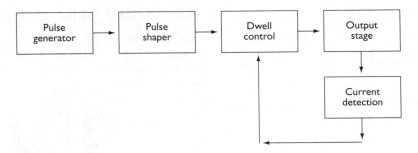

Figure 8.8 Constant energy ignition

performance criteria. Figure 8.8 is a block diagram of a closed loop constant energy ignition system. The earlier open loop systems are the same but without the current detection feedback section.

Due to the high energy nature of constant energy ignition coils, the coil cannot be allowed to remain switched on for more than a certain time. This is not a problem when the engine is running, as the variable dwell or current limiting circuit prevents the coil overheating. Some form of protection must be provided for, however, when the ignition is switched on but the engine is not running. This is known as the 'stationary engine primary current cut off'.

8.2.4 Hall effect pulse generator

The operating principle of the Hall effect is discussed in Chapter 2. The Hall effect distributor has become very popular with many manufacturers. Figure 8.9 shows a typical distributor with a Hall effect sensor.

As the central shaft of the distributor rotates, the vanes attached under the rotor arm alternately covers and uncovers the Hall chip. The number of vanes corresponds to the number of cylinders. In constant dwell systems the dwell is determined by the width of the vanes. The vanes cause the Hall chip to be alternately in and out of a magnetic field. The result of this is that the device will produce almost a square wave output, which can then easily be used to switch further electronic circuits. The three terminals on the distributor are marked '+, 0, –', the terminals + and –, are for a voltage supply and terminal '0' is the output signal. Typically, the output from a Hall effect sensor will switch between 0 V and about 7 V as shown in Figure 8.10. The supply voltage is taken from the ignition ECU and, on some systems, is stabilized at about 10 V to prevent changes to the output of the sensor when the engine is being cranked.

Hall effect distributors are very common due to the accurate signal produced and long term reliability. They are suitable for use on both

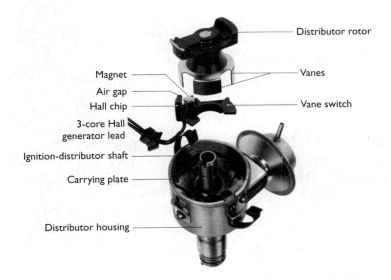

Figure 8.9 Ignition distributor with Hall generator

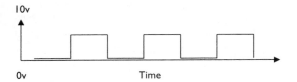

Figure 8.10 A Hall effect sensor output will switch between 0 V and about 7 V

constant dwell and constant energy systems. Operation of a Hall effect pulse generator can easily be tested with a DC voltmeter or a logic probe. Note that tests must not be carried out using an ohmmeter as the voltage from the meter can damage the Hall chip.

8.2.5 Inductive pulse generator

Inductive pulse generators use the basic principle of induction to produce a signal typical of the one shown in Figure 8.11. Many forms exist but all are based around a coil of wire and a permanent magnet.

The example distributor shown in Figure 8.12 has the coil of wire wound on the pick-up and, as the reluctor rotates, the magnetic flux varies due to the peaks on the reluctor. The number of peaks, or teeth, on the reluctor corresponds to the number of engine cylinders. The gap between the reluctor and pick-up can be important and manufacturers have recommended settings.

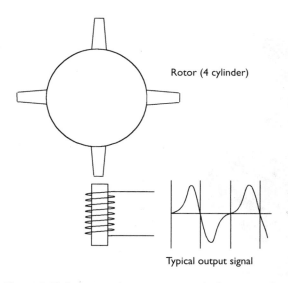

Figure 8.11 Inductive pulse generators use the basic principle of induction to produce a signal

8.2.6 Other pulse generators

Early systems were known as transistor assisted contacts (TAC) where the contact breakers were used as the trigger. The only other technique,

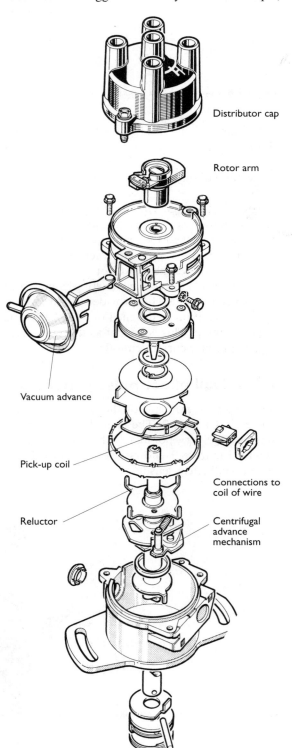

Figure 8.12 Inductive pulse generator in a distributor

which has been used on a reasonable scale, is the optical pulse generator. This involved a focused beam of light from a light emitting diode (LED) and a phototransistor. The beam of light is interrupted by a rotating vane, which provides a switching output in the form of a square wave. The most popular use for this system is in the after-market as a replacement for conventional contact breakers. Figure 8.13 shows the basic principle of an optical pulse generator; note how the beam is focused to ensure accurate switching.

8.2.7 Dwell angle control (open loop)

Figure 8.14 shows a circuit diagram of a transistorized ignition module. For the purposes of explaining how this system works, the pulse generator is the inductive type. To understand how

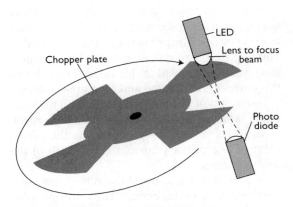

Figure 8.13 Basic principle of an optical pulse generator

the dwell is controlled, an explanation of the whole circuit is necessary.

The first part of the circuit is a voltage stabilizer to prevent damage to any components and to allow known voltages for charging and discharging the capacitors. This circuit consists of ZD_1 and R_1.

The alternating voltage coming from the inductive-type pulse generator must be reshaped into square-wave type pulses in order to have the correct effect in the trigger box. The reshaping is done by an electronic threshold switch known as a Schmitt trigger. This circuit is termed a pulse shaping circuit because of its function in the trigger box.

The pulse shaping circuit starts with D_4, a silicon diode which, due to its polarity, will only allow the negative pulses of the alternating control voltage to reach the base of transistor T_1. The induction-type pulse generator is loaded only in the negative phase of the alternating control voltage because of the output of energy. In the positive phase, on the other hand, the pulse generator is not loaded. The negative voltage amplitude is therefore smaller than the positive amplitude.

As soon as the alternating control voltage, approaching from negative values, exceeds a threshold at the pulse shaping circuit input, transistor T_1 switches off and prevents current passing. The output of the pulse-shaping circuit is currentless for a time (anti-dwell?). This switching state is maintained until the alternating control voltage, now approaching from positive

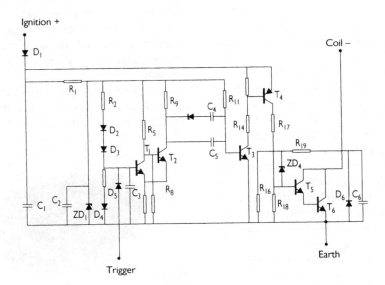

Figure 8.14 Circuit diagram of Bosch transistorized ignition module

values, drops below the threshold voltage. Transistor T_1 now switches off. The base of T_2 becomes positive via R_5 and T_2 is on. This alternation – T_1 on / T_2 off or T_1 off / T_2 on – is typical of the Schmitt trigger and the circuit repeats this action continuously. Two series-connected diodes, D_2 and D_3, are provided for temperature compensation. The diode D_1 is for reverse polarity protection.

The energy stored in the ignition coil can be put to optimum use with the help of the dwell section in the trigger box. The result is that sufficient high voltage is available for the spark at the spark plug under any operating condition of the engine. The dwell control specifies the start of the dwell period. The beginning of the dwell period (when T_3 switches on), is also the beginning of a rectangular current pulse that is used to trigger the transistor T_4, which is the driver stage. This in turn switches on the output stage.

A timing circuit using RC elements is used to provide a variable dwell. This circuit alternately charges and discharges capacitors by way of resistors. This is an open loop dwell control circuit because the combination of the resistors and capacitors provides a fixed time relationship as a function of engine speed.

The capacitor C_5 and the resistors R_9 and R_{11} form the RC circuit. When transistor T_2 is switched off, the capacitor C_5 will charge via R_9 and the base emitter of T_3. At low engine speed the capacitor will have time to charge to almost 12 V. During this time T_3 is switched on and, via T_4, T_5 and T_6, so is the ignition coil. At the point of ignition T_2 switches on and capacitor C_5 can now discharge via R_{11} and T_2. T_3 remains switched off all the time C_5 is discharging. It is this discharge time (which is dependent on how much C_5 had been charged), that delays the start of the next dwell period. Capacitor C_5 finally begins to be charged, via R_{11} and T_2, in the opposite direction and, when it reaches about 12 V, T_3 will switch back on. T_3 remains on until T_2 switches off again. As the engine speed increases, the charge time available for capacitor C_5 decreases. This means it will only reach a lower voltage and hence will discharge more quickly. This results in T_3 switching on earlier and hence a longer dwell period is the result.

The current from this driver transistor drives the power output stage (a Darlington pair). In this Darlington circuit the current flowing into the base of transistor T_5 is amplified to a considerably higher current, which is fed into the base of the transistor T_6. The high primary cur-

rent can then flow through the ignition coil via transistor T_6. The primary current is switched on the collector side of this transistor. The Darlington circuit functions as one transistor and is often described as the power stage.

Components not specifically mentioned in this explanation are for protection against back EMF (ZD_4, D_6) from the ignition coil and to prevent the dwell becoming too small (ZD_2 and C_4). A trigger box for Hall effect pulse generators functions in a similar manner to the above description. The hybrid ignition trigger boxes are considerably smaller than those utilizing discrete components. Figure 8.15 is a picture of a typical complete unit.

8.2.8 Current limiting and closed loop dwell

Primary current limiting ensures no damage can be caused to the system by excessive primary current, but also forms a part of a constant energy system. The primary current is allowed to build up to its pre-set maximum as soon as possible and then be held at this value. The value of this current is calculated and then pre-set during construction of the amplifier module. This technique, when combined with dwell angle control, is known as closed loop control as the actual value of the primary current is fed back to the control stages.

A very low resistance, high power precision resistor is used in this circuit. The resistor is connected in series with the power transistor and the ignition coil. A voltage sensing circuit connected across this resistor will be activated at a pre-set voltage (which is proportional to the current), and will cause the output stage to hold the current at a constant value. Figure 8.16 shows a block diagram of a closed loop dwell control system.

Stationary current cut-off is for when the ignition is on but the engine is not running. This is achieved in many cases by a simple timer

Figure 8.15 Transistorized ignition module

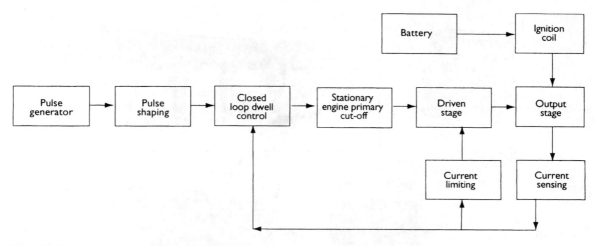

Figure 8.16 Closed loop dwell control system

circuit, which will cut the output stage after about one second.

8.2.9 Capacitor discharge ignition

Capacitor discharge ignition (CDI) has been in use for many years on some models of the Porsche 911 and some Ferrari models.

Figure 8.17 shows a block diagram of the CDI system. The CDI works by first stepping up the battery voltage to about 400 V (DC), using an oscillator and a transformer, followed by a rectifier. This high voltage is used to charge a capacitor. At the point of ignition the capacitor is discharged through the primary winding of a coil, often by use of a thyristor. This rapid discharge through the coil primary will produce a very high voltage output from the secondary winding. This voltage has a very fast rise time compared with a more conventional system. Typically, the rise time for CDI is 3–10 kV/μs as compared with the pure inductive system, which is 300–500 V/μs. This very fast rise time and high voltage will ensure that even a carbon- or oil-fouled plug will be fired. The disadvantage, however, is that the spark duration is short, which can cause problems particularly during starting.

This is often overcome by providing the facility for multi-sparking. However, when used in conjunction with direct ignition (one coil for each plug) the spark duration is acceptable.

It is also used on the Saab direct ignition system as shown in Figure 8.18.

8.3 Programmed ignition

8.3.1 Overview

'Programmed ignition' is the term used by some manufacturers, while others call it 'electronic spark advance' (ESA). Constant energy electronic ignition was a major step forwards and is still used on countless applications. However, its limitations lay in still having to rely upon mechanical components for speed and load advance characteristics. In many cases these did not match ideally the requirements of the engine.

Programmed ignition systems have a major difference compared with earlier systems, in that they operate digitally. Information about the operating requirements of a particular engine is programmed into the memory inside the electronic control unit. The data for storage in ROM

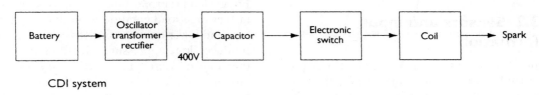

CDI system

Figure 8.17 CDI system

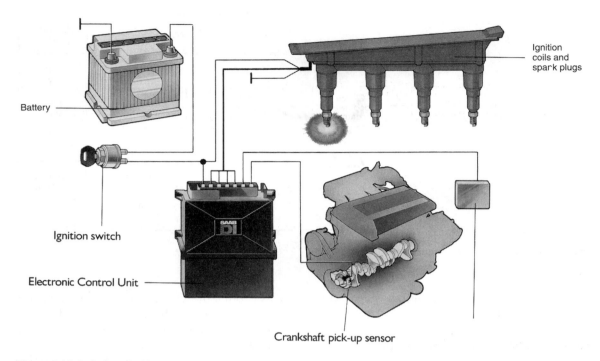

Battery

Ignition coils and spark plugs

Ignition switch

Electronic Control Unit

Crankshaft pick-up sensor

Figure 8.18 Saab direct ignition system

are obtained from rigorous testing on an engine dynamometer and from further development work on the vehicle under various operating conditions.

Programmed ignition has several advantages.

- The ignition timing can be accurately matched to the individual application under a range of operating conditions.
- Other control inputs can be utilized such as coolant temperature and ambient air temperature.
- Starting is improved and fuel consumption is reduced, as are emissions, and idle control is better.
- Other inputs can be taken into account such as engine knock.
- The number of wearing components in the ignition system is considerably reduced.

Programmed ignition, or ESA, can be a separate system or be included as part of the fuel control system.

8.3.2 Sensors and input information

Figure 8.19 shows the layout of the Rover programmed ignition system. In order for the ECU to calculate suitable timing and dwell outputs, certain input information is required.

Engine speed and position – crankshaft sensor

This sensor is a reluctance sensor positioned as shown in Figure 8.20. The device consists of a permanent magnet, a winding and a soft iron core. It is mounted in proximity to a reluctor disc. The disc has 34 teeth, spaced at 10° intervals around the periphery of the disc. It has two teeth missing, 180° apart, at a known position before TDC (BTDC). Many manufacturers use this technique with minor differences. As a tooth from the reluctor disc passes the core of the sensor, the reluctance of the magnetic circuit is changed. This induces a voltage in the winding, the frequency of the waveform being proportional the engine speed. The missing tooth causes a 'missed' output wave and hence the engine position can be determined.

Engine load – manifold absolute pressure sensor

Engine load is proportional to manifold pressure in that high load conditions produce high pressure and lower load conditions – such as cruise – produce lower pressure. Load sensors are therefore pressure transducers. They are either mounted in the ECU or as a separate unit, and are connected to the inlet manifold with a pipe. The pipe often incorporates a restriction to

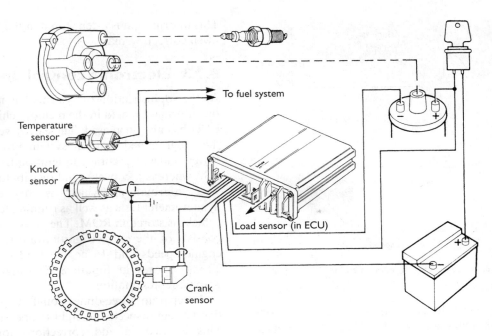

Figure 8.19 Programmed ignition system

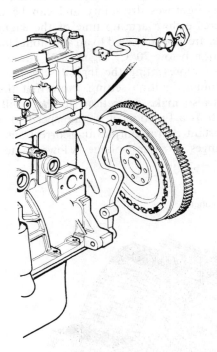

Figure 8.20 Position of a programmed ignition crankshaft sensor

damp out fluctuations and a vapour trap to prevent petrol fumes reaching the sensor.

Engine temperature – coolant sensor

Coolant temperature measurement is carried out by a simple thermistor, and in many cases the same sensor is used for the operation of the temperature gauge and to provide information to the fuel control system. A separate memory map is used to correct the basic timing settings. Timing may be retarded when the engine is cold to assist in more rapid warm up.

Detonation – knock sensor

Combustion knock can cause serious damage to an engine if sustained for long periods. This knock, or detonation, is caused by over-advanced ignition timing. At variance with this is that an engine will, in general, run at its most efficient when the timing is advanced as far as possible. To achieve this, the data stored in the basic timing map will be as close to the knock limit of the engine as possible (see Figure 8.21). The knock sensor provides a margin for error. The sensor itself is an accelerometer often of the piezoelectric type. It is fitted in the engine block between cylinders two and three on in-line four-cylinder engines. Vee engines require two sensors, one on each side. The ECU responds to signals from the knock sensor in the engine's knock window for each cylinder – this is often just a few degrees each side of TDC. This prevents clatter from the valve mechanism being interpreted as knock. The signal from the sensor is also filtered in the ECU to remove unwanted noise. If detonation is detected, the ignition timing is retarded on the fourth ignition pulse after detection (four-cylinder engine) in steps until

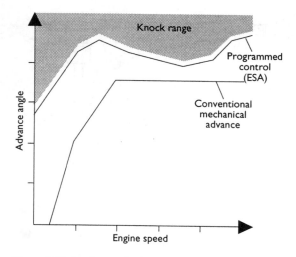

Figure 8.21 Ideal timing angle for an engine

knock is no longer detected. The steps vary between manufacturers, but about 2° is typical. The timing is then advanced slowly in steps of, say 1°, over a number of engine revolutions, until the advance required by memory is restored. This fine control allows the engine to be run very close to the knock limit without risk of engine damage.

Battery voltage

Correction to dwell settings is required if the battery voltage falls, as a lower voltage supply to the coil will require a slightly larger dwell figure.

This information is often stored in the form of a dwell correction map.

8.3.3 Electronic control unit

As the sophistication of systems has increased, the information held in the memory chips of the ECU has also increased. The earlier versions of the programmed ignition system produced by Rover achieved accuracy in ignition timing of ± 1.8° whereas a conventional distributor is ±8°. The information, which is derived from dynamometer tests as well as running tests in the vehicle, is stored in ROM. The basic timing map consists of the correct ignition advance for 16 engine speeds and 16 engine load conditions. This is shown in Figure 8.22 using a cartographic representation.

A separate three-dimensional map is used that has eight speed and eight temperature sites. This is used to add corrections for engine coolant temperature to the basic timing settings. This improves drivability and can be used to decrease the warm-up time of the engine. The data are also subjected to an additional load correction below 70°C. Figure 8.23 shows a flow chart representing the logical selection of the optimum ignition setting. Note that the ECU will also make corrections to the dwell angle, both as a function of engine speed to provide constant energy output and corrections due to changes in battery voltage. A lower battery volt-

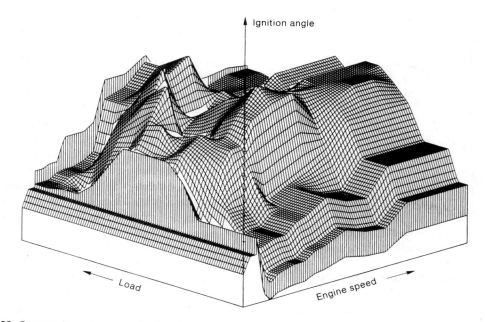

Figure 8.22 Cartographic map representing how ignition timing is stored in the ECU

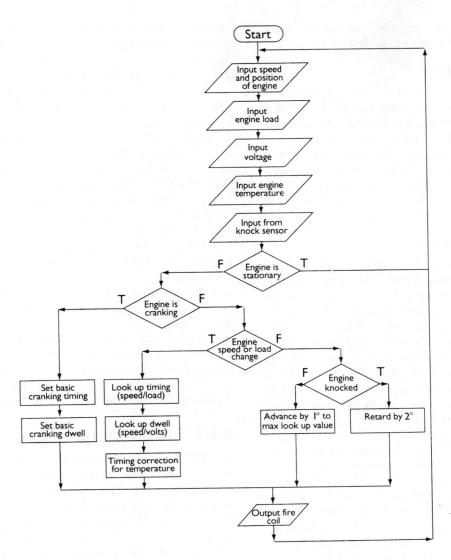

Figure 8.23 Ignition calculation flow diagram

age will require a slightly longer dwell and a higher voltage a slightly shorter dwell.

Typical of most 'computer' systems, a block diagram as shown in Figure 8.24 can represent the programmed ignition ECU. Input signals are processed and the data provided are stored in RAM. The program and pre-set data are held in ROM. In these systems a microcontroller is used to carry out the fetch execute sequences demanded by the program. Information, which is collected from the sensors, is converted to a digital representation in an A/D circuit. Rover, in common with many other manufacturers, use an on-board pressure sensor consisting of an aneroid chamber and strain gauges to indicate engine load.

A flow chart used to represent the program held in ROM, inside the ECU, is shown in Fig-

ure 8.23. A Windows 95/98/2000 shareware program that simulates the ignition system (as well

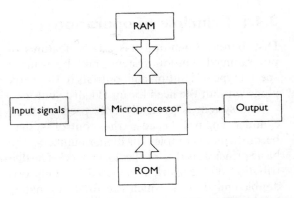

Figure 8.24 Typical of most 'computer' systems, the programmed ignition ECU can be represented by a block diagram

as many other systems) is available for downloading from my web site (details in Preface).

Ignition output

The output of a system, such as this programmed ignition, is very simple. The output stage, in common with most electronic ignitions, consists of a heavy-duty transistor that forms part of, or is driven by, a Darlington pair. This is simply to allow the high ignition primary current to be controlled. The switch off point of the coil will control ignition timing and the switch on point will control the dwell period.

HT distribution

The high tension distribution is similar to a more conventional system. The rotor arm however is mounted on the end of the camshaft with the distributor cap positioned over the top. The material used for the cap is known as Velox, which is similar to the epoxy type but has better electrical characteristics – it is less prone to tracking, for example. The distributor cap is mounted on a base plate made of Crasline which, as well as acting as the mounting point, prevents any oil that leaks from the camshaft seal fouling the cap and rotor arm. Another important function of the mounting plate is to prevent the build-up of harmful gases such as ozone and nitric oxide by venting them to the atmosphere. These gases are created by the electrolytic action of the spark as it jumps the air gap between the rotor arm and the cap segment. The rotor arm is also made of Crasline and is reinforced with a metal insert to relieve fixing stresses.

8.4 Distributorless ignition

8.4.1 Principle of operation

Distributorless ignition has all the features of programmed ignition systems but, by using a special type of ignition coil, outputs to the spark plugs without the need for an HT distributor.

The system is generally only used on four-cylinder engines because the control system becomes more complex for higher numbers. The basic principle is that of the 'lost spark'. The distribution of the spark is achieved by using two double-ended coils, which are fired alternately by the ECU. The timing is determined from a crankshaft speed and position sensor as well as load and other corrections. When one of the coils is fired, a spark is delivered to two engine cylinders, either 1 and 4, or 2 and 3. The spark delivered to the cylinder on the compression stroke will ignite the mixture as normal. The spark produced in the other cylinder will have no effect, as this cylinder will be just completing its exhausted stroke.

Because of the low compression and the exhaust gases in the 'lost spark' cylinder, the voltage used for the spark to jump the gap is only about 3 kV. This is similar to the more conventional rotor arm to cap voltage. The spark produced in the compression cylinder is therefore not affected.

An interesting point here is that the spark on one of the cylinders will jump from the earth electrode to the spark plug centre. Many years ago this would not have been acceptable, as the spark quality when jumping this way would not have been as good as when it jumps from the centre electrode. However, the energy available from modern constant energy systems will produce a spark of suitable quality in either direction. Figure 8.25 shows the layout of the distributorless ignition system (DIS) system.

8.4.2 System components

The DIS system consists of three main components: the electronic module, a crankshaft position sensor and the DIS coil. In many systems a manifold absolute pressure sensor is integrated in the module. The module functions in much the same way as has been described for the previously described electronic spark advance system.

The crankshaft position sensor is similar in

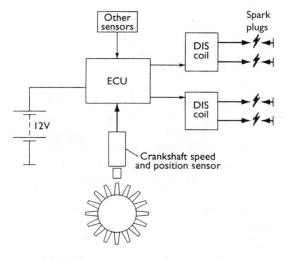

Figure 8.25 DIS ignition system

operation to the one described in the previous section. It is again a reluctance sensor and is positioned against the front of the flywheel or against a reluctor wheel just behind the front crankshaft pulley. The tooth pattern consists of 35 teeth. These are spaced at 10° intervals with a gap where the 36th tooth would be. The missing tooth is positioned at 90° BTDC for cylinders number 1 and 4. This reference position is placed a fixed number of degrees before top dead centre, in order to allow the timing or ignition point to be calculated as a fixed angle after the reference mark.

The low tension winding is supplied with battery voltage to a centre terminal. The appropriate half of the winding is then switched to earth in the module. The high tension windings are separate and are specific to cylinders 1 and 4, or 2 and 3. Figure 8.26 shows a typical DIS coil.

8.5 Direct ignition

8.5.1 General description

Direct ignition is, in a way, the follow-on from distributorless ignition. This system utilizes an inductive coil for each cylinder. These coils are mounted directly on the spark plugs. Figure 8.27 shows a cross-section of the direct ignition coil. The use of an individual coil for each plug ensures that the rise time for the low inductance primary winding is very fast. This ensures that a very high voltage, high energy spark is produced. This voltage, which can be in excess of 40 kV, provides efficient initiation of the combustion process under cold starting conditions and with weak mixtures. Some direct ignition systems use capacitor discharge ignition.

In order to switch the ignition coils, igniter units are used. These can control up to three coils and are simply the power stages of the control unit but in a separate container. This allows less interference to be caused in the main ECU due to heavy current switching and shorter runs of wires carrying higher currents.

8.5.2 Control of ignition

Ignition timing and dwell are controlled in a manner similar to the previously described programmed system. The one important addition to this on some systems is a camshaft sensor to provide information as to which cylinder is on the compression stroke. A system that does not require a sensor to determine which cylinder is on compression (engine position is known from a crank sensor) determines the information by initially firing all of the coils. The voltage across the plugs allows measurement of the current for each spark and will indicate which cylinder is on its combustion stroke. This works because a burning mixture has a lower resistance. The cylinder with the highest current at this point will be the cylinder on the combustion stroke.

A further feature of some systems is the case when the engine is cranked over for an excessive time, making flooding likely. The plugs are all

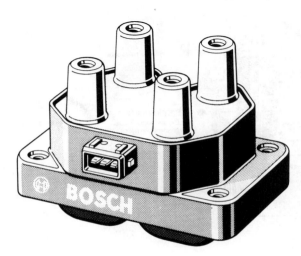

Figure 8.26 DIS coil

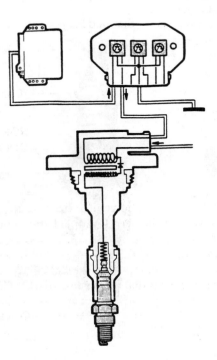

Figure 8.27 Direct ignition system

fired with multisparks for a period of time after the ignition is left in the on position for 5 seconds. This will burn away any excess fuel.

During difficult starting conditions, multi-sparking is also used by some systems during 70° of crank rotation before TDC. This assists with starting and then, once the engine is running, the timing will return to its normal calculated position.

8.6 Spark plugs

8.6.1 Functional requirements

The simple requirement of a spark plug is that it must allow a spark to form within the combustion chamber, to initiate burning. In order to do this the plug has to withstand a number of severe conditions. Consider, as an example, a four-cylinder four-stroke engine with a compression ratio of 9:1, running at speeds up to 5000 rev/min. The following conditions are typical. At this speed the four-stroke cycle will repeat every 24 ms.

- End of induction stroke – 0.9 bar at 65°C.
- Ignition firing point – 9 bar at 350°C.
- Highest value during power stroke – 45 bar at 3000°C.
- Power stroke completed – 4 bar at 1100°C.

Besides the above conditions, the spark plug must withstand severe vibration and a harsh chemical environment. Finally, but perhaps most important, the insulation properties must withstand voltage pressures up to 40 kV.

8.6.2 Construction

Figure 8.28 shows a standard and a resistor spark plug. The centre electrode is connected to the top terminal by a stud. The electrode is constructed of a nickel-based alloy. Silver and platinum are also used for some applications. If a copper core is used in the electrode this improves the thermal conduction properties.

The insulating material is ceramic-based and of a very high grade. Aluminium oxide, Al_2O_3 (95% pure), is a popular choice, it is bonded into the metal parts and glazed on the outside surface. The properties of this material, which make it most suitable, are as follows:

- Young's modulus:
  340 kN/mm².
- Coefficient of thermal expansion:
  $7.8 \times 10^{-6} K^{-1}$.
- Thermal conductivity:
  15–5 W/m K (Range 200–900°C).
- Electrical resistance:
  $>10^{13} \Omega/m$.

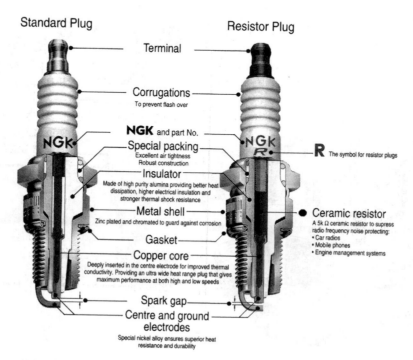

Figure 8.28 Spark-plug construction

The above list is intended as a guide only, as actual values can vary widely with slight manufacturing changes. The electrically conductive glass seal between the electrode and terminal stud is also used as a resistor. This resistor has two functions. First, to prevent burn-off of the centre electrode, and secondly to reduce radio interference. In both cases the desired effect is achieved because the resistor damps the current at the instant of ignition.

Flash-over, or tracking down the outside of the plug insulation, is prevented by ribs that effectively increase the surface distance from the terminal to the metal fixing bolt, which is of course earthed to the engine.

8.6.3 Heat range

Due to the many and varied constructional features involved in the design of an engine, the range of temperatures a spark plug is exposed to can vary significantly. The operating temperature of the centre electrode of a spark plug is critical. If the temperature becomes too high then pre-ignition may occur as the fuel–air mixture may become ignited due to the incandescence of the plug electrode. On the other hand, if the electrode temperature is too low then carbon and oil fouling can occur as deposits are not burnt off. Fouling of the plug nose can cause shunts (a circuit in parallel with the spark gap). It has been shown through experimentation and experience that the ideal operating temperature of the plug electrode is between 400 and 900 °C. Figure 8.29 shows how the temperature of the electrode changes with engine power output.

The heat range of a spark plug then is a measure of its ability to transfer heat away from the centre electrode. A hot running engine will require plugs with a higher thermal loading ability than a colder running engine. Note that hot and cold running of an engine in this sense refers to the combustion temperature and not to the efficiency of the cooling system.

The following factors determine the thermal capacity of a spark plug.

- Insulator nose length.
- Electrode material.
- Thread contact length.
- Projection of the electrode.

All these factors are dependent on each other and the position of the plug in the engine also has a particular effect.

It has been found that a longer projection of the electrode helps to reduce fouling problems due to low power operation, stop–go driving and high altitude conditions. In order to use greater projection of the electrode, better quality thermal conduction is required to allow suitable heat transfer at higher power outputs. Figure 8.30 shows the heat conducting paths of a spark plug together with changes in design for heat ranges. Also shown are the range of part numbers for NGK plugs.

8.6.4 Electrode materials

The material chosen for the spark plug electrode must exhibit the following properties:

- High thermal conductivity.
- High corrosion resistance.
- High resistance to burn-off.

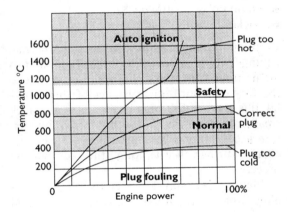

Figure 8.29 Temperature of a spark plug electrode changes with engine power output

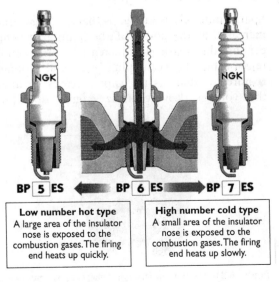

Figure 8.30 Heat conducting paths of a spark plug

For normal applications, alloys of nickel are used for the electrode material. Chromium, manganese, silicon and magnesium are examples of the alloying constituents. These alloys exhibit excellent properties with respect to corrosion and burn-off resistance. To improve on the thermal conductivity, compound electrodes are used. These allow a greater nose projection for the same temperature range, as discussed in the last section. A common example of this type of plug is the copper-core spark plug.

Silver electrodes are used for specialist applications as silver has very good thermal and electrical properties. Again, with these plugs nose length can be increased within the same temperature range. The thermal conductivity of some electrode materials is listed for comparison.

- Silver   407 W/m K
- Copper   384 W/m K
- Platinum   70 W/m K
- Nickel   59 W/m K

Compound electrodes have an average thermal conductivity of about 200 W/m K. Platinum tips are used for some spark plug applications due to the very high burn-off resistance of this material. It is also possible because of this to use much smaller diameter electrodes, thus increasing mixture accessibility. Platinum also has a catalytic effect, further accelerating the combustion process.

Figure 8.31 shows a semi-surface sparkplug, which, because of its design, has good anti-fouling properties.

8.6.5 Electrode gap

Spark plug electrode gaps have, in general, increased as the power of the ignition systems driving the spark has increased. The simple relationship between plug gap and voltage required is that, as the gap increases so must the voltage (leaving aside engine operating conditions). Furthermore, the energy available to form a spark at a fixed engine speed is constant, which means that a larger gap using higher voltage will result in a shorter duration spark. A smaller gap will allow a longer duration spark. For cold starting an engine and for igniting weak mixtures, the duration of the spark is critical. Likewise the plug gap must be as large as possible to allow easy access for the mixture in order to prevent quenching of the flame.

The final choice is therefore a compromise reached through testing and development of a particular application. Plug gaps in the region of 0.6–1.2 mm seem to be the norm at present.

8.6.6 V-grooved spark plug

The V-grooved plug is a development by NGK designed to reduce electrode quenching and allow the flame front to progress more easily from the spark. This is achieved by forming the electrode end into a 'V' shape, as shown in Figure 8.32.

This allows the spark to be formed at the side of the electrode, giving better propagation of the flame front and less quenching due to contact with the earth and centre electrodes. Figure 8.33 shows a V-grooved plug firing together with a graphical indication of the potential improvements when compared with the conventional plug.

8.6.7 Choosing the correct plug

Two methods are often used to determine the best spark plug for a given application. In the main it is the temperature range that is of prime importance. The first method of assessing plug temperature is the thermocouple spark plug, as shown in Figure 8.34. This allows quite accurate measurement of the temperature but does not

Figure 8.31 The semi-surface spark plug has good anti-fouling characteristics

Figure 8.32 V-grooved plug

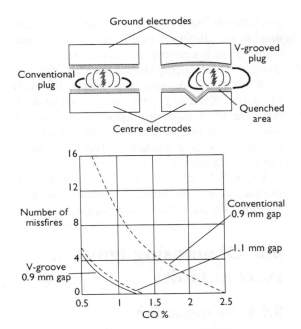

Engine at 550 rev/min, timing 11°BTDC

Figure 8.33 V-grooved spark plug firing, together with a graph indicating potential improvements when compared with the conventional plug

allow the test to be carried out for all types of plug.

A second method is the technique of ionic current measurement. When combustion has been initiated, the conductivity and pattern of current flow across the plug gap is a very good indication of the thermal load on the plug. This process allows accurate matching of the spark plug heat range to every engine, as well as providing data on the combustion temperature of a test engine. This technique is starting to be used as feedback to engine management systems to assist with accurate control.

In the after-market, choosing the correct plug is a matter of using manufacturers' parts catalogues.

8.7 Case studies

8.7.1 Introduction

Most modern ignition systems are combined with the fuel management system. For this reason I have chosen older case studies. I have even included contact breakers, for fear that we forget how they work!

8.7.2 Integrated ignition assembly (Toyota)

Figure 8.35 shows the components of an integrated ignition assembly. The pulse generator, ignition coil and igniter (module) are all mounted on the distributor. The unit contains conventional advance weights and a

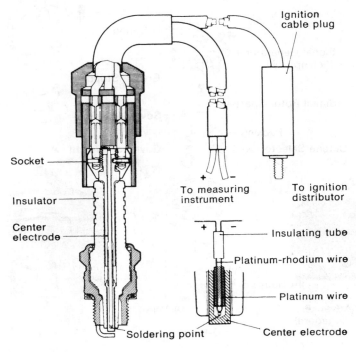

Figure 8.34 Thermocourse spark plug

vacuum/load sensitive advance unit. This also doubles as an octane selector.

The circuit diagram is shown in Figure 8.36. This shows how the inductive rotor triggers a Darlington pair in the igniter unit to operate the coil primary.

Mounting all the components as one unit can cause overheating problems. If the system is dismantled then any heat sink grease disturbed must be replaced.

8.7.3 Contact breaker ignition (lots of older cars)

Figure 8.4, at the start of this chapter, shows the circuit of a typical contact breaker ignition system. The distributor rotates at half engine speed, and a cam causes the contacts to open and close. This switching action turns the current flow in the coil primary on and off which, by mutual induction, creates a high voltage in the secondary winding. This voltage is distributed in the form of a spark via the cap and rotor arm.

A distributor is shown in Figure 8.37 complete with the centrifugal advance weights and vacuum capsule. As the engine speed increases, the weights fly outwards under the control of springs. This movement causes the cam on the

top central shaft of the distributor to rotate against the direction of rotation of the lower shaft. This opens the contacts earlier in the cycle, thus advancing the ignition timing.

A vacuum advance unit moves the base plate on which the contacts are secured, in response to changes in engine load. This has most effect during cruising due to the advance needed to burn a weaker mixture used under these conditions.

Figure 8.21 shows the advance characteristics of this type of distributor. The straight lines are normally described as the advance curve.

8.8 Diagnosing ignition system faults

8.8.1 Introduction

As with all systems, the six stages of fault-finding should be followed.

1. Verify the fault.
2. Collect further information.
3. Evaluate the evidence.
4. Carry out further tests in a logical sequence.
5. Rectify the problem.
6. Check all systems.

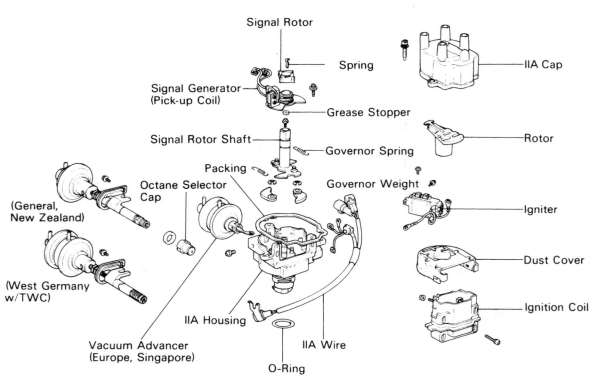

Figure 8.35 Toyota integrated ignition assembly

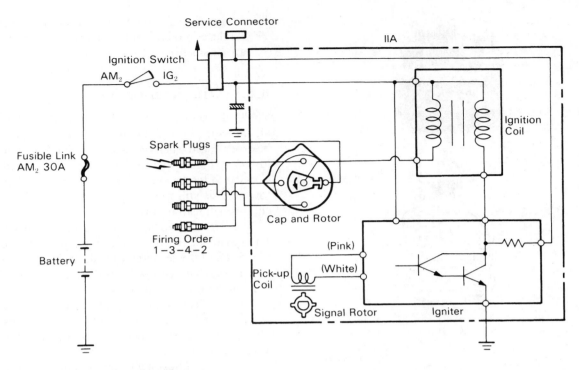

Figure 8.36 Toyota integrated ignition circuit

The procedure outlined in the next section is related primarily to Stage 4 of the process. Table 8.3 lists some common symptoms of an ignition system malfunction together with suggestions for the possible fault.

8.8.2 Testing procedure

Caution/Achtung/Attention – high voltages can seriously damage your health!
The following procedure is generic and with a little adaptation can be applied to any ignition system. Refer to the manufacturer's recommendations if in any doubt.

1. Check battery state of charge (at least 70%).
2. Hand and eye checks (all connections secure and clean).
3. Check supply to ignition coil (within 0.5 V of battery).
4. Spark from coil via known good HT lead (jumps about 10 mm, but do not try more).
5. If good spark then check HT system for tracking and open circuits. Check plug condition (leads should be a maximum resistance of about 30 kΩ/m and per lead) – stop here in this procedure.
6. If no spark, or it will only jump a short distance, continue with this procedure (colour of spark is not relevant).

7. Check continuity of coil windings (primary 0.5–3 Ω, secondary *several* kΩ.

Table 8.3 Common symptoms of an ignition system malfunction and possible faults

Symptom	Possible fault
Engine rotates but does not start	• Damp ignition components • Spark plugs worn to excess • Ignition system open circuit
Difficult to start when cold	• Spark plugs worn to excess • High resistance in ignition circuit
Engine starts but then stops immediately	• Ignition wiring connection intermittent • Ballast resistor open circuit (older cars)
Erratic idle	• Incorrect plug gaps • Incorrect ignition timing
Misfire at idle speed	• Ignition coil or distributor cap tracking • Spark plugs worn to excess
Misfire through all speeds	• Incorrect plugs or plug gaps • HT leads breaking down
Lack of power	• Ignition timing incorrect • HT components tracking
Backfires	• Incorrect ignition timing • Tracking
Runs on when switched off	• Ignition timing incorrect • Carbon build-up in engine
Pinking or knocking under load	• Ignition timing incorrect • Ignition system electronic fault • Knock sensor not working

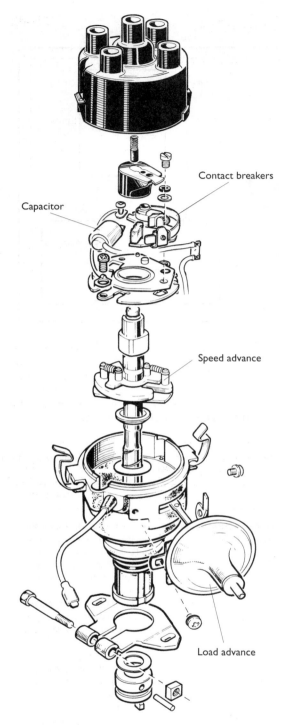

Figure 8.37 Contact breaker distributor

8. Supply and earth to 'module' (12 V minimum supply, earth drop 0.5 V maximum).
9. Supply to pulse generator if appropriate (10–12 V).
10. Output of pulse generator (inductive about 1 V AC when cranking, Hall type switches 0 V to 8 V DC).

11. Continuity of LT wires (0–0.1 Ω).
12. Replace 'module' but only if all tests above are satisfactory.

8.8.3 DIS diagnostics

The DIS system is very reliable due to the lack of any moving parts. Some problems, however, can be experienced when trying to examine HT oscilloscope patterns due to the lack of a king lead. This can often be overcome with a special adaptor but it is still necessary to move the sensing clip to each lead in turn.

The DIS coil can be tested with an ohmmeter. The resistance of each primary winding should be 0.5 Ω and the secondary windings between 11 and 16 kΩ. The coil will produce in excess of 37 kV in an open circuit condition.

The plug leads have integral retaining clips to prevent water ingress and vibration problems. The maximum resistance for the HT leads is 30 kΩ per lead.

No service adjustments are possible with this system, with the exception of octane adjustment on some models. This involves connecting two pins together on the module for normal operation, or earthing one pin or the other to change to a different fuel. The actual procedure must be checked with the manufacturer for each particular model.

8.8.4 Spark plug diagnostics

Examination of the spark plugs is a good way of assessing engine and associated systems condition. Figure 8.38 is a useful guide as provided by NGK plugs.

8.9 Advanced ignition technology

8.9.1 Ignition coil performance

The instantaneous value of the primary current in the inductive circuit of the ignition coil is determined by a number of factors. The HT produced is mainly dependent on this value of primary current. The rate of increase of primary current is vital because this determines the value of current when the circuit is 'broken' in order to produce the collapse of the magnetic field.

If the electrical constants of the primary ignition system are known it is possible to calculate

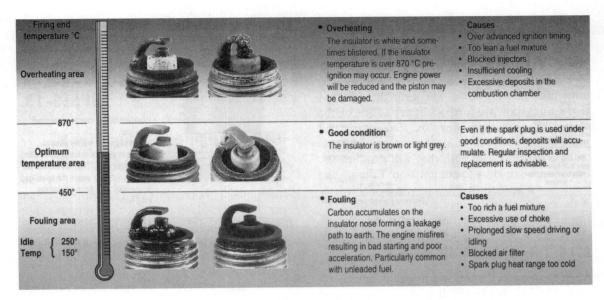

Figure 8.38 Assessing spark plug condition

the instantaneous primary current. This requires the exponential equation:

$$i = \frac{V}{R}(1 - e^{-Rt/L})$$

where i = instantaneous primary current, R = total primary resistance, L = inductance of primary winding, t = time the current has been flowing, e = base of natural logs.

Some typical values for comparison are given in Table 8.4.

Using, as an example, a four-cylinder engine running at 3000 rev/min, 6000 sparks per minute are required (four sparks during the two revolutions to complete the four-stroke cycle). This equates to 6000/60 or 100 sparks per second. At this rate each spark must be produced and used in 10 ms.

Taking a typical dwell period of say 60%, the time t, at 3000 rev/min on a four-cylinder engine, is 6 ms. At 6000 rev/min, t will be 3 ms. Employing the exponential equation above, the instantaneous current for each system is:

	3000 rev/min	6000 rev/min
Conventional system =	3.2 A	2.4 A
Electronic system =	10.9 A	7.3 A

This gives a clear indication of how the energy stored in the coil is much increased by the use of low resistance and low inductance igni-

Table 8.4 Comparison of conventional and electronic ignitions

Conventional ignition	Electronic ignition
$R = 3–4\,\Omega$	$R = 1\,\Omega$
$V = 14\,V$	$V = 14\,V$
$L = 10\,mH$	$L = 4\,mH$

tion coils. It is important to note that the higher current flowing in the electronic system would have been too much for the conventional contact breakers.

The energy stored in the magnetic field of the ignition coil is calculated as shown:

$$E = \frac{1}{2}(L \times i^2)$$

where E = energy, L = inductance of primary winding, and i = instantaneous primary current. The stored energy of the electronic system at 6000 rev/min is 110 mJ; the energy in the conventional system is 30 mJ. This clearly shows the advantage of electronic ignition as the spark energy is directly related to the energy stored in the coil.

8.10 New developments in ignition systems

8.10.1 Spark plugs

Most developments in spark plug technology are incremental. Recent trends have been towards the use of platinum plugs and the development of a plug that will stay within acceptable parameters for long periods (i.e. in excess of 50 000 miles/80 000 km).

Multiple electrode plugs are a contribution to long life and reliability. Do note though that these plugs only produce one spark at one of the electrodes each time they fire. The spark will jump across the path of least resistance and this will normally be the path that will produce the best ignition or start to combustion. Equally, the wear rate is spread over two or more electrodes. A double electrode plug is shown as Figure 8.31. Figure 8.39 shows a platinum spark plug.

8.10.2 Ignition coil cores

Most ignition coil cores are made of laminated iron. The iron is ideal as it is easily magnetized and demagnetized. The laminations reduce eddy currents, which cause inefficiency due to the heating effect (iron losses). If thinner laminations or sheets are used, then the better the performance.

Powder metal is now possible for use as coil cores. This reduces eddy currents to a minimum

Figure 8.39 Platinum spark plug

but the density of the magnetism is decreased. Overall, however, this produces a more efficient and higher output ignition coil. Developments are continuing and the flux density problem is about to be solved, giving rise to even more efficient components.

8.10.3 Engine management

Most serious developments in ignition are now linked with the full control of all engine functions. This means that the ignition system *per se* is not likely to develop further in its own right. Ignition timing, however, is being used to a greater extent for controlling idle speed, traction control and automatic gearbox surge control.

We have come a long way since 'hot tube' ignition!

8.11 Self-assessment

8.11.1 Questions

1. Describe the purpose of an ignition system.
2. State five advantages of electronic ignition compared with the contact breaker system.
3. Draw the circuit of a programmed ignition system and clearly label each part.
4. Explain what is meant by ignition timing and why certain conditions require it to be advanced or retarded.
5. Make a sketch to show the difference between a hot and cold spark plug.
6. Describe what is meant by 'mutual induction' in the ignition coil.
7. Explain the term 'constant energy' in relation to an ignition system.
8. Using a programmed ignition system fitted with a knock sensor as the example, explain why knock control is described as closed loop.
9. Make a clearly labelled sketch to show the operation of an inductive pulse generator.
10. List all the main components of a basic (not ESA) electronic ignition system and state the purpose of each component.

8.11.2 Assignment

Draw an 8 × 8 look-up table (grid) for a digital ignition system. The horizontal axis should represent engine speed from zero to 5000 rev/min, and the vertical axis engine load from zero to 100%. Fill in all the boxes with realistic figures and explain why you have chosen these figures. You should explain clearly the trends and not each individual figure.

Download the 'Automotive Technology – Electronics' simulation program from my web site and see if your figures agree with those in the program. Discuss reasons why they may differ.

9
Electronic fuel control

9.1 Combustion

9.1.1 Introduction

The process of combustion in spark and compression ignition engines is best considered for petrol and diesel engines in turn. The knowledge of the more practical aspects of combustion has been gained after years of research and is by no means complete even now. For a complete picture of the factors involved, further reference should be made to appropriate sources. However, the combustion section here will give enough details to allow considered opinion about the design and operation of electronic fuel control systems.

9.1.2 Spark ignition engine combustion process

A simplified description of the combustion process within the cylinder of a spark ignition engine is as follows. A single high intensity spark of high temperature passes between the electrodes of the spark plug leaving behind it a thin thread of flame. From this thin thread combustion spreads to the envelope of mixture immediately surrounding it at a rate that depends mainly on the flame front temperature, but also, to a lesser degree, on the temperature and density of the surrounding envelope.

In this way, a bubble of flame is built up that spreads radially outwards until the whole mass of mixture is burning. The bubble contains the highly heated products of combustion, while ahead of it, and being compressed by it, lies the still unburnt mixture.

If the cylinder contents were at rest this bubble would be unbroken, but with the air turbulence normally present within the cylinder, the filament of flame is broken up into a ragged front, which increases its area and greatly increases the speed of advance. While the rate of advance depends on the degree of turbulence,

the direction is little affected, unless some definite swirl is imposed on the system. The combustion can be considered in two stages.

1. Growth of a self-propagating flame.
2. Spread through the combustion chamber.

The first process is chemical and depends on the nature of the fuel, the temperature and pressure at the time and the speed at which the fuel will oxidize or burn. Shown in Figure 9.1, it appears as the interval from the spark (A) to the time when an increase in pressure due to combustion can first be detected (B).

This ignition delay period can be clearly demonstrated. If fuel is burned at constant volume, having been compressed to a self-ignition temperature, the pressure–time relationship is as shown in Figure 9.2. The time interval occurs with all fuels but may be reduced with an increase of compression temperature. A similar result can be demonstrated, enabling the effect of mixture strength on ignition delay to be investigated.

Returning to Figure 9.1, with the combustion under way, the pressure rises within the engine

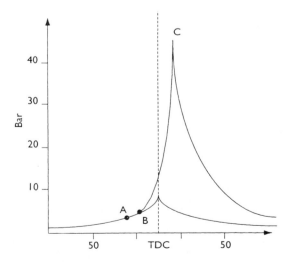

Figure 9.1 The speed at which fuel will oxidize or burn

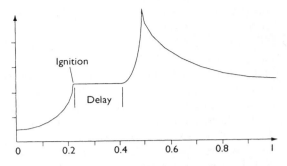

Figure 9.2 Fuel is burned at constant volume having been compressed to a self-ignition temperature. The pressure–time relationship is shown

cylinder from (B) to (C), very rapidly approaching the 'constant volume' process of the four-stroke cycle. While (C) represents the peak cylinder pressure and the completion of flame travel, all available heat has not been liberated due to re-association, and what can be referred to as after-burning continues throughout the expansion stroke.

9.1.3 Range and rate of burning

The range and rate of burning can be summarized by reference to the following graphs.

Figure 9.3 shows the approximate relation between flame temperature and the time from spark to propagation of flame for a hydrocarbon fuel.

Figure 9.4 shows the relation between the flame temperature and the mixture strength.

Figure 9.5 shows the relationship between mixture strength and rate of burning.

These graphs show that the minimum delay time (A to B) is about 0.2 ms with the mixture slightly rich.

While the second stage (B to C) is roughly dependent upon the degree of the turbulence (and on the engine speed), the initial delay necessitates ignition advance as the engine speed increases.

Figure 9.6 shows the effects of incorrect ignition timing. As the ignition is advanced there is

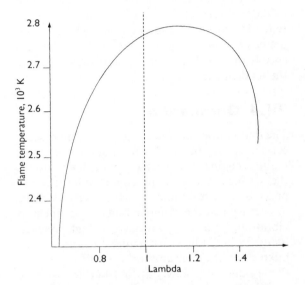

Figure 9.4 Relationship between flame temperature and mixture strength

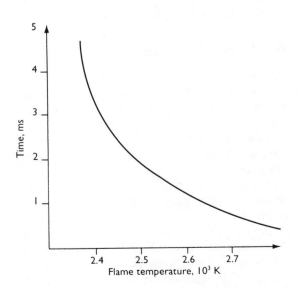

Figure 9.3 Approximated relationship between flame temperature and the time from spark to propagation of flame for a hydrocarbon fuel

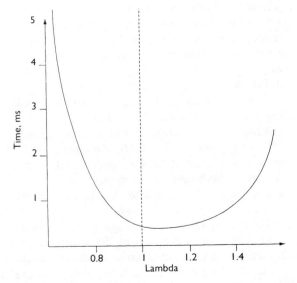

Figure 9.5 Relationship between mixture strength and rate of burning

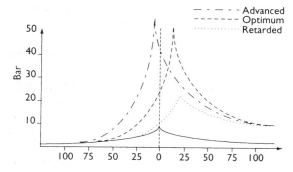

Figure 9.6 Effects of faulty ignition timing on fuel burn

an increase in firing pressure (or maximum cylinder pressure) generally accompanied by a reduction in exhaust temperature. The effect of increasing the range of the mixture strength speeds the whole process up and thus increases the tendency to detonate.

9.1.4 Detonation

The detonation phenomenon is the limiting factor on the output and efficiency of the spark ignition engine. The mechanism of detonation is the setting up within the engine cylinder of a pressure wave travelling at such velocity as, by its impact against the cylinder walls, to set them in vibration, and thus produce a high pitched 'ping'. When the spark ignites a combustible mixture of the fuel and air, a small nucleus of flame builds up, slowly at first but accelerating rapidly. As the flame front advances it compresses the remaining unburned mixture ahead of it. The temperature of the unburned mixture is raised by compression and radiation from the advancing flame until the remaining charge ignites spontaneously. The detonation pressure wave passes through the burning mixture at a very high velocity and the cylinder walls emit the ringing knock.

Detonation is seldom dangerous in small engines since it is usually avoided at the first warning by easing the load, but at higher speeds, where the noise level is high, the characteristic noise can and often does go undetected. It can be extremely dangerous, prompting pre-ignition and possibly the complete destruction of the engine.

High compression temperature and pressure tend to promote detonation. In addition, the ability of the unburnt mixture to absorb or get rid of the heat radiated to it by the advancing flame front is also important. The latent enthalpy of the mixture and the design of the combustion chamber affect this ability. The latter must be arranged for adequate cooling of the unburnt mixture by placing it near a well-cooled feature such as an inlet valve.

The length of flame travel should be kept as short as possible by careful positioning of the point of ignition. Other factors include the time (hence the ignition timing), since the reaction in the unburnt mixture must take some time to develop; the degree of turbulence (in general, higher turbulence tends to reduce detonation effects); and, most importantly, the tendency of the fuel itself to detonate.

Some fuels behave better in this respect. Fuel can be treated by additives (e.g. tetra-ethyl lead) to improve performance. However, this aggravates an already difficult pollution problem. A fuel with good anti-knock properties is iso-octane, and a fuel that is susceptible to detonation is normal heptane.

To obtain the octane number or the anti-knock ratings of a particular blend of fuel, a test is carried out on an engine run under carefully monitored conditions, and the onset of detonation is compared with those values obtained from various mixtures of iso-octane and normal heptane. If the performance of the fuel is identical to, for example, a mixture of 90% iso-octane and 10% heptane, then the fuel is said to have an octane rating of 90.

Mixing water, or methanol and water, with the fuel can reduce detonation. A mainly alcohol-based fuel, which enables the water to be held in solution, is also helpful so that better use can be made of the latent enthalpy of the water.

9.1.5 Pre-ignition

Evidence of the presence of pre-ignition is not so apparent at the onset as detonation, but the results are far more serious. There is no characteristic 'ping'. In fact, if audible at all, it appears as a dull thud. Since it is not immediately noticeable, its effects are often allowed to take a serious toll on the engine. The process of combustion is not affected to any extent, but a serious factor is that control of ignition timing can be lost.

Pre-ignition can occur at the time of the spark with no visible effect. More seriously, the 'auto-ignition' may creep earlier in the cycle. The danger of pre-ignition lies not so much in development of high pressures but in the very great increase in heat flow to the piston and cylinder walls. The maximum pressure does not, in fact, increase appreciably although it may occur a little early.

In a single-cylinder engine, the process is not dangerous since the reduction usually causes the engine to stall. In a multiple-cylinder engine the remaining cylinders (if only one is initially affected), will carry on at full power and speed, dragging the pre-igniting cylinder after them. The intense heat flow in the affected cylinder can result in piston seizure followed by the breaking up of the piston with catastrophic results to the whole engine.

Pre-ignition is often initiated by some form of hot spot, perhaps red-hot carbon or some poorly cooled feature of combustion space. In some cases, if the incorrect spark plug is used, over-heated electrodes are responsible, but often detonation is the prime cause. The detonation wave scours the cylinder walls of residual gases present in a film on the surface with the result that the prime source of resistance to heat flow is removed and a great release of heat occurs. Any weaknesses in the cooling system are tested and any hot spots formed quickly give rise to pre-ignition.

9.1.6 Combustion chamber design

To avoid the onset of detonation and pre-ignition, a careful layout of the valves and spark plugs is essential. Smaller engines, for automotive use, are firmly tied to the poppet valve. This, together with the restriction of space involved with high compression ratios, presents the designer with interesting problems.

The combustion chamber should be designed bearing in mind the following factors:

- The compression ratio should be $9:1$ for normal use, 11 or $12:1$ for higher performance.
- The plug or plugs should be placed to minimize the length of flame travel. They should not be in pockets or otherwise shrouded since this reduces effective cooling and also increases the tendency toward cyclical variations.

Experimental evidence shows a considerable variation in pressure during successive expansion stokes. This variation increases, as the mixture becomes too weak or too rich. Lighter loads and lower compression ratios also aggravate the process. While the size and position of the point of maximum pressure changes, the mean effective pressure and engine output is largely unaffected.

9.1.7 Stratification of cylinder charge

A very weak mixture is difficult to ignite but has great potential for reducing emissions and improving economy. One technique to get around the problem of igniting weak mixtures is stratification.

It is found that if the mixture strength is increased near the plug and weakened in the main combustion chamber an overall reduction in mixture strength results, but with a corresponding increase in thermal efficiency. To achieve this, petrol injection is used – stratification being very difficult with a conventional carburation system. A novel approach to this technique is direct mixture injection, which, it is claimed, can allow a petrol engine to run with air-to-fuel ratios in the region of $150:1$. This is discussed in a later section. The gasoline direct injection (GDi) engine from Mitsubishi is interesting in this area and is again discussed in a later section.

9.1.8 Mixture strength and performance

The effect of varying the mixture strength while maintaining the throttle position, engine speed and ignition timing constant is shown in Figure 9.7.

Figure 9.8 shows the effect of operating at part throttle with varying mixture strength. The chemically correct mixture of approximately $14.7:1$ lies between the ratio that provides

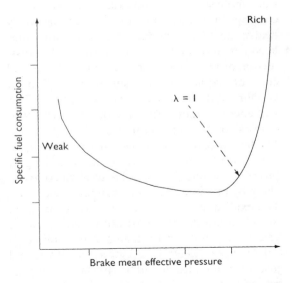

Figure 9.7 Effect of varying mixture strength while maintaining throttle, engine speed and ignition timing constant

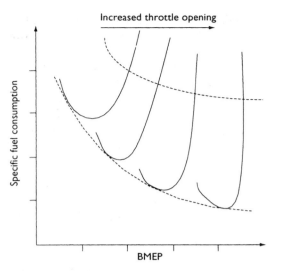

Figure 9.8 Effect of operating at part throttle with varying mixture strength

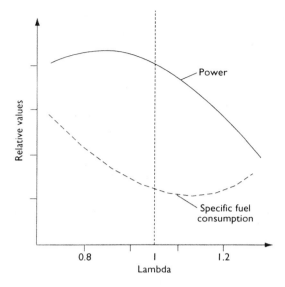

Figure 9.9 Comparison of engine power output and fuel consumption, with changes in air–fuel ratio

maximum power (12:1), and minimum consumption (16:1). The stoichiometric ratio of 14.7:1 is known as a lambda value of one.

Figure 9.9 shows a comparison between engine power output and fuel consumption with changes in air–fuel ratio.

9.1.9 Compression ignition engines

The process of combustion in the compression ignition engine differs from that in a spark ignition engine. In this case the fuel is injected in a liquid state, into a highly compressed, high-temperature air supply in the engine cylinder. Each minute droplet is quickly surrounded by an envelope of its own vapour as it enters the highly heated air. This vapour, after a certain time, becomes inflamed on the surface. A cross-section of any one droplet would reveal a central core of liquid, a thin surrounding film of vapour, with an outer layer of flame. This sequence of vaporization and burning persists as long as combustion continues.

The process of combustion (oxidization of the hydrocarbon fuel), is in itself a lengthy process, but one that may be accelerated artificially by providing the most suitable conditions. The oxidization of the fuel will proceed in air at normal atmospheric temperatures, but it will be greatly accelerated if the temperature is raised. It will take years at 20°C, a few days at 200°C and just a few minutes at 250°C. In these cases, the rate of temperature rise due to oxidization is less than the rate at which the heat is being lost due to convection and radiation. Ultimately, as the temperature is raised, a critical stage is reached where heat is being generated by oxidization at a greater rate than it is being dissipated.

The temperature then proceeds to rise automatically. This, in turn, speeds up the oxidization process and with it the release of heat. Events now take place very rapidly, a flame is established and ignition takes place. The temperature at which this critical change takes place is usually termed the self-ignition temperature of the fuel. This, however, depends on many factors such as pressure, time and the ability to transmit heat from the initial oxidization.

We will now look at the injection of the fuel as a droplet into the heated combustion chamber. At a temperature well above the ignition point, the extreme outer surface of the droplet immediately starts to evaporate, surrounding the core with a thin film of vapour. This involves a supply of heat from the air surrounding the droplet in order to supply the latent enthalpy of evaporation. This supply is maintained by continuing to draw on the main supply of heat from the mass of hot air.

Ignition can and will occur on the vapour envelope even with the core of the droplet still liquid and relatively cold. Once the flame is established, the combustion proceeds at a more rapid rate. This causes a delay period, after injection commences and before ignition takes place. The delay period therefore depends on:

- Excess of air temperature over and above the self-ignition temperature of the fuel.

- Air pressure, both from the point of view of the supply of oxygen and improved heat transfer between the hot air and cold fuel.

Once the delay period is over, the rate at which each flaming droplet can find fresh oxygen to replenish its consumption controls the rate of further burning. The relative velocity of the droplet to the surrounding air is thus of considerable importance. In the compression ignition engine, the fuel is injected over a period of perhaps 40–50° of crank angle. This means that the oxygen supply is absorbed by the fuel first injected, with a possible starvation of the last fuel injected.

This necessitates a degree of turbulence of the air so that the burnt gases are scavenged from the injector zone and fresh air is brought into contact with the fuel. It is clear that the turbulence should be orderly and not disorganized, as in a spark ignition engine, where it is only necessary in order to break up the flame front.

In a compression ignition engine the combustion can be regarded as occurring in three distinct phases as shown in Figure 9.10.

- Delay period.
- Rapid pressure rise.
- After-burning, i.e. the fuel is burning as it leaves the injector.

The longer the delay, the greater and more rapid the pressure rise since more fuel will be present in the cylinder before the rate of burning comes under direct control of the rate of injection. The aim should be to reduce the delay as much as possible, both for the sake of smooth running,

the avoidance of knock and also to maintain control over the pressure change. There is, however, a lower limit to the delay since, without delay, all the droplets would burn as they leave the nozzle. This would make it almost impossible to provide enough combustion air within the concentrated spray and the delay period also has its use in providing time for the proper distribution of the fuel. The delay period therefore depends on:

- The pressure and temperature of the air.
- The cetane rating of the fuel.
- The volatility and latent enthalpy of the fuel.
- The droplet size.
- Controlled turbulence.

The effect of droplet size is important, as the rate of droplet burning depends primarily on the rate at which oxygen becomes available. It is, however, vital for the droplet to penetrate some distance from the nozzle around which burning will later become concentrated. To do this, the size of the droplets must be large enough to obtain sufficient momentum at injection. On the other hand, the smaller the droplet the greater the relative surface area exposed and the shorter the delay period. A compromise between these two effects is clearly necessary.

With high compression ratios (15 : 1 and above) the temperature and pressure are raised so that the delay is reduced, which is an advantage. However, high compression ratios are a disadvantage mechanically and also inhibit the design of the combustion chamber, particularly in small engines where the bumping clearance consumes a large proportion of the clearance volume.

9.1.10 Combustion chamber design – diesel engine

The combustion chamber must be designed to:

- Give the necessary compression ratio.
- Provide the necessary turbulence.
- Position for correct and optimum operation of the valves and injector.

These criteria have effects that are interrelated. Turbulence is normally obtained at the expense of volumetric efficiency. Masked inlet valves (which are mechanically undesirable) or 'tangent' directional ports restrict the air flow and therefore are restrictive to high-speed engines.

To assist in breathing, four or even six valves per cylinder can be used. This arrangement has

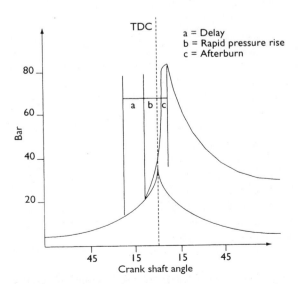

Figure 9.10 Phases of diesel combustion

the advantage of keeping the injector central, a desirable aim for direct injection engines. Large valves and their associated high lift, in addition to providing mechanical problems often require heavy piston recesses, which disturb squish and orderly movement of the air.

A hemispherical combustion chamber assists with the area available for valves, at the expense of using an offset injector. Pre-combustion chambers, whether of the air cell or 'combustion swirl' type have the general disadvantage of being prone to metallurgical failure or at least are under some stress since, as they are required to produce a 'hot spot' to assist combustion, the temperature stresses in this region are extremely high. There is no unique solution and the resulting combustion chamber is always a compromise.

9.1.11 Summary of combustion

Section 9.1 has looked at some of the issues of combustion, and is intended to provide a background to some of the other sections in this book. The subject is very dynamic and improvements are constantly being made. Some of the key issues this chapter has raised so far include points such as the time to burn a fuel–air mixture, the effects of changes in mixture strength and ignition timing, the consequences of detonation and other design problems.

Accurate control of engine operating variables is one of the keys to controlling the combustion process. This is covered in other chapters.

9.2 Engine fuelling and exhaust emissions

9.2.1 Operating conditions

The ideal air–fuel ratio is about 14.7 : 1. This is the theoretical amount of air required to burn the fuel completely. It is given a 'lambda (λ)' value of 1.

λ = actual air quantity $\div$ theoretical air quantity

The air–fuel ratio is altered during the following operating conditions of an engine to improve its performance, drivability, consumption and emissions.

- Cold starting – a richer mixture is needed to compensate for fuel condensation and improves drivability.

- Load or acceleration – a richer mixture to improve performance.
- Cruise or light loads – a weaker mixture for economy.
- Overrun – very weak mixture (if any) to improve emissions and economy.

The more accurately the air–fuel ratio is controlled to cater for external conditions, then the better the overall operation of the engine.

9.2.2 Exhaust emissions

Figure 9.11 shows, firstly, the theoretical results of burning a hydrocarbon fuel and, secondly, the actual combustion results. The top part of the figure is ideal but the lower part is the realistic result under normal conditions. Note that this result is prior to any further treatment, for example by a catalytic converter.

Figure 9.12 shows the approximate percentages of the various exhaust gas emissions. The volume of pollutants is small but, because they are so poisonous, they are undesirable and strong legislation now exists to encourage their reduction. The actual values of these emissions varies depending on engine design, operating conditions, temperature and smooth running, to name just a few variables.

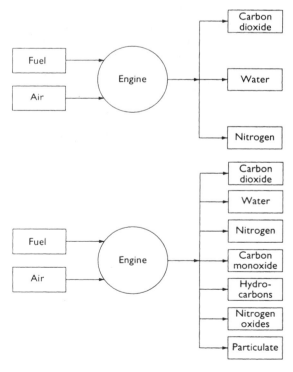

Figure 9.11 Theoretical results of burning a hydrocarbon fuel and actual combustion results

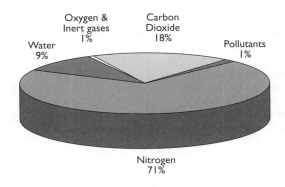

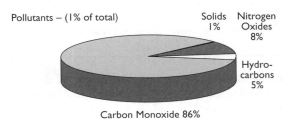

Figure 9.12 Composition of exhaust

Table 9.1 lists the four main emissions that are hazardous to health, together with a short description of each.

9.2.3 Other sources of emissions

The main source of vehicle emissions is the exhaust, but other areas of the vehicle must also come under scrutiny.

As well as sulphur in fuel, another area of contention between car manufacturers and oil companies is the question of who should bear the cost of collecting fuel vapour at filling stations. The issue of evaporative fuel emissions (EFEs) has become a serious target for environmentalists. Approximately 10% of EFEs escape during refuelling.

In the US, the oil companies have won the battle. All cars manufactured from the start of 1998 must be fitted with 10 litre canisters filled with carbon to catch and absorb the vapours. The outcome in Europe is not certain and there is considerable debate as to whether it should be the responsibility of the oil companies to collect this vapour at the pump.

This still leaves the matter of preventing evaporation from the fuel line itself, another key problem for car manufacturers. Technological advances in design actually increase fuel evaporation from within the fuelling system. This is because of the increasing use of plastics, rather than metal, for manufacturing fuel lines. Plastics allow petrol vapour to permeate through into the atmosphere. The proximity of catalytic converters, which generate tremendous heat to the fuel tank and the under-body shielding, contributes to making the fuel hotter and therefore more liable to evaporate.

Table 9.2 describes this issue further and also looks at crankcase emissions.

Evaporative emissions are measured in a 'shed'! This Sealed Housing for Evaporative Determination (SHED) is used in two ways:

- The vehicle with 40% of its maximum fuel is warmed up (from about 14–28 °C) in the shed

Table 9.1 Main health hazard emissions

Substance	Description
Carbon monoxide (CO)	This gas is very dangerous even in low concentrations. It has no smell or taste and is colourless. When inhaled it combines in the body with the red blood cells, preventing them from carrying oxygen. If absorbed by the body it can be fatal in a very short time.
Nitrogen oxides (NOx)	Oxides of nitrogen are colourless and odourless when they leave the engine but as soon as they reach the atmosphere and mix with more oxygen, nitrogen oxides are formed. They are reddish brown and have an acrid and pungent smell. These gases damage the body's respiratory system when inhaled. When combined with water vapour, nitric acid can be formed which is very damaging to the windpipe and lungs. Nitrogen oxides are also a contributing factor to acid rain.
Hydrocarbons (HC)	A number of different hydrocarbons are emitted from an engine and are partly burnt or unburnt fuel. When they mix with the atmosphere they can contribute to form smog. It is also believed that hydrocarbons may be carcinogenic.
Particulate matter (PM)	This heading mainly covers lead and carbon. Lead was traditionally added to petrol to slow its burning rate in order to reduce detonation. It is detrimental to health and is thought to cause brain damage, especially in children. Lead will eventually be phased out as all new engines now run on unleaded fuel. Particles of soot or carbon are more of a problem on diesel-fuelled vehicles and these now have limits set by legislation.

Table 9.2 Fuel evaporation and crankcase emissions

Source	Comments
Fuel evaporation from the tank and system	Fuel evaporation causes hydrocarbons to be produced. The effect is greater as temperature increases. A charcoal canister is the preferred method for reducing this problem. The fuel tank is usually run at a pressure just under atmospheric by a connection to the intake manifold, drawing the vapour through the charcoal canister. This must be controlled by the management system, however, as even a 1% concentration of fuel vapour would shift the lambda value by 20%. This is done by using a 'purge valve', which under some conditions is closed (full-load and idle, for example) and can be progressively opened under other conditions. The system monitors the effect through the use of the lambda sensor signal.
Crankcase fumes (blow by)	Hydrocarbons become concentrated in the crankcase mostly due to pressure blowing past the piston rings. These gases must be conducted back into the combustion process. This usually happens via the air intake system. This is described as positive crankcase ventilation.

and the increased concentration of the hydrocarbons measured.

- The vehicle is first warmed up over the normal test cycle and then placed in the shed. The increase in HC concentration is measured over one hour.

9.2.4 Leaded and unleaded fuel

Tetra-ethyl lead was first added to petrol in the 1920s to slow down the rate of burning, improve combustion and increase the octane rating of the fuel. All this at less cost than further refining by the petrol companies.

The first real push for unleaded fuel was from Los Angeles in California. To reduce this city's severe smog problem, the answer at the time seemed to be to employ catalytic converters. However, if leaded fuel is used, the 'cat' can be rendered inoperative. A further study showing that lead causes brain damage in children sounded the death knell for leaded fuel. This momentum spread worldwide and still exists.

New evidence is now coming to light showing that the additives used instead of lead were ending up in the environment. The two main culprits are benzene, which is strongly linked to leukaemia, and MTBE, which poisons water and is very toxic to almost all living things. This is potentially a far worse problem than lead, which is now not thought to be as bad as the initial reaction suggested.

It is important, however, to note that this is still in the 'discussion' stage; further research is necessary for a fully reasoned conclusion. Note though how any technological issue has far more to it than first meets the eye.

Modern engines are now designed to run on unleaded fuel, with one particular modification being hardened valve seats. In Europe and other places, leaded fuel has now been phased out completely. This is a problem for owners of classic vehicles. Many additives are available but these are not as good as lead. Here is a list of comments I have collated from a number of sources.

- All engines with cast iron heads and no special hardening of the exhaust valve seats will suffer some damage running on unleaded. The extent of the damage depends on the engine and on the engine revs.
- No petrol additives prevent valve seat recession completely. Some are better than others but none replace the action of lead.
- The minimum critical level of lead in the fuel is about 0.07 g Pb/l. Current levels in some leaded fuel are 0.15 g Pb/l and so mixing alternate tanks of leaded and unleaded is likely to be successful.
- It is impossible to predict wear rates accurately and often wear shows up predominantly in only one cylinder.
- Fitting hardened valve seats or performing induction hardening on the valve seats is effective in engines where either of these processes can be done.
- Tests done by Rover appear to back up the theory that, although unleaded petrol does damage all iron heads, the less spirited driver will not notice problems until a high mileage has been covered on entirely lead-free fuel.
- When unprotected engines are bench-tested on unleaded fuel and then stripped down, damage will always be evident. However, drivers seldom complain of trouble running on unleaded, perhaps because they are not over-revving the engine or are not covering high mileage.

I will leave you to make your own mind up about these matters.

9.2.5 Exhaust emission regulations

At the time of publication, the current emission regulations can be summarized by the following tables: firstly, MOT regulations for UK vehicles (Table 9.3) and, secondly, limits set for new vehicles produced in or imported into the EC (Table 9.4). The MOT tests are carried out with a warm engine at the recommended idle speed. However, the hydrocarbon figure can also be checked at 1200 rev/min if out of setting at idle speed.

Diesel engined vehicles are also tested for particulate emissions at idle and full load. At present, the full load test which involved revving an engine to its governed maximum speed has been suspended due to a number of engines suffering serious damage during the test! European regulations are applicable to new vehicles. The current and proposed European standards are summarized in Table 9.4.

The stages 1 and 2 directives have served to ensure that all cars are required to be fitted with a three-way catalytic converter to meet the standards. Stages 3 and 4 proposals require further technology in all areas of engine control. Manufacturers are working hard to introduce these measures. The present EEC test cycle has been under scrutiny as it was felt that it did not reflect the conditions for normal driving in northern Europe. Figure 9.13 shows current and future test cycles.

The soak, or original, temperature of the vehicle is currently 20–30°C, and an engine idle time of 40 s is not included in the test. Proposals are currently being discussed to include the idle period and use a soak temperature of between 0 and 5°C. This will require further methods of reducing emissions to be included on the vehicle. One further problem is the measurement of hydrocarbons. At present, the two main types – non-methane (NMHC) and methane hydrocarbons (MHC) – are included in the test. However, as catalytic conversion of MHC is difficult, if these stay in the test it will make further proposals more difficult to achieve.

Table 9.3 UK MOT regulations (introduced on 1 November 1991)

Vehicles first used	Regulations
Before 1.8.1975	Visual check – excessive emissions only
On or after 1.8.1975	Carbon monoxide – 4.5% maximum Hydrocarbons – 1200 ppm maximum
On or after 1.8.1986	Carbon monoxide – 3.5% maximum Hydrocarbons – 1200 ppm maximum

On or after 1.8.1994
(Minimum oil temperature 60°C)
	At idle 450–1500 rev/min:
	  Carbon monoxide – 0.5% maximum
	At fast idle 2500–3000 rev/min:
	  Carbon monoxide – 0.3% maximum
	  Hydrocarbons – 200 ppm maximum
	Lambda: 0.97–1.03

Table 9.4 UK/EEC regulations for new vehicles

Standard (g/km)	CO	HC/NOx	PM
Stage 1. 1992/1993			
Petrol	2.72	0.97	NA
Diesel	2.72	0.97	0.14
To 1994 diesel DI	2.72	1.36	0.19
Stage 2. 1995/1996/1997			
Petrol	2.2	0.5	NA
Diesel	1.0	0.7	0.08
To 1999 diesel DI	1.0	0.9	0.10
Stage 3. 1999/2000/2001			
Petrol	2.3	0.2/0.15	NA
Diesel	0.64	0.56/0.5	0.05
Stage 4 proposal. 2005			
Petrol	1.0	0.1/0.08	NA
Diesel	0.5	0.3/0.25	0.025

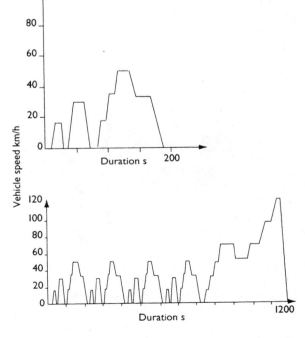

Figure 9.13 EC current and possible future test cycles

9.3 Electronic control of carburation

9.3.1 Basic carburation

Figure 9.14 shows a simple fixed choke carburettor, in order to describe the principles of operation of this device. The float and needle valve assembly ensure a constant level of petrol in the float chamber. The Venturi causes an increase in air speed and hence a drop in pressure in the area of the outlet. The main jet regulates how much fuel can be forced into this intake air stream by the higher pressure now apparent in the float chamber. The basic principle is that as more air is forced into the engine then more fuel will be mixed into the air stream.

Figure 9.15 shows the problem with this very simple system; the amount of fuel forced into the air stream does not linearly follow the increase in air quantity. This means further compensation fuel and air jets are required to meet all operating requirements.

Figure 9.16 shows a variable Venturi carburettor, which keeps the air pressure in the Venturi constant, and uses a tapered needle to control the amount of fuel.

9.3.2 Areas of control

One version of the variable Venturi carburettor (Figure 9.17) is used with electronic control. In

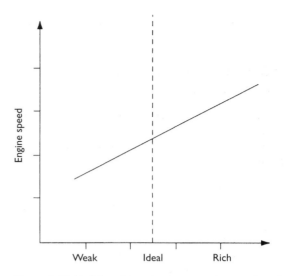

Figure 9.15 Fuel forced into the air stream does not linearly follow the increase in air quantity with a simple fixed choke carburettor

general, electronic control of a carburettor is used in the following areas.

Idle speed

Controlled by a stepper motor to prevent stalling but still allow a very low idle speed to improve economy and reduce emissions. Idle speed may also be changed in response to a signal from an automatic gearbox to prevent either the engine from stalling or the car from trying to creep.

Fast idle

The same stepper motor as above controls fast idle in response to a signal from the engine temperature sensor during the warm up period.

Choke (warm up enrichment)

A rotary choke or some other form of valve or flap operates the choke mechanism depending on engine and ambient temperature conditions.

Overrun fuel cut off

A small solenoid operated valve or similar cuts off the fuel under particular conditions. These are often that the engine temperature is above a set level, the engine speed is above a set level and that the accelerator pedal is in the off position.

The main control of the air–fuel ratio is a function of the mechanical design and is very difficult to control by electrical means. Some systems have used electronic control of a needle and jet but this did not prove to be very popular.

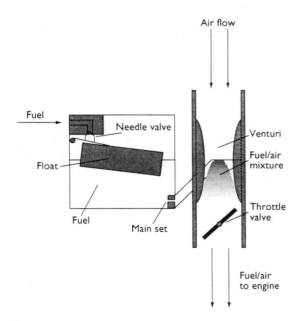

Figure 9.14 Simple fixed choke carburettor

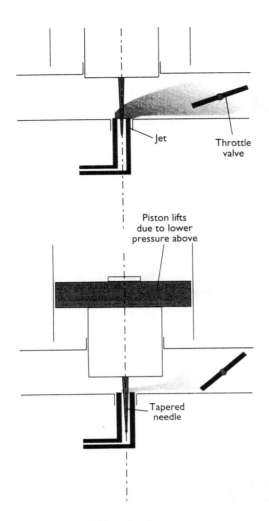

Figure 9.16 Variable Venturi carburettor

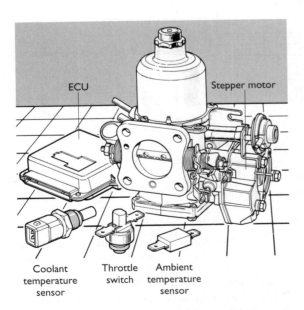

Figure 9.17 HIF variable Venturi carburettor with electronic control components

9.4 Fuel injection

9.4.1 Advantages of fuel injection

The major advantage of any type of fuel injection system is accurate control of the fuel quantity injected into the engine. The basic principle of fuel injection is that if petrol is supplied to an injector (electrically controlled valve), at a constant differential pressure, then the amount of fuel injected will be directly proportional to the injector open time.

Most systems are now electronically controlled even if containing some mechanical metering components. This allows the operation of the injection system to be very closely matched to the requirements of the engine. This matching process is carried out during development on test beds and dynamometers, as well as development in the car. The ideal operating data for a large number of engine operating conditions are stored in a read only memory in the ECU. Close control of the fuel quantity injected allows the optimum setting for mixture strength when all operating factors are taken into account (see the air–fuel ratio section).

Further advantages of electronic fuel injection control are that overrun cut off can easily be implemented, fuel can be cut at the engines rpm limit and information on fuel used can be supplied to a trip computer.

Fuel injection systems can be classified into two main categories:

- Single-point injection – see Figure 9.18.
- Multipoint injection – see Figure 9.19.

Both of these systems are discussed in more detail in later sections of this chapter.

9.4.2 System overview

Figure 9.20 shows a typical control layout for a fuel injection system. Depending on the sophistication of the system, idle speed and idle mixture adjustment can be either mechanically or electronically controlled.

Figure 9.21 shows a block diagram of inputs and outputs common to most fuel injection systems. Note that the two most important input sensors to the system are speed and load. The basic fuelling requirement is determined from these inputs in a similar way to the determination of ignition timing, as described in a previous section.

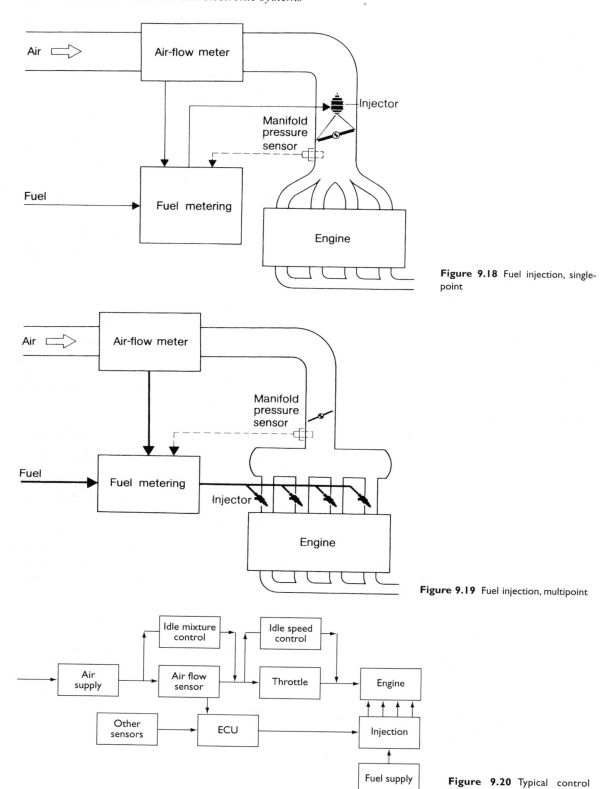

Figure 9.18 Fuel injection, single-point

Figure 9.19 Fuel injection, multipoint

Figure 9.20 Typical control layout for a fuel injection system

A three-dimensional cartographic map, shown in Figure 9.22, is used to represent how the information on an engine's fuelling requirements is stored. This information forms part of a read only memory (ROM) chip in the ECU. When the ECU has determined the look-up value of the fuel required (injector open time), corrections to this figure can be added for bat-

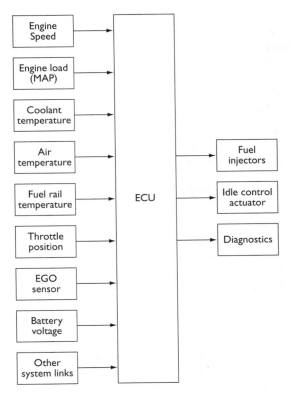

Figure 9.21 Block diagram of inputs and outputs common to most fuel injection systems

tery voltage, temperature, throttle change or position and fuel cut off.

Idle speed and fast idle are also generally controlled by the ECU and a suitable actuator. It is also possible to have a form of closed loop control with electronic fuel injection. This involves a lambda sensor to monitor exhaust gas oxygen content. This allows very accurate control of the mixture strength, as the oxygen content of the exhaust is proportional to the air–fuel ratio. The signal from the lambda sensor is used to adjust the injector open time.

Figure 9.23 is a flow chart showing one way in which the information from the sensors could

be processed to determine the best injector open duration as well as control of engine idle speed.

9.4.3 Components of a fuel injection system

The following parts with some additions, are typical of the Bosch 'L' Jetronic systems. These components are only briefly discussed, as most are included in other sections in more detail.

Flap type air flow sensor (Figure 9.24)

A Bosch vane-type sensor is shown which moves due to the air being forced into the engine. The information provided to the ECU is air quantity and engine load.

Engine speed sensor

Most injection systems, which are not combined directly with the ignition, take a signal from the coil negative terminal. This provides speed data but also engine position to some extent. A resistor in series is often used to prevent high voltage surges reaching the ECU.

Temperature sensor (Figure 9.25)

A simple thermistor provides engine coolant temperature information.

Throttle position sensor (Figure 9.26)

Various sensors are shown consisting of the two-switch types, which only provide information that the throttle is at idle, full load or anywhere else in between; and potentiometer types, which give more detailed information.

Lambda sensor (Figure 9.27)

This device provides information to the ECU on exhaust gas oxygen content. From this informa-

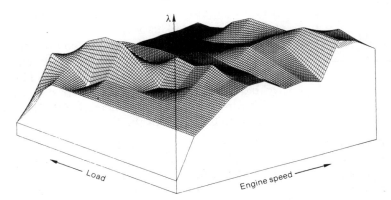

Figure 9.22 Cartographic map used to represent how the information on an engine's fuelling requirements are stored

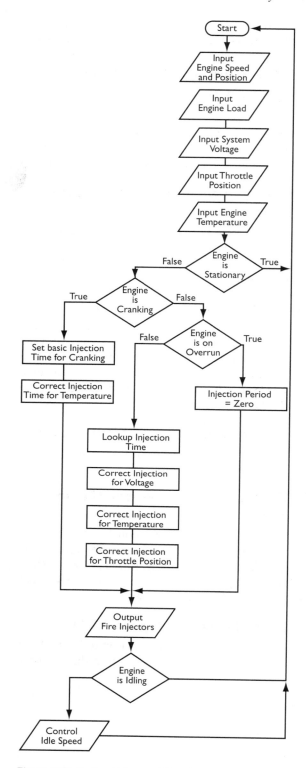

Figure 9.23 Fuel and idle speed flow diagram

tion, corrections can be applied to ensure the engine is kept at or very near to stoichiometry. Also shown in this figure is a combustion chamber pressure sensor.

Idle or fast idle control actuator (Figure 9.28)

Bimetal or stepper motor actuators are used but the one shown is a pulsed actuator. The air that it allows through is set by its open/close ratio.

Fuel injector(s) (Figure 9.29)

Two types are shown–the pintle and disc injectors. They are simple solenoid-operated valves designed to operate very quickly and produce a finely atomized spray pattern.

Injector resistors

These resistors were used on some systems when the injector coil resistance was very low. A lower inductive reactance in the circuit allows faster operation of the injectors. Most systems now limit injector maximum current in the ECU in much the same way as for low resistance ignition on coils.

Fuel pump (Figure 9.30)

The pump ensures a constant supply of fuel to the fuel rail. The volume in the rail acts as a swamp to prevent pressure fluctuations as the injectors operate. The pump must be able to maintain a pressure of about 3 bar.

Fuel pressure regulator (Figure 9.31)

This device ensures a constant differential pressure across the injectors. It is a mechanical device and has a connection to the inlet manifold.

Cold start injector and thermo-time switch (Figure 9.32)

An extra injector was used on earlier systems as a form of choke. This worked in conjunction with the thermo-time switch to control the amount of cold enrichment. Both engine temperature and a heating winding heat it. This technique has been replaced on newer systems, which enrich the mixture by increasing the number of injector pulses or the pulse length.

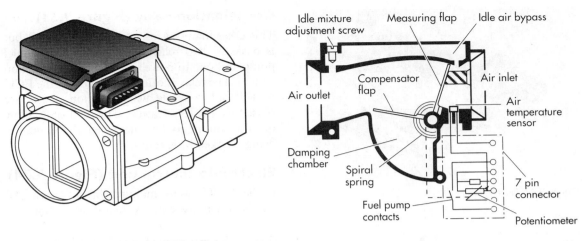

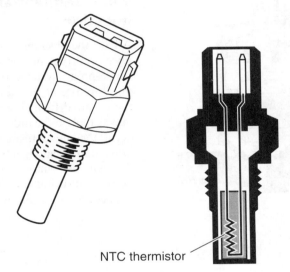

Figure 9.24 Airflow meter

Figure 9.25 Coolant temperature sensor

NTC thermistor

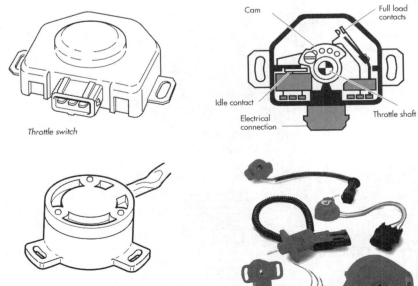

Throttle switch

Throttle potentiometer

Figure 9.26 Throttle position sensors

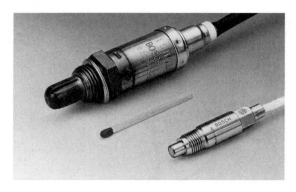

Figure 9.27 Lambda sensor

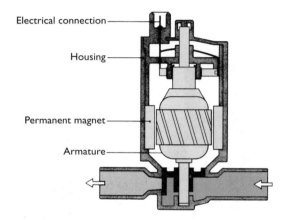

Figure 9.28 Rotary idle actuator

Combination relay (Figure 9.33)

This takes many forms on different systems but is basically two relays, one to control the fuel pump and one to power the rest of the injection system. The relay is often controlled by the ECU or will only operate when ignition pulses are sensed as a safety feature. This will only allow the fuel pump to operate when the engine is being cranked or is running.

Electronic control unit (Figure 9.34)

Earlier ECUs were analogue in operation. All ECUs now in use employ digital processing.

9.4.4 Sequential multipoint injection

All of the systems discussed previously either inject the fuel in continuous pulses, as in the single-point system, or all of the multipoint injectors fire at the same time, injecting half of the required fuel. A sequential injection system injects fuel on the induction stroke of each cylinder in the engine firing order. This system, while more complicated, allows the stratification of the cylinder charge to be controlled to some extent, allowing an overall weaker charge. Sequential injection is normally incorporated

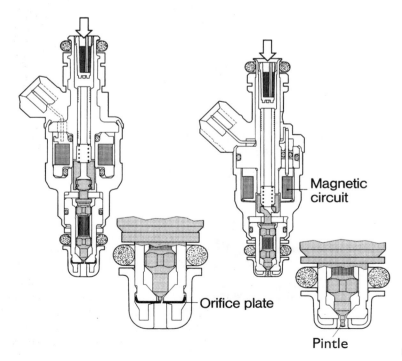

Figure 9.29 Fuel injector

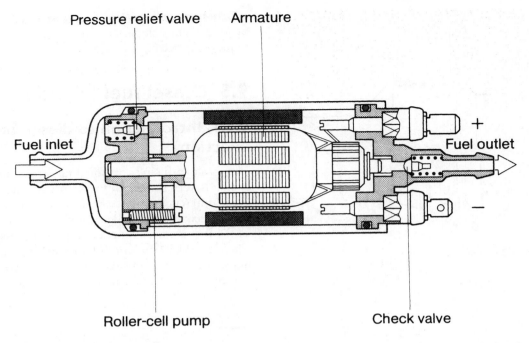

Figure 9.30 Fuel pump (high pressure)

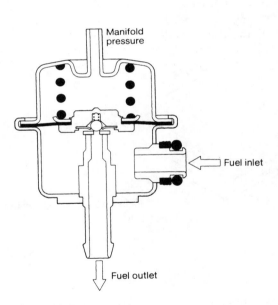

Figure 9.31 Pressure regulator

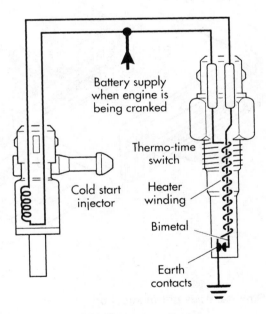

Figure 9.32 Typical cold start arrangement

9.4.5 Summary

The developments of fuel injection in general, and the reduced complexity of single-point systems in particular, have now started to make the carburettor obsolete. As emission regulations continue to become more stringent, manufacturers are being forced into using fuel injection, even on lower priced models. This larger market

with full engine management, which is discussed further in Chapter 10. Figure 9.35 shows a comparison between normal and sequential injection.

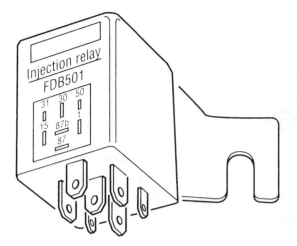

Figure 9.33 Combination relay

will, in turn, pull the price of the systems down, making them comparable to carburation techniques on price but superior in performance.

9.5 Diesel fuel injection

9.5.1 Introduction to diesel fuel injection

The basic principle of the four-stroke diesel engine is very similar to the petrol system. The main difference is that the mixture formation takes place in the cylinder combustion chamber as the fuel is injected under very high pressure. The timing and quantity of the fuel injected is important from the usual viewpoints of performance, economy and emissions.

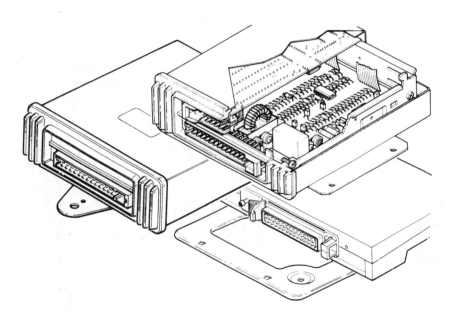

Figure 9.34 Electronic control units

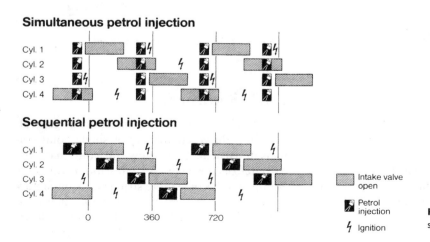

Figure 9.35 Simultaneous and sequential petrol injection

Fuel is metered into the combustion chamber by way of a high pressure pump connected to injectors via heavy duty pipes. When the fuel is injected it mixes with the air in the cylinder and will self-ignite at about 800°C. See the section on diesel combustion for further details. The mixture formation in the cylinder is influenced by the following factors.

Start of delivery and start of injection (timing)

The timing of a diesel fuel injection pump to an engine is usually done using start of delivery as the reference mark. The actual start of injection, in other words when fuel starts to leave the injector, is slightly later than start of delivery, as this is influenced by the compression ratio of the engine, the compressibility of the fuel and the length of the delivery pipes. This timing increases the production of carbon particles (soot) if too early, and increases the hydrocarbon emissions if too late.

Spray duration and rate of discharge (fuel quantity)

The duration of the injection is expressed in degrees of crankshaft rotation in milliseconds. This clearly influences fuel quantity but the rate of discharge is also important. This rate is not constant due to the mechanical characteristics of the injection pump.

Injection pressure

Pressure of injection will affect the quantity of fuel, but the most important issue here is the effect on atomization. At higher pressures, the fuel will atomize into smaller droplets with a corresponding improvement in the burn quality. Indirect injection systems use pressures up to about 350 bar, while direct injection systems can be up to about 1000 bar. Emissions of soot are greatly reduced by higher pressure injection.

Injection direction and number of jets

The direction of injection must match very closely the swirl and combustion chamber design. Deviations of only 2° from the ideal can greatly increase particulate emissions.

Excess air factor (air–fuel ratio)

Diesel engines do not, in general, use a throttle butterfly as the throttle acts directly on the injection pump to control fuel quantity. At low speeds in particular, the very high excess air factor ensures complete burning and very low emissions. Diesel engines operate where possible with an excess air factor even at high speeds.

Figure 9.36 shows a typical diesel fuel injection system. Detailed operation of the components is beyond the scope of this book. The principles and problems are the issues under consideration here, in particular, the way electronics can be employed to solve some of these problems.

9.5.2 Diesel exhaust emissions

Overall, the emissions from diesel combustion are far lower than emission from petrol combustion. Figure 9.37 shows the comparison between petrol and diesel emissions. The CO, HC and NOx emissions are lower, mainly due to the higher compression ratio and excess air factor. The higher compression ratio improves the

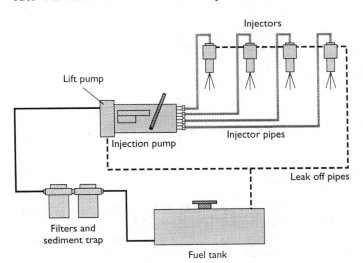

Figure 9.36 Diesel fuel injection system

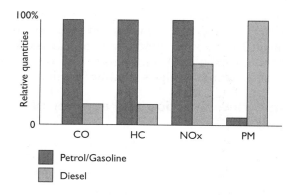

Figure 9.37 Comparison between petrol and diesel emissions

thermal efficiency and thus lowers the fuel consumption. The excess air factor ensures more complete burning of the fuel.

The main problem area is that of particulate emissions. These particle chains of carbon molecules can also contain hydrocarbons, mostly aldehydes. The effect of this emission is a pollution problem but the possible carcinogenic effect of this soot also a gives cause for concern. The diameter of these particles is only a few ten thousandths of a millimetre – consequently they float in the air and can be inhaled.

9.5.3 Electronic control of diesel injection

The advent of electronic control over the diesel injection pump has allowed many advances over the purely mechanical system. The production of high pressure and injection is, however, still mechanical with all current systems. The following advantages are apparent over the non-electronic control system.

- More precise control of fuel quantity injected.
- Better control of start of injection.
- Idle speed control.
- Control of exhaust gas recirculation.
- Drive by wire system (potentiometer on throttle pedal).
- An antisurge function.
- Output to data acquisition systems etc.
- Temperature compensation.
- Cruise control.

Figure 9.38 shows a distributor-type injection pump used with electronic control. Because fuel must be injected at high pressure, the hydraulic head, pressure pump and drive elements are still used. An electromagnetic moving iron actuator adjusts the position of the control collar, which

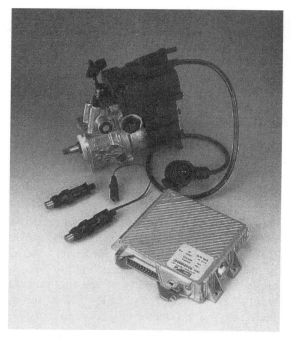

Figure 9.38 Distributor type injection pump with electronic control

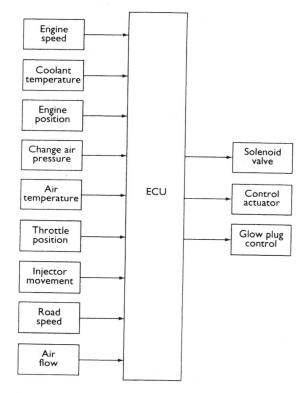

Figure 9.39 Block diagram of typical electronic diesel control system

in turn controls the delivery stroke and therefore the injected quantity of fuel. Fuel pressure is applied to a roller ring and this controls the start of injection. A solenoid-operated valve controls the supply to the roller ring. These actuators together allow control of the start of injection and injection quantity.

Figure 9.39 shows a block diagram of a typical electronic diesel control system. Ideal values for fuel quantity and timing are stored in memory maps in the electronic control unit. The injected fuel quantity is calculated from the accelerator position and the engine speed. The start of injection is determined from the following:

- Fuel quantity.
- Engine speed.
- Engine temperature.
- Air pressure.

The ECU is able to compare start of injection with actual delivery from a signal produced by the needle motion sensor in the injector. Figure 9.40 shows a typical injector complete with a needle motion sensor.

Control of exhaust gas recirculation is by a simple solenoid valve. This is controlled as a function of engine speed, temperature and injected quantity. The ECU is also in control of the stop solenoid and glow plugs (Figure 9.41) via a suitable relay.

Figure 9.42 is the complete layout of an electronic diesel control system.

9.6 Case studies

9.6.1 Bosch 'L' Jetronic – variations

Owing to continued demands for improvements, the 'L' Jetronic system has developed and

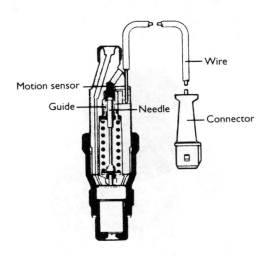

Figure 9.40 Diesel injector complete with needle motion sensor

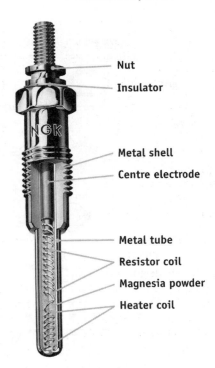

Figure 9.41 A typical diesel glow plug

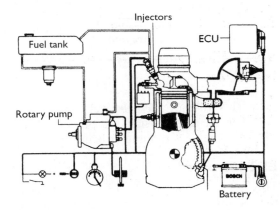

Figure 9.42 Layout of an electrical diesel control system

changed over the years. This section will highlight the main changes that have taken place. The 'L' variation is shown in Figure 9.43.

L2 Jetronic

This system is changed little except for the removal of the injector series resistors as the ECU now limits the output current to the injectors. The injector resistance is $16\,\Omega$.

LE1 Jetronic

No current resistors are used and the throttle switch is adjustable. The fuel pump does not have safety contacts in the air flow sensor. The safety circuit is incorporated in the electronic

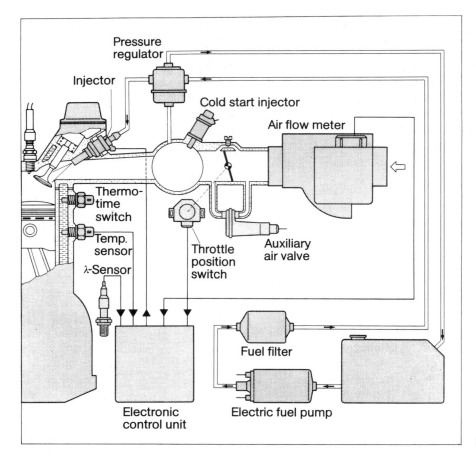

Figure 9.43 L-Jetronic

relay. This will only allow the fuel pump to operate when an ignition signal is present; that is, when the engine is running or being cranked.

LE2 Jetronic

This is very similar to the LE1 systems except the thermo-time switch and cold start injector are not used. The ECU determines cold starting enrichment and adjusts the injector open period accordingly.

LU Jetronic

This system is a further refinement of the LE systems but also utilizes closed loop lambda control.

L3 Jetronic

The ECU for the L3 Jetronic forms part of the air flow meter installation, as shown in Figure 9.44. The ECU now includes a 'limp home' facility. The system can be operated with or without lambda closed loop control. The air–fuel ratio can be adjusted by a screw-operated potentiometer on the side of the ECU.

LH Jetronic

The LH system incorporates most of the improvements noted above. The main difference is that a hot-wire type of air flow meter is used. The component layout is shown in Figure 9.45. Further developments are continuing but, in general, most systems have now developed into combined fuel and ignition control systems as discussed in the next chapter.

Figure 9.44 L3 Jetronic

9.6.2 Lucas hot wire – multipoint injection

The Lucas hot-wire fuel injection system is a multipoint, indirect and intermittent injection system. In line with many other systems, the basic fuelling requirements are determined from engine speed and rate of air flow. Engine load, engine temperature and air temperature are the three main correction factors. The calculation for the fuel injection period is a digital process and the look-up values are stored in a memory chip in the ECU. It is important with this and other systems that no unmetered air enters the engine except via the idle mixture screw on the throttle body. Figure 9.46 is the schematic arrangement of the hot wire system.

All the major components of the system are shown in Figure 9.47. The ECU acts on the signals received from sensors and adjusts the length of pulse supplied to the injectors. The ECU also controls the time at which the injector pulses occur relative to signals from the coil negative terminal. During normal running conditions, the injectors on a four-cylinder engine are all fired at the same time and inject half of the required amount, twice during the complete engine cycle.

The fuel tank contains a swirl pot as part of the pick-up pipe. This is to ensure that the pick-up pipe is covered in fuel at all times, thus preventing air being drawn into the fuel lines. A permanent-magnet electric motor is used for the fuel pump, which incorporates a roller-cell-type pumping assembly. An eccentric rotor on the motor shaft has metal rollers in cut-outs around its edge. These rollers are forced out by centrifugal force as the motor rotates. This traps the fuel and forces it out of the pressure side of the system. The motor is always filled with fuel and the pump is able to self-prime. A non-return valve and a pressure relief valve are fitted. These will cause a pressure to be held in the system and prevent excessive pressure build up, respectively. The pump is controlled by the ECU via a relay. When the ignition is first switched on, the pump runs for a short time to ensure the system is at the correct pressure. The pump will then only run when the engine is being cranked or is running. A 1 Ω ballast resistor is often fitted in the

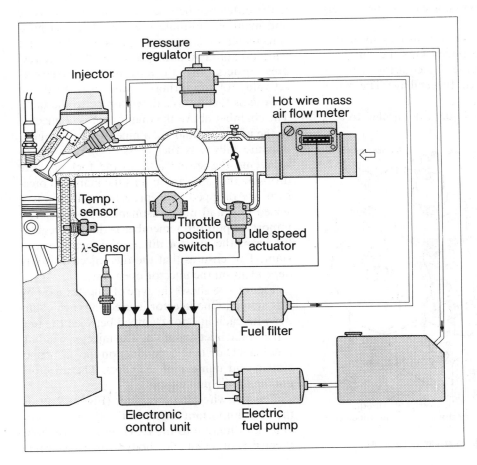

Figure 9.45 LH-Jetronic

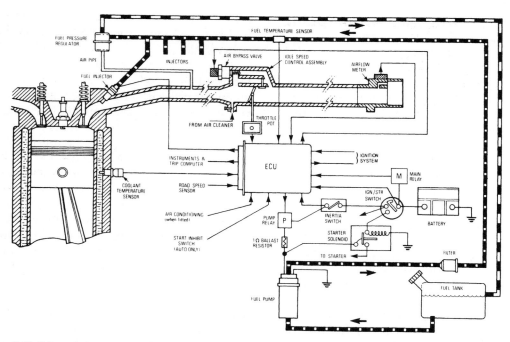

Figure 9.46 Schematic arrangement of hot-wire electronic fuel injection system

supply to the pump. This will cut down on noise but is also bypassed when the engine is being cranked to ensure the pump runs at a 'normal' speed even when cranking causes the battery voltage to drop.

An inertia switch, which is usually located in the passenger compartment, cuts the supply to the fuel pump in the case of a collision. This is a safety feature to prevent fuel spillage. The switch can be reset by hand.

In order for the fuel quantity injected to be a

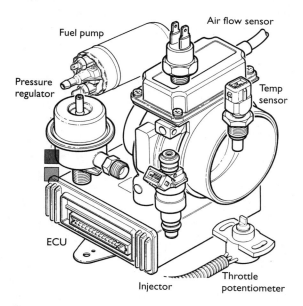

Figure 9.47 Hot-wire injection system components

function of the injection pulse length, the fuel pressure across the injector must be constant. This fuel pressure, which is in the region of 3 bar, is the difference between absolute fuel pressure and manifold absolute pressure. The fuel pressure regulator is a simple pressure relief valve with a diaphragm and spring on which the fuel pressure acts. When the pressure exceeds the preset value (of the spring), a valve is opened and the excess fuel returns down a pipe to the tank. The chamber above the diaphragm is connected to the inlet manifold via a pipe. As the manifold pressure falls, less fuel pressure is required to overcome the spring and so the fuel pressure drops by the same amount as the manifold pressure has dropped. The pressure regulator is a sealed unit and no adjustment is possible. The important point to remember is that the regulator keeps the injector differential pressure constant. This ensures that the fuel injected is only dependent on the injector open time.

Figure 9.48 shows the type of injector used by this system. One injector is used for each cylinder with each injector clamped between the fuel rail and the inlet manifold. The injector winding is either $4\,\Omega$ or $16\,\Omega$ depending on the particular system and number of cylinders. The injectors are the needle/pintle types.

The hot-wire air flow meter (Figure 9.49) is the most important sensor in the system. It provides information to the ECU on air mass flow. It consists of a cast alloy body with an electronic

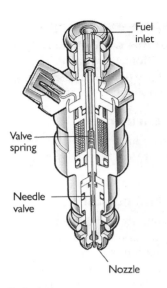

Fuel
inlet

Valve
spring

Needle
valve

Nozzle

Figure 9.48 Fuel injector

module on the top. Air drawn into the engine
passes through the main opening, with a small
proportion going through a bypass in which two
small wires are fixed. These two wires are a sensing wire and a compensation wire. The compensation wire reacts only to the air temperature.
The sensing wire is heated with a small current
from the module. The quantity of air drawn over
this wire will cause a cooling effect and alter its
resistance, which is sensed by the module. The
air flow meter has just three wires, a positive and
negative supply and an output that varies
between about 0 and 5 V depending on air mass
flow rate. This system can react very quickly to
changes and also automatically compensates for
changes in altitude. Chapter 2 gives further
details of the operation of this sensor. Each air
flow meter is matched to its module, therefore
repair is not normally possible.

A throttle potentiometer is used to provide
the ECU with information on throttle position
and rate of change of throttle position. The
device is a simple three-wire variable resistor
using a carbon track, it is attached to the main
throttle butterfly spindle. A stable supply of 5 V
allows a variable output voltage depending on
throttle position. At idle, the output should be
325 mV and, at full load, 4.8 V. The rate of
change indicates the extent of acceleration or
deceleration. This is used to enrich the mixture
or implement overrun fuel cut-off as may be
appropriate.

The throttle body is an alloy casting bolted to
the inlet manifold and connected to the air flow
sensor by a flexible trunking. This assembly contains the throttle butterfly and potentiometer,
and also includes the stepper motor, which controls the air bypass circuit. Heater pipes and
breather pipes are also connected to the throttle
body.

The stepper motor is a four-terminal, two-
coil, permanent magnet motor. It is controlled
by the ECU to regulate idle speed and fast idle
speed during the warm-up period. The valve is
located in an airway, which bypasses the throttle
valve. A cut-away section can be seen in Figure
9.50. The rotary action of the stepper motor acts
on a screw thread. This causes the cone section
at the head of the valve to move linearly, progressively opening or closing an aperture. An
idle mixture screw is also incorporated in the
throttle body which allows a small amount of air
to bypass the air flow sensor.

The coolant sensor is a simple thermistor and
provides information on engine temperature.
The fuel temperature sensor is a switch on earlier
vehicles, and a thermistor on later models. The

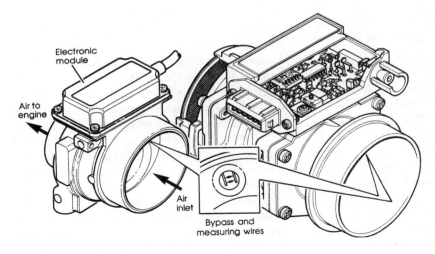

Electronic
module

Air to
engine

Air
inlet

Bypass and
measuring wires

Figure 9.49 Hot-wire air flow
meters

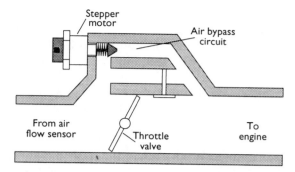

Figure 9.50 Idle control on the hot-wire system is by stepper motor

information provided allows the ECU to determine when hot start enrichment is required. This is to counteract the effects of fuel evaporation.

The heart of the system is the electronic control unit. It contains a map of the ideal fuel settings for 16 engine speeds and eight engine loads. The figure from the memory map is the basic injector pulse width. Corrections are then added for a number of factors, the most important being engine temperature and throttle position. Corrections are also added for some or all of the following when appropriate.

Voltage correction

Pulse length is increased if battery voltage falls, this is to compensate for the slower reaction time of the injectors.

Cranking enrichment

The injectors are fired every ignition pulse instead of every other pulse for cranking enrichment.

After-start enrichment

This is to ensure smooth running after starting. This is provided at all engine temperatures, and it decays over a set time. It is, however, kept up for a longer period at lower temperatures. The ECU increases the pulse length to achieve this enrichment.

Hot-start enrichment

A short period of extra enrichment, which decays gradually, is used to assist with hot starting.

Acceleration enrichment

When the ECU detects a rising voltage from the throttle sensor the pulse length is increased to achieve a smoother response. The extra fuel is needed as the rapid throttle opening causes a sudden inrush of air and, without extra fuel, a weak mixture would cause a flat spot.

Deceleration weakening

The ECU detects this condition from a falling throttle potentiometer voltage. The pulse length is shortened to reduce fuel consumption and exhaust emissions.

Full load enrichment

This is again an increase in pulse length but by a fixed percentage of the look-up and corrected value.

Overrun fuel cut-off

This is an economy and emissions measure. The injectors do not operate at all during this condition. This situation will only occur with a warm engine, throttle in the closed position and the engine speed above a set level. If the throttle is pressed or the engine falls below the threshold speed the fuel is reinstated gradually to ensure smooth take up.

Overspeed fuel cut-off

To prevent the engine from being damaged by excess speed, the ECU can switch off the injectors above a set speed. The injectors are reinstated once engine speed falls below the threshold figure.

Hot-wire fuel injection is a very adaptable system and will remain current in various forms for some time. By way of summary, Figure 9.51 is a typical circuit diagram of the hot wire system.

9.6.3 Bosch Mono Jetronic – single point injection

The Mono Jetronic is an electronically controlled system utilizing just one injector positioned above the throttle butterfly valve. The throttle body assembly is similar in appearance to a carburettor. A low pressure (1 bar) fuel supply pump, as shown in Figure 9.52, is used to supply the injector, which injects the fuel intermittently into the inlet manifold. In common with most systems, sensors measuring engine variables supply the operating data. The ECU computes the ideal fuel requirements and outputs to the injector. The width of the injector pulses determines the quantity of fuel introduced.

The injector for the system is a very fast-acting valve. Figure 9.53 shows the injector in section. A pintle on the needle valve is used and a conical spray pattern is produced. This ensures

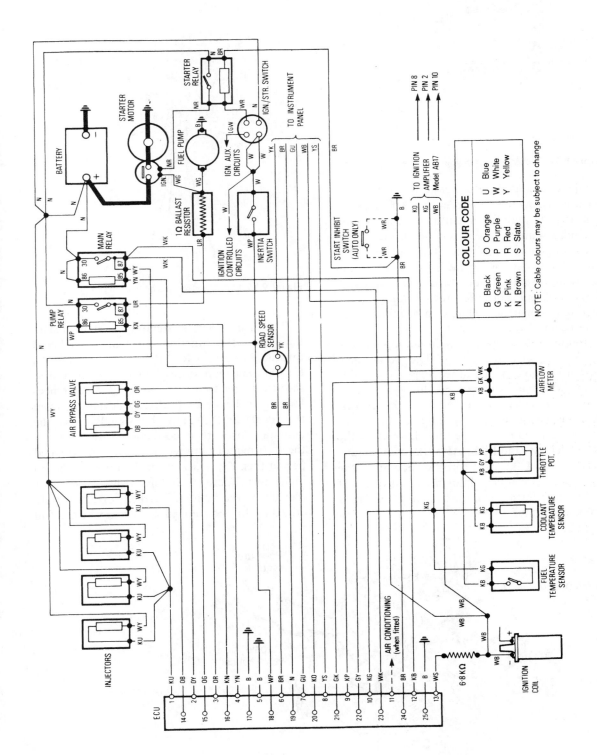

Figure 9.51 Circuit diagram of a hot-wire system

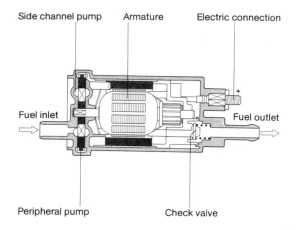

Figure 9.52 Electric fuel pump (low pressure)

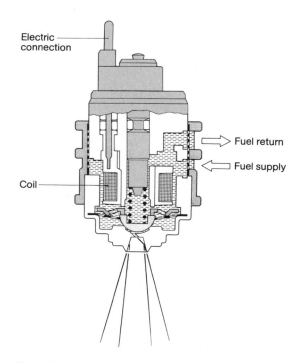

Figure 9.53 Low pressure injector (mini-injector)

excellent fuel atomization and hence a better 'burn' in the cylinder. In order to ensure accurate metering of small fuel quantities the valve needle and armature have a very small mass. This permits opening and closing times of less than 1 ms. The fuel supply to the injector is continuous, this prevents air locks and a constant supply of cool fuel. This also provides for good hot starting performance, which can be inhibited by evaporation if the fuel is hot.

Figure 9.54 shows the main components of the Mono Jetronic system. The component most noticeable by its absence is an air flow sensor which is not used by this system. Air mass and load are calculated from the throttle position sensor, engine speed and air intake temperature. This is sometimes known as the speed density method. At a known engine speed with a known throttle opening, the engine will 'consume' a known volume of air. If the air temperature is known then the air mass can be calculated.

The basic injection quantity is generated in the ECU as a function of engine speed and throttle position. A ROM chip, represented by a cartographic map, stores data at 16 speed and 16 throttle angle positions, giving 256 references altogether. If the ECU detects deviations from the ideal air–fuel ratio by signals from the lambda sensor then corrections are made. If these corrections are required over an extended period then the new corrected values are stored in memory. These are continuously updated over the life of the system. Further corrections are added to this look-up value for temperature, full

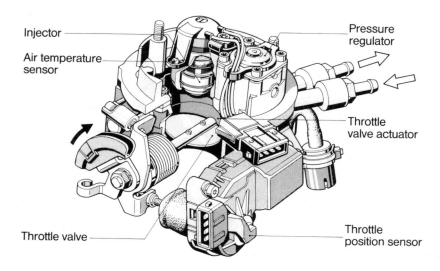

Figure 9.54 Central injection unit of the Mono Jetronic

load and idle conditions. Overrun fuel cut-off and high engine speed cut-off are also implemented when required.

The Bosch Mono Jetronic system also offers adaptive idle control. This is to allow the lowest possible smoothed idle speed to reduce fuel consumption and exhaust emissions. A throttle valve actuator changes the position of the valve in response to a set speed calculated in the ECU, which takes into account the engine temperature and electrical loads on the alternator. The required throttle angle is computed and placed in memory. The adaptation capability of this system allows for engine drift during its life and also makes corrections for altitude.

The electronic control unit checks all signals for plausibility during normal operation. If a signal deviates from the normal, this fault condition is memorized and can be output to a diagnostic tester or read as a blink code from a fault lamp.

9.6.4 Toyota Computer Controlled System (TCCS)

The EFi system as shown in Figure 9.55 is composed, as are most such systems, of three basic sub-systems:

- Fuel.
- Air.
- Electronic control.

Fuel is supplied under constant pressure to the injectors by an electric fuel pump. The injectors inject a metered quantity of fuel into the intake manifold under the control of the ECU.

The air induction system is via an air filter and provides sufficient air under all operating conditions.

The central operation of injection is by microcomputer control. The TCCS controls the injectors in response to signals relating to:

- Intake air volume.
- Intake air temperature.
- Coolant temperature.
- Engine speed.
- Acceleration/deceleration.
- Exhaust oxygen content.

The ECU detects any malfunctions and stores them in the memory. The codes can be read as flashes of the check engine warning light. In the event of serious malfunction, a back-up circuit takes over to provide minimal drivability.

9.7 Diagnosing fuel control system faults

9.7.1 Introduction

As with all systems, the six stages of fault-finding should be followed.

1. Verify the fault.
2. Collect further information.
3. Evaluate the evidence.
4. Carry out further tests in a logical sequence.
5. Rectify the problem.
6. Check all systems.

The procedure outlined in the next section is related primarily to stage 4 of the process. Table 9.5 lists some common symptoms of a fuel system malfunction together with suggestions for

Table 9.5 Common symptoms of a fuel system malfunction and possible faults

Symptom	Possible fault
Engine rotates but does not start	• No fuel in the tank! • Air filter dirty or blocked. • Fuel pump not running. • No fuel being injected.
Difficult to start when cold	• Air filter dirty or blocked. • Fuel system wiring fault. • Enrichment device not working (choke or injection circuit).
Difficult to start when hot	• Air filter dirty or blocked. • Fuel system wiring fault.
Engine starts but then stops immediately	• Fuel system contamination. • Fuel pump or circuit fault (relay). • Intake system air leak.
Erratic idle	• Air filter blocked. • Inlet system air leak. • Incorrect CO setting. • Fuel injectors not spraying correctly.
Misfire through all speeds	• Fuel filter blocked. • Fuel pump delivery low. • Fuel tank ventilation system blocked.
Engine stalls	• Idle speed incorrect. • CO setting incorrect. • Fuel filter blocked. • Air filter blocked. • Intake air leak. • Idle control system not working.
Lack of power	• Fuel filter blocked. • Air filter blocked. • Low fuel pump delivery. • Fuel injectors blocked.
Backfires	• Fuel system fault (air flow sensor on some cars).

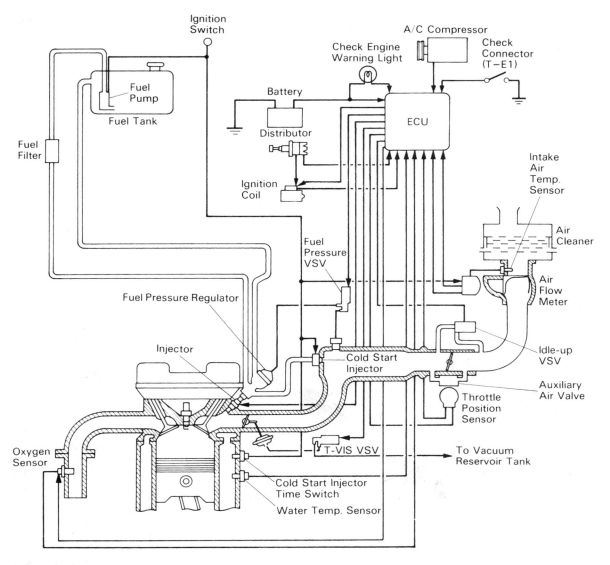

Figure 9.55 Toyota computer control system (TCCS)

the possible fault. Note that when diagnosing engine fuel system faults, the same symptoms may indicate an ignition problem.

9.7.2 Testing procedure

Caution/Achtung/Attention – Burning fuel can seriously damage your health!

The following procedure is generic and, with a little adaptation, can be applied to any fuel injection system. Refer to manufacturer's recommendations if in any doubt. It is assumed the ignition system is operating correctly. Most tests are carried out while cranking the engine.

1. Check battery state of charge (at least 70%).
2. Hand and eye checks (all fuel and electrical connections secure and clean).
3. Check fuel pressure supplied to rail (in multi-point systems it will be about 2.5 bar but check specifications).
4. If the pressure is *not* correct jump to stage 10.
5. Is injector operation OK? – continue if not (suitable spray pattern or dwell reading across injector supply).
6. Check supply circuits from main relay (battery volts minimum).
7. Continuity of injector wiring (0–0.2 Ω and note that many injectors are connected in parallel).
8. Sensor readings and continuity of wiring (0–0.2 Ω for the wiring sensors will vary with type).
9. If no fuel is being injected and all tests so far are OK, suspect ECU.

10. Fuel supply – from stage 4.
11. Supply voltage to pump (within 0.5 V battery – pump fault if supply is OK).
12. Check pump relay and circuit (note in most cases the ECU closes the relay but this may be bypassed on cranking).
13. Ensure all connections (electrical and fuel) are remade correctly.

9.8 Advanced fuel control technology

9.8.1 Air–fuel ratio calculations

The ideal ratio by mass of air to fuel for complete combustion is 14.7:1. This is given the lambda value 1, which is known as stoichiometry. This figure can be calculated by working out the exact number of oxygen atoms, that are required to oxidize completely the particular number of hydrogen and carbon atoms in the hydrocarbon fuel, then multiplying by the atomic mass of the respective elements.

Petrol consists of a number of ingredients, these are known as fractions and fall into three chemical series.

- Paraffins   e.g. octane C_8H_{18}
- Napthenes   e.g. cyclohexane C_6H_{12}
- Aromatics   e.g. benzene C_6H_6

The ideal air–fuel ratio for each of these can be calculated from the balanced chemical equation and the atomic mass of each atom. The atomic masses of interest are:

- Carbon (C) = 12
- Hydrogen (H) = 1
- Oxygen (O) = 16

The balanced chemical equation for complete combustion of octane is as follows:

$$2C_8H_{18} + 25O_2 \rightarrow 16CO_2 + 18H_2O$$

The molecular mass of $2C_8H_{18}$ is:

$$(2 \times 12 \times 8) + (2 \times 1 \times 18) = 228$$

The molecular mass of $25 \times O_2$ is:

$$(25 \times 16 \times 2) = 800$$

Therefore the oxygen to octane ratio is 800:228 or 3.5:1; in other words 1 kg of fuel uses 3.5 kg of oxygen. Air contains 23% of oxygen by mass (21% by volume), which means 1 kg of air contains 0.23 kg of oxygen. Further, there is 1 kg of oxygen in 4.35 kg of air.

The ideal air–fuel (A/F) ratio for complete combustion of octane is $3.5 \times 4.35 = 15.2:1$.

Octane: $2C_8H_{18} + 25O_2 \rightarrow 16CO_2 + 18H_2O$
A/F ratio = 15.2:1

If a similar calculation is carried out for cyclohexane and benzene, the results are as follows.

Cyclohexane: $C_6H_{12} + 9O_2 \rightarrow 6CO_2 + 6H_2O$
A/F ratio = 14.7:1

Benzene: $C_6H_6 + 15O_2 \rightarrow 6CO_2 + 3H_2O$
A/F ratio = 13.2:1

The above examples serve to explain how the air–fuel ratio is calculated and how petrol/gasoline, being a mixture of a number of fractions, has an ideal air–fuel ratio of 14.7:1.

This figure is, however, only the theoretical ideal and takes no account of pollutants produced and the effect the air–fuel ratio has on engine performance. With modern engine fuel control systems it is possible to set the air–fuel ratio exactly at this stoichiometric ratio if desired. As usual though, a compromise must be sought as to the ideal setting. Figure 9.9 shows a graph comparing engine power output and fuel consumption, with changes in air–fuel ratio.

Figure 9.56 shows the influence of air–fuel ratio on the three main pollutants created from a spark ignition, internal combustion engine. A ratio slightly weaker than the lambda value of 1 (or about 15.5:1 ratio) is often an appropriate compromise.

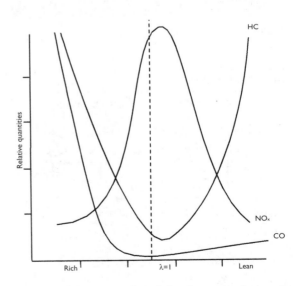

Figure 9.56 Influence of air–fuel ratio on the three main pollutants created from a spark ignition engine (no catalyst in use)

9.9 New developments

9.9.1 Mazda lean burn technology

The optimum air–fuel ratio is 14.7 : 1 to ensure complete combustion. Increasing this ratio (introducing more air) results in what is known as lean burn. Fuel economy is maximized when the ratio is in the 20 to 22 : 1 range. Running leaner mixtures also reduces NOx emissions. However, the potential for unstable combustion increases. Reducing NOx emissions under lean burn conditions is difficult because the normal catalytic converter needs certain conditions to work properly. Mazda have produced a 'Z-lean engine' that offers both a wide lean burn range and good power output at normal rev/min. Figure 9.57 shows a cutaway view of this engine.

Introducing more air into the cylinder necessarily results in a lower fuel density in the mixture and thus a lower combustion temperature. This in turn means that less heat energy is lost from the combustion chamber to the surrounding parts of the engine. In addition to reduced heat loss, pumping loss is also lower because one can open the throttle wider when adjusting air input. These two effects contribute to the higher fuel economy of lean burn engines. Figure 9.58 shows these features.

The tumble swirl control (TSC) valve and its effects are shown as Figure 9.59. The Z-lean engine uses a feature known as a TSX (tumble swirl multiplex) port to control the vortex inside the cylinder. Combining this with an air mixture type injector which turns the fuel into a very fine spray and a high-energy ignition system ensures that it can operate on very lean mixtures up to 25 : 1. A special catalytic converter combines the NOx and HC into H_2O, CO_2 and N.

9.9.2 In-cylinder catalysts

A novel approach to reducing hydrocarbon emissions has been proposed and investigated by a team from Brunel University (SAE paper 952419). The unburned hydrocarbons in spark ignition engines arise primarily from sources near the combustion chamber walls.

A platinum-rhodium coating was deposited on the top and side surfaces of the piston crown and its effects were examined under a variety of operating conditions.

The results were as follows:

- HC emissions were reduced by about 20%.
- NOx emissions did not change appreciably.

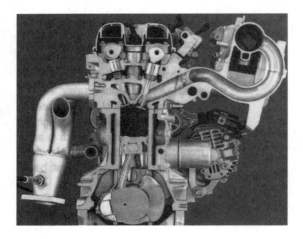

Figure 9.57 Cylinder-head and inlet path of a lean burn engine

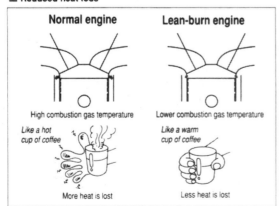

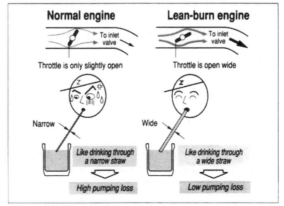

Figure 9.58 Features of the lean burn system

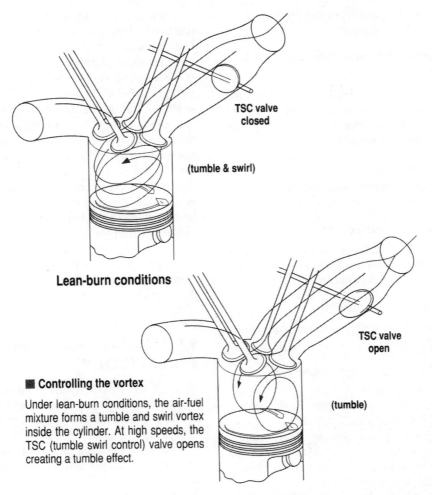

TSC valve closed

(tumble & swirl)

Lean-burn conditions

■ **Controlling the vortex**

Under lean-burn conditions, the air-fuel mixture forms a tumble and swirl vortex inside the cylinder. At high speeds, the TSC (tumble swirl control) valve opens creating a tumble effect.

TSC valve open

(tumble)

High-speed conditions   **Figure 9.59** Tumble swirl control

The catalyst caused a slightly faster initial flame development but no evident effect on the burning rate.

9.9.3 Electronic unit injection (EUI) – diesel fuel

The advantages of electronic unit injection are as follows.

Lower emissions

Through the use of higher injection pressures (up to 2000 bar), lower emissions of particulates and NOx are achieved, together with a reduction in the levels of noise traditionally associated with diesel engines.

Electronic fuel quantity and timing control

Precise electronic control also assists in the reduction of emissions.

Shot to shot fuel adjustment

This feature also provides a very quick transient response, improving vehicle drivability.

Control of all engine functions

Through a series of sensors connected to the electronic control unit (ECU), the EUI system ensures that all the engine functions consistently operate at optimum performance.

Electronically controlled pilot injection

A new feature developed to meet tighter NOx emissions standards, without loss of fuel consumption. Pilot injection also reduces combustion noise.

Communication with other vehicle systems

Linked to the ECU, the EUI system can communicate with other vehicle systems such as ABS, transmission and steering, making further systems development possible.

Cylinder cut-out

This is used as a diagnostic aid and offers potential for fuel economy at idling and low loads.

Reliability and durability

The EUI's reliability is proven under field conditions. Experience in the truck market indicates a service life of at least 800 000 km.

Full diagnostics capability

Fault codes can be stored and diagnostic equipment connected.

Further development potential

EUI technology is currently only at the beginning of its life cycle; it has significant further development potential which will enable the system to meet future tough emissions legislation.

In the EUI system, the fuel injection pump, the injector and a solenoid valve are combined in one, single unit; these unit injectors are located in the cylinder-head, above the combustion chamber. The EUI is driven by a rocker arm, which is in turn driven by the engine camshaft. This is the most efficient hydraulic and mechanical layout, giving the lowest parasitic losses. The fuel feed and spill pass through passages integrated in the cylinder-head.

The EUI uses sensors and an electronic control unit (ECU) to achieve precise injection timing and fuel quantities. Sensors located on the engine pass information to the ECU on all the relevant engine functions. This evaluates the information and compares it with optimum values stored in the ECU to decide on the exact injection timing and fuel quantity required to realize optimum performance. Signals are then sent to the unit injector's solenoid-actuated spill valve system to deliver fuel at the timing required to achieve this performance.

Injection is actuated by switching the integrated solenoid valve. The closing point of the valve marks the beginning of fuel delivery, and the duration of closing determines the fuel quantity. The operating principle is as follows.

Each plunger moves through a fixed stroke, actuated by the engine camshaft. On the upward (filling) stroke, fuel passes from the cylinder-head through a series of integrated passages and the open spill valve into a chamber below the plunger. The ECU then sends a signal to the solenoid stator, which results in the closure of the spill control valve. The plunger continues its downward stroke causing pressure to build in the high pressure passages. At a pre-set pressure the nozzle opens and fuel injection begins. When the solenoid stator is de-energized the spill control valve opens, causing the pressure to collapse, which allows the nozzle to close, resulting in a very rapid termination of injection.

Lucas electronic unit injectors (Figure 9.60) have been developed in a range of sizes to suit all engines, and can be fitted to light- and heavy-duty engines suitable for small cars and the largest premium trucks.

9.9.4 Lucas diesel common rail system (LDCR)

To meet the future stringent emissions requirements, and offering further improvements in fuel economy, the common rail fuel injection system is becoming popular.

Fuel injection equipment with the capability of operating at very high pressures is required to achieve the ultra low emissions and low noise demands of the future. The advantages of a system developed by Lucas are summarized below.

Compact design

The compact design of the injector outline enables the LDCR system to be used on 2 or 4 valves per cylinder engines.

Figure 9.60 Unit injectors (Lucas)

Modular system

With one electronically driven injector per engine cylinder, the system is modular and can be used on 3, 4, 5 and 6 cylinder engines.

Low drive torque

As the pumping of the pressure rail is not phased with the injection, the common rail system requires a low drive torque from the engine.

Independent injection pressure

The injection pressure is independent of the engine speed and load, so enabling high injection pressures at low speed if required.

Lower NOx emissions

Injection sequences that include periods both pre and post the main injection can be utilized to reduce emissions, particularly NOx, enabling the system to meet the stringent emissions levels required by EURO-III and US–98 legislation and beyond.

Noise reduction and NOx control

The inclusion of pilot injection results in a significant reduction in engine noise.

Full electronic control

The common rail offers all the benefits of full electronic control for vehicles, including extremely accurate fuel metering and timing, as well as the option to interface with other vehicle functions.

The common rail can be easily adapted for different engines. The main components are as follows.

- Common pressure accumulator (the 'Rail').
- High pressure regulator.
- High pressure supply pump.
- Injectors.
- Electronic solenoids.
- Electronic Control Unit.
- Filter unit.

Figure 9.61 shows the layout of a common rail injection system. The system consists of a common pressure accumulator, called the 'rail', which is mounted along the engine block, and fed by a high pressure pump. The pressure level of the rail is electronically regulated by a combination of metering on the supply pump and fuel discharge by a high-pressure regulator. The pressure accumulator operates independently of

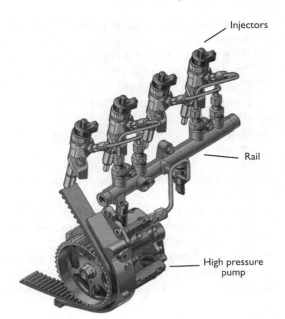

Figure 9.61 Diesel common rail injection

engine speed or load, so that high injection pressure can be produced at low speeds if required. A series of injectors is connected to the rail, and each injector is opened and closed by a solenoid, driven by the Electronic Control Unit.

A feed pump delivers the fuel through a filter unit to the high-pressure pump. The high-pressure pump delivers fuel to the high pressure rail. The injectors inject fuel into the combustion chamber when the solenoid valve is actuated.

Because the injection pressure is independent of engine speed and load, the actual start of injection, the injection pressure, and the duration of injection can be freely chosen from a wide range. The introduction of pilot injection, which is adjusted depending on engine needs, results in significant engine noise reduction, together with a reduction in NOx emissions. The actuator controls the pressure in the system.

The Lucas system has been designed for use on future HSDI engines for passenger cars, which will be required to meet the EURO-III and US–98 emissions legislation and beyond.

9.10 Self-assessment

9.10.1 Questions

1. Explain what is meant by a lambda (λ) value of 1.
2. State five advantages of fuel injection.

3. With reference to the combustion process, describe the effects of ignition timing.
4. With reference to the combustion process, describe the effects of mixture strength.
5. Draw a block diagram of a fuel injection system. Describe briefly the purpose of each component.
6. Explain the combustion process in a diesel engine.
7. Describe how electronic control of diesel fuel injection is achieved and state the advantages of EUI.
8. List all the main components of an electronic carburation control system and state the purpose of each component.
9. Make a clearly labelled sketch to show the operation of a fuel injector.
10. State six sources of emissions from a vehicle and describe briefly how manufacturers are tackling each of them

9.10.2 Assignment

Draw an 8×8 look-up table (grid) for a digital fuel control system. The horizontal axis should represent engine speed from zero to 5000 rev/min, and the vertical axis engine load from zero to 100%. Fill in all the boxes with realistic figures and explain why you have chosen these figures. You should explain the trends and not each individual figure.

Download the 'Automotive Technology – Electronics' simulation program from my web site and see if your figures agree with those in the program. Discuss reasons why they may differ.

10
Engine management

10.1 Combined ignition and fuel management

10.1.1 Introduction

As the requirements for lower and lower emissions continue, together with the need for better performance, other areas of engine control are constantly being investigated. This control is becoming even more important as the possibility of carbon dioxide emissions being included in future regulations increases. Some of the current and potential areas for further control of engine operation are included in this section. Although some of the common areas of 'control' have been covered in the previous two chapters, this chapter will cover some aspects in more detail and introduce further areas of engine control. Some of the main issues are:

● Ignition timing.
● Dwell angle.
● Fuel quantity.
● EGR (exhaust gas recirculation).
● Canister purge.
● Idle speed.

An engine management system con be represented by the standard three-stage model as shown in Figure 10.1. This representation shows closed loop feedback, which is a common feature, particularly related to:

● lambda control,
● knock,
● idle speed.

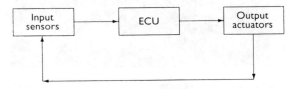

Figure 10.1 Representation of complete engine control as the standard functional system

The block diagram shown as Figure 10.2 can further represent an engine management system. This series of 'inputs' and 'outputs' is a good way of representing a complex system. This section continues with a look at some of the less common 'inputs and outputs'.

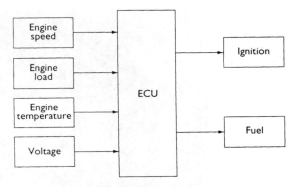

Figure 10.2 General block diagram of an ignition and fuel control system

10.1.2 Variable inlet tract

For an engine to operate at its best, volumetric efficiency is not possible with fixed manifolds. This is because the length of the inlet tract determines the velocity of the intake air and, in particular, the propagation of the pressure waves set up by the pumping action of the cylinders. These standing waves can be used to improve the ram effect of the charge as it enters the cylinder but only if they coincide with the opening of the inlet valves. The length of the inlet tract has an effect on the frequency of these waves. One method of changing the length of the inlet tract is shown in Figure 10.3. The control valves move, which changes the effective length of the inlet.

Figure 10.4 shows how the design of the inlet manifold is a significant feature of the Volvo S80 engine.

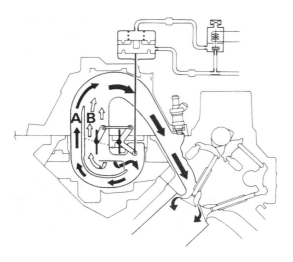

Figure 10.3 Variable length inlet manifold. A = long tract; B = short tract

10.1.3 Variable valve timing

With the widespread use of twin cam engines, one cam for the inlet valves and one for the exhaust valves, it is possible to vary the valve overlap while the engine is running. Honda has a system that noticeably improves the power and torque range by only opening both of the inlet valves at higher speed. This system is shown as Figure 10.5.

A system of valves using oil pressure to turn the cam with respect to its drive gear controls the cam positions on the BMW system shown in Figure 10.6. The position of the cams is determined from a suitable map held in ROM in the control unit.

A system that not only allows changes in valve timing but also valve open periods is also

Figure 10.4 Volvo engine showing the feature of the inlet manifold design

Figure 10.5 Honda's valve control system. At low revs the VTEC-E engine opens only one inlet valve per cylinder fully, so just 12 valves control the mixture and combustion of air and fuel. This delivers maximum efficiency with the lowest possible emissions. At higher engine speeds, hydraulic pins activate the extra valves to give 16 valve performance

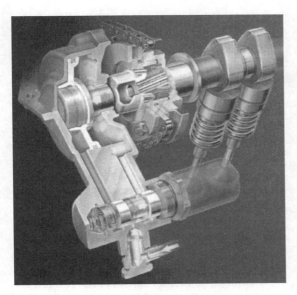

Figure 10.6 Variable valve timing from BMW

starting to be used. The system is known as active valve train (AVT) and was intended to be a development tool for the design of fixed camshafts. However, production versions are being developed. The opening of the inlet and exhaust valves will be by hydraulic actuators working at up to 200 bar with a high-speed servo valve controlling flow to the actuators.

10.1.4 Combustion flame and pressure sensing

Research is ongoing in the development of cost effective sensors for determining combustion pressure and combustion flame quality. These sensors are used during development but currently are prohibitivly expensive for use in production. When available, these sensors will provide instantaneous closed loop feedback about the combustion process. This will be particularly important with lean burning engines.

10.1.5 Wide range lambda sensors

Most lambda sensors provide excellent closed-control of the air–fuel ratio at or very near to stoichiometry (14.7:1). A sensor is now available that is able to provide a linear output between air–fuel ratios of 12:1 and about 24:1. This allows closed loop feedback over a much wider range of operating conditions.

10.1.6 Injectors with air shrouding

If high-speed air is introduced at the tip of an injector, the dispersal of the fuel is considerably improved. Droplet size can be reduced to below 50 μm during idle conditions. Figure 10.7 shows an injector with air shrouding.

Figure 10.8 shows the effect of this air shrouding as two photographs, one with the feature and one without. The improved dispersal and droplet size is clear.

10.1.7 On-board diagnostics (OBD)

Figure 10.9 shows the Bosch Motronic M5 with the OBD 2 system. On-board diagnostics are

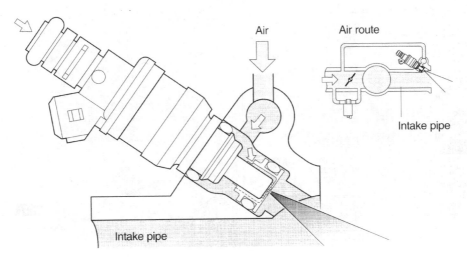

Figure 10.7 Injection valve with air shrouding

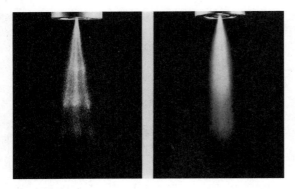

Figure 10.8 Better fuel preparation through injection with air shrouding. Left: injection valve without air shrouding. Right: injection valve with air shrouding

becoming essential for the longer term operation of a system in order for it to produce a clean exhaust. Many countries now require a very comprehensive diagnosis of all components which affect the exhaust. Any fault detected will be indicated to the driver by a warning light. The OBD 2 system is intended to standardize the many varying methods used by different manufacturers. It is also thought that an extension to total vehicle diagnostics through a common interface is possible in the near future.

Digital electronics allow both sensors and actuators to be monitored. Allocating values to all operating states of the sensors and actuators achieves this. If a deviation from these figures is detected, it is stored in memory and can be output in the workshop to assist with fault-finding.

Monitoring of the ignition system is very important as misfiring not only produces more emissions of hydrocarbons, but the unburned fuel can enter the catalytic converter and burn there. This can cause higher than normal temperatures and may damage the catalytic converter.

An accurate crankshaft speed sensor is used to monitor ignition and combustion in the cylinders. Misfiring alters the torque of the crankshaft for an instant, which causes irregular rotation. This can be monitored, thus allowing a misfire to be recognized instantly.

A number of further sensors are required for the functions of the OBD 2 system. Another lambda sensor, placed after the catalytic converter, monitors the operation of the OBD 2. An intake pressure sensor and a valve are needed to control the activated charcoal filter to reduce and monitor evaporative emissions from the fuel tank. A differential pressure sensor also monitors the fuel tank permeability. As well as the driver's fault lamp a considerable increase in the electronics is required in the control unit in order to operate an OBD system. A better integral-monitoring system will have a superior effect in

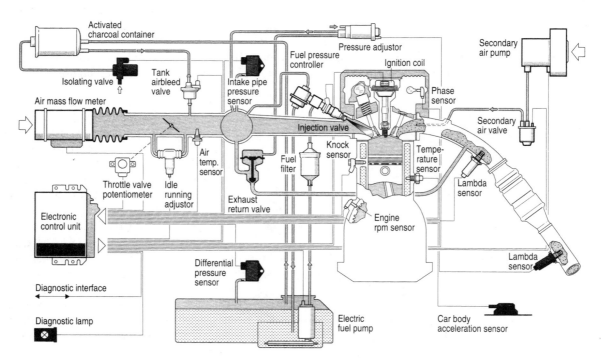

Figure 10.9 Motronic M5 with OBD 2

reducing vehicle emissions than tighter MOT regulations.

The diagnostic socket used by systems conforming to OBD 2 standards should have the following pin configuration.

1. Manufacturer's discretion.
2. Bus + Line, SAE J1850.
3. Manufacturer's discretion.
4. Chassis ground.
5. Signal ground.
6. Manufacturer's discretion.
7. K Line, ISO 9141.
8. Manufacturer's discretion.
9. Manufacturer's discretion.
10. Bus – Line, SAE J1850.
11. Manufacturer's discretion.
12. Manufacturer's discretion.
13. Manufacturer's discretion.
14. Manufacturer's discretion.
15. L line, ISO 9141.
16. Vehicle battery positive.

It is hoped that with future standards and goals set it will be beneficial for vehicle manufacturers to begin implementation of at least the common connector in the near term. Many diagnostic system manufacturers would welcome this move.

If the current lack of standardization continues, it will become counter-productive for all concerned.

10.2 Exhaust emission control

10.2.1 Engine design

Many design details of an engine have a marked effect on the production of pollutant emissions. With this in mind, it will be clear that the final design of an engine is a compromise between conflicting interests. The major areas of interest are as discussed in the following sections.

10.2.2 Combustion chamber design

The main source of hydrocarbon emissions is unburnt fuel that is in contact with the combustion chamber walls. For this reason the surface area of the walls should be kept as small as possible and with the least complicated shape. A theoretical ideal is a sphere but this is far from

practical. Good swirl of the cylinder charge is important, as this facilitates better and more rapid burning. Perhaps more important is to ensure a good swirl in the area of the spark plug. This ensures a mixture quality that is easier to ignite. The spark plug is best positioned in the centre of the combustion chamber as this reduces the likelihood of combustion knock by reducing the distance the flame front has to travel.

10.2.3 Compression ratio

The higher the compression ratio, the higher, in general, the thermal efficiency of the engine and therefore the better the performance and fuel consumption. The two main drawbacks to higher compression ratios are the increased emissions and the increased tendency to knock. The problem with emissions is due to the high temperature, which in turn causes greater production of NOx. The increase in temperature makes the fuel and air mixture more likely to self-ignite, causing a higher risk of combustion knock. Countries which have had stringent emission regulations for some time, such as the USA and Japan, have tended to develop lower compression engines. However, with the changes in combustion chamber design and the more widespread introduction of four valves per cylinder, together with greater electronic control and other methods of dealing with emissions, compression ratios have increased over the years.

10.2.4 Valve timing

The effect of valve timing on exhaust emissions can be quite considerable. One of the main factors is the amount of valve overlap. This is the time during which the inlet valve has opened but the exhaust valve has not yet closed. The duration of this phase determines the amount of exhaust gas left in the cylinder when the exhaust valve finally closes. This has a significant effect on the reaction temperature (the more exhaust gas the lower the temperature), and hence has an effect on the emissions of NOx. The main conflict is that, at higher speeds, a longer inlet open period increases the power developed. The downside is that this causes a greater valve overlap and, at idle, this can greatly increase emissions of hydrocarbons. This has led to the successful introduction of electronically controlled valve timing.

10.2.5 Manifold designs

Gas flow within the inlet and exhaust manifolds is a very complex subject. The main cause of this complexity is the transient changes in flow that are due not only to changes in engine speed but also to the pumping action of the cylinders. This pumping action causes pressure fluctuations in the manifolds. If the manifolds and both induction and exhaust systems are designed to reflect the pressure wave back at just the right time, great improvements in volumetric efficiency can be attained. Many vehicles are now fitted with adjustable length induction tracts. Longer tracts are used at lower engine speeds and shorter tracts at higher speed.

10.2.6 Charge stratification

If the charge mixture can be inducted into the cylinder in such a way that a richer mixture is in the proximity of the spark plug, then overall the cylinder charge can be much weaker. This can bring great advantages in fuel consumption, but the production of NOx can still be a problem. The later section on direct mixture injection development is a good example of the use of this technique. Many lean-burn engines use a form of stratification to reduce the chances of misfire and rough running.

10.2.7 Warm up time

A significant quantity of emissions produced by an average vehicle is created during the warm-up phase. Suitable materials and care in the design of the cooling system can reduce this problem. Some engine management systems even run the ignition timing slightly retarded during the warm-up phase to heat the engine more quickly.

10.2.8 Exhaust gas recirculation

This technique is used primarily to reduce peak combustion temperatures and hence the production of nitrogen oxides (NOx). Exhaust gas recirculation (EGR) can be either internal as mentioned above, due to valve overlap, or external via a simple arrangement of pipes and a valve (Figure 10.10). A proportion of exhaust gas is simply returned to the inlet side of the engine.

This EGR is controlled electronically as determined by a ROM in the ECU. This ensures that drivability is not affected and also that the rate of EGR is controlled. If the rate is too high,

then the production of hydrocarbons increases. Figure 10.11 shows the effect of various rates of EGR.

One drawback of EGR systems is that they can become restricted by exhaust residue over a period of time, thus changing the actual percentage of recirculation. However, valves are now available that reduce this particular problem.

10.2.9 Ignition system

The ignition system can affect exhaust emissions in two ways; first, by the quality of the spark produced, and secondly, the timing of the spark. The quality of a spark will determine its ability to ignite the mixture. The duration of the spark in particular is significant when igniting weaker mixtures. The stronger the spark the less the likelihood of a misfire, which can cause massive increases in the production of hydrocarbons.

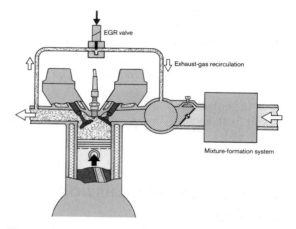

Figure 10.10 Exhaust-gas recirculation system

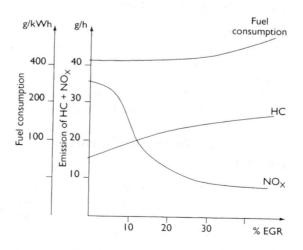

Figure 10.11 Effect of various rates of EGR

The timing of a spark is clearly critical but, as ever, is a compromise with power, drivability, consumption and emissions. Figure 10.12 is a graph showing the influence of ignition timing on emissions and fuel consumption. The production of carbon monoxide is dependent almost only on fuel mixture and is not significantly affected by changes in ignition timing. Electronic and programmed ignition systems have made significant improvements to the emission levels of today's engines.

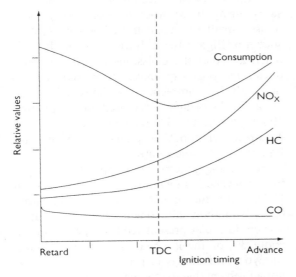

Figure 10.12 Influence of ignition timing on emissions and fuel consumption

10.2.10 Thermal after-burning

Prior to the more widespread use of catalytic converters, thermal after-burning was used to reduce the production of hydrocarbons. In fact, hydrocarbons do continue to burn in the exhaust manifold and recent research has shown that the type of manifold used, such as cast iron or pressed steel, can have a noticeable effect on the reduction of HC. At temperatures of about 600°C, HC and CO are burnt or oxidized into H_2O and CO_2. If air is injected into the exhaust manifold just after the valves, then the after-burning process can be encouraged.

10.2.11 Catalytic converters

Stringent regulations in most parts of the world have made the use of a catalytic converter almost indispensable. The three-way catalyst (TWC) is used to great effect by most manufacturers. It is a very simple device and looks similar to a standard exhaust box. Note that, in order to operate correctly, however, the engine must be run at – or very near to – stoichiometry. This is to ensure that the right 'ingredients' are available for the catalyst to perform its function.

Figure 10.13 shows a view of the inside of a catalytic converter. There are many types of hydrocarbons but the following example illustrates the main reaction. Note that the reactions rely on some CO being produced by the engine

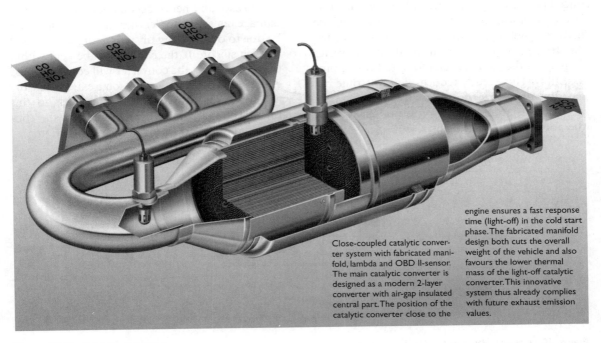

Close-coupled catalytic converter system with fabricated manifold, lambda and OBD II-sensor. The main catalytic converter is designed as a modern 2-layer converter with air-gap insulated central part. The position of the catalytic converter close to the engine ensures a fast response time (light-off) in the cold start phase. The fabricated manifold design both cuts the overall weight of the vehicle and also favours the lower thermal mass of the light-off catalytic converter. This innovative system thus already complies with future exhaust emission values.

Figure 10.13 Catalytic converter

in order to reduce the NOx. This is one of the reasons that manufacturers have been forced to run engines at stoichiometry. This legislation has tended to stifle the development of lean burn techniques. The fine details of the emission regulations can in fact, have a very marked effect on the type of reduction techniques used. The main reactions in the 'cat' are as follows:

- $2CO + O_2 \rightarrow 2CO_2$
- $2C_2H_6 + 2CO \rightarrow 4CO_2 + 6H_2O$
- $2NO + 2CO \rightarrow N_2 + 2CO_2$

The ceramic monolith type of base, when used as the catalyst material, is a magnesium aluminium silicate and, due to the several thousand very small channels, provides a large surface area. This area is coated with a wash coat of aluminium oxide, which further increases its effective surface area by a factor of about seven thousand. Noble metals are used for the catalysts. Platinum promotes the oxidation of HC and CO, and rhodium helps the reduction of NOx. The converter shown is the latest metal substrate type with a built-in manifold. The whole three-way catalytic converter only contains about 3–4 g of the precious metals.

The ideal operating temperature range is from about 400 to 800°C. A serious problem to counter is the delay in the catalyst reaching this temperature. This is known as the 'catalyst light-off time'. Various methods have been used to reduce this time as significant emissions are produced before 'light-off' occurs. Electrical heating is one solution, as is a form of burner, which involves lighting fuel inside the converter. Another possibility is positioning the converter as part of the exhaust manifold and down pipe assembly. This greatly reduces light-off time but gas flow problems, vibration and excessive temperature variations can be problems that reduce the potential life of the unit.

Catalytic converters can be damaged in two ways. The first is by the use of leaded fuel, which causes lead compounds to be deposited on the active surfaces, thus reducing the effective area, and, secondly, by engine misfire, which can cause the catalytic converter to overheat due to burning inside the unit. BMW, for example, uses a system on some vehicles where a sensor monitors the output of the ignition HT system and, if the spark is not present, will not allow fuel to be injected.

A further possible technique to reduce emissions during the warm-up time of the catalyst is to use a small electrically heated pre-converter as shown in Figure 10.14. Initial tests of this system show that the emissions of hydrocarbons during the warm-up phase can be reduced significantly. The problem yet to be solved is that about 30 kW of heat is required during the first 30 s to warm up the pre-converter. This will require a current in the region of 250 A; an extra battery may be one solution.

For a catalytic converter to operate at its optimum conversion rate in order to oxidize CO and HC whilst reducing NOx, a narrow band within 0.5% of lambda value one is essential. Lambda sensors in use at present tend to operate within about 3% of the lambda mean value. When a catalytic converter is in prime condition this is not a problem due to storage capacity within the converter for CO and O_2. Damaged converters, however, cannot store a sufficient quantity of these gases and hence become less efficient. The damage, as suggested earlier in this section, can be due to overheating or 'poisoning' due to lead or even silicon. If the control can be kept within 0.5% of lambda the converter will continue to be effective even if damaged to some extent. Sensors are becoming available that can work to this tolerance. A second sensor fitted after the converter can be used to ensure ideal operation.

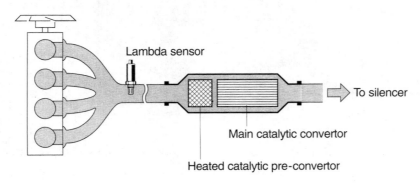

Lambda sensor

To silencer

Main catalytic convertor

Heated catalytic pre-convertor

Figure 10.14 Electrically heated catalytic pre-converter

10.2.12 Closed loop lambda control

Current regulations have almost made mandatory closed-loop control of the air–fuel mixture in conjunction with a three-way catalytic converter. It was under discussion that a lambda value of 1 should become compulsory for all operating conditions, but this was not agreed.

Lambda control is a closed loop feedback system in that the signal from a lambda sensor in the exhaust can directly affect the fuel quantity injected. The lambda sensor is described in more detail in Chapter 2. Figure 10.15 shows a block diagram of the lambda control system.

A graph to show the effect of lambda control and a three-way catalyst (TWC) is shown in Figure 10.16. The principle of operation is as follows: the lambda sensor produces a voltage that is proportional to the oxygen content of the exhaust, which is in turn proportional to the air–fuel ratio. At the ideal setting, this voltage is about 450 mV. If the voltage received by the ECU is below this value (weak mixture) the quantity of fuel injected is increased slightly. If the signal voltage is above the threshold (rich mixture) the fuel quantity is reduced. This alteration in the air–fuel ratio must not be too sudden as it could cause the engine to buck. To prevent this, the ECU contains an integrator, which changes the mixture over a period of time.

A delay also exists between the mixture formation in the manifold and the measurement of the exhaust gas oxygen. This is due to the engine's working cycle and the speed of the inlet mixture, the time for the exhaust to reach the sensor and the sensor's response time. This is sometimes known as 'dead time' and can be as much as one second at idle speed but only a few hundred milliseconds at higher engine speeds.

Due to the dead time the mixture cannot be controlled to an exact value of $\lambda = 1$. If the integrator is adjusted to allow for engine speed then it is possible to keep the mixture in the lambda window (0.97–1.03), which is the region in which the TWC is at its most efficient.

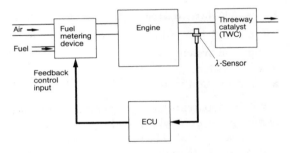

Figure 10.15 Fuel metering with closed loop control

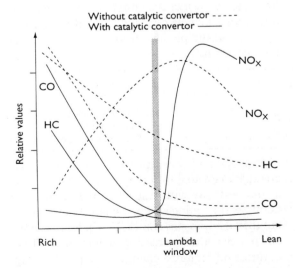

Figure 10.16 The effect of lambda control and a three-way catalyst (TWC)

10.3 Control of diesel emissions

10.3.1 Introduction

Exhaust emissions from diesel engines have been reduced considerably by changes in the design of combustion chambers and injection techniques. More accurate control of start of injection and spill timing has allowed further improvements to be made. Electronic control has also made a significant contribution. A number of further techniques can be employed to control emissions.

10.3.2 Exhaust gas recirculation

In much the same way as with petrol engines, exhaust gas recirculation (EGR) is employed primarily to reduce NOx emissions by reducing the reaction temperature in the combustion chamber. However, if the percentage of EGR is too high, increased hydrocarbons and soot are produced.

10.3.3 Intake air temperature

This is appropriate to turbocharged engines such that if the air is passed through an intercooler and there are improvements in volumetric efficiency, lower temperature will again reduce the production of NOx. The intercooler is fitted in the same area as the cooling system radiator.

10.3.4 Catalytic converter

On a diesel engine, a catalyst can be used to reduce the emission of hydrocarbons but will have less effect on nitrogen oxides. This is because diesel engines are always run with excess air to ensure better and more efficient burning of the fuel. A normal catalyst therefore will not strip the oxygen off the NOx to oxidize the hydrocarbons because the excess oxygen will be used instead. Special NOx converters are becoming available.

10.3.5 Filters

To reduce the emission of particulate matter (soot), filters can be used. These can vary from a fine grid design made from a ceramic material, to centrifugal filters and water trap techniques. The problem to overcome is that the filters can get blocked, which adversely affects the overall performance. Several techniques are employed, including centrifugal filters.

10.4 Complete vehicle control systems

10.4.1 Introduction

The possibility of a complete vehicle control system has been around since the first use of digital control. Figure 10.17 shows a representation of a full vehicle control system. In principle, it involves one ECU, which is capable of controlling all aspects of the vehicle.

Figure 10.18 shows one way in which a number of ECUs can be linked. In reality, however, rather than one control unit, separate ECUs are used that are able to communicate with each other via a controller area network (CAN) data bus.

10.4.2 Advantages of central control

The advantages of central control come under two main headings, inputs and outputs. On the input side, consider all the inputs required to operate each of the following:

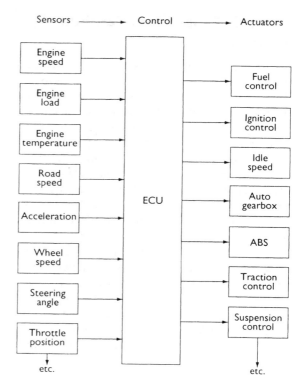

Figure 10.17 Representation of a full vehicle control system

- Ignition system.
- Fuel system.
- Transmission system.

It will be apparent that there are many common requirements even with just three possible areas of vehicle control. Having one central control system can potentially decrease the complexity of the wiring whilst increasing the possibilities for control. This is, in fact, the advantage of the 'outputs'. Consider the common operating condition for a vehicle of a sudden and hard acceleration and the possible responses from each of the systems listed.

System	Possible action
Ignition	Retard the timing
Fuel	Inject extra fuel
Transmission	Change down a gear

If each system is operating in its own right, it is possible that each, to some extent, will not react in the best way with respect to the others. For example, the timing and fuel quantity may be set but then the transmission ECU decides to change down a gear thus increasing engine speed. This, in turn, will require a change in fuel and timing. During the transition stage, a decrease in efficiency and an increase in emissions are likely.

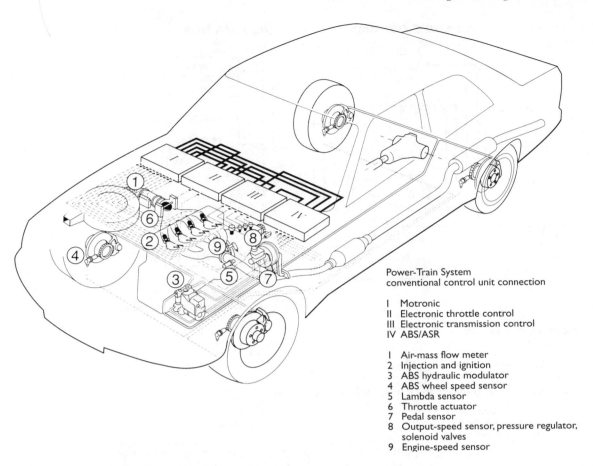

Power-Train System
conventional control unit connection

I   Motronic
II Electronic throttle control
III Electronic transmission control
IV ABS/ASR

1   Air-mass flow meter
2   Injection and ignition
3   ABS hydraulic modulator
4   ABS wheel speed sensor
5   Lambda sensor
6   Throttle actuator
7   Pedal sensor
8   Output-speed sensor, pressure regulator, solenoid valves
9   Engine-speed sensor

Figure 10.18 Linking ECUs

With a single control unit, or at least communication between them, the ideal actions could all take place at the most appropriate time. The complexity of the programming, however, requires much increased computing power. This is particularly apparent if other vehicle systems are considered, such as traction control, ABS, active suspension and steering. These systems are discussed individually in other sections of this book.

10.4.3 Bosch Cartronic system

The complexity of combining systems as suggested above is increasing. Bosch has a system involving a hierarchy of vehicle electronics. Improvements in performance, emissions, driver safety and comfort require increased interconnection of various electronic systems. In the previous section, a simple example highlighted the need for separate electronic systems to communicate with each other. Bosch uses a hierarchical signal structure to solve this problem. Figure 10.19. shows two ways in which the systems can

be linked. The first using conventional wiring and the second using a **Controller Area Network** (CAN).

Figure 10.20 shows the difference between the data flow in a stand-alone system and the data flow in a hierarchical system. The Cartronic system works on the principle that each system can only be controlled by a system placed above it in the hierarchy. As an example, the integrated transmission control systems of engine control and gearbox control do not communicate directly but via the hierarchically superior transmission control system.

10.4.4 Summary

Research is continuing into complete control systems for vehicles. As more and more systems are integrated then the cost of the electronics necessary will reduce. The computing power required for these types of developments is increasing, and 32- (or even 64-) bit high-speed microcontrollers will soon become the norm. The down-side of using a single ECU to control

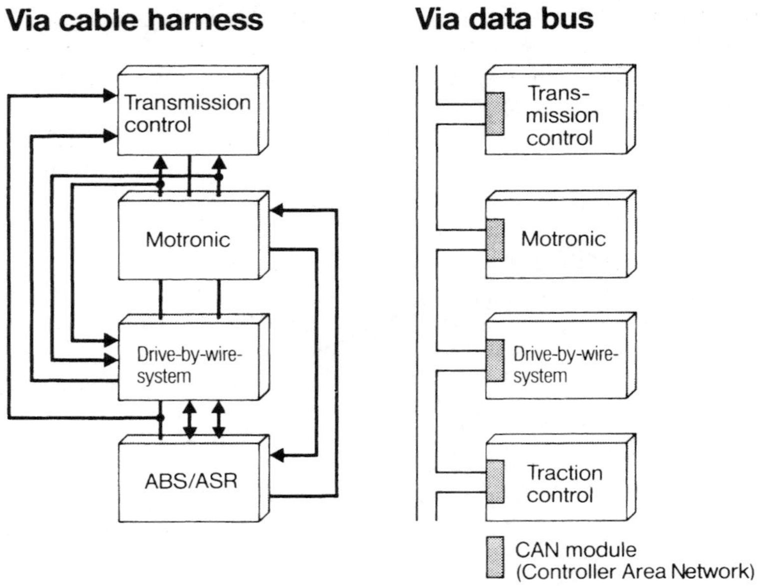

Via cable harness
Via data bus
Transmission control
Motronic
Drive-by-wire-system
ABS/ASR
Trans-mission control
Motronic
Drive-by-wire-system
Traction control
CAN module (Controller Area Network)

Figure 10.19 System link

Distribution and interconnection of functions
Data flow between stand-alone systems
Data flow in hierarchical system
Driver intention
Driver intention
Driving dynamics
Wheel torques
Drive shift
Steering
Motor torque
Engine
Gearbox
Brake
Control commands
Control commands

Figure 10.20 Cartronic system

the entire vehicle is the replacement cost of the unit. At present prices, even a single system ECU can cost a significant amount. Overall though, the cost of vehicle manufacture may be less.

Full central control has other possible advantages such as allowing the expansion of on-board diagnostics (OBD) to cover the whole vehicle, potentially saving repair time and running costs.

10.5 Case study – Mitsubishi GDI

10.5.1 Introduction

I am grateful to Mitsubishi for the information in this section.

For many years, innovative engine technology has been a development priority of Mitsubishi Motors. In particular, Mitsubishi has sought to improve engine efficiency in an endeavour to

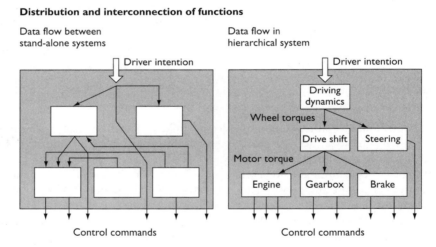

meet growing environmental demands – such as those for energy conservation and the reduction of CO_2 emissions in order to limit the negative impact of the greenhouse effect.

In Mitsubishi's endeavour to design and build ever more efficient engines, it has devoted significant resources to developing a gasoline direct injection engine. For years, automotive engineers have believed this type of engine has the greatest potential to optimize fuel supply and combustion, which in turn can deliver better performance and lower fuel consumption. Until now, however, no one has successfully designed an in-cylinder direct injection engine for use on production vehicles. A result of Mitsubishi's engine development capabilities, Mitsubishi's advanced Gasoline Direct Injection 'GDI' engine is the realization of an engineering dream.

For the fuel supply, conventional engines use a fuel injection system which replaced the carburation system. MPI or Multi-Point Injection, where the fuel is injected to each intake port, is currently one of the most widely used systems. However, even in MPI engines there are limits to the fuel supply response and the combustion control because the fuel mixes with air before entering the cylinder. Mitsubishi set out to push these limits by developing an engine where gasoline is directly injected into the cylinder as in a diesel engine, and, moreover, where injection timings are precisely controlled to match load conditions. The GDI engine achieved the following outstanding characteristics.

- Extremely precise control of fuel supply to achieve fuel efficiency that exceeds that of diesel engines by enabling combustion of an ultra-lean mixture supply.
- A very efficient intake and a relatively high compression ratio unique to the GDI engine deliver both high performance and a response that surpass those of conventional MPI engines.

Figure 10.21 shows the progress towards higher output and efficiency. For Mitsubishi, the technology realized for this GDI engine will form the cornerstone of the next generation of high efficiency engines and, in its view, the technology will continue to develop in this direction.

Figure 10.22 shows the transition of the fuel supply system.

Figure 10.23 is the Mitsubishi Gasoline Direct Injection (GDI) engine.

10.5.2 Major objectives of the GDI engine

- Ultra-low fuel consumption that is even better than that of diesel engines.
- Superior power to conventional MPI engines.

Technical features

- Upright straight intake ports for optimal air-flow control in the cylinder.
- Curved-top pistons for better combustion.
- High-pressure fuel pump to feed pressurized fuel into the injectors.
- High-pressure swirl injectors for optimum air–fuel mixture.

The major characteristics of the GDI engine are considered in the next few sections.

10.5.3 Lower fuel consumption and higher output

Optimal fuel spray for two combustion modes

Using methods and technologies unique to Mitsubishi, the GDI engine provides both lower fuel consumption and higher output. This seemingly contradictory and difficult feat is achieved with

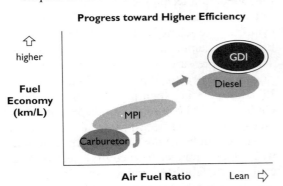

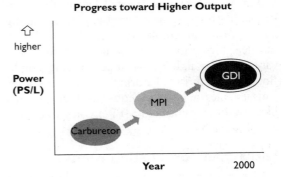

Figure 10.21 Progress towards higher output and efficiency

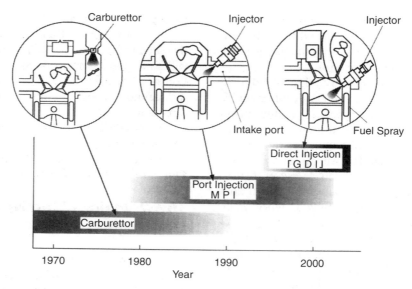

Figure 10.22 Transition of fuel supply system

Figure 10.23 Mitsubishi gasoline direct injection 'GDI' engine

the use of two combustion modes. Put another way, injection timings change to match engine load.

For the load conditions required in average urban driving, fuel is injected late in the compression stroke, as in a diesel engine. By doing so, an ultra-lean combustion is achieved due to an ideal formation of a stratified air–fuel mixture. During high performance driving conditions, fuel is injected during the intake stroke. This enables a homogeneous air–fuel mixture, like that in conventional MPI engines, to deliver a higher output.

Ultra-lean combustion mode

Under most normal driving conditions, up to speeds of 120 km/h, the Mitsubishi GDI engine operates in ultra-lean combustion mode, resulting in less fuel consumption. In this mode, fuel injection occurs at the latter stage of the compression stroke and ignition occurs at an ultra-lean air–fuel ratio of 30 : 40 (35 : 55, including EGR).

Superior Output Mode

When the GDI engine is operating with higher loads or at higher speeds, fuel injection takes place during the intake stroke. This optimizes combustion by ensuring a homogeneous, cooler air–fuel mixture which minimizes the possibility of engine knocking.

These two modes are represented in Figure 10.24.

10.5.4 The GDI engine's foundation technologies

There are four technical features that make up the foundation technology. The 'upright straight intake port' supplies optimal airflow into the cylinder. The 'curved top piston' controls combustion by helping to shape the air–fuel mixture. The 'high-pressure fuel pump' supplies the high-pressure needed for direct in-cylinder injection. In addition, the 'high-pressure swirl injector' controls the vaporization and dispersion of the fuel spray.

These fundamental technologies, combined with other unique fuel control technologies, enabled Mitsubishi to achieve both development objectives – fuel consumption lower than that of diesel engines and output higher than that of

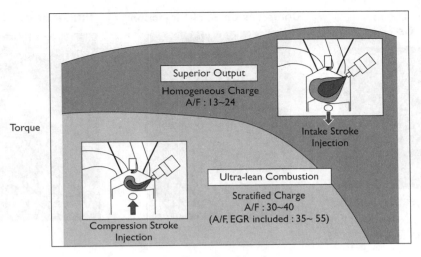

Torque

Engine Speed (rpm)

Figure 10.24 Two combustion modes

conventional MPI engines. The methods are shown below.

In-cylinder airflow

The GDI engine has upright straight intake ports rather than the horizontal intake ports used in conventional engines. The upright straight intake ports efficiently direct the airflow down at the curved-top piston, which redirects the airflow into a strong reverse tumble for optimal fuel injection, as shown in Figure 10.25.

Fuel spray

Newly developed high-pressure swirl injectors provide the ideal spray pattern to match each engines operational modes. This is shown as Figure 10.26. At the same time, by applying highly swirling motion to the entire fuel spray, the injectors enable sufficient fuel atomization that is mandatory for the GDI even with a relatively low fuel pressure of 50 kg/cm.

Optimized configuration of the combustion chamber

The curved-top piston controls the shape of the air–fuel mixture as well as the airflow inside the combustion chamber and has an important role in maintaining a compact air–fuel mixture. The mixture, which is injected late in the compression stroke, is carried towards the spark plug before it can disperse.

Mitsubishi's advanced in-cylinder observation techniques, including laser-methods, have been utilized to determine the optimum piston shape shown in Figure 10.27.

10.5.5 Realization of lower fuel consumption

Basic concept

In conventional gasoline engines, dispersion of an air–fuel mixture with the ideal density around

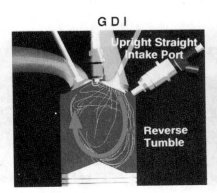
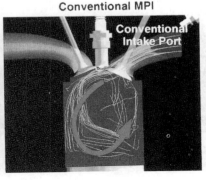

Figure 10.25 Upright straight intake ports

**Fuel Spray Locus
(Bottom view)**

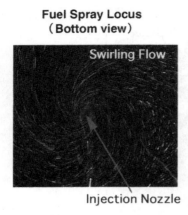

Compression Stroke Injection

Compact Spray
(Realization of)
Stratified Charge

Intake Stroke Injection

Cone Shape Spray
(Realization of)
Homogeneous Charge

Figure 10.26 Swirl injectors

Figure 10.27 Optimum piston shape

the spark plug was very difficult. However, this is possible in the GDI engine. Furthermore, extremely low fuel consumption is achieved because ideal stratification enables fuel injected late in the compression stroke to maintain an ultra-lean air–fuel mixture.

An engine for analysis purposes has proved that an air–fuel mixture with the optimum density gathers around the spark plug in a stratified charge. This is also borne out by analysing the behaviour of the fuel spray immediately before ignition and analysing the air–fuel mixture itself.

As a result, extremely stable combustion of an ultra-lean mixture with an air–fuel ratio of 40 (55, EGR included) is achieved as shown in Figure 10.28.

Combustion of ultra-lean mixture

In conventional MPI engines, there were limits to the mixture's leanness due to large changes in combustion characteristics. However, the stratified mixture of the GDI enabled greatly decreasing the air–fuel ratio without leading to poorer combustion. For example, during idling, when combustion is most inactive and unstable, the GDI engine maintains a stable and fast combustion even with an extremely lean mixture of 40 : 1 air–fuel ratio (55 : 1, EGR included). Figure

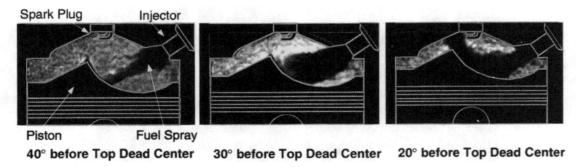

40° before Top Dead Center   30° before Top Dead Center   20° before Top Dead Center

Figure 10.28 Behaviour of fuel spray (injection in compression stroke) – Schlieren photo method

10.29 shows a comparison between GDI and a conventional multipoint system.

Vehicle fuel consumption

Fuel consumption is considered under idling, cruising and city driving conditions.

Fuel consumption during idling

The GDI engine maintains stable combustion even at low idle speeds. Moreover, it offers greater flexibility in setting the idle speed. Compared with conventional engines, its fuel consumption during idling is 40% less, as represented in Figure 10.30.

Fuel consumption during cruising

At 40 km/h, the GDI engine uses 35% less fuel than a comparably sized conventional engine (Figure 10.31).

Fuel consumption in city driving

In Japanese 10.15 mode tests (representative of Japanese urban driving), the GDI engine used 35% less fuel than comparably sized conventional gasoline engines. Moreover, these results indicate that the GDI engine uses less fuel than even diesel engines (Figure 10.32).

Emission control

Previous efforts to burn a lean air–fuel mixture have resulted in difficulty in controlling NOx emissions. However, for the GDI engine, 97% NOx reduction is achieved by utilizing a high-rate EGR (Exhaust Gas Recirculation) such as

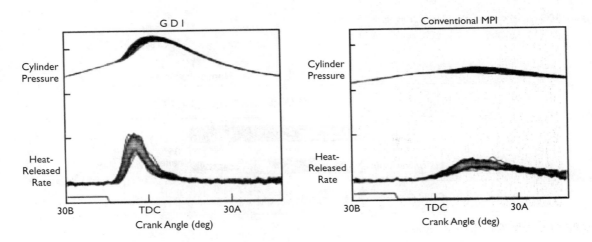

Figure 10.29 Comparison between GDI and a conventional multipoint system

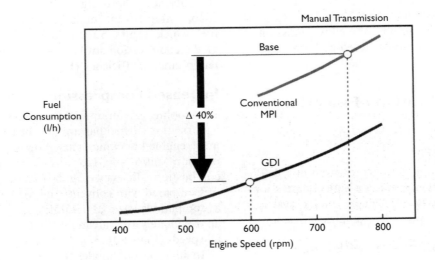

Figure 10.30 Fuel consumption during idling

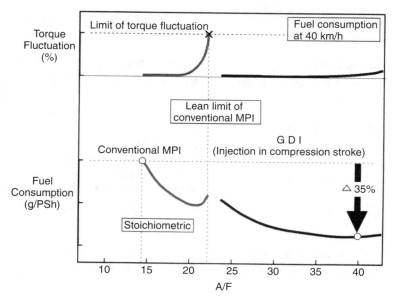

Figure 10.31 Fuel consumption during cruising

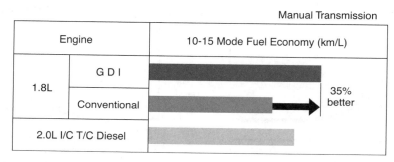

Figure 10.32 Fuel consumption in city driving

30%, which is allowed by the stable combustion unique to the GDI, as well as by the use of a newly developed lean-NOx catalyst. Figure 10.33 shows a graph of NOx emissions. Figure 10.34 is a newly developed lean NOx catalyst.

10.5.6 Realization of superior output

Basic concept

To achieve power superior to conventional MPI engines, the GDI engine has a high compression ratio and a highly efficient air intake system, which result in improved volumetric efficiency.

Improved volumetric efficiency

Compared with conventional engines, the Mitsubishi GDI engine provides better volumetric efficiency. The upright straight intake ports enable smoother air intake. The vaporization of fuel, which occurs in the cylinder at a late stage of the compression stroke, cools the air for better volumetric efficiency (Figure 10.35).

Increased compression ratio

The cooling of air inside the cylinder by the vaporization of fuel has another benefit to minimize engine knocking. This allows a high compression ratio of 12, and thus improved combustion efficiency (Figure 10.36).

Compared with conventional MPI engines of a comparable size, the GDI engine provides approximately 10% greater output and torque at all speeds (Figure 10.37).

In high-output mode, the GDI engine provides outstanding acceleration. Figure 10.38

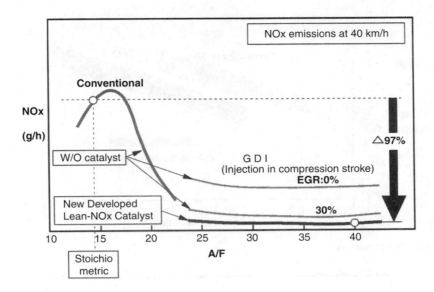

Figure 10.33 NOx emissions

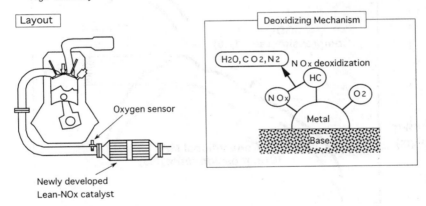

Figure 10.34 Newly developed lean NOx catalyst (HC selective deoxidization type)

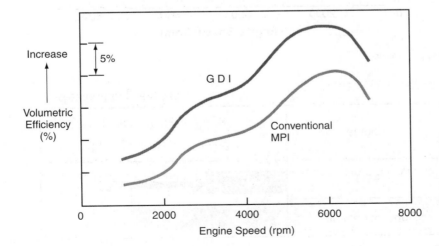

Figure 10.35 Improved volumetric efficiency

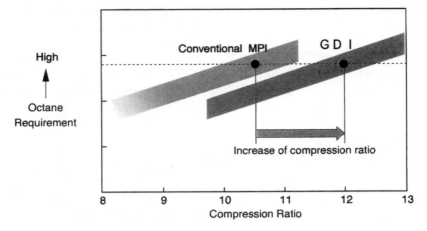

Figure 10.36 Increased compression ratio

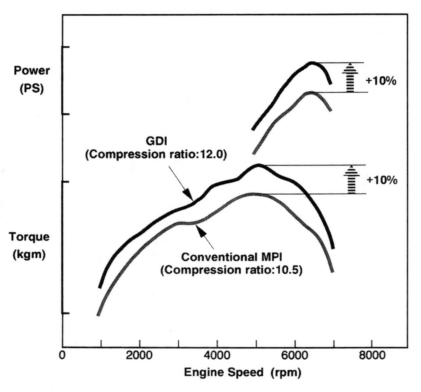

Figure 10.37 Engine performance

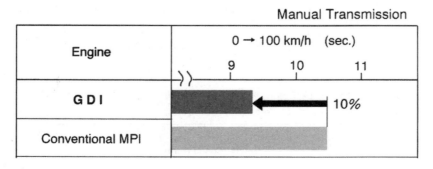

Figure 10.38 Vehicle acceleration

compares the performance of the GDI engine with a conventional MPI engine.

10.6 Case study – Bosch Motronic M3

10.6.1 Introduction

The combination of ignition and injection control has several advantages. The information received from various sensors is used for computing both fuelling and ignition requirements. Perhaps more importantly, ignition and injection are closely linked. The influence they have on each other can easily be taken into account to ensure that the engine is working at its optimum, under all operation conditions.

Overall, this type of system is less complicated than separate fuel and ignition systems and, in many cases, the ECU is able to work in an emergency mode by substituting missing information from sensors with pre-programmed values. This will allow limited but continued operation in the event of certain system failures.

The ignition system is integrated and is operated without a high tension distributor. The ignition process is controlled digitally by the ECU. The data for the ideal characteristics are stored in ROM from information gathered during both prototyping and development of the engine. The main parameters for ignition advance are engine speed and load, but greater accuracy can be achieved by taking further parameters into account, such as engine temperature. This provides both optimum output and close control of anti-pollution levels. Performance and pollution level control means that the actual ignition point must, in many cases, be a trade-off between the two.

The injection system is multipoint and, as is the case for all fuel systems, the amount of fuel delivered is primarily determined by the amount of air drawn into the engine. The method for measuring these data is indirect in the case of this system as a pressure sensor is being used to determine the air quantity.

Electromagnetic injectors control the fuel supply into the engine. The injector open period is determined by the ECU. This will obtain very accurate control of the air–fuel mixture under all operating conditions of the engine. The data for this are stored in ROM in the same way as for the ignition.

Figure 10.39 shows the components of this system.

10.6.2 Ignition system operation

The main source of reference for the ignition system is from the crankshaft position sensor. This is a magnetic inductive pick-up sensor positioned next to a flywheel ring containing 58

Figure 10.39 Bosch Motronic system components

teeth. Each tooth takes up a 6° angle of the fly-wheel with one 12° gap positioned 114° before top dead centre (TDC) for the number one cylinder.

Typical resistance of the sensor coil is 800 Ω. The air gap between the sensor and flywheel ring is about 1 mm. The signal produced by the fly-wheel sensor is shown in Figure 10.40. It is essentially a sine wave with one cycle missing, which corresponds to the gap in the teeth of the reluctor plate.

The information provided to the ECU is engine speed from the frequency of the signal, and engine position from the number of pulses before or after the missed pulses.

The block diagram in Figure 10.41 shows a block diagram layout of how the ignition system is controlled. At ignition system level the ECU must be able to:

- Determine and create advance curves.
- Establish constant energy.
- Transmit the ignition signal direct to the ignition coil.

The basic ignition advance angle is obtained from a memorized cartographic map. This is held in a ROM chip within the ECU. The parameters for this are:

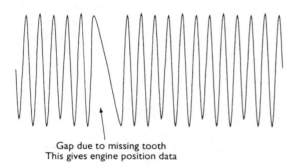

Gap due to missing tooth
This gives engine position data

Figure 10.40 Crankshaft sensor signal

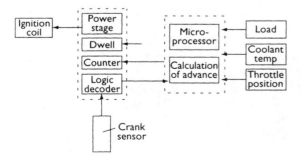

Figure 10.41 Simplified layout of the control of the ignition system

- Engine rev/min – given by the flywheel sensor.
- Inlet air pressure – given by the manifold absolute pressure sensor.

The above two parameters (speed and load) give the basic setting but to ensure optimum advance angle the timing is corrected by:

- Coolant temperature.
- Air temperature.
- Throttle butterfly position.

The ignition is set to a predetermined advance during the starting phase. Figure 10.42 shows a typical advance map and a dwell map used by the Motronic system. These data are held in ROM. For full ignition control, the electronic control unit has first to determine the basic timing for three different conditions.

- Under idling conditions, ignition timing is often moved very quickly by the ECU in order to control idle speed. When timing is advanced, engine speed will increase within certain limits.
- Full load conditions require careful control of ignition timing to prevent combustion knock. When a full load signal is sensed by the ECU (high manifold pressure) the ignition advance angle is reduced.
- Partial throttle is the main area of control and, as already stated, the basic timing is set initially by a programme as a function of engine speed and manifold pressure.

Corrections are added according to:

- Operational strategy.
- Knock protection.
- Phase correction.

The ECU will also control ignition timing variation during overrun fuel cut-off and reinstatement and also ensure anti-jerk control. When starting, the ignition timing plan is replaced by a specific starting strategy. Phase correction is when the ECU adjusts the timing to take into account the time taken for the HT pulse to reach the spark plugs. To ensure good drivability the ECU can limit the variations between the two ignition systems to a maximum value, which varies according to engine speed and the basic injection period.

The anti-jerk function operates when the basic injection period is less than 2.5 ms and the engine speed is between 720 and 3200 rev/min. This function operates to correct the pro-grammed ignition timing in relation to the

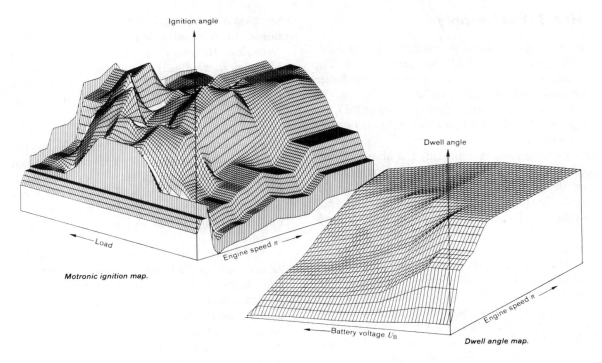

Figure 10.42 Engine timing and dwell maps

instantaneous engine speed and a set filtered speed; this is done to stabilize the engine rotational characteristics as much as possible.

In order to maintain constant high tension (HT) energy, the dwell period must increase in line with engine speed. To ensure the ignition primary current reaches its maximum at the point of ignition, the ECU controls the dwell by the use of another memory map, which takes battery voltage into account.

The signal from the flywheel sensor is virtually a sinusoid created as the teeth pass the winding. The zero value of this signal occurs as the sensor 'sees' the apex of each tooth. A circuit within the ECU (a Schmitt trigger) converts the signal into a square wave. The passage of the missing teeth gives a longer duration signal. The ECU detects the gap in the teeth and, from this, can determine the first TDC. The second TDC in the cycle is the determined by counting 29 teeth, which is half a revolution. The ECU, having determined the ignition angle then controls the coil every half engine revolution. Using the reference signal, the ECU switches the coil on at a point determined by a number of teeth corresponding to the dwell, before the point determined by timing value, where the coil is switched off.

The ignition module is only used as a simple switch to control the coil primary windings. It consists of a Darlington-type amplifier. This switching function is carried out within the ECU on some systems, this choice very much depends on the location of the ECU compared with the ignition coil. Also, the heat generated by the switching of heavy current may be better separate from the main ECU. A final consideration is whether the interference caused by the switching could cause problems within the ECU.

The 'distributorless' ignition coil is made up of two primary windings and two secondary windings. The primary windings have a common 12 V supply and are switched to earth in turn in the normal manner. The primary resistance is of the order of 0.5 Ω and the secondary resistance is 14.5 kΩ. The system works on the lost spark principle in that cylinders 1 and 4 fire together as do 2 and 3. The disadvantage of this system is that one cylinder of each pair has the spark jumping from the plug earth electrode to the centre. However, owing to the very high energy available for the spark, this has no significant effect on performance.

The HT cables used are resistive. Spark plugs used for this system are standard but vary between types of engine. A gap of around 0.8 mm is the norm.

10.6.3 Fuel supply

Fuel is collected from the tank by a pump either immersed in it or outside, but near the tank. The immersed type is quieter in operation has better cooling and no internal leaks. The fuel is directed forwards to the fuel rail or manifold, via a paper filter. Figure 10.43 shows the fuel supply system.

Fuel pressure is maintained at about 2.5 bar above manifold pressure by a regulator mounted on the fuel rail. Excess fuel is returned to the tank. The fuel is usually picked up via a swirl pot in the tank to prevent aeration of the fuel. Each of the four inlet manifold tracts has its own injector.

The fuel pump is a high-pressure type and is a two-stage device. A low-pressure stage, created by a turbine, draws fuel from the tank and a high-pressure stage, created by a gear pump, delivers fuel to the filter. It is powered by a 12 V supply from the fuel pump relay, which is controlled by the ECU as a safety measure.

The fuel pump characteristics are:

- Delivery – 120 litres per hour at 3 bars.
- Resistance – 0.8 Ω (static).
- Voltage – 12 V.
- Current – 10.5 A.

The rotation of the turbine draws fuel in via the inlet. The fuel passes through the turbine and enters the pump housing where it is pressurized by rotation of the pump and the reduction of the volume in the gear chambers. This pressure opens a residual valve and fuel passes to the fil-

ter. When the pump stops, pressure is maintained by this valve, which prevents the fuel returning. If, due to a faulty regulator or a blockage in the line, fuel pressure rises above 7 bar an over-pressure valve will open, releasing fuel back to the tank. Figure 9.30 shows this type of pump.

The fuel filter is placed between the fuel pump and the fuel rail. It is fitted to ensure that the outlet screen traps any paper particles from the filter element. The filter will stop contamination down to between 8 and 10 μm. Replacement of the filter varies between manufacturers but 80 000 km (50 000 miles) is often recommended.

The fuel rail, in addition to providing a uniform supply to the injectors, acts as an accumulator. Depending on the size of the fuel rail some systems also use an extra accumulator. The volume of the fuel rail is large enough to act as a pressure fluctuation damper, ensuring that all injectors are supplied with fuel at a constant pressure.

10.6.4 Injectors and associated components

One injector is used for each cylinder although very high performance vehicles may use two. The injectors are connected to the fuel rail by a rubber seal. The injector is an electrically operated valve manufactured to a very high precision. The injector comprises a body and needle attached to a magnetic core. When the winding in the injector housing is energized, the core or

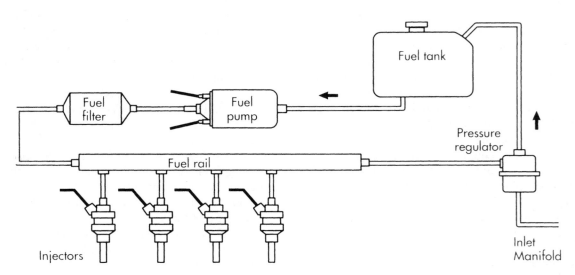

Figure 10.43 The main components in the fuel supply system

armature is attracted and the valve opens, compressing a return spring. The fuel is delivered in a fine spray to wait behind the closed inlet valve until the induction stroke begins. Providing the pressure across the injector remains constant, the quantity of fuel admitted is related to the open period, which in turn is determined by the time the electromagnetic circuit is energized. The injectors typically have the following characteristics (Figure 9.29 shows typical fuel injectors):

- Supply voltage – 12 V.
- Resistance – 16 Ω.
- Static output – 150 cc per minute at 3 bar.

The purpose of the fuel pressure regulator is to maintain differential pressure across the injectors at a pre-determined constant. This means the regulator must adjust the fuel pressure in response to changes in manifold pressure. It is made of two compressed cases containing a diaphragm, spring and a valve. Figure 9.31 is a fuel pressure regulator similar to those used on this and many other injection systems.

The calibration of the regulator valve is determined by the spring tension. Changes in manifold pressure vary the basic setting. When the fuel pressure is sufficient to move the diaphragm, the valve opens and allows fuel to return to the tank. The decrease in pressure in the manifold, also acting on the diaphragm for example, idle speed, will allow the valve to open more easily, hence maintaining a constant differential pressure between the fuel rail and the inlet manifold. This is a constant across the injectors and hence the quantity of fuel injected is determined only by the open time of the injectors. The differential pressure is maintained at about 2.5 bar.

The air supply circuit will vary considerably between manufacturers but an individual manifold from a collector housing, into which the air is fed via a simple butterfly valve, essentially supplies each cylinder. The air is supplied from a suitable filter. A supplementary air circuit is utilized during the warm-up period after a cold start and to control idle speed.

10.6.5 Fuel mixture calculation

The quantity of fuel to be injected is determined primarily by the quantity of air drawn into the engine. This is dependent on two factors:

- Engine rpm.
- Inlet manifold pressure.

This speed load characteristic is held in the ECU memory in ROM look-up tables.

A sensor connected to the manifold by a pipe senses the manifold absolute pressure. It is a piezoelectric-type sensor, where the resistance varies with pressure. The sensor is fed with a stabilized 5 V supply and transmits an output voltage according to the pressure. The sensor is fitted away from the manifold and hence a pipe is required to connect it. A volumetric capacity is usually fitted in this line to damp down pressure fluctuations. The output signal varies between about 0.25 V at 0.17 bar to about 4.75 V at 1.05 bar. Figure 10.44 shows a pressure sensor and its voltage output.

The density of air varies with temperature such that the information from the MAP sensor on air quantity will be incorrect over wide temperature variations. An air temperature sensor is used to inform the ECU of the inlet air temperature such that the ECU may correct the quantity of fuel injected. As the temperature of air

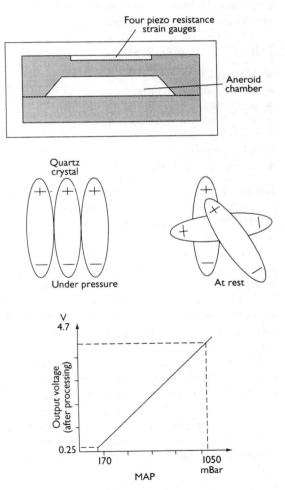

Figure 10.44 Pressure sensor and its voltage output

decreases its density increases and hence the quantity of fuel injected must also be increased.

The sensor is a negative temperature coefficient (NTC) resistor. The resistance value decreases as temperature increases and vice versa. The output characteristic of this sensor is non-linear. Further details about this type of sensor and one solution to the non-linear response problem are examined in Chapter 2.

In order to operate the injectors, the ECU needs to know – in addition to air pressure – the engine speed to determine the injection quantity. The same flywheel sensor used by the ignition system provides this information. All four injectors operate simultaneously, once per engine revolution, injecting half of the required fuel. This helps to ensure balanced combustion. The start of injection varies according to ignition timing.

A basic open period for the injectors is determined by using the ROM information relating to manifold pressure and engine speed. Two corrections are then made, one relative to air temperature and another depending on whether the engine is idling, at full or partial load.

The ECU then carries out another group of corrections, if applicable:

- after-start enrichment,
- operational enrichment,
- acceleration enrichment,
- weakening on deceleration,
- cut-off on overrun,
- reinstatement of injection after cut-off,
- correction for battery voltage variation.

Under starting conditions, the injection period is calculated differently. This is determined mostly from a set figure, which is varied as a function of temperature.

The coolant temperature sensor is a thermistor and is used to provide a signal to the ECU relating to engine coolant temperature. The ECU can then calculate any corrections to fuel injection and ignition timing. The operation of this sensor is the same as the air temperature sensor.

The throttle potentiometer is fixed on the throttle butterfly spindle and informs the ECU of the throttle position and rate of change of throttle position. The sensor provides information on acceleration, deceleration and whether the throttle is in the full load or idle position. Figure 10.45 shows the throttle potentiometer and its electrical circuit. It comprises a variable resistance and a fixed resistance. As is common with many sensors, a fixed supply of 5 V is pro-

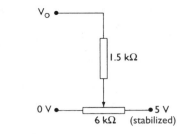

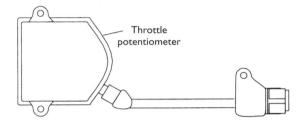

Figure 10.45 Throttle potentiometer and its electrical circuit

vided and the return signal will vary approximately between 0 and 5 V. The voltage increases as the throttle is opened.

10.6.6 Operating functions

The operation functions employed by this system can be examined under a number of headings or phases, as follows.

Starting phase

Entry to the starting phase occurs as soon as the ECU receives a signal from the flywheel sensor. The ignition advance is determined relative to the engine speed and the water temperature. The ECU operates the injectors four times per engine cycle (twice per crankshaft revolution) in order to obtain the most uniform mixture and to avoid wetting the plugs during the starting phase. Figure 10.46 shows the injection and ignition timing relative to engine position. Injection ceases 24° after the flywheel TDC signal. The ECU sets an appropriate injection period, corrected in relation to water temperature if starting from cold and air temperature if starting from hot. Exit from this starting phase is when the engine speed passes a threshold determined by water temperature.

After-start enrichment phase

Enrichment is necessary to avoid stalling after starting. The amount of enrichment is determined by water and air temperature and decreases under control of the ECU. If the engine is cold or an intermediate temperature,

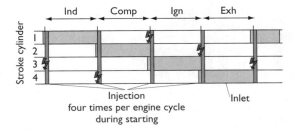

Figure 10.46 Injection and ignition timing relative to engine position

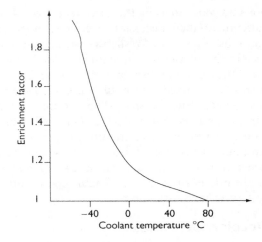

Figure 10.48 Enrichment factor during warm up

the initial mixture is a function of water temperature. If the engine is hot, the initial mixture is a function of air and water temperature. Figure 10.47 is a representation of the decreasing mixture enrichment after a cold start. If the engine happens to stop within a certain period of time just after a cold start, the next post-start enrichment will be reduced slightly.

Engine cold running phase

During warm up, the ignition timing is corrected in relation to water temperature. Timing will also alter depending on engine speed and load. During the warm-up phase, the injector open period is increased by the coolant temperature signal to make up for fuel losses and to prevent the engine speed dropping. The enrichment factor is reduced as the resistance of the temperature sensor falls, finally ceasing at 80°C. Figure 10.48 shows the enrichment factor during warm up. The enrichment factor is determined by engine speed and temperature at idle and at other times by the programmed injection period relative to engine speed as well as the water temperature. To overcome the frictional resistance of a cold engine it is important to increase the mixture supply. This is achieved by using a supplementary air control device, which allows air to bypass the throttle butterfly.

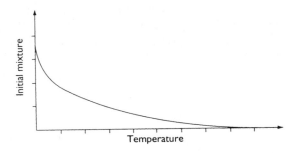

Figure 10.47 Decreasing mixture enrichment after a cold start

Idling phase

Air required for idling bypasses the throttle butterfly by a passage in the throttle housing. A volume screw is fitted for adjustment of idle speed. Idle mixture adjustment is carried out electronically in response to the adjusting of a potentiometer, either on the ECU or as a separate unit. The ignition and injection functions for idle condition are set using information from the throttle potentiometer that the throttle is at the idle position, and engine speed is set by information from the flywheel sensor.

Full load phase

Under full load conditions the ignition timing is related to engine speed and full load information from the throttle potentiometer. The injection function in order to achieve maximum power must be set such that the mixture ratio is increased to 1:1. The information from the throttle potentiometer triggers a programme in the ECU to enrich the mixture in relation to engine speed in order to ensure maximum power over the speed range but also to minimize the risk of knocking. It is also important not to increase fuel consumption unnecessarily and not to allow significant increases in exhaust emissions.

Acceleration phase

When a rapid acceleration is detected by the ECU from the rate of change of the throttle potentiometer signal, enrichment occurs over a certain number of ignitions. The enrichment value is determined from water temperature and pressure variations in the inlet manifold. The enrichment then decreases over a number of ignitions. Figure 10.49 shows the acceleration enrichment phase. The enrichment is applied for the calibrated number of ignitions and then reduced at a fixed rate until it is non-existent. Acceleration enrichment will not occur if the engine speed is above 5000 rev/min or at idle. Under very strong acceleration it is possible to have unsynchronized injection. This is determined from the water temperature, a ROM map of throttle position against engine speed and a battery voltage correction.

Deceleration phase

If the change in manifold pressure is greater than about 30 mbar the ECU causes the mixture to be weakened relative to the detected pressure change.

Injection cut-off on deceleration phase

This is designed to improve fuel economy and to reduce particular emissions of hydrocarbons. It will occur when the throttle is closed and when the engine speed is above a threshold related to water temperature (about 1500 rev/min). When the engine speed falls to about 1000 rev/min, injection recommences with the period rising to the value associated with the current engine speed and load. Figure 10.50 shows the strategy used to control injection cut-out and reinstatement.

Knock protection phase

Ignition timing is also controlled to reduce jerking and possible knocking during cut-off and reinstatement. The calculated advance is reduced to keep the ignition just under the knock limit. The advance correction against knock is a programme relating to injection period, engine speed and water/air temperature.

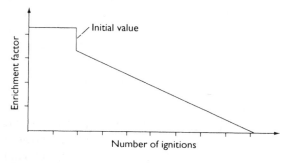

Figure 10.49 Acceleration enrichment phase

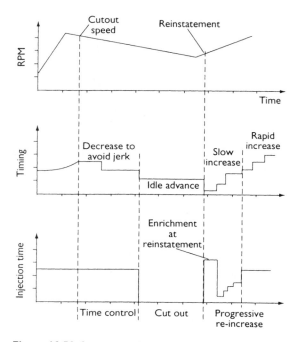

Figure 10.50 Strategy used to control injection cut-out and reinstatement

Engine speed limitation

Injection is cut-off when the engine speed rises above 6900 rev/min and is reinstated below this figure. This is simply to afford some protection against overrevving of the engine and the damage that may be caused.

Battery voltage correction

This is a correction in addition to all other functions in order to compensate for changes in system voltage. The voltage is converted every TDC and the correction is then applied to all injection period calculations. On account of the time taken for full current to flow in the injector winding and the time taken for the current to cease, a variation exists depending on applied voltage. Figure 10.51 shows how this delay can occur; if S_1 is greater than S_2 a correction is required. $S_1 - S_2 = S$ where S represents the time delay due to the inductance of the injector winding.

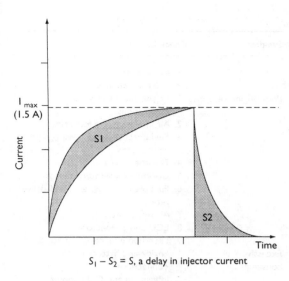

Figure 10.51 Injector operation time curve

$S_1 - S_2 = S$, a delay in injector current

10.7 Diagnosing engine management system faults

10.7.1 Introduction

As with all systems the six stages of fault finding should be followed.

1. Verify the fault.
2. Collect further information.
3. Evaluate the evidence.
4. Carry out further tests in a logical sequence.
5. Rectify the problem.
6. Check all systems.

The procedure outlined in the next section is related primarily to stage 4 of the process. Table 10.1 is based on information available from 'Autodata' in its excellent range of books. It relates in particular to the Bosch LH-Jetronic fuel system but it is also a good guide to many other systems. The numbers relate to the order in which the systems should be checked.

10.7.2 ECU auto-diagnostic function

Most ECUs are equipped to advise the driver of a fault in the system and to aid the repairer in detection of the problem. The detected fault is first notified to the driver by a dashboard warning light. A code giving the details is held in RAM within the ECU. The repairer can read this fault code as an aid to fault-finding.

Table 10.1 Common symptoms of an engine malfunction and checks for possible faults

Symptom	Check for possible fault
Engine will not start	1. Engine and battery earth connections.
	2. Fuel filter and fuel pump.
	3. Air intake system for leaks.
	4. Fuses/fuel pump/system relays.
	5. Fuel injection system wiring and connections.
	6. Coolant temperature sensor.
	7. Auxiliary air valve/idle speed control valve.
	8. Fuel pressure regulator and delivery rate.
	9. ECU and connector.
	10. Limp home function – if fitted.
Engine difficult to start when cold	1. Engine and battery earth connections.
	2. Fuel injection system wiring and connections.
	3. Fuses/fuel pump/system relays.
	4. Fuel filter and fuel pump.
	5. Air intake system for leaks.
	6. Coolant temperature sensor.
	7. Auxiliary air valve/idle speed control valve.
	8. Fuel pressure regulator and delivery rate.
	9. ECU and connector.
	10. Limp home function – if fitted.
Engine difficult to start when warm	1. Engine and battery earth connections.
	2. Fuses/fuel pump/system relays.
	3. Fuel filter and fuel pump.
	4. Air intake system for leaks.
	5. Coolant temperature sensor.
	6. Fuel injection system wiring and connections.
	7. Air mass meter.
	8. Fuel pressure regulator and delivery rate.
	9. Air sensor filter.
	10. ECU and connector.
	11. Knock control – if fitted.
Engine starts then stops	1. Engine and battery earth connections.
	2. Fuel filter and fuel pump.
	3. Air intake system for leaks.
	4. Fuses/fuel pump/system relays.
	5. Idle speed and CO content.
	6. Throttle potentiometer.
	7. Coolant temperature sensor.
	8. Fuel injection system wiring and connections.
	9. ECU and connector.
	10. Limp home function – if fitted.
Erratic idling speed	1. Engine and battery earth connections.
	2. Air intake system for leaks.
	3. Auxiliary air valve/idle speed control valve.
	4. Idle speed and CO content.

Table 10.1 Continued

Symptom	Check for possible fault	Symptom	Check for possible fault
	5. Fuel injection system wiring and connections.		3. Air mass meter.
	6. Coolant temperature sensor.		4. ECU and connector.
	7. Knock control – if fitted.	Poor engine response	1. Engine and battery earth connections.
	8. Air mass meter.		2. Air intake system for leaks.
	9. Fuel pressure regulator and delivery rate.		3. Fuel injection system wiring and connections.
	10. ECU and connector.		4. Throttle linkage.
	11. Limp home function – if fitted.		5. Coolant temperature sensor.
Incorrect idle speed	1. Air intake system for leaks.		6. Fuel pressure regulator and delivery rate.
	2. Vacuum hoses for leaks.		7. Air mass meter.
	3. Auxiliary air valve/idle speed control valve.		8. ECU and connector.
	4. Idle speed and CO content.		9. Limp home function – if fitted.
	5. Coolant temperature sensor.	Excessive fuel consumption	1. Engine and battery earth connections.
Misfire at idle speed	1. Engine and battery earth connections.		2. Idle speed and CO content.
	2. Air intake system for leaks.		3. Throttle potentiometer.
	3. Fuel injection system wiring and connections.		4. Throttle valve/housing/sticking/initial position.
	4. Coolant temperature sensor.		5. Fuel pressure regulator and delivery rate.
	5. Fuel pressure regulator and delivery rate.		6. Coolant temperature sensor.
	6. Air mass meter.		7. Air mass meter.
	7. Fuses/fuel pump/system relays.		8. Limp home function – if fitted.
Misfire at constant speed	1. Air flow sensor.	CO level too high	1. Limp home function – if fitted.
			2. ECU and connector.
Hesitation when accelerating	1. Engine and battery earth connections.		3. Emission control and EGR valve – if fitted.
	2. Air intake system for leaks.		4. Fuel injection system wiring and connections.
	3. Fuel injection system wiring and connections.		5. Air intake system for leaks.
	4. Vacuum hoses for leaks.		6. Coolant temperature sensor.
	5. Coolant temperature sensor.		7. Fuel pressure regulator and delivery rate.
	6. Fuel pressure regulator and delivery rate.	CO level too low	1. Engine and battery earth connections.
	7. Air mass meter.		2. Air intake system for leaks.
	8. ECU and connector.		3. Idle speed and CO content.
	9. Limp home function – if fitted.		4. Coolant temperature sensor.
Hesitation at constant speed	1. Engine and battery earth connections.		5. Fuel injection system wiring and connections.
	2. Throttle linkage.		6. Injector valves.
	3. Vacuum hoses for leaks.		7. ECU and connector.
	4. Auxiliary air valve/idle speed control valve.		8. Limp home function – if fitted.
	5. Fuel lines for blockage.		9. Air mass meter.
	6. Fuel filter and fuel pump.		10. Fuel pressure regulator and delivery rate.
	7. Injector valves.	Poor performance	1. Engine and battery earth connections.
	8. ECU and connector.		2. Air intake system for leaks.
	9. Limp home function – if fitted.		3. Throttle valve/housing/sticking/initial position.
Hesitation on overrun	1. Air intake system for leaks.		4. Fuel injection system wiring and connections.
	2. Fuel injection system wiring and connections.		5. Coolant temperature sensor.
	3. Coolant temperature sensor.		6. Fuel pressure regulator/fuel pressure and delivery rate.
	4. Throttle potentiometer.		7. Air mass meter.
	5. Fuses/fuel pump/system relays.		8. ECU and connector.
	6. Air sensor filter.		9. Limp home function – if fitted.
	7. Injector valves.		
	8. Air mass meter.		
Knock during acceleration	1. Knock control – if fitted.		
	2. Fuel injection system wiring and connections.		

Each fault detected is memorized as a numerical code and can only be erased by a voluntary action. Often, if the fault is not detected again for 50 starts of the engine, the ECU erases the code automatically. Only serious faults will light the lamp but minor faults are still recorded in memory. The faults are memorized in the order of occurrence. Certain major faults will cause the ECU to switch over to an emergency mode. In this mode, the ECU substitutes alternative values in place of the faulty signal. This is called a 'limp home facility'.

Faults can be read as two digit numbers from the flashing warning light by shorting the diagnostic wire to earth for more than 2.5 s but less than 10 s. Earthing this wire for more than 10 s will erase the fault memory, as does removing the ECU constant battery supply. Earthing a wire to read fault codes should only be carried out in accordance with the manufacturer's recommendations. The same coded signals can be more easily read on many after-sales service testers. On some systems it is not possible to read the fault codes without a code reader.

10.7.3 Testing procedure

Caution/Achtung/Attention – Burning fuel can seriously damage your health!
Caution/Achtung/Attention – High voltages can seriously damage your health!

The following procedure is very generic but with a little adaptation can be applied to any fuel injection system. Refer to the manufacturer's recommendations if in any doubt.

1. Check battery state of charge (at least 70%).
2. Hand and eye checks (all fuel and electrical connections secure and clean).
3. Check for spark at plug lead (if poor or no spark jump to stage 15).
4. Check fuel pressure supplied to rail (for multipoint systems it will be about 2.5 bar but check specifications).
5. If the pressure is NOT correct jump to stage 11.
6. Is injector operation OK? – continue if NOT (suitable spray pattern or dwell reading across injector supply).
7. Check supply circuits from main relay (battery volts minimum).
8. Continuity of injector wiring ($0–0.2\,\Omega$ and note that many injectors are connected in parallel).
9. Sensor readings and continuity of wiring ($0–0.2\,\Omega$ for the wiring sensors will vary with type).
10. If no fuel is being injected and all tests so far are OK (suspect ECU).
11. Fuel supply – from stage 5.
12. Supply voltage to pump (within 0.5 V of battery – pump fault if supply is OK).
13. Check pump relay and circuit (note that, in most cases, the ECU closes the relay but this may be bypassed on cranking).
14. Ensure all connections (electrical and fuel) are remade correctly.
15. Ignition section (if appropriate).
16. Check supply to ignition coil (within 0.5 V of battery).
17. Spark from coil via known good HT lead (jumps about 10 mm, but do not try more).
18. If good spark then check HT system for tracking and open circuits. Check plug condition (leads should be a maximum resistance of about $30\,k\Omega/m$ per lead) – stop here in this procedure.
19. If no spark, or it will only jump a short distance, continue with this procedure (colour of spark is not relevant).
20. Check continuity of coil windings (primary $0.5–3\,\Omega$, secondary *several* $k\Omega$).
21. Supply and earth to 'module' (12 V minimum supply, earth drop 0.5 V maximum).
22. Supply to pulse generator if appropriate (10–12 V).
23. Output of pulse generator (inductive about 1 V AC when cranking, Hall-type switches 0–8 V DC).
24. Continuity of LT wires ($0–0.1\,\Omega$).
25. Suspect ECU but only if all of the above tests are satisfactory.

10.7.4 Injection duration signals

Figure 10.52 shows typical injector signals as would be shown on an oscilloscope during a test procedure. These will vary depending on the particular system but, in principle, are the same. The most important parts of the traces are marked. These are the open time or dwell, current limiting phase and the back EMF produced when the injector is switched off. The traces showing variations in the dwell represent how the quantity of fuel injected is varied. The difference in how the dwell is varied is due to the method of injector switching. If a simple on/off technique is used then the trace will be as shown in the first two sketches; if current limiting is

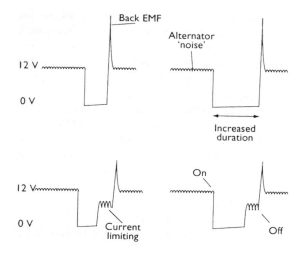

Figure 10.52 Injector signals as would be shown on an oscilloscope during a test procedure

used then the trace will be slightly different, as shown by the lower two sketches.

These traces are very useful for diagnosing faults – it is possible to see how the trace changes under the engine operating conditions, for example:

- Does the trace width extend under acceleration?
- Does the trace cut off on overrun?
- Does the trace width reduce as the engine warms up?

10.8 Advanced engine management technology

10.8.1 Speed density and fuel calculations

Engine management systems that do not use an air flow sensor rely on the speed-density method for determining the required fuel quantity. Accurate measurement of the manifold absolute pressure (MAP) and intake air temperature are essential with this technique.

The volume flow rate of air taken into an engine at a given speed can be calculated by:

$$A_v = \left[\left(\frac{RPM}{60} \right) \left(\frac{D}{2} \right) V_e \right] - EGR_v$$

where A_v = air volume flow rate (litres/s); EGR_v = exhaust gas recirculation volume (litres/s); D = displacement of the engine (litres); and V_e = volumetric efficiency (as a percentage from look-up tables).

The density of air in the inlet manifold is related to its temperature and pressure. If these are measured accurately then density can be calculated. A basic gas law states that, in a fixed volume:

$$d_a = d_o \left(\frac{p_i}{p_o} \times \frac{T_o}{T_i} \right)$$

where d_a = density; p_i = intake pressure; and T_i = intake temperature.

p_o, d_o and T_o are known values relating to pressure, density and temperature under 'sea level standard day' (SLSD) conditions

The mass of the air can be calculated by:

$$M_a = d_a \times V$$

where M_a = mass of air (kg); d_a = density of the air (kg/litre); and V = volume of air (litres).

The mass flow rate can now be calculated by:

$$A_m = d_a \times A_v$$

where A_m = air mass flow rate (kg/s).

Finally, by substitution and simplification, air mass flow can be calculated by:

$$A_m = d_a \left[\left(\frac{RPM.D.V_e}{120} \right) - EGR_v \right]$$

Further to this calculation, the basic fuel quantity can be determined as follows:

$$F = \frac{A_m}{AFR}$$

where F = fuel quantity (kg) and AFR = desired air–fuel ratio.

To inject the required quantity of fuel, the final calculation is that of the injector pulse width:

$$T = \frac{F}{R_f}$$

where T = time and R_f = fuel injector(s) delivery rate.

Note that the actual injection period will also depend on a number of other factors such as temperature and throttle position. The total fuel quantity may also be injected in two halves.

10.8.2 Ignition timing calculation

Data relating to the ideal ignition timing for a particular engine are collected from dynamometer tests and operational tests in the vehicle. These data are stored in the form of look-up tables in ROM. These look-up tables hold data relative to the speed and load of the engine. The number of look-up values is determined by the computing power of the microcontroller, in other words the number of bits, as this determines the size of memory that can be addressed.

Inputs from speed and load sensors are converted to digital numbers and these form the reference to find the ideal timing value. A value can also be looked up for the temperature correction. These two digital numbers are now added to give a final figure. Further corrections can be added in this way for conditions such as overrun and even barometric pressure if required.

This 'timing number' is used to set the point at which the coil is switched off; that is, the actual ignition point. The ECU receives a timing pulse from the 'missing flywheel tooth' and starts a 'down counter'. The coil is fired (switched off) when the counter reaches the 'timing number'. The computing of the actual 'timing number' is represented by Figure 10.53.

To prevent engine damage caused by detonation or combustion knock, but still allow the timing to be set as far advanced as possible, a knock sensor is used. The knock sensor (accelerometer) detects the onset of combustion knock, but the detection process only takes place in a 'knock window'. This window is just a few degrees of crankshaft rotation either side of top dead centre compression for each cylinder. This window is the only time knock can occur and is also a quiet time as far as valve opening and closing is concerned. The sensor is tuned to respond to a particular frequency range of about 5–10 kHz, which also helps to eliminate erroneous signals. The resonant frequency of this type of accelerometer is greater than about 25 kHz.

The signal from the knock sensor is filtered and integrated in the ECU. A detection circuit determines a yes/no answer to whether the engine knocked or not. When knock is detected on a particular cylinder, the timing for that cylinder is retarded by a set figure, often 2°, each

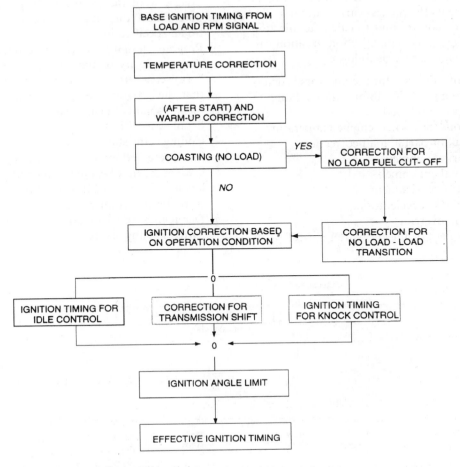

Figure 10.53 Determination of effective ignition timing

time the cylinder fires, until the knocking stops. The timing is then advanced more slowly back towards the look-up value. Figure 10.54 represents this process in more detail.

10.8.3 Dwell calculation

In order for an ignition system to produce constant energy the dwell angle must increase as the engine speed increases. Ideal dwell values are held in a look-up table; engine speed forms one axis, and battery voltage the other. If battery voltage falls, the dwell angle is increased to compensate. The 'dwell number' is used in a similar way to the 'timing number' in the previous section except that this time, the 'dwell number' is used to determine the switch-on point of the coil during operation of the down counter.

10.8.4 Injection duration calculation

The main criteria for the quantity of fuel required for injection are engine speed and load. Further corrections are then added. Figure 10.55 represents the process carried out in a digital electronic control unit to calculate injection duration. The process of injection duration calculation is summarized as follows.

- A basic open period for the injectors is determined from the ROM information relating to engine speed and load.
- Corrections for air and engine temperature.
- Idling, full or partial load corrections.
- After-start enrichment.
- Operational enrichment.
- Acceleration enrichment.
- Weakening on deceleration.
- Cut-off on overrun.
- Reinstatement of injection after cut-off.

- Correction for battery voltage variation.

Under starting conditions the injection period is calculated differently. This is determined from a set figure varied as a function of temperature.

10.8.5 Developing and testing software

There is, of course, more than one way of producing a 'computer' program. Most programs used in the electronic control unit of a vehicle digital control system are specialist applications and, as such, are one-off creations. The method used to create the final program is known as the 'top down structured programming technique'. Following on from a 'need' for the final product, the process can be seen to pass through six definable stages.

1. Requirement analysis seeks to answer the question as to whether a computerized approach is the best solution. It is, in effect, a feasibility study.
2. Task definition is a process of deciding exactly what the software will perform. The outcome of this stage will be a set of functional specifications.
3. Program design becomes more important as the complexity of the task increases. This is because, where possible, it is recommended that the program be split into a number of much smaller tasks, each with its own detailed specification.
4. Coding is the stage at which the task begins to be represented by a computer language. This is when the task becomes more difficult to follow as the language now used is to be understood by the 'computer'.
5. Debugging and validation is the process of correcting any errors or a bug in the program

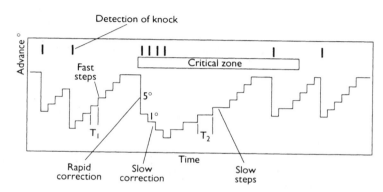

Figure 10.54 How timing is varied in response to combustion knock

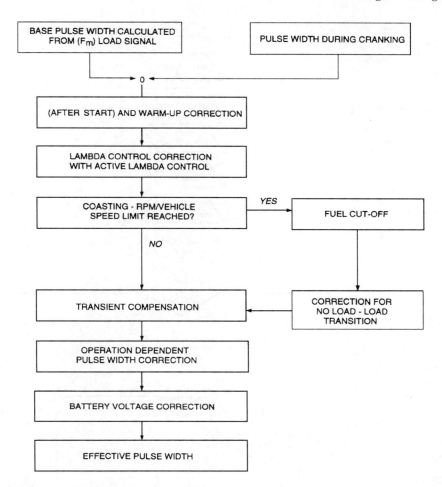

Figure 10.55a Determination of effective injector pulse width

code and then finally ensuring that it is valid. This means checking that the desired outputs appear in response to appropriate inputs. In other words, does it work? (As a slight aside, did you know that the original computer bug was actually a moth trapped between the contacts of a relay?) Note that it is very important to get the program right at this stage as it is likely to be incorporated into tens of thousands of specially produced microcontrollers. A serious error can be very expensive to rectify.

6. Operation and maintenance is the stage when the program is actually in use. Occasionally slight errors do not come to light until this stage, such as a slight hesitation during acceleration at high altitude or some other obscure problem. These can be rectified by program maintenance for inclusion in later models.

This section has been included with the intention of filling in the broader picture of what is involved in producing a program for, say, an electronic spark advance system. Many good books are available for further reading on this subject.

10.8.6 Example programmes

A much simplified BASIC program is included in the Appendix to illustrate the principle of digital engine control. It is not intended to illustrate actual ECU programs although the principle is much the same. In reality, faster programming languages and more complicated algorithms are used. The programs listed in my website should run on suitable computers using a BASIC interpreter. Please feel free to copy and try them out. They are available to download free of charge from my web site.

- Programmed ignition control.
- Full engine management.

The programs in this section are for illustration

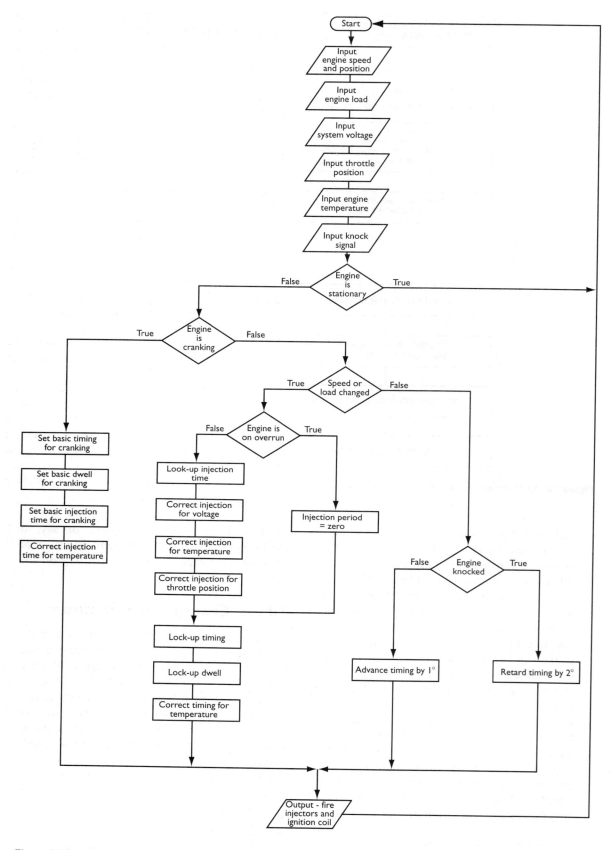

Figure 10.55b Engine management fuel and ignition calculation flow diagram

purposes only. They are not 'error trapped' or even fully tested and debugged. That is your problem!

10.8.7 Hot chipping!

Hot chipping is the name often given to the fitting of new processors/memory to improve the performance of a vehicle. It should be noted that the improvements are at the expense of economy, emissions and engine life! Fitting a 'Power Processor', which is a programmable computer specifically designed for high performance engines, is the first step. The fuel map, engine ignition timing map, acceleration fuel and all parameters for fuel management are programmable using an IBM compatible PC or laptop computer. Note that a new ECU is needed in most cases but this does allow improvement of other features.

The software even allows changes to be made while you are driving the vehicle. This system is appropriate for virtually any fuel injected engine. A basic calibration is used to get the engine started and running. The user then performs fine tuning. The systems are capable of closed or open loop operation. Some systems even feature control of nitrous injection with automatic engagement based on throttle position and rev/min. Ignition timing is automatically retarded with pre-set parameters.

CalMap Software is a well respected system for developing custom calibrations for high performance engines. The software allows 'online' and 'offline' adjustments to be made to the 'ACCEL Digital Fuel Injection Power Processors' (contact information is available in Chapter 18). The software kit comes with an interface cable, a user manual and a floppy disk, which contains the software. The software is user friendly and arguably should be considered a must for all modern performance shops. Setting and adjusting the spark curve for your distributor from a laptop computer in the vehicle is possible – as you are driving down the road. Snapshots can be taken using the software. For example, you could record a set distance run and review the engine performance in order to determine how the engine is running.

Tuning a fuel-injected engine requires experience, time and patience. One mistake with the laptop keyboard and your engine can easily be turned into a pile of junk from detonation or a lean condition! When determining the size of the base fuel map's rev/min resolution, the cell widths should be as small as possible. This gives the most tuning set points in the operating range of the engine. If the map is configured to 5000 rev/min, any resolution above that figure would be lost, but resolution would be gained where the engine spends its most time, i.e. below 5000 rev/min.

If the fuel map is calibrated to 5000 rev/min and the calibrated pulse width at that speed is 12 ms, the ECU will keep issuing pulses of 12 ms at any speed above this value. It is beneficial to use as many of the 256 (16×16 look-up table or 2^8 relating to 8 bits) set points as possible during tuning. This is established by setting the rev/min between cells. The largest fuel commands should be at the peak torque and, as the engine speed escalates above peak torque, the pulse width reduces. Most values from the ECU's inputs and outputs will be available 'on-screen', as if from the serial data link on a production ECU.

Most systems use 'interpolative' software, meaning the cells surrounding the actual chosen cell in the fuel map will affect the issued pulse width. Getting the fuel calculations as near to the stoichiometric set point as possible and using very little, if any, oxygen sensor trim is a good technique. This is the approach that the original equipment manufacturers use. While working on the base fuel map, note that with injector pulse widths below 2 ms, you are entering an unstable range. Work with all of the cells around the chosen idle cell because the surrounding cell values are used for interpolation. Large variations in matrix values around the idle cell can lead to surging.

The resolution of the ignition map is referenced from the fuel table and is scaled at a rate of 1.5 to the fuel table. The same theory applies to the spark table, as to the fuel table, in regard to keeping the same timing command beyond its rev/min resolution. The amount of retardation required to stop detonation once it is started in the combustion chamber is greater than the amount that would be needed never to allow detonation to start. A trial and error method is required for the best results. The amount of spark advance is affected by engine criteria such as:

- Cylinder-head combustion chamber design.
- Mixture movement.
- Piston design.
- Intake manifold length and material.
- Compression ratio.

- Available fuel.
- Thermal transfer from the cylinder-head to the cooling system.

10.8.8 Artificial Intelligence

Artificial intelligence (AI) is the ability of an artificial mechanism to exhibit intelligent behaviour. The term invites speculation about what constitutes the mind or intelligence. Such questions can be considered separately, but the endeavour to construct and understand increasingly sophisticated mechanisms continues.

AI has shown great promise in the area of expert systems, or knowledge-based expert programs, which, although powerful when answering questions within a specific domain, are nevertheless incapable of any type of adaptable, or truly intelligent, reasoning.

No generally accepted theories have yet emerged within the field of AI, due in part to AI being a new science. However, it is assumed that on the highest level, an AI system must receive input from its environment, determine an action or response and deliver an output to its environment. This requires techniques of expert reasoning, common sense reasoning, problem solving, planning, signal interpretation and learning. Finally, the system must construct a response that will be effective in its environment.

The possibilities for AI in vehicle use are unlimited. In fact, it becomes more a question of how much control the driver would be willing to hand over to the car. If, for example, the vehicle radar detects that you tend to follow the car in front too closely, should it cause the brakes to be applied? The answer would probably be no, but if the question was, as the engine seems to surge at idle should the idle speed be increased slightly, then the answer would most likely be yes.

It is not just the taking in of information and then applying a response as this is carried out by all electronic systems to some extent, but in being able to adapt and change. For example, if the engine was noticed to surge when the idle speed was set to 600 rev/min, then the ECU would increase the speed to, say, 700 rev/min. The adaptability, or a very simple form of AI, comes in deciding to set the idle speed at 700 rev/min on future occasions. This principle of modifying the response is the key. Many systems use a variation of this idea to control idle speed and also to adapt air–fuel ratios in response to a lambda sensor signal.

An adaptive ignition system has the ability to adapt the ignition point to the prevailing conditions. Programmed ignition has precise values stored in the memory appropriate for a particular engine. However, due to manufacturing tolerances, engine wear with age and road conditions means that the ideal timing does not always correspond to that held in the ECU memory.

The adaptive ignition ECU has a three-dimensional memory map as normal for looking up the basic timing setting, but it also has the ability to alter the spark timing rapidly, either retarding or advancing, and to assess the effect this has on engine torque. The ECU monitors engine speed by the crankshaft sensor, and if it sees an increase in speed after a timing alteration, it can assume better combustion. If this is the case, the appropriate speed load site on the memory map is updated. The increase in speed detected is for one cylinder at a time; therefore, normal engine speed changes due to the throttle operation do not affect the setting.

The operation of the adaptive ignition system is such as to try and achieve a certain slope on the timing versus torque curve as shown by Figure 10.56. Often the slope is zero (point A) for maximum economy but is sometimes non-zero (point B), to avoid detonation and reduce emissions.

Figure 10.57 shows the adaptive ignition block diagram. The fixed spark timing map produces a 'non-adapted' timing setting. A variation is then added or subtracted from this point and the variation is also sent to the slope detector. The slope detector determines whether the engine torque was increased or decreased from

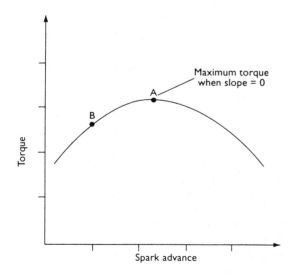

Figure 10.56 Timing versus torque curve

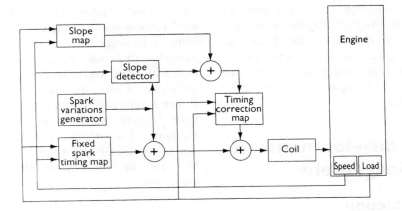

Figure 10.57 Adaptive ignition block diagram

the measure of the slope on the torque/timing curve compared with data from the slope map. The difference is used to update the timing correction map. The correction map can be updated every time a spark variation occurs, allowing very fast adaptation even during rapid changes in engine operation. The slope map can be used to aim for either maximum torque or minimum emissions.

10.8.9 Neural computing

The technology behind neural computing is relatively new and is expanding rapidly. The exciting aspect is that neural networks have the capacity to learn rather than having to be programmed. This form of artificial intelligence does not require specific instructions on how a problem can be solved. The user allows the computer to adapt itself during a training period, based on examples of similar problems. After training, the computer is able to relate the problem to the solution, inputs to outputs, and thus offer a viable answer to the 'question'.

The main part of a neural computer is the neural network, a schematic representation of which is shown in Figure 10.58. In this representation the circles represent neurons and the lines represent links between them. A neuron is a simple processor, which takes one or more inputs and produces an output. Each input has an associated 'weight', which determines its intensity or strength. The neuron simply has to determine the weight of its inputs and produce a suitably weighted output. The number of neurons in a network can range from tens to many thousands.

The way the system learns is by comparing its actual output with an expected output. This pro-duces an error value, which in turn changes the relative weights of the links back through the whole network. This eventually results in an ideal solution, as connections leading to the correct answer are strengthened. This, in principle, is similar to the way a human brain works. The neural computing system has a number of advantages over the conventional method.

- Very fast operation due to 'parallel processing'.
- Reduced development time.
- Ability to find solutions to problems that are difficult to define.
- Flexible approach to a solution, which can be adapted to changing circumstances.
- More robust, as it can handle 'fuzzy' data or unexpected situations. An adaptive fuzzy system acts like a human expert. It learns from experience and uses new data to fine-tune its knowledge.

The advantages outlined make the use of

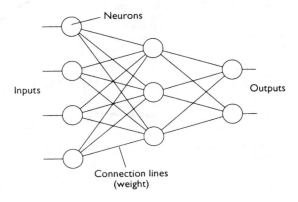

Figure 10.58 Neural network

neural nets on automobile systems almost inevitable. Some are even starting to be used in such a way that the engine control system is able to learn the driver's technique and anticipate the next most likely action. It can then set appropriate system parameters before the action even happens!

10.9 New developments in engine management

10.9.1 Introduction

Research is going on all the time into different ways of reducing emissions in order to keep within the current and expected regulations. In a way, the European market has an advantage as the emission laws in the USA and, in particular, the State of California, are very stringent and set to become more so. It is reasonable to expect that EC regulations will broadly follow the same route. The one potential difference is if CO_2 is included in the legislation. This will, in effect, make fuel consumption as big an issue as noxious emissions. Some of the current areas of development are briefly mentioned below. It is becoming clear that nitrogen oxides are the most difficult gases to reduce in line with future legislation. The technology for a NOx reducing catalyst has just started to reach production stage.

10.9.2 Lean burn engines

Any engine running at a lambda value greater than one is a form of lean burn. In other words,

the combustion takes place with an excess of air. Fuel consumption is improved and CO_2 emissions are lower than with a conventional 'lambda equals one and catalyst system'. However, with the same comparison, NOx emissions are higher. This is due to the excess air factor. Rough running can also be a problem with lean burn (Figure 10.59), due to the problems encountered lighting lean mixtures. A form of charge stratification is a way of improving this. Note also the case studies in this and the previous chapters.

10.9.3 Direct mixture injection

A new technique called DMI, or direct mixture injection, shows a potential 30% saving in fuel. This system involves loading a small mixing chamber above the cylinder-head with a suitable quantity of fuel during the compression stroke and start of combustion. This may be by a normal injector. The heat of the chamber ensures total fuel evaporation.

During an appropriate point in the next cycle the mixture is injected into the combustion chamber. This is one of the key advances because it is injected in such a way that the charge is in the immediate vicinity of the spark plug. This stratification is controlled by the mixture injection valve opening, the in-cylinder pressure and the mixing chamber pressure. Figure 10.60 shows the layout of a DMI system. The lambda values possible with this system range from 8 to 10 at idle and from 0.9 to 1 at full load. Compare this with the lean limits of a homogeneous mixture, which is typically $\lambda = 1.6–1.8$.

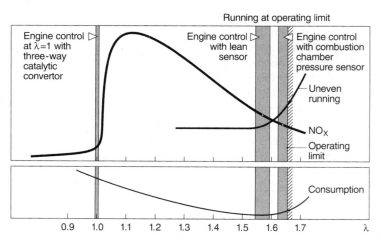

Figure 10.59 Lean-burn engine

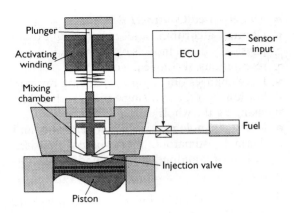

Figure 10.60 Layout of a direct mixture injection system

10.9.4 Two stroke engines

The two-stroke engine could be the answer to emission problems, but experts have differing views. The main reason for this is that the potential improvements for the four-stroke system have by no means been exhausted. The claimed advantages of the two-stroke engine are lower weight, lower fuel consumption and higher power density. These, however, differ depending on engine design. The major disadvantages are less smooth running, shorter life and higher NOx emissions. An Australian company, Orbital, have made a considerable contribution to two-stroke technology. A simple shutter control is used in their system and, in a published paper, a one litre two-stroke engine was compared with a one litre four-stroke engine. The two-stroke engine weighs 30% less, has lower consumption and low NOx levels while being comparable in all other ways. The engine can use direct injection to stratify the charge.

10.9.5 Alternative fuels

Engines using alcohol (e.g. ethanol) do not require major design changes. The fuel supply components would need to withstand corrosion and slightly different cold start strategies are needed. Other than this, changes to the engine 'maps' are all that is required. If an alcohol sensor is used in the fuel tank, the management system could adapt to changes in the percentage of alcohol used, if mixed with petrol. Some advantages in emissions are apparent with ethanol–petrol mixtures. It is said that the use of alcohol fuels is a political, not a technical issue.

Gas powered engines have been used for some time but storage of suitable quantities is a problem. These engines, however, do produce lower CO, HC and CO_2 emissions. Hydrogen powered vehicles offer the potential to exceed the ultra low emission vehicle (ULEV) limits, but are still in the early stages. Many manufacturers do, however, have prototypes. Electric powered vehicles, which meet the zero emission vehicle (ZEV) limits, are discussed in Chapter 17.

When all alternatives are considered it is clear that the petrol/gasoline and diesel engines are not easily replaceable. Indeed there are still many possible areas for further improvements.

10.9.6 Delphi's 'building block' approach to advanced engine management systems

This section is included as an example of how the 'current thinking' is going with regard to engine management systems in general. Delphi is a well respected company in this area.

The following is taken from a Press Release – Delphi Energy & Engine Management Systems, Presentation to the SAE, 1998.

'Engine management is the science of equipping and calibrating an engine to achieve the cleanest possible exhaust stream while maintaining top performance and fuel economy, and continuously diagnosing system faults. However, the focus on those priorities often varies around the world, reflecting differing governmental regulations, customer expectations and driving conditions and a host of vehicle types and content levels.

Typically, an engine management system integrates numerous elements, including:

- An engine control module (ECM).
- Control and diagnostics software.
- An air induction and control subsystem.
- A fuel handling module.
- A fuel injection module.
- An ignition subsystem.
- A catalytic converter.
- A subsystem to handle evaporative emissions.
- A variety of sensors and solenoids.'

Delphi states the following: 'We don't start at ground zero with each customer, in each market, with each vehicle. We use modular systems architecture, rapid calibration development tools and controls based on real world models. We use off-the-shelf interchangeable hardware whenever possible and software that will work in most systems and most processors. We use "plug and play" tools, like auto-code generation, so we do not have to recalibrate the whole system when we modify a piece of it.

Highlighted advanced engine management systems include the following:

- Modular systems architecture.
- Delphi's building block approach to engine management selects from sets of 'commonized', interchangeable software and electronics in the engine or powertrain control modules.
- Allowing OEMs to custom-build systems for widely differing markets.
- Software has expansion/deletion capabilities.
- Systems are designed with a minimum number of basic electronic controllers, which can be expanded if desired.
- Component hardware is interchangeable among systems.
- Software can be used across a variety of systems.
- Rapid Calibration Development Tools (RapidCal).
- Rapid prototyping permits immediate evaluation of the performance of new systems developments.
- Results can be benchmarked against plant/control models and rapid prototypes to verify correct implementation.

- Model-Based Controls (MBC).
- Control algorithms are redesigned around physically based models or mathematical representations of 'the real world'.
- Piece changes only require changing the calibration data for that single piece, rather than changing the whole system.
- MBC technologies include pneumatic and thermal estimators, model-based transient fuel control and individual cylinder fuel control.

Benefits of the building block approach include the following:

- Saves development costs.
- Offers flexibility to manufacturers.
- Adapts easily to the needs of a variety of customers, from emerging markets to high-end applications.
- Allows use of off-the-shelf components with minimal recalibration after modification.
- Enables compliance with varying emissions regulations over a wide range of driving conditions, driving habits, customer expectations and vehicle types.
- Saves fuel, reduces emissions.
- Reduces time-to-market for vehicle manufacturers.'

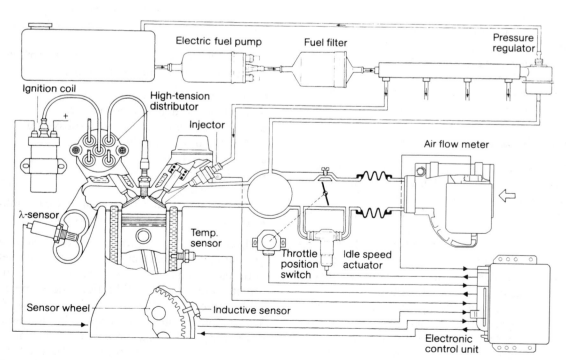

Figure 10.61 Motronic

10.9.7 Video link diagnostics

Some manufacturers have introduced hand-held video cameras to aid with diagnosing faults. This is relevant to all areas of the vehicle as well as engine management systems.

The camera is linked via an Internet/modem line from the dealers to the manufacturers. The technician is therefore able to show what tests have been done as well as describe the problem to the engineer/specialist.

10.10 Self-assessment

10.10.1 Questions

1. Describe what is meant by 'Engine management'.
2. State what the term 'light off' refers to in connection with catalytic converters.
3. Explain the stages of calculating 'fuel quantity' that take place in an ECU.
4. Make a clearly labelled sketch to show an exhaust gas recirculation system.
5. Draw a block diagram of an engine management system showing all the main inputs and outputs.
6. Describe the purpose of on-board diagnostics (OBD).
7. Make a simple sketch to show a variable length inlet manifold system.
8. State the information provided by a throttle potentiometer.
9. State four methods of reducing diesel engine emissions.
10. Explain the operation of a gasoline direct injection (GDI) system.

10.10.2 Assignment

1. Research the current state of development of 'lean-burn' technology. Produce an essay discussing current progress. Consider also the advantages and disadvantages of this method of engine operation. Make a reasoned prediction of the way in which this technology will develop.
2. Figure 10.61 shows an early version of the Motronic system. Compare this with the Motronic M5 or other systems and report on where, and why, changes have been made.

11 Lighting

11.1 Lighting fundamentals

11.1.1 Introduction

Vehicle lighting systems are very important, particularly where road safety is concerned. If headlights were suddenly to fail at night and at high speed, the result could be catastrophic. Many techniques have been used, ranging from automatic changeover circuits to thermal circuit breakers, which pulse the lights rather than putting them out as a blown fuse would. Modern wiring systems fuse each bulb filament separately and even if the main supply to the headlights failed, it is likely that dim-dip would still work.

We have come a long way since lights such as the Lucas 'King of the road' were in use. These were acetylene lamps! A key point to remember with vehicle lights is that they must allow the driver to:

- See in the dark.
- Be seen in the dark (or conditions of poor visibility).

Sidelights, tail lights, brake lights and others are relatively straightforward. Headlights present the most problems, namely that, on dipped beam they must provide adequate light for the driver but without dazzling other road users. Many techniques have been tried over the years and great advances have been made, but the conflict between seeing and dazzling is very difficult to overcome. One of the latest developments, ultraviolet (UV) lighting, which is discussed later, shows some promise.

11.1.2 Bulbs

Joseph Swan in the UK demonstrated the first light bulb in 1878. Much incremental development has taken place since that time. The number, shape and size of bulbs used on vehicles is

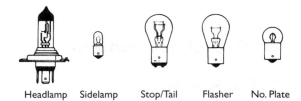

Headlamp   Sidelamp   Stop/Tail   Flasher   No. Plate

Figure 11.1 Selection of bulbs

increasing all the time. Figure 11.1 shows a common selection. Most bulbs for vehicle lighting are generally either conventional tungsten filament bulbs or tungsten halogen.

In the conventional bulb the tungsten filament is heated to incandescence by an electric current. In a vacuum the temperature is about 2300°C. Tungsten is a heavy metallic element and has the symbol W; its atomic number is 74; and its atomic weight 2.85. The pure metal is steel grey to tin white in colour. Its physical properties include the highest melting point of all metals: 3410°C. Pure tungsten is easily forged, spun, drawn and extruded, whereas in an impure state it is brittle and can be fabricated only with difficulty. Tungsten oxidizes in air, especially at higher temperatures, but it is resistant to corrosion and is only slightly attacked by most mineral acids. Tungsten or its alloys are therefore ideal for use as filaments for electric light bulbs. The filament is normally wound into a 'spiralled spiral' to allow a suitable length of thin wire in a small space and to provide some mechanical strength. Figure 11.2 shows a typical bulb filament.

If the temperature mentioned above is exceeded even in a vacuum, then the filament will become very volatile and break. This is why the voltage at which a bulb is operated must be kept within tight limits. The vacuum in a bulb prevents the conduction of heat from the filament but limits the operating temperature.

Gas-filled bulbs are more usual, where the glass bulb is filled with an inert gas such as

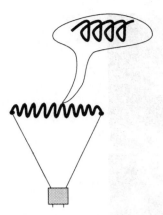

Figure 11.2 A bulb filament is like a spiralled spiral

argon under pressure. This allows the filament to work at a higher temperature without failing and therefore produce a whiter light. These bulbs will produce about 17 lm/W compared with a vacuum bulb, which will produce about 11 lm/W.

Almost all vehicles now use tungsten halogen bulbs for their headlights as these are able to produce about 24 lm/W (more for some modern designs). The bulb has a long life and will not blacken over a period of time like other bulbs. This is because in normal gas bulbs, over a period of time, about 10% of the filament metal evaporates and is deposited on the bulb wall. The gas in halogen bulbs is mostly iodine. The name halogen is used because there are four elements within group VIIA of the periodic table, known collectively as the halogens. The name, derived from the Greek hal- and -gen, means 'salt-producing'. The four halogens are bromine, chlorine, fluorine and iodine. They are highly reactive and are not found free in nature. The gas is filled to a pressure of several bar.

The glass envelope used for the tungsten halogen bulb is made from fused silicon or quartz. The tungsten filament still evaporates but, on its way to the bulb wall, the tungsten atom combines with two or more halogen atoms forming a tungsten halide. This will not be deposited on to the bulb because of its temperature. The convection currents will cause the halide to move back towards the filament at some point and it then splits up, returning the tungsten to the filament and releasing the halogen. Because of this the bulb will not become blackened, the light output will therefore remain constant throughout its life. The envelope can also be made smaller as can the filament, thus allowing better focusing. Figure 11.3 shows a tungsten halogen headlight bulb.

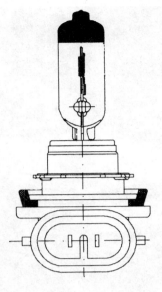

Figure 11.3 Halogen bulb

Next, some common bulbs are discussed further.

Festoon

The glass envelope has a tubular shape, with the filament stretched between brass caps cemented to the tube ends. This bulb was commonly used for number-plate and interior roof lighting.

Miniature centre contact (MCC)

This bulb has a bayonet cap consisting of two locating pins projecting from either side of the cylindrical cap. The diameter of the cap is about 9 mm. It has a single central contact (SCC), with the metal cap body forming the second contact, often the earth connection. It is made with various power ratings ranging from 1 to 5 W.

Capless bulb

These bulbs have a semi-tubular glass envelope with a flattened end, which provides the support for the terminal wires, which are bent over to form the two contacts. The power rating is up to 5 W, and these bulbs are used for panel lights, sidelights and parking. They are now very popular due to the low cost of manufacture.

Single contact, small bayonet cap (SBC)

These bulbs have a bayonet cap with a diameter of about 15 mm with a spherical glass envelope enclosing a single filament. A single central contact (SCC) uses the metal cap body to form the

(a)

(b)

second contact. The size or wattage of the bulb is normally 5 W or 21 W. The small 5 W bulb, is used for side or tail lights and the larger 21 W bulb is used for indicators, hazard, reversing and rear fog-lights.

Double contact, small bayonet cap

Similar in shape and size to the large SCC 15 mm SBC bulb, as described above. It has two filaments, one end of each being connected to an end contact, and both of the other ends are joined to the cap body forming a third contact,

which is usually the earth. These caps have offset bayonet pins so that the two filaments, which are of different wattage, cannot be connected the wrong way around. One filament is used for the stop light and the other for the tail light. They are rated at 21 and 5 W (21/5 W) respectively.

11.1.3 External lights

Regulations exist relating to external lights, the following is a simplified interpretation and amalgamation of current regulations, the range

(c)

(d)

Figure 11.4 Vehicle lighting designs. (a) Ford Focus; (b) Jaguar S-Type; (c) Mercedes-Benz S-class; (d) the Hyundai XG

of permissible luminous intensity is given in brackets after each sub heading.

Sidelights (up to 60 cd)

A vehicle must have two sidelights each with wattage of less than 7 W. Most vehicles have the sidelights incorporated as part of the headlight assembly.

Rear lights (up to 60 cd)

Again, two must be fitted each with wattage not less than 5 W. Lights used in Europe must be 'E'

marked and show a diffused light. Their position must be within 400 mm from the vehicle edge and over 500 mm apart, and between 350 and 1500 mm above the ground.

Brake lights (40–100 cd)

There two lights are often combined with the rear lights. They must be between 15 and 36 W each, with diffused light and must operate when any form of first line brake is applied. Brake lights must be between 350 and 1500 mm above the ground and at least 500 mm apart in a symmetrical position. High-level brake lights are now allowed and, if fitted, must operate with the primary brake lights.

Reversing lights (300–600 cd)

No more than two lights may be fitted with a maximum wattage each of 24 W. The light must not dazzle and either be switched automatically from the gearbox or with a switch incorporating a warning light. Safety reversing 'beepers' are now often fitted in conjunction with this circuit, particularly on larger vehicles.

Day running lights (800 cd max)

Volvo use day running lights as these are in fact required in Sweden and Finland. These lights come on with the ignition and must only work in conjunction with the rear lights. Their function is to indicate that the vehicle is moving or about to move. They switch off when parking or headlights are selected.

Rear fog lights (150–300 cd)

One or two may be fitted but, if only one, then it must be on the offside or centre line of the vehicle. They must be between 250 and 1000 mm above the ground and over 100 mm from any brake light. The wattage is normally 21 W and they must only operate when either the sidelights, headlights or front fog lights are in use.

Front spot and fog lights

If front spot lights are fitted (auxiliary driving lights), they must between 500 and 1200 mm above the ground and more than 400 mm from the side of the vehicle. If the lights are non-dipping then they must only operate when the headlights are on main beam. Front fog lamps are fitted below 500 mm from the ground and may only be used in fog or falling snow. Spot lamps are designed to produce a long beam of light to illuminate the road in the distance. Fog lights are designed to produce a sharp cut off line such as to illuminate the road just in front of the vehicle but without reflecting back or causing glare.

Figure 11.4 shows a selection of vehicle light designs and some of the groupings used.

11.1.4 Headlight reflectors

Light from a source, such as the filament of a bulb, can be projected in the form of a beam of varying patterns by using a suitable reflector and a lens. Reflectors used for headlights are usually para-bolic, bifocal or homifocal. Lenses, which are also used as the headlight cover glass, are used to direct the light to the side of the road and in a downward direction. Figure 11.5 shows how lenses and reflectors can be used to direct the light.

The object of the headlight reflector is to direct the random light rays produced by the bulb into a beam of concentrated light by applying the laws of reflection. Bulb filament position relative to the reflector is important, if the desired beam direction and shape are to be obtained. This is demonstrated in Figure 11.5(a). First, the light source (the light filament) is at the focal point, so the reflected beam will be parallel to the principal axis. If the filament is between the focal point and the reflector, the reflected beam will diverge – that is, spread outwards along the principal axis. Alternatively, if the filament is positioned in front of the focal point the reflected beam will converge towards the principal axis.

A reflector is basically a layer of silver, chrome or aluminium deposited on a smooth and polished surface such as brass or glass. Consider a mirror reflector that 'caves in' – this is called a concave reflector. The centre point on the reflector is called the pole, and a line drawn perpendicular to the surface from the pole is known as the principal axis. If a light source is moved along this line, a point will be found where the radiating light produces a reflected beam parallel to the principal axis. This point is known as the focal point, and its distance from the pole is known as the focal length.

Parabolic reflector

A parabola is a curve similar in shape to the curved path of a stone thrown forward in the air. A parabolic reflector (Figure 11.5(a)) has the property of reflecting rays parallel to the principal axis when a light source is placed at its focal point, no matter where the rays fall on the reflector. It therefore produces a bright parallel reflected beam of constant light intensity. With a parabolic reflector, most of the light rays from the light-bulb are reflected and only a small amount of direct rays disperses as stray light.

The intensity of reflected light is strongest near the beam axis, except for light cut-off by the bulb itself. The intensity drops off towards the outer edges of the beam. A common type of reflector and bulb arrangement is shown in Figure 11.6 where the dip filament is shielded. This gives a nice sharp cut-off line when on dip beam and is used mostly with asymmetric headlights.

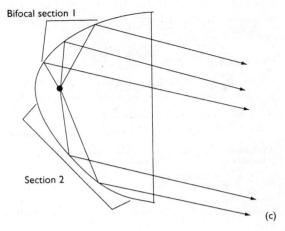

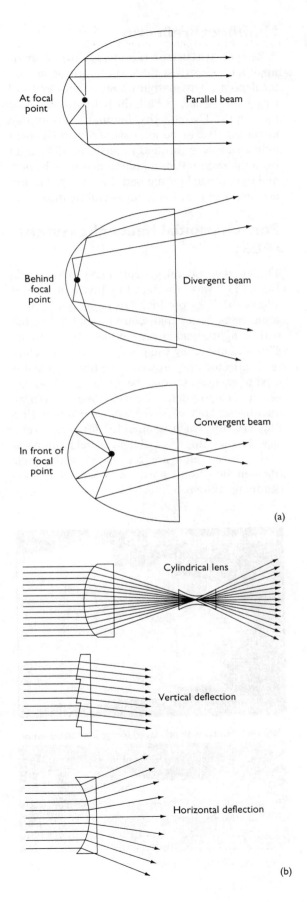

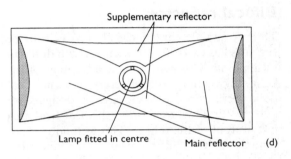

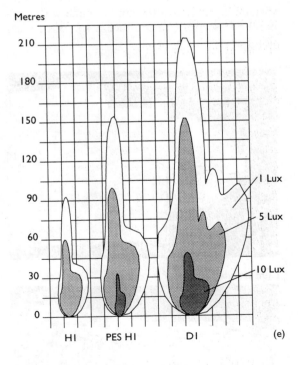

Figure 11.5 Headlight patterns are produced by careful use of lenses and reflectors

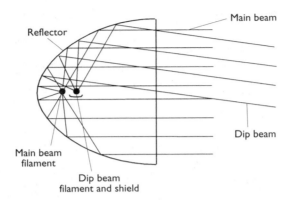

Figure 11.6 Creating a dip beam with a twin filament shielded bulb

Bifocal reflector

The bifocal reflector (Figure 11.5(c)) as its name suggests has two reflector sections with different focal points. This helps to take advantage of the light striking the lower reflector area. The parabolic section in the lower area is designed to reflect light down to improve the near field area just in front of the vehicle. This technique is not suitable for twin filament bulbs, it is therefore only used on vehicles with a four-headlight system. With the aid of powerful CAD programs, variable focus reflectors can be made with non-parabolic sections to produce a smooth transition between each area.

Homifocal reflector

A homifocal reflector (Figure 11.5(d)) is made up of a number of sections each with a common focal point. This design allows a shorter focal length and hence, overall, the light unit will have less depth. The effective luminous flux is also increased. It can be used with a twin filament bulb to provide dip and main beam. The light from the main reflector section provides the normal long range lighting and the auxiliary reflectors improve near field and lateral lighting.

Poly-ellipsoidal headlight system (PES)

The poly-ellipsoidal system (PES) as shown in Figure 11.7 was introduced by Bosch in 1983. It allows the light produced to be as good, or in some cases better than conventional lights, but with a light-opening area of less than $30\,cm^2$. This is achieved by using a CAD designed elliptical reflector and projection optics. A shield is used to ensure a suitable beam pattern. This can be for a clearly defined cut-off line or even an intentional lack of sharpness. The newer PES Plus system, which is intended for larger vehicles, further improves the near-field illumination. These lights are only used with single filament bulbs and must form part of a four-headlamp system.

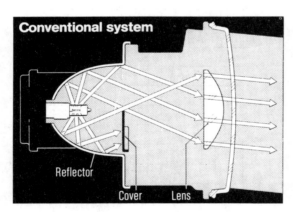

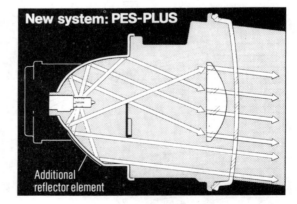

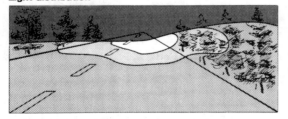

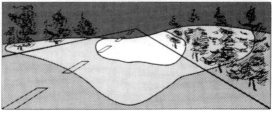

Figure 11.7 Improved poly-ellipsoid low beam

11.1.5 Headlight lenses

A good headlight should have a powerful far-reaching central beam, around which the light is distributed both horizontally and vertically in order to illuminate as great an area of the road surface as possible. The beam formation can be considerably improved by passing the reflected light rays through a transparent block of lenses. It is the function of the lenses partially to redistribute the reflected light beam and any stray light rays, so that a better overall road illumination is achieved with the minimum of glare. A block prism lens is shown as Figure 11.5(b).

Lenses work on the principle of refraction – that is, the change in the direction of light rays when passing into or out of a transparent medium, such as glass (plastic on some very recent headlights). The headlight front cover and glass lens, is divided up into a large number of small rectangular zones, each zone being formed optically in the shape of a concave flute or a combination of flute and prisms. The shape of these sections is such that, when the roughly parallel beam passes through the glass, each individual lens element will redirect the light rays to obtain an improved overall light projection or beam pattern.

The flutes control the horizontal spread of light. At the same time the prisms sharply bend the rays downwards to give diffused local lighting just in front of the vehicle. The action of lenses is shown as Figure 11.5(b).

Many headlights are now made with clear lenses, which means that all the light directionality is performed by the reflector (see Figure 11.4).

11.1.6 Headlight levelling

The principle of headlight levelling is very simple, the position of the lights must change depending on the load in the vehicle. Figure 11.8 shows a simple manual aiming device operated by the driver.

An automatic system can be operated from sensors positioned on the vehicle suspension. This will allow automatic compensation for whatever the load distribution on the vehicle. Figure 11.9 shows the layout of this system. The actuators, which actually move the lights, can vary from hydraulic devices to stepper motors.

The practicality of headlight aiming is represented by Figure 11.10. Adjustment is by moving two screws positioned on the headlights, such that one will cause the light to move up and down the other will cause side-to-side movement.

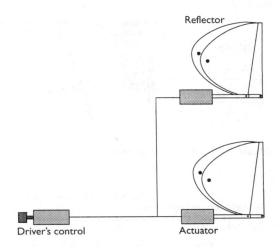

Figure 11.8 Manual headlight levelling

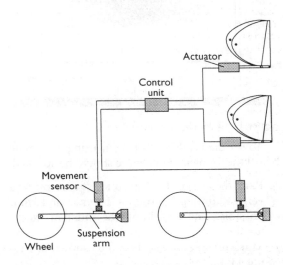

Figure 11.9 Automatic headlight adjustment

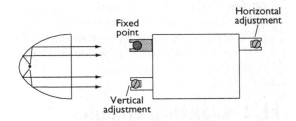

Figure 11.10 Principle of headlight aiming

11.1.7 Headlight beam setting

Many types of beam-setting equipment are available and most work on the same principle, which is represented in Figure 11.11. The method is the same as using an aiming board but is more convenient and accurate due to easier working and less room being required.

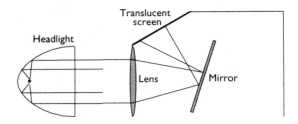

Figure 11.11 Beam setter principle

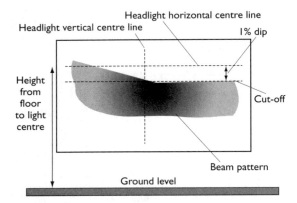

Figure 11.12 Headlight aiming board

To set the headlights of a car using an aiming board the following procedure should be adopted.

1. Park the car on level ground, square on to a vertical aiming board at a distance of 10 m if possible. The car should be unladen except for the driver.
2. Mark out the aiming board as shown in Figure 11.12.
3. Bounce the suspension to ensure it is level.
4. With the lights set on dip beam, adjust the cut-off line to the horizontal mark, which will be 1 cm* below the height of the headlight centre, for every 1 m the car is away from the board. The break-off point should be adjusted to the centre line of each light in turn.

11.2 Lighting circuits

11.2.1 Basic lighting circuit

Figure 11.13 shows a simple lighting circuit. Whilst this representation helps to demonstrate the way in which a lighting circuit operates, it is not now used in this simple form. The circuit does, however, help to show in a simple way how various lights in and around the vehicle

* or whatever the manufacturer recommends

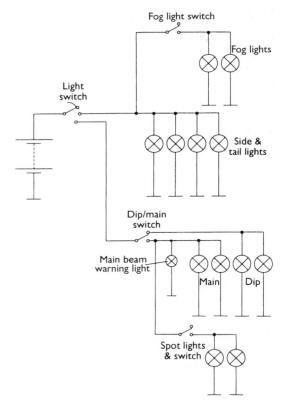

Figure 11.13 Simplified lighting circuit

operate with respect to each other. For example, fog lights can be wired to work only when the sidelights are on. Another example is how the headlights cannot be operated without the side-lights first being switched on.

11.2.2 Dim-dip circuit

Dim-dip headlights are an attempt to stop drivers just using sidelights in semi-dark or poor visibility conditions. The circuit is such that when sidelights and ignition are on together, then the headlights will come on automatically at about one-sixth of normal power.

If there is any doubt as to the visibility or conditions, switch on dipped headlights. If your vehicle is in good order it will not discharge the battery.

Dim-dip lights are achieved in one of two ways. The first uses a simple resistor in series with the headlight bulb and the second is to use a 'chopper' module, which switches the power to the headlights on and off rapidly. In either case the 'dimmer' is bypassed when the driver selects normal headlights. Figure 11.14 is a simplified circuit of dim-dip lights using a

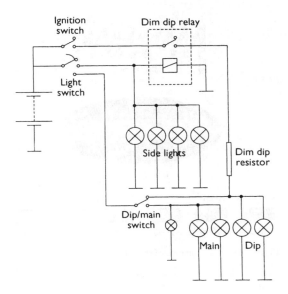

Figure 11.14 Simplified circuit of dim-dip lights using a series resistor

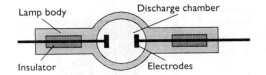

Figure 11.15 Operating principle of a gas discharge bulb

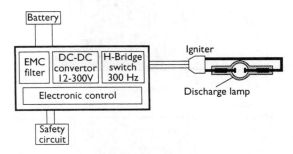

Figure 11.16 Ballast system to control a GDL

series resistor. This is the most cost-effective method but has the problem that the resistor (about $1\,\Omega$) gets quite hot and hence has to be positioned appropriately.

11.3 Gas discharge and LED lighting

11.3.1 Gas discharge lamps

Gas discharge headlamps (GDL) are now being fitted to vehicles. They have the potential to provide more effective illumination and new design possibilities for the front of a vehicle. The conflict between aerodynamic styling and suitable lighting positions is an economy/safety trade-off, which is undesirable. The new headlamps make a significant contribution towards improving this situation because they can be relatively small. The GDL system consists of three main components.

Lamp

This operates in a very different way from conventional incandescent bulbs. A much higher voltage is needed. Figure 11.15 illustrates the operating principle of a GD bulb.

Ballast system

This contains an ignition and control unit and converts the electrical system voltage into the operating voltage required by the lamp. It controls the ignition stage and run up as well as regulating during continuous use and finally monitors operation as a safety aspect. Figure 11.16 shows the lamp circuit and components.

Headlamp

The design of the headlamp is broadly similar to conventional units. However, in order to meet the limits set for dazzle, a more accurate finish is needed, hence more production costs are involved.

The source of light in the gas discharge lamp is an electric arc, and the actual discharge bulb is only about 10 mm across. Two electrodes extend into the bulb, which is made from quartz glass. The gap between these electrodes is 4 mm. The distance between the end of the electrode and the bulb contact surface is 25 mm – this corresponds to the dimensions of the standardized H1 bulb.

At room temperature, the bulb contains a mixture of mercury, various metal salts and xenon under pressure. When the light is switched on, the xenon illuminates at once and evaporates the mercury and metal salts. The high luminous efficiency is due to the metal vapour mixture. The mercury generates most of the light and the metal salts affect the colour spectrum. Figure 11.17 shows the spectrum of light produced by the GDL compared with that from a halogen H1 bulb. Table 11.1 highlights the difference in output between the D1 and H1 bulbs (the figures are approximate and for comparison only).

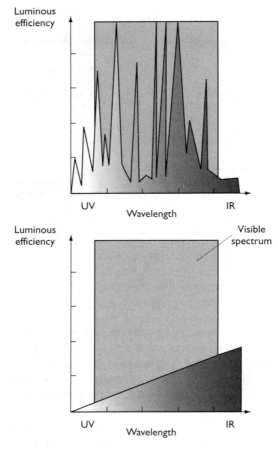

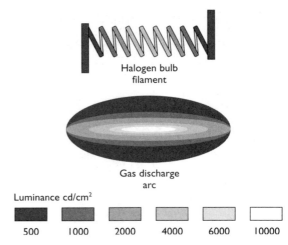

Figure 11.18 Luminance of the GDL compared with a halogen light bulb

Figure 11.17 Spectrum of light produced by the GDL (top) compared with that from a halogen H1 bulb

Table 11.1 Comparison of H1 and D1 bulbs

Bulb	Light	Heat	UV radiation
H1	8%	92%	<1%
D1	28%	58%	14%

The high output of UV radiation from the GDL means that for reasons of safety, special filters are required. Figure 11.18 shows the luminance of the GDL again compared with an H1 bulb. The average output of the GDL is three times greater.

To start the D1 lamp, the following four stages are run through in sequence.

● Ignition – a high voltage pulse causes a spark to jump between the electrodes, which ionizes the gap. This creates a tubular discharge path.
● Immediate light – the current flowing along the discharge path excites the xenon, which then emits light at about 20% of its continuous value.
● Run-up – the lamp is now operated at increased wattage, the temperature rises rapidly and the mercury and metal salts evaporate. The pressure in the lamp increases as the luminous flux increases and the light shifts from the blue to the white range.
● Continuous – the lamp is now operated at a stabilized power rating of 35 W. This ensures that the arc remains still and the output does not flicker. The luminous flux (28 000 lm) and the colour temperature (4500 K) are reached.

In order to control the above stages of operation, a ballast system is required. A high voltage, which can be as much as 20 kV, is generated to start the arc. During run-up, the ballast system limits the current and then also limits voltage. This wattage control allows the light to build up very quickly but prevents overshoot, which would reduce the life of the bulb. The ballast unit also contains radio suppression and safety circuits.

The complete headlamp can be designed in a different way, as the D1 bulb produces 2.5 times the light flux and at less than half the temperature of the conventional H1 bulb. This allows far greater variation in the styling of the headlamp and hence the front end of the vehicle.

If the GDL system is used as a dip beam, the self-levelling lights are required because of the high luminous intensities. However, use as a main beam may be a problem because of the on/off nature. A GDL system for dip beam, which stays on all the time and is supplemented by a conventional main beam (four headlamp system), may be the most appropriate use. Figure 11.5(e) shows the light distribution of the D1 and H1 bulbs used in headlamps.

11.3.2 Ultraviolet headlights

The GDL can be used to produce ultraviolet (UV) lights. Since UV radiation is virtually invisible it will not dazzle oncoming traffic but will illuminate fluorescent objects such as specially treated road markings and clothing. These glow in the dark much like a white shirt under some disco lights. The UV light will also penetrate fog and mist, as the light reflected by water droplets is invisible. It will even pass through a few centimetres of snow.

Cars with UV lights use a four-headlamp system. This consists of two conventional halogen main/dip lights and two UV lights. The UV lights come on at the same time as the dipped beams, effectively doubling their range but without dazzling.

Two-stage blue filters are used to eliminate visible light. Precise control of the filter colour is needed to ensure UVB and UVC are filtered out, as these can cause eye damage and skin cancer. This leaves UVA, which is just beyond the visible spectrum and is used, for example, in suntan lamps. However, some danger still exists; for example, if a child were to look directly and at close range into the faint blue glow of the lights. To prevent this, the lights will only operate when the vehicle is moving. This is a very promising contribution to road safety.

11.3.3 LED lighting

Light emitting diode (LED) displays were first produced commercially in 1968. Almost from this time there has been speculation as to possible vehicle applications. Such LEDs have certainly found applications in the interior vehicle, particularly in dashboard displays. However, until recently, legislation has prevented the use of LEDs for exterior lighting. A simple change in the legislative language from 'incandescent lamp' to 'light source', has at last made it possible to use lighting devices other than filament bulbs. Figure 11.19 shows a light unit containing LEDs.

The advantages of LED lighting are clear, the greatest being reliability. LEDs have a typical rated life of over 50000 hours, compared with just a few thousand for incandescent lamps. The environment in which vehicle lights have to survive is hostile to say the least. Extreme variations in temperature and humidity as well as serious shocks and vibration have to be endured.

Figure 11.19 Light units with LEDs

LEDs are more expensive than bulbs but the potential savings in design costs due to sealed units being used and the greater freedom of design could outweigh the extra expense. A further advantage is that they turn on quicker than ordinary bulbs. This turn-on time is important; the times are about 130 ms for the LEDs, and 200 ms for bulbs. If this is related to a vehicle brake light at motorway speeds, then the increased reaction time equates to about a car length. This is also potentially a major contribution to road safety.

Most of the major manufacturers are undertaking research into the use of LED lighting. Much time is being spent looking at the use of LEDs as high-level brake lights. This is because of their shock resistance, which will allow them to be mounted on the boot lid. In convertible cars, which have no rear screen as such, this application is ideal. Many manufacturers are designing rear spoilers with lights built in, and this is a good development as a safety aspect.

Heavy vehicle side marker lights are an area of use where LEDs have proved popular. Many lighting manufacturers are already producing lights for the after-market. Being able to use sealed units will greatly increase the life expectancy. Side indicator repeaters are a similar issue due to the harsh environmental conditions.

11.4 Case studies

11.4.1 Rover lighting circuit

The circuit shown in Figure 11.20 is the complete lighting system of a Rover vehicle. Operation of the main parts of this circuit is as follows.

Sidelights

Operation of the switch allows the supply on the N or N/S wire (colour codes are discussed on page 80) to pass to fuses 7 and 8 on an R wire. The two fuses then supply left sidelights and right sidelights as well as the number plate light.

Dipped beam

When the dip beam is selected, a supply is passed to fuse 9 on a U wire and then to the dim-dip unit, which is now de-energized. This then allows a supply to fuses 10 and 11 on the O/U wire. This supply is then passed to the left light on a U/K wire and the right light on a U/B wire.

Main beam

Selecting main beam allows a supply on the U/W wire to the main/dip relay, thus energizing it. A supply is therefore placed on fuses 21 and 22 and hence to each of the headlight main beam bulbs.

Dim-dip

When sidelights are on there is a supply to the dim-dip unit on the R/B wire. If the ignition supplies a second feed on the G wire then the unit will allow a supply from fuse 5 to the dim-dip resistor on an N/S wire and then on to the dim-dip unit on an N/G wire. The unit then links this supply to fuses 10 and 11 (dip beam fuses).

11.4.2 Generic lighting circuit – Bosch

Figure 11.21 shows a typical lighting circuit using the 'flow diagram' or schematic technique. The identifiers are listed in the Table 11.2. Note that, when following this circuit, the wires do not pass directly through the 'lamp check module' from top to bottom. There is a connection to the appropriate lamp but this will be through for example, a sensing coil.

Also, note how codes are used to show connections from some components to others rather than a line representing the wire. This is to reduce the number of wires in general but also to reduce cross-over points.

11.4.3 Xenon lighting – Hella

The risk of being injured or killed in a traffic accident on the roads is much higher at night than during the day, in spite of the smaller volumes of traffic. Although only about 33% of accidents occur at dusk or in the dark, the number of persons seriously injured increases by 50%, and the number of deaths by 136% compared with accidents that occur during the day.

Alongside factors such as self-dazzling caused by wet road surfaces, higher speeds because of the reduced traffic density and a reduction of about 25% of the distance maintained to the vehicle in front, causes relating to eye physiology play a very important role.

The eyes age faster than any other sensory organ, and the human eye's powers of vision begin to deteriorate noticeably from as early an age as 30! The consequence of this – a reduction in visual acuity and contrast sensitivity when the light begins to fade – is a situation that is very rarely noticed by the motorist, as these functional deficits develop only slowly.

However, the vision – even of a person with healthy eyes – is considerably reduced at night. The associated risk factors include delayed adjustment to changes between light and dark, impaired colour vision and the slow transition from day to night, which, through the habituation effect, can lull the motorist into a false sense of security.

Hella – for the past 100 years a forerunner in the development and production of innovative headlamp and lighting systems – is therefore giving increasing backing to xenon technology, the only system that offers more light than conventional tungsten bulbs – and that is daylight quality.

However, a good xenon headlamp alone is not enough to translate the additional light quantity and quality into increased safety. In order, for example, to avoid the hazard of being dazzled by oncoming traffic, the legally required range of additional equipment includes such items as headlamp cleaning equipment and automatic beam levellers. Only the system as whole is able to provide the clear advantage of higher safety for all road-users, even under the most adverse weather conditions. This means that even in rain, fog and snow, spatial vision is improved and the motorist's orientation abilities are less restricted.

Already today, according to a survey, 94% of

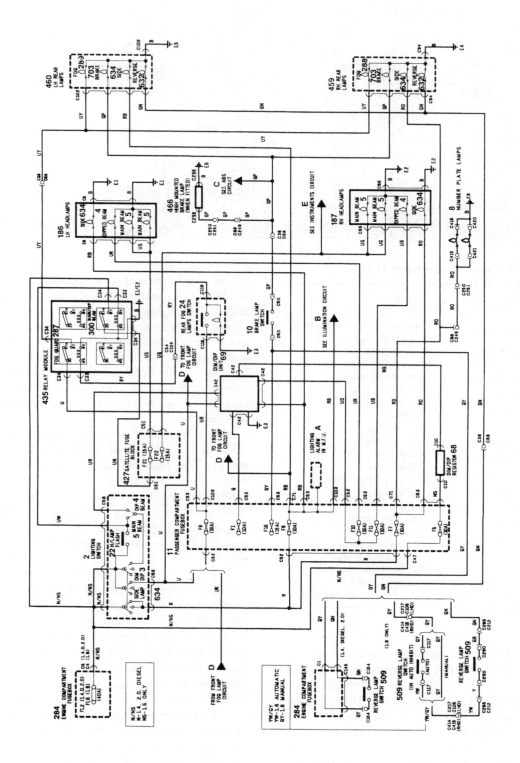

Figure 11.20 Complete vehicle lighting circuit

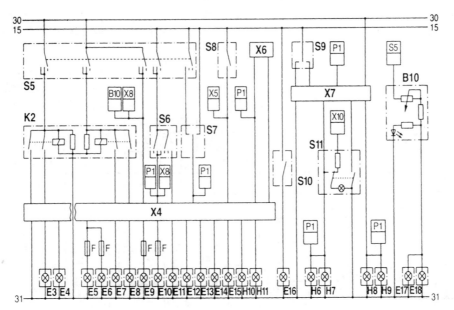

Figure 11.21 Lighting circuit flow diagram

Table 11.2 Identifiers for Figure 11.21

Identifier	Device
B10	Dimmer for instrument lighting
E3,4	Fog warning lamps
E5,6	Main beam headlamps
E7,8	Fog lamps
E9, 10	Dip beam headlamps
E11, 12	Side-marker lamps
E13	Number plate lamp
E14, 15	Tail lamps
E16	Reverse lamp
E17, 18	Instrument lighting
H6, 7, 8, 9	Indicator lamps
K2	Lighting relay
S5	Headlamp switch
S6	Fog lamp switch
S7	Dip switch
S8	Stop lamp switch
S9	Turn signal switch
S10	Back-up lamp switch
S11	Hazard warning switch
X4	Plug, lamp check module
X6	Plug, check control
X7	Socket, hazard warning relay

xenon headlamp users are convinced of their positive benefits. Night vision is improved claim 85% of users – in the case of the over-50s this figure is even increased to 90%. Visibility in rain is also judged by 80% to be better, while 75% of those surveyed have perceived an increase in safety for cyclists and pedestrians owing to the wider illumination of the road. The same percentage maintains that, thanks to xenon light, obstacles on the road are more easily recognized.

In order to make this increase in active safety available to as many road users as possible, the automobile industry – whether as standard equipment or as an optional accessory – is laying more emphasis on xenon headlamps. The annual requirement for xenon headlamps in Europe is estimated to rise to over two million units by the year 2000. Today, more than 600 000 cars have already been equipped with xenon headlamps.

The xenon bulb is a micro-discharge bulb filled with a mixture of noble gases including xenon. The bulb has no filament, as is the case with a halogen bulb, but the light arc is created between two electrodes. As is the case with other gas discharge bulbs, the xenon bulb has an electronic starter for quick ignition, and requires an electronic ballast to function properly.

The xenon bulb provides more than twice the amount of light of a halogen bulb, while only consuming half the power. Therefore, the driver can see more clearly, and the car has more power for other functions. Moreover, it is environmentally friendly, as less power means less fuel consumption. The clear white light produced by the xenon bulb is similar to daylight. Research has shown that this enables drivers to concentrate better. Furthermore, this particular light colour reflects the road markings and signs better that conventional lighting. The xenon bulb also delivers a marked contribution to road safety in the event of limited visibility due to weather conditions. In practical terms, the life span of the bulb is equal to that of the car, which means that the bulb need only be replaced in exceptional cases.

The light produced by a xenon bulb is, in fact, not blue but white, falling well within the international specifications for white light – the light only appears blue in comparison to the warmer 'yellow' light produced by halogen. However, it clearly appears white in comparison to daylight. Technically speaking, it is possible to adapt the light colour produced, but this would lead to a substantial loss of intensity, thereby cancelling out the particular advantages.

The international regulations governing light distribution and intensity on the road are very strict. Xenon light falls well within these boundaries. In addition, technically speaking, xenon lighting is less irritating than conventional light. As the light–darkness borders are much more clearly defined, less light is reflected into the eyes of oncoming drivers. The increased amount (double) of light produced is mainly used to achieve higher intensity and better distribution of light on the road. Moreover, the verges are also better lit. There are three conditions that must be met. These are contained in the international regulations concerning the use of xenon light: the headlamps must be aligned according to regulations; the vehicle must be fitted with an automatic headlamp levelling system, so that when the load is increased the headlight beams are automatically adjusted; the headlamp must be fitted with an automatic cleaning system, as dirt deposits on the lens act as a diffuser, thereby projecting the light beyond the prescribed range. These three conditions together with the extensive life span of the xenon bulb greatly reduce the risk of incorrectly aligned headlamps. The use of halogen bulbs entails a much higher risk.

Xenon light sometimes appears to irritate oncoming drivers. In normal circumstances drivers look straight ahead; however, due to the conspicuous colour of xenon light, drivers are more inclined to look into the headlamps. The same phenomenon was experienced during the introduction of halogen headlamps in the 1960s. In those days people also spoke of 'that irritating white light'. The introduction of xenon headlamps will therefore entail a period in which everybody will become accustomed. Figure 11.22 shows the xenon lamp from Hella.

Figure 11.22 Hella xenon lighting

11.5 Diagnosing lighting system faults

11.5.1 Introduction

As with all systems the six stages of faultfinding should be followed.

1. Verify the fault.
2. Collect further information.
3. Evaluate the evidence.
4. Carry out further tests in a logical sequence.
5. Rectify the problem.
6. Check all systems.

The procedure outlined in the next section is related primarily to stage 4 of the process. Table 11.3 lists some common symptoms of a lighting system malfunction together with suggestions for the possible fault. The faults are very generic but will serve as a good reminder.

Table 11.3 Common symptoms and possible faults of a lighting system malfunction

Symptom	Possible fault
Lights dim	• High resistance in the circuit
	• Low alternator output
	• Discoloured lenses or reflectors
Headlights out of adjustment	• Suspension fault
	• Loose fittings
	• Damage to body panels
	• Adjustment incorrect
Lights do not work	• Bulbs blown
	• Fuse blown
	• Loose or broken wiring/connections/fuse
	• Relay not working
	• Corrosion in light units
	• Switch not making contact

11.5.2 Testing procedure

The process of checking a lighting system circuit is broadly as follows:

1. Hand and eye checks (loose wires, loose switches and other obvious faults) – all connections clean and tight.
2. Check battery (see Chapter 5) – must be 70% charged.
3. Check bulb(s) – visual check or test with ohmmeter.
4. Fuse continuity – (do not trust your eyes) voltage at both sides with a meter or a test lamp.
5. If used, does the relay click (if yes, jump to stage 8) – this means the relay has operated, it is not necessarily making contact.
6. Supply to switch – battery volts.
7. Supply from the switch – battery volts.
8. Supplies to relay – battery volts.
9. Feed out of the relay – battery volts.
10. Voltage supply to the light – within 0.5 V of the battery.
11. Earth circuit (continuity or voltage) – $0\,\Omega$ or 0 V.

11.6 Advanced lighting technology

11.6.1 Lighting terms and definitions

Many unusual terms are used when relating to lighting, this section aims to give a simplified description of those used when dealing with vehicle lighting. First, terms associated with the light itself are given, and then terms relating more particularly to vehicle lights. The definitions given are generally related to the construction and use of headlights.

Luminous flux (φ)

The unit of luminous flux is the lumen (lm). Luminous flux is defined as the amount of light passing through an area in one second. The lumen is defined as the light falling on a unit area at a unit distance from a light source, which has a luminous intensity of one candela.

Luminous intensity *I*

This is the power to produce illumination at a distance. The unit is the candela (cd); it is a mea-

sure of the brightness of the light rather than the amount of light falling on an object.

Illumination intensity *E*

This can be defined on a surface as the luminous flux reaching it per unit area. The luminous intensity of a surface such as the road will be reduced if the light rays are at an angle. The unit is the lux (lx), it is equivalent to one lumen per square metre or to the illuminance of a surface one metre from a point source of light of one candela. In simple terms it depends on the brightness, distance from, and angle to, a light source.

Brightness or luminance *L*

This should not be confused with illumination. For example when driving at night the illumination from the vehicle lights will remain constant. The brightness or luminance of the road will vary depending on its surface colour. Luminance therefore depends not just on the illumination but also on the light reflected back from the surface.

Range of a headlight

The distance at which the headlight beam still has a specified luminous intensity.

Geometric range

This is the distance to the cut-off line on the road surface when the dip beam is set at an inclination below the horizontal.

Visual range

This is affected by many factors so cannot be expressed in units but it is defined broadly as the distance within the luminous field of vision, at which an object can still be seen.

Signal identification range

The distance at which a light signal can be seen under poor conditions.

Glare or dazzle

This is again difficult to express, as different people will perceive it in different ways. A figure is used, however, and that is if the luminous intensity is 1 lx at a distance of 25 m, in front of a dipped headlight at the height of the light centre, then the light is said not to glare or dazzle. The old British method stated that the lights must not dazzle a person on the same horizontal plane as the vehicle at a distance over 25 feet, whose eye level is more than 3 ft 6 in above the plane (I presume s/he is sitting down.)! In general, headlights when on dipped beam must fall below a horizontal line by 1% (1.2% or more in some cases) or 1 cm/m.

11.6.2 Expert Lighting Systems

The Expert Lighting System is a new Valeo technology developed to adapt the headlamp beam to various road and traffic conditions. The low beam is adapted to curves and the high beam to the vehicle's speed. These lighting functions provide drivers with:

- enhanced comfort due to the increased quantity of light and quality of the beam,
- improved safety, particularly in difficult driving conditions such as winding mountain roads.

This function is achieved by additional moving reflectors, which rotate according to the position of the steering wheel (in line with the direction of the driver's sight). The additional beam illuminates the area beyond or at the curve that is not normally illuminated by a traditional low beam function.

High beam adaptation to speed is based on the translation of 'additional mirrors' within the high beam reflector. The high beam is automatically adapted for beam width and range according to vehicle speed. This function is not subject to the introduction of new regulations.

11.6.3 Intelligent front lighting – Hella

The lighting of modern vehicles has improved continually in the past few decades. The halogen technology developed by Hella in particular set new standards after it was introduced early in the 1970s, as has xenon technology in the 1990s. The advantages of these systems were, and still are, their high lighting performance and their precise light distribution. The intelligent lighting systems of the future, however, will have to offer even more than this in order to make driving safer and more enjoyable.

In cooperation with the motor industry, Hella is masterminding a project for the development of an intelligent front lighting system for future generations of motor vehicles. Market research surveys conducted all over Europe first enabled an analysis to be made of the requirements drivers make on their vehicle lighting.

European drivers, according to this study, would like the front lighting to respond to the various different light conditions they encounter such as daylight, twilight, night-time, and driving in and out of tunnels, and to such weather situations as rain, fog, or falling snow. They would also like better illumination on bends. Drivers would also like better light on motorways. Their list of requirements also includes better light along the edge of the road, and additional light for parking in a narrow space and when reversing.

For Hella's lighting experts, turning these requirements into an intelligent front lighting system means comprehensive detail work and the development of totally new lighting technologies that can respond in various different ways to all these different situations, some of which call for contradictory patterns of light distribution.

For instance, direct lighting of the area immediately in front of the car is desirable when the roadway is dry, but can dazzle oncoming traffic if the road is wet. Light emitted above the cut-off line in fog dazzles the driver him/herself. And a long-range, narrow pattern of light distribution for high-speed motorway driving is unsuitable on twisting country roads, where the need is for a broad illumination in front of the car, possibly augmented by special headlamps for bends or a 'dynamic' long-range lighting system. Despite the wide diversity of all these light-distribution patterns, none must be allowed to dazzle oncoming drivers.

Another theme is the idea of lights that switch on automatically. Unlit vehicles keep turning up at night, for instance in city-centre traffic, because the street lighting is so good that some drivers fail to notice that they are driving without lights. The same phenomenon can be seen where cars drive through tunnels. In both cases, the unlit vehicles represent a major safety risk because other road users can hardly see them.

With the aid of the sensors that are already installed on some vehicles, an intelligent lighting system can recognize the ever-changing light situation and give the appropriate assistance to the driver. For instance, the sunlight sensors that already exist for controlling air-conditioning systems, or speed sensing devices, could also deliver data to an intelligent lighting system.

Additional sensors for ambient light and light density in the field of vision, for identifying a dry or wet road, fog, and whether the road ahead is straight or curved, could also deliver important data. In modern vehicles with digital electronic systems and bus interfaces, these data will not only be useful to the lighting system but also to the other electronically controlled systems, such as ABS or ASR, and give the driver vital assistance particularly in the most difficult driving situations.

The data transmitted by the various sensors on a vehicle can only be put to use if the vehicle has a 'dynamic' headlamp system that is capable of producing various different light-distribution patterns. This could begin with an automatic, dynamic height-adjustment and headlamps that automatically swivel sideways and could even include variable reflectors providing a whole range of light-distribution patterns.

11.7 New developments in lighting systems

11.7.1 Blue lights!

Philips 'BlueVision' white light stimulates driver concentration and makes night-time driving less tiring and reflects much better on road markings and signs. The new headlight and sidelight bulbs meet all the European safety legislation. The bulbs are directly interchangeable with existing bulbs.

With the introduction of BlueVision, Philips Automotive Lighting is illuminating the way ahead to the future of enhanced headlamp performance. The future is ... white light BlueVision. For the simple reason that the Blue Vision lamps reproduce day-light type light ... in night-time conditions!

Using the UV cut quartz developed by Philips for halogen lamps means that BlueVision can safely be used for all headlamps. However, it should be noted that halogen technology is not comparable to the xenon discharge technology, fitted as original equipment to more and more of the world's cars.

11.7.2 New signalling and lighting technologies

Valeo Lighting Systems has developed new signal lighting technologies to provide more variety and innovation to signal lamp concepts, which are a key styling feature on cars.

Jewel aspect signal lamps

Jewel aspect signal lamps are based on the complex shape technology widely used in headlamps. Beam pattern is no longer completely controlled by the lens but by the reflector which, in some cases, may be in conjunction with an intermediary filter. Conventional lens optics using prisms is minimized, giving the impression of greater depth and brightness.

Mono-colour signal lamps

With mono-colour technology, in addition to the traditional red functions (stop, tail lamp and fog), the reverse and turn signal functions appear red when not in use, but emit white and amber light respectively when functioning. Several technologies make this possible. In the case of subtractive synthesis lamps, coloured screens are placed in front of the bulb. Their colours are selected so that, in conjunction with the red of the external lens, they colour the light emitted by the lamp in line with the regulations: white for reverse, amber for the turn signal. Complementary colour technology uses a two-colour external lens, which combines red (dominant) and its complementary colour (yellow for the turn signal, blue for reverse). The combination of these two lights – red and yellow for the turn signal, red and blue for reverse – produces the colour of light (white or amber) stipulated by the regulations.

Linear lighting

Linear tail lamps can easily be harmonized with the design of the vehicle by introducing the aspect of very elongated lamps. Each function light is narrow, (35 mm), and can be up to 400 mm long. The lamps use optical intermediary screens, which are so precise that they not only fulfil legal photometric requirements but also create a harmonious overall aspect and very distinct separations between the function lights. This new technology is particularly well suited for the rear of mini-vans and light trucks.

New light sources for signal lamps

LED (light emitting diode) and neon combination lamps are a unique way to combine style and safety. Innovative style: thanks to their compactness, LED and neon offer enhanced design flexibility, notably for highlighting the lines of the vehicle and illuminating the bumper. Their homogeneous or pointillist appearance accentuates the differentiation and high-tech aspect of these signal lamps. Increased safety: the response time of these new sources, approximately 0.2 s faster than incandescent bulbs, allows danger to be anticipated as it provides the equivalent of 5 m extra braking distance for a vehicle travelling behind at 120 km/h.

Centre high mounted stop lamps (CHMSLs)

An LED CHMSL illuminates 0.2 s faster than conventional incandescent lamps, improving driver response time and providing extra braking distance of 5 m at 120 km/h. Owing to their low height and reduced depth, LED CHMSLs can be easily harmonized with all vehicle designs, whether they are mounted inside or integrated into the exterior body or spoiler. The lifetime of an LED CHMSL is greater than 2 000 hours, exceeding the average use of the light during the life of the vehicle. Each new LED generation feature enhances photometric performance and allows a reduction in the number of LEDs required for the CHMSL function. This number has already decreased from 16 to 12 in some configurations and should decrease even further over the next few years.

Neon technology

As with LED technology, neon lamps have an almost instantaneous response time (increased safety), take up little space (design flexibility) and last more than 2000 hours, thus exceeding the average use of a CHMSL during the life of the vehicle. Moreover, the neon CHMSL is very homogeneous in appearance and offers unmatched lateral visibility.

11.7.3 Electric headlamp levelling actuators

The primary function of a levelling actuator is to adjust the low beam in accordance with the load carried by the car and thereby avoid dazzling oncoming traffic. Manual electric levelling actuators are connected up to a control knob on the dashboard so allowing the driver to adjust beam height.

In addition to its range of manual electric headlamp levelling actuators, Valeo now also offers a new range of automatic actuators. As their name implies, these products do not require any driver adjustment. They are of two types.

- Automatic static actuators adjust beam height to the optimum position in line with vehicle load conditions. The system includes two sensors (front and rear) which measure the attitude of the vehicle. An electronic module converts data from the sensors and drives two electric gear motors (or actuators) located at the rear of the headlamps, which are mechanically attached to the reflectors. Beam height is adjusted every 10–30 s.
- Automatic dynamic adjusters have two sensors, an electronic module and two actuators. The sensors are the same as in the static system but the electronic module is more sophisticated in that it includes electronics that control rapid response actuator stepper motors. Response time to changes in vehicle attitude due to acceleration or deceleration is measured in tenths of a second. Corrective action is continuous and provides enhanced driving comfort, as the beam aim is optimized. In line with regulations, automatic dynamic levelling actuators are mandatory on all vehicles equipped with high intensity discharge (HID) lighting systems.

11.7.4 Baroptic styling concept

The Baroptic concept provides flexibility in the front-end styling of vehicles for the year 2000 and beyond while optimizing aerodynamics. The Baroptic lighting system's volume is significantly reduced as compared with complex shape technology. The volume benefits allow enhanced management of 'under hood' packaging. The product is a breakthrough both in terms of volume and shape. The futuristic elongated appearance of Baroptic headlamps, illuminated or not, sets them apart from conventional headlamps which tend to be oval or circular-shaped.

The Baroptic uses a new optical concept. Traditionally, the luminous flux emitted by the source is reflected by the surface of the reflector (parabolic or complex shape) and the beam is spread by a striated outer lens or refocused by the inner lenses (elliptical reflector), which then projects this flux onto the road.

In the Baroptic system, the luminous flux generated either by a halogen or a HID lamp is projected into an optical guide with reflecting facets. It is then focused through lenses and, positioned along the optical guide, which defines, in conjunction with shields, the desired beam characteristics: spread, width, length, cutoff and homogeneity.

The benefit of this total reflection system is that photometric performance is similar to

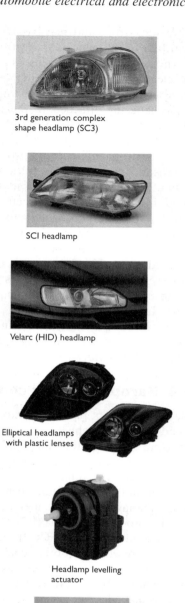

3rd generation complex
shape headlamp (SC3)

SCI headlamp

Velarc (HID) headlamp

Elliptical headlamps
with plastic lenses

Headlamp levelling
actuator

Fog lamp

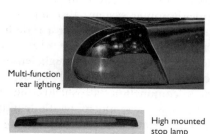

Multi-function
rear lighting

High mounted
stop lamp

Figure 11.23 SC3 and other lighting products from 'Valeo'

normal-sized headlamps. The spread of light is also optimized, which serves to enhance visual comfort when driving at night. The Baroptic system is currently under development.

11.7.5 Complex shape reflectors

The surface of the reflector is calculated through advanced computer analysis using a minimum of 50 000 individual points, each specific to the headlamp model under design. The third generation of complex shape reflectors (SC3) combines the benefits of the first two developments and controls both beam cut-off and pattern as well as homogeneity. SC3 headlamp lenses can be perfectly clear or with striations purely for decorative purposes. The lens is there to enhance aesthetic appeal and aerodynamics. Figure 11.23 shows a headlamp using this technique together with some other lighting components.

11.7.6 Infrared lights

Thermal-imaging technology promises to make night driving less hazardous. Infrared thermal-imaging systems are going to be fitted to cars. The Cadillac division of General Motors is now offering a system called 'Night Vision' as an option. After 'Night Vision' is switched on, 'hot' objects, including animals and people show up as white in the thermal image, as shown in Figure 11.24.

The infrared end of the light spectrum was discovered as long ago as 1800 by William Herschel. When investigating light passing through a prism, Herschel found heat was being emitted by rays he could not see. This part of the spectrum is called infrared (from the Latin infra, meaning 'below') because the rays are below the frequency of red light. The infrared spectrum begins at a wavelength of about 0.75 µm and

Figure 11.24 Night vision system in use

extends up to 1 mm. Every object at a temperature above absolute zero (−273°C) emits some kind of infrared radiation.

On the vehicle system a camera unit sits on headlamp-type mountings in the centre of the car, behind the front grille. Its aim is adjusted just like that of headlamps. The mid-grille position was chosen because most front collisions involve offset rather than full head-on impacts. However, the sensor is claimed to be tough enough to withstand 9 mph (14.5 kph) bumper impacts anyway. The sensor is focused 125 m ahead of the car as shown in Figure 11.25.

The outer lens of the sensor is coated with silicon to protect it against scratching. Behind this are two lenses made of black glass called tecalgenite. This is a composite material that transmits infrared easily but visible light will not pass through it.

The device looks a bit like a conventional camera, but instead of film it houses a bank of ferroelectric barium-strontium-titanate (BST) sensor elements; 76 800 of them can be packed onto a substrate measuring 25 mm square. Each element is a temperature dependent capacitor, the capacitance of which changes in direct proportion to how much infrared radiation it senses. This is termed an uncooled focal plane array (UFPA). An electrically-heated element maintains a temperature of 10°C inside the UFPA, enabling it to operate between ambient temperatures of −40 and +85°C.

Between the lens and the bank of UFPA sensor elements there is a thin silicon disc rotated by an electric motor at 1800 rev/min. Helical swirls are etched on some segments of the disc. Infrared radiation is blocked by the swirls but passes straight through the plain segments. The UFPA elements respond to the thermal energy of the objects viewed by the lens. Each sensor's reading switches on and off every 1/30 of a second, thus providing video signals for the system's head-up display (HUD).

The display, built into the dashboard, projects a black-and-white image, which the driver sees near the front edge of the car's bonnet. Objects in the image are the same size as viewed by the UFPA, helping the driver judge distances to them.

11.7.7 RGB lights

The reliability of the LED is allowing designers to integrate lights into the vehicle body in ways that have so far not been possible. The colour of light emitted by LEDs is red, orange, amber,

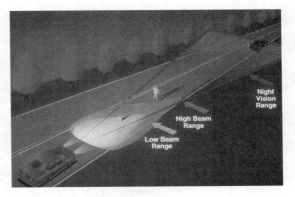

Figure 11.25 Night vision system range

yellow or green. Developments are progressing to produce a blue LED which, when combined with red and green, will allow white light from a solid state device. Red, green and blue are the primary colours of light and can be mixed to produce any other colour. This is how the combinations of pixels (RGB), on a colour monitor or television screen operate.

The possibilities as the technology develops are very wide. The type of lights used and the possible position of the lights on the vehicle are limitless. Rear lights in particular could be changed depending on what the requirements were. For example, when travelling normally, the rear lights would be red but when reversing all of the light could be white.

11.7.8 Single light-source lighting

It is now possible to use a gas discharge lamp (GDL) as a central source for vehicle lighting. Development of this new headlamp system allows a reduction in headlamp dimensions for the same output or improved lighting with the same dimensions. Using a GDL as a central light source for all the vehicle lights is shown in Figure 11.26.

The principle is that light from the 'super light source', is distributed to the headlamps and other lamps by a light-guide or fibre-optic link. The light from the GDL enters the fibre-optics via special lenses and leaves the light-guide in a similar manner as shown in Figure 11.27. A patterned covered lens provides the required light distribution. Shields can provide functions such as indicators, or electro-chromatic switches may even become available.

Heat build-up can be a problem in the fibre-optics but an infrared permeable coating on the reflector will help to alleviate this issue. The light-guide system has a very low photometric efficiency (10–20% at best), but the very efficient

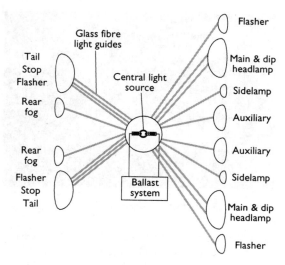

Figure 11.26 Using a GDL as central light source for all the vehicle lights

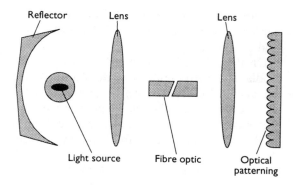

Figure 11.27 The light from the gas discharge lamp (GDL) enters and leaves the light guide via a special lens

light source still makes this technique feasible. One of the main advantages is being able to improve the light distribution of the main headlamp. Due to the legal limits with regard to dazzle, conventional lights do not intensely illuminate the area just under the cut-off line. Consequently, several glass fibre bundles can be used to direct the light in an even distribution onto the desired areas of the road.

The central light source can be placed anywhere in the vehicle. Only one source is required but it is thought that a second would be used for safety reasons. A vehicle at present uses some 30 to 40 bulbs, and this number could be reduced markedly. A single light source could be utilized for rear lights on the vehicle, which would allow rear lights with an overall depth of only about 15 mm. This could be supplied with light from a single conventional bulb.

11.7.9 Xenon headlamps for retro-fitting

Hella can now supply headlamps with xenon light for retro-fitting to the current BMW 5-series and the Mercedes Benz E-series models. In addition to the headlamps themselves, the package also includes an automatic headlamp-range adjustment system and, if it is not already fitted to the car, a headlamp cleaning system. These components are mandatory in many countries for xenon headlamp systems.

The automatic headlamp-range adjustment system ensures that the xenon headlamps are always aligned correctly, regardless of the load the car is carrying. An axle sensor developed by Hella with an integral control unit is mounted on the rear axle, from where it can measure the degree to which the vehicle is being pressed down on its suspension. Changes are communicated to an actuator attached to the headlamps, which makes sure that the xenon light is always aligned correctly and does not dazzle.

When the driver operates the windscreen-washing system, when the headlamps are switched on, they are also cleaned at the same time. The headlamp cleaning system works on the water-jet principle; a rotary pump generates the necessary water pressure, and the kinetic effect of the drops of water, directed through special swirl-chamber nozzles, forces the particles of dirt off the headlamp and rinses them away. Clean headlamps are best for lighting up the road and prevent the dazzle that might be caused by stray light reflected from the particles of dirt. It is estimated that installing the whole package will require about 5 man-hours, or 3 man-hours without the headlamp cleaning system.

11.8 Self-assessment

11.8.1 Questions

1. Describe briefly the reasons for fitting vehicle lights.
2. State four methods of converting electrical energy into light energy.
3. Explain the reason why headlights are fused independently.
4. Draw a simplified circuit of a lighting system

showing the side- and headlight bulbs, light switch, dip switch and main beam warning light.

5. Make a clearly labelled sketch to show the 'aiming board' method of setting headlight alignment.
6. Describe the operation of a gas discharge lamp.
7. List the advantages and disadvantages of gas discharge lamps.
8. Explain the operation of infrared lighting and sketch a block diagram of the system components.
9. Define the term 'Expert or Intelligent lighting'.
10. Draw a typical dim-dip circuit and state the reason why it is used.

11.8.2 Assignment

Design a vehicle lighting system using technology described in this chapter. Decide which techniques you are going to use and justify your choices. For example, you may choose to use a single light source for all lights or you may decide to use neon lights for the rear and gas discharge for the front. Whatever the choice, it should be justified with sound reasons such as cost, safety, aerodynamics, styling, reliability and so on.

Make sketches to show exterior views. Circuit diagrams are not necessary but you should note where components would be located. State whether the vehicle is standard or 'top of the range' etc.

12
Auxiliaries

12.1 Windscreen washers and wipers

12.1.1 Functional requirements

The requirements of the wiper system are simple. The windscreen must be clean enough to provide suitable visibility at all times. To do this, the wiper system must meet the following requirements.

- Efficient removal of water and snow.
- Efficient removal of dirt.
- Operate at temperatures from −30 to 80°C.
- Pass the stall and snow load test.
- Service life in the region of 1 500 000 wipe cycles.
- Resistant to corrosion from acid, alkali and ozone.

In order to meet the above criteria, components of good quality are required for both the wiper and washer system. The actual method used by the blades in cleaning the screen can vary, providing the legally prescribed area of the screen is cleaned. Figure 12.1 shows five such techniques.

Figure 12.2 shows how the front screen is split into 'zones' and how a 'non-circular wiping' technique is applied.

12.1.2 Wiper blades

The wiper blades are made of a rubber compound and are held on to the screen by a spring in the wiper arm. The aerodynamic properties of the wiper blades have become increasingly important due to the design of the vehicle as different air currents flow on and around the screen area. The strip on top of the rubber element is often perforated to reduce air drag. A good quality blade will have a contact width of about 0.1 mm. The lip wipes the surface of the screen at an angle of about 45°. The pressure of the blade on the screen is also important as the coefficient of friction between the rubber and glass

can vary from 0.8 to 2.5 when dry and 0.1 to 0.6 when wet. Temperature and velocity will also affect these figures.

12.1.3 Wiper linkages

Most wiper linkages consist of series or parallel mechanisms. Some older types use a flexible

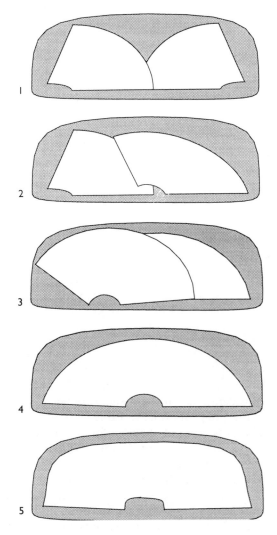

Figure 12.1 Five techniques of moving wiper blades on the screen

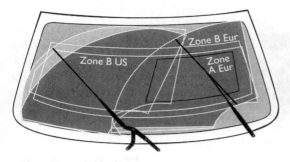

Figure 12.2 Non-circular wiping

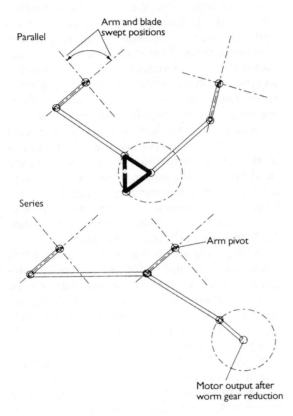

Figure 12.3 Two typical wiper linkage layouts

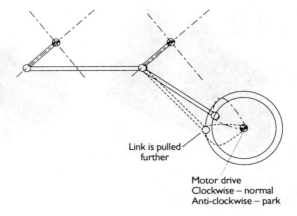

Figure 12.4 Wiper linkage used on some vehicles, together with the cam link which allows off-screen reverse parking

rack and wheel boxes similar to the operating mechanism of many sunroofs. One of the main considerations for the design of a wiper linkage is the point at which the blades must reverse. This is because of the high forces on the motor and linkage at this time. If the reverse point is set so that the linkage is at its maximum force transmission angle then the reverse action of the blades puts less strain on the system. This also ensures smoother operation. Figure 12.3 shows two typical wiper linkage layouts, the first figure is shown at the reverse point. Note that the position of the rotary link and the angles of the rods

are designed to reduce the loading on the motor at this point.

Figure 12.4 shows one method used on some vehicles together with the cam linkage, which allows off-screen parking.

12.1.4 Wiper motors

Most, if not all, wiper motors now in use are the permanent magnet motors. The drive is taken via a worm gear to increase torque and reduce speed. Three brushes may be used to allow two-speed operation. The normal speed operates through two-brushes placed in the usual positions opposite to each other. For a fast speed, the third brush is placed closer to the earth brush. This reduces the number of armature windings between them, which reduces resistance and hence increases current and therefore speed. Figure 12.5 shows two typical wiper motors. Typical specifications for wiper motor speed and hence wipe frequency are 45 rev/min at normal speed and 65 rev/min at fast speed. The motor must be able to overcome the starting friction of each blade at a minimum speed of 5 rev/min.

The characteristics of a typical car wiper motor are shown in Figure 12.6. The two sets of curves indicate fast and slow speed.

Wiper motors, or the associated circuit, often have some kind of short circuit protection. This is to protect the motor in the event of stalling, if frozen to the screen for example. A thermal trip of some type is often used or a current sensing circuit in the wiper ECU, if fitted. The maximum time a motor can withstand stalled current is normally specified. This is usually in the region of about 15 minutes.

Rear motor with
electronic components

Front wiper motor

Figure 12.5 Wiper motors

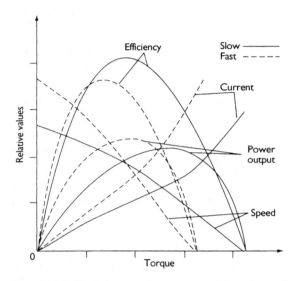

Figure 12.6 Characteristics of a wiper motor; the two sets of curves indicate fast and slow speed

12.1.5 Windscreen washers

The windscreen washer system usually consists of a simple DC permanent magnet motor driving a centrifugal water pump. The water, preferably with a cleaning additive, is directed onto an appropriate part of the screen by two or more jets. A non-return valve is often fitted in the line to the jets to prevent water siphoning back to the reservoir. This also allows 'instant' operation when the washer button is pressed. The washer circuit is normally linked to the wiper circuit such that when the washers are operated the wipers start automatically and will continue for several more sweeps after the washers have stopped. The circuit is shown in the next section.

12.1.6 Washer and wiper circuits

Figure 12.7 shows a circuit for fast, slow and intermittent wiper control. The switches are shown in the off position and the motor is stopped and in its park position. Note that the two main brushes of the motor are connected together via the limit switch, delay unit contacts and the wiper switch. This causes regenerative braking because of the current generated by the motor due to its momentum after the power is switched off. Being connected to a very low resistance loads up the 'generator' and it stops instantly when the park limit switch closes.

When either the delay contacts or the main switch contacts are operated the motor will run at slow speed. When fast speed is selected the third brush on the motor is used. On switching off, the motor will continue to run until the park limit switch changes over to the position shown. This switch is only in the position shown when the blades are in the parked position.

A simple capacitor-resistor (CR) timer circuit often based around a 555 IC or similar integrated circuit is used to control intermittent wipe. The charge or discharge time of the capacitor causes a delay in the operation of a transistor, which in turn operates a relay with change-over contacts.

Figure 12.8 shows the circuit of a programmed wiper system. The ECU contains two change-over relays to enable the motor to be reversed. Also contained in the ECU is a circuit to switch off the motor supply in the event of the blades stalling. To reset this the driver's switch must be returned to the off position.

12.1.7 Electronic control of windscreen wipers

Further control of wipers other than just delay is possible with appropriate electronic control. Manufacturers have used programmed electronic control of the windscreen wipers for a number of years now. One system consists of a two-speed motor with two limit switches, one for the park position and one that operates at the top limit of the sweep. A column switch is utilized that has positions for wash/wipe, fast

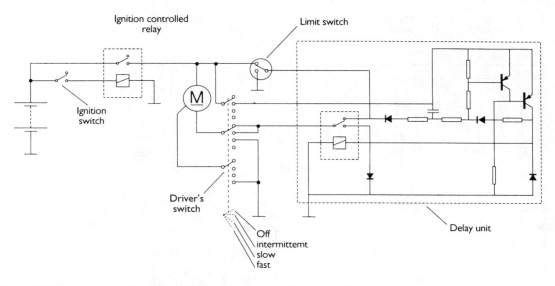

Figure 12.7 Wiper circuit with intermittent/delay operation as well as slow and fast speed

speed, slow speed, flick wipe and delay, and which has several settings. The heart of this system is the programmed wiper control unit. An innovative feature is that the wiper blades may be parked below the screen. This is achieved by utilizing the top limit switch to signal the ECU to reverse the motor for parking. The switch is normally closed and switches open circuit when the blades reach the 'A' post. Due to the design of the linkage, the arms move further when working in reverse and pull the blades off the screen. The normal park limit switch stops the motor, via the ECU in this position.

Some vehicles use a similar system with even more enhanced facilities. This is regulated by either a central control unit (CCU) or a multi-function unit (MFU). These units can often control other systems as well as the wipers, thus allowing reduced wiring bulk under the dash area. Electric windows, headlights and a heated rear window, to name just a few, are now often controlled by a central unit. A CCU allows the following facilities for the wipers (front and rear).

Front wash/wipe

The CCU activates the wipers when the washer switch is pressed and keeps them going for a further six seconds when the switch is released.

Intermittent wipe

When the switch is moved to this position, the CCU operates the wipers for one sweep. When back in the rest position, the CCU waits for a set time and then operates another sweep and so on.

This continues until the switch is moved to the off position. The time delay can be set by the driver – as one of five settings of a variable resistor. This changes the delay from about 3 s with a resistance of $500\,\Omega$, to a delay of about 20 s with a resistance of $5400\,\Omega$.

Rear wiper system

When the switch is operated, the CCU operates the rear wipers for three sweeps by counting the signal from the park switch. The wiper will then be activated once every six seconds until switched off by the driver.

Rear wash/wipe

When the rear washer switch is pressed, the CCU will operate the rear wiper and then continue its operation for three sweeps after the washer switch is released. If the rear wiper is not switched on the CCU will operate the blades for one more sweep after about 18 s. This is commonly known as the 'dribble wipe'!

Rear wiper when reverse gear is selected

If the front wipers are switched on and reverse gear is selected the CCU will operate the rear wiper continuously. This will stop when either the front wipers are switched off or reverse gear is deselected.

Stall protection

When the rear wiper is operated, the CCU starts a timer. If no movement is detected within

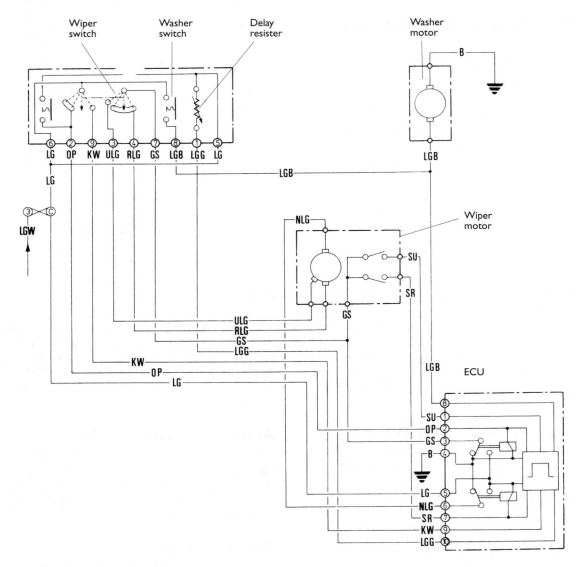

Figure 12.8 Programmed washer wipe and variable intermittent wipe circuit

15 s the power to the motor is removed. This is reset when the driver's switch is moved to the off position.

12.1.8 Microprocessor controlled wipers

A problem facing car manufacturers is that of fitting a suitable wiper linkage into the minimal space available with modern body styles. One solution is to use a separate motor for each blade. This leaves another problem, and that is how to synchronize the operation of each motor. In order to allow synchronization, a datum

point and a way of measuring distance from this point is needed. The solution to this is to utilize a normal park limit switch as the datum and to count the revolutions of the motor armature to imply distance moved.

A computer program can then be used to control the motors. The inputs to the program are from the driver's switch, the motor limit switches and the motor armature revolution counters. Fully programmed operation in this way will allow more sophisticated facilities to be used if required. A slight delay in the start and reverse point of each motor can be used to reduce high current draw.

12.2 Signalling circuits

12.2.1 Introduction

Direction indicators have a number of statutory requirements. The light produced must be amber, but the indicators may be grouped with other lamps. The flashing rate must be between one and two per second with a relative 'on' time of between 30 and 57%. If a fault develops, this must be apparent to the driver by the operation of a warning light on the dashboard. The fault can be indicated by a distinct change in frequency of operation or the warning light remaining on. If one of the main bulbs fails then the remaining lights should continue to flash perceptibly.

Legislation exists as to the mounting position of the exterior lamps, such that the rear indicator lights must be within a set distance of the tail lights and within a set height. The wattage of the indicator light bulbs is normally 21 W at 6, 12 or 24 V as appropriate.

Brake lights fall under the heading of auxiliaries or 'signalling'. A circuit is examined later in this section.

12.2.2 Flasher units

Figure 12.9 shows the internal circuit of an electronic flasher unit. The operation of this unit is based around an integrated circuit. The type shown can operate at least four 21 W bulbs (front and rear) and two 5 W side repeaters when operating in hazard mode. This will continue for several hours if required. Flasher units are rated by the number of bulbs they are capable of operating. When towing a trailer or caravan the unit must be able to operate at a higher wattage. Most units use a relay for the actual switching as this is not susceptible to voltage spikes and also provides an audible signal.

The electronic circuit is constructed together with the relay, on a printed circuit board. Very few components are used as the integrated circuit is specially designed for use as an indicator timer. The integrated circuit itself has three main sections. The relay driver, an oscillator and a bulb failure circuit. A Zener diode is built in to the IC to ensure constant voltage such that the frequency of operation will remain constant in the range 10–15 V. The timer for the oscillator is controlled by R_1 and C. The values are normally set to give an on–off ratio of 50% and an operating frequency of 1.5 Hz (90 per minute).

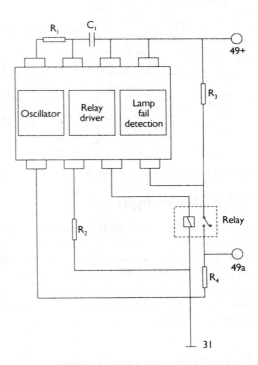

Figure 12.9 Circuit diagram of an electronic flasher unit

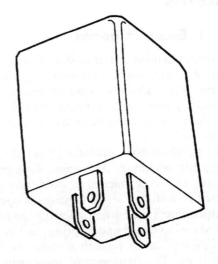

Figure 12.10 Electronic flasher unit

The on–off signals produced by the oscillator are passed to a driver circuit, which is a Darlington pair with a diode connected to protect it from back-EMF as the relay coil is switched on and off. Bulb failure is recognized when the volt drop across the low value resistor R_2 falls. The bulb failure circuit causes the oscillator to double the speed of operation. Extra capacitors can be used for added protection against transient voltages and for interference suppression. Figure 12.10 shows the normal 'packaging' for a flasher unit.

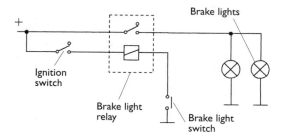

Figure 12.11 Typical brake light circuit

12.2.3 Brake lights

Figure 12.11 shows a typical brake light circuit. Most incorporate a relay to switch the lights, which is in turn operated by a spring-loaded switch on the brake pedal. Links from this circuit to cruise control may be found. This is to cause the cruise control to switch off as the brakes are operated.

12.3 Other auxiliary systems

12.3.1 Electric horns

Regulations in most countries state that the horn (or audible warning device) should produce a uniform sound. This consequently makes sirens and melody-type fanfare horns illegal! Most horns draw a large current, so are switched by a suitable relay.

The standard horn operates by simple electromagnetic switching. As current flow causes an armature that is attached to a tone disc to be attracted to a stop, a set of contacts is opened. This disconnects the current allowing the armature and disc to return under spring tension. The whole process keeps repeating when the horn switch is on. The frequency of movement and hence the fundamental tone is arranged to lie between 1.8 and 3.5 kHz. This gives good penetration through traffic noise. Twin horn systems, which have a high and low tone horn, are often used. This produces a more pleasing sound but is still very audible in both town and higher speed conditions. Figure 12.12 shows a typical horn together with its associated circuit.

12.3.2 Engine cooling fan motors

Most engine cooling fan motors (radiator cooling) are simple permanent magnet types. Figure 12.13 shows a typical example. The fans used

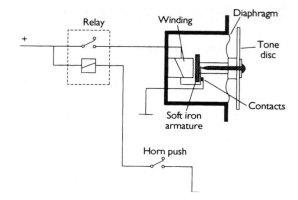

Figure 12.12 Horn and circuit

Figure 12.13 Engine cooling motor

often have the blades placed asymmetrically (balanced but not in a regular pattern) to reduce noise when operating.

When twin cooling fans and motors are fitted, they can be run in series or parallel. This is often the case when air conditioning is used as the condenser is usually placed in front of the radiator and extra cooling air speed may be needed.

A circuit for series or parallel operation of cooling fans is shown in Figure 12.14.

12.3.3 Headlight wipers and washers

There are two ways in which headlights are cleaned, first by high pressure jets, and secondly by small wiper blades with low pressure water supply. The second method is, in fact, much the same as windscreen cleaning but on a smaller scale. The high pressure system tends to be

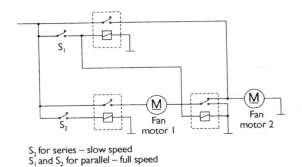

S₂ for series – slow speed
S₁ and S₂ for parallel – full speed

Figure 12.14 Circuit for series or parallel operation of cooling fans

Figure 12.15 Headlight washers in action

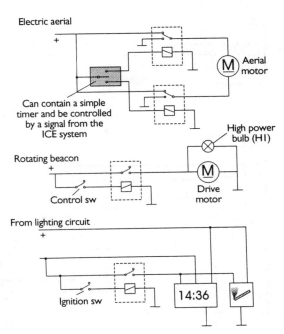

Figure 12.16 Electric aerial, rotating beacon, cigar lighter and clock circuit

favoured but can suffer in very cold conditions due to the fluid freezing. It is expected that the wash system should be capable of about 50 operations before refilling of the reservoir is necessary. Figure 12.15 shows the pressure wash technique.

Headlight cleaners are often combined with the windscreen washers. They operate each time the windscreen washers are activated, if the headlights are also switched on.

A retractable nozzle for headlight cleaners is often used. When the water pressure is pumped to the nozzle it pushes the nozzle from its retracted position, flush with the bodywork.

When the washing is completed the jet is retracted back into the housing.

Some minor vehicle electrical systems, which are not covered elsewhere, are shown in Figure 12.16. Cigar lighter, clock, rotating beacon and electric aerial are all circuits that could be used by many other systems.

12.4 Case studies

12.4.1 Indicators and hazard circuit – Rover

The circuit diagram shown in Figure 12.17 is part of the circuit from a Rover car and shows the full layout of the indicator and hazard lights wiring. Note how the hazard switch, when operated, disconnects the ignition supply from the flasher unit and replaces it with a constant supply. The hazard system will therefore operate at any time but the indicators will only work when the ignition is switched on. When the indicator switch is operated left or right, the front, rear and repeater bulbs are connected to the output terminal of the flasher unit, which then operates and causes the bulbs to flash.

When the hazard switch is operated, five sets of contacts are moved. Two sets connect left and right circuits to the output of the flasher unit.

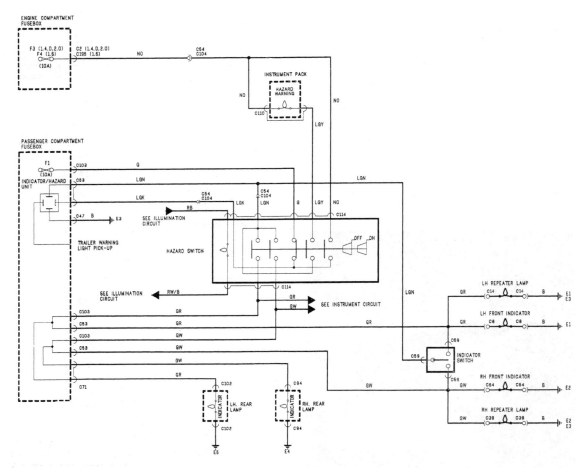

Figure 12.17 Indicator and hazard circuit – Rover

One set disconnects the ignition supply and another set connects the battery supply to the unit. The final set of contacts causes a hazard warning light to be operated. On this and most vehicles the hazard switch is illuminated when the sidelights are switched on.

When operating in hazard mode the bulbs would draw 7.8 A (94 W/12 V).

However, this current will peak much higher due to the cold resistance of the bulbs. In the circuit shown, the top fuse is direct from the battery and the other is ignition controlled.

With the ignition switched on, fuse 1 in the passenger compartment fusebox provides a feed to the hazard warning switch on the G wire. Provided the hazard warning switch is in the off position the feed crosses the switch and supplies the flasher unit on the LG/K wire. When the switch control is moved for a right turn, the switch makes contact when the LG/N wire from the flasher unit is connected to the G/W wire, allowing a supply to pass the right-hand front and rear indicator lights and then to earth on the

B wire. When the switch control is moved for a left turn, the switch makes contact with the G/R wire, which allows the supply to pass to the left-hand front and rear indicator lights and then to earth on the B wire. The action of the flasher unit causes the circuit to 'make and break'.

By pressing the hazard warning switch a battery supply on the N/O from fuse 3 (1.4, 2.0 and diesel models) or 4 (1.6 models) in the engine bay fusebox crosses the switch and supplies the flasher unit on the LG/K wire. At the same time contacts are closed to connect the hazard warning light and the flasher unit to both the G/W and GIR wires, the right-hand and left-hand indicators and the warning light flash alternately.

12.4.2 Wiper circuit – Ford

The circuit shown in Figure 12.18 is similar to that used on many Ford vehicles. Note that the two sets of switch contacts are mechanically linked together. The switches are shown in the 'off' position. A link is shown to a headlamp

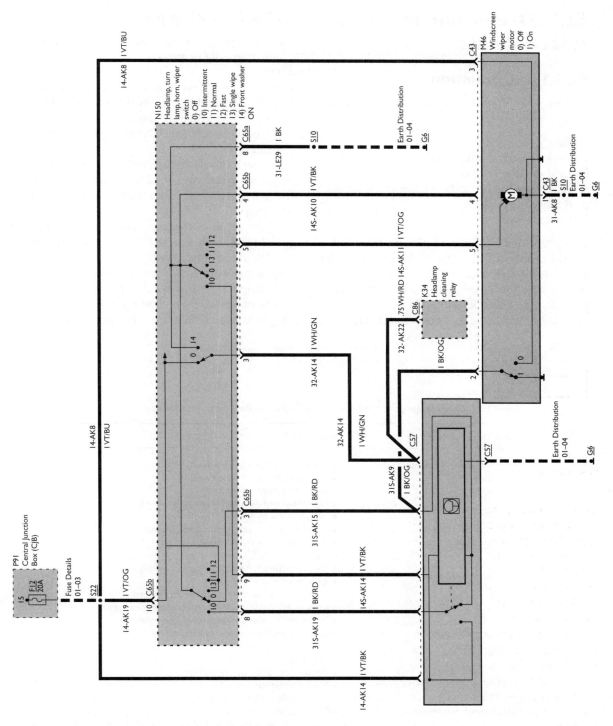

Figure 12.18 Wiper/washer control circuit used by Ford

cleaning relay (if fitted) to allow operation of the headlamp washers as the screen washers are used. This will only occur if the headlamps are also switched on.

The wire codes follow the convention outlined in Chapter 3. The motor is a three-brush PM type and contains a parking switch. Following the top terminal of the motor, as shown, results in a connection to earth via the control switch and the limit switch. This is to achieve regenerative braking.

12.5 Diagnosing auxiliary system faults

12.5.1 Introduction

As with all systems the six stages of fault-finding should be followed.

1. Verify the fault.
2. Collect further information.
3. Evaluate the evidence.
4. Carry out further tests in a logical sequence.
5. Rectify the problem.
6. Check all systems.

The procedure outlined in the next section is related primarily to stage 4 of the process. Table 12.1 lists some common symptoms of an auxiliary system malfunction together with suggestions for the possible fault. The faults are very generic but will serve as a good reminder.

12.5.2 Testing procedure

The process of checking an auxiliary system circuit is broadly as follows.

1. Hand and eye checks (loose wires, loose switches and other obvious faults) – all connections clean and tight.
2. Check battery (see Chapter 5) – must be 70% charged.
3. Check motor linkage/bulbs – visual check.
4. Fuse continuity – (do not trust your eyes) voltage at both sides with a meter or a test lamp.
5. If used does the relay click (if yes, jump to stage 8) – this means the relay has operated, but it is not necessarily making contact.
6. Supply to switch – battery volts.
7. Supply from the switch – battery volts.
8. Supplies to relay – battery volts.
9. Feed out of the relay – battery volts.

Table 12.1 Common symptoms and possible faults of an auxiliary system malfunction

Symptom	Possible fault
Horn not working or poor sound quality	• Loose or broken wiring/connections/fuse. • Corrosion in horn connections. • Switch not making contact. • High resistance contact on switch or wiring. • Relay not working.
Wipers not working or poor operation	• Loose or broken wiring/connections/fuse. • Corrosion in wiper connections. • Switch not making contact. • High resistance contact on switch or wiring. • Relay/timer not working. • Motor brushes or slip ring connections worn. • Limit switch contacts open circuit or high resistance. • Blades and/or arm springs in poor condition.
Washers not working or poor operation	• Loose or broken wiring/connections/fuse. • Corrosion in washer motor connections. • Switch not making contact. • Pump motor poor or not working. • Blocked pipes or jets. • Incorrect fluid additive used.
Indicators not working or incorrect operating speed	• Bulb(s) blown. • Loose or broken wiring/connections/fuse. • Corrosion in horn connections. • Switch not making contact. • High resistance contact on switch or wiring. • Relay not working.
Heater blower not working or poor operation	• Loose or broken wiring/connections/fuse. • Switch not making contact. • Motor brushes worn. • Speed selection resistors open circuit.

10. Voltage supply to the motor – within 0.5 V of the battery.
11. Earth circuit (continuity or voltage) – 0 Ω or 0 V.

12.6 Advanced auxiliary systems technology

12.6.1 Wiper motor torque calculations

The torque required to overcome starting friction of each wiper blade can be calculated as follows:

$$T = F\mu_{max}f_sf_tl\left(\frac{w_a}{w_m}\right)\left(\frac{1}{e}\right)\left(\frac{R_h}{R_c}\right)$$

where

T = torque to move one wiper arm;
F = force of one blade onto the screen;
μ_{max} = maximum dry coefficient of friction (e.g. 2.5);
f_s = multiplier for joint friction (e.g. 1.15);
f_t = tolerance factor (e.g. 1.12);
l = wiper arm length;
w_a = maximum angular velocity of arm;
w_m = mean angular velocity of motor crank;
e = efficiency of the motor gear unit (e.g. 0.8);
R_h = motor winding resistance – hot;
R_c = motor winding resistance – cold.

12.6.2 PM Motor – electronic speed control

The automotive industry uses permanent magnet (PM) motors because they are economical to produce and provide good performance. A simple current limiting resistor or a voltage regulator can vary the motor's speed. This simple method is often used for motors requiring variable speed control. However, to control the speed of a motor that draws 20 A at full speed and about 10 A at half speed is a problem.

At full speed, the overall motor control system's efficiency is around 80%. If the speed is reduced to half the system's, then efficiency drops to 40%. This is because there would be a heat loss of 70 W in the series resistor and 14 W lost in the motor. A more efficient speed control system is therefore needed.

One way is to interrupt the motor's voltage at a variable duty cycle using a switching power supply. A system known as pulse width modulation (PWM) has been developed. An introduction to this technique follows.

Because the armature of the PM motor acts as a flywheel, the voltage interruption rate can be 1 kHz or slower, without causing the motor's speed to pulsate. A problem at this or other audible frequencies is the noise generated from within the motor. At higher frequencies, 16 kHz for example, the audible noise is minimized. A further noise problem is significant EMR (electromagnetic radiation). This is generated by the fast switching speeds. This can be improved by slowing down the switching edge of the operating signal. A compromise has to be made between the edge speeds and power device heat loss.

When the EMR problems are safely contained, the stalled motor condition must be considered. The motor's copper windings have a positive temperature coefficient of 0.00393 Ω/°C. Therefore, a 0.25 Ω motor resistance value at 25°C would be about 0.18 Ω at –40°C. Using a typical 20 A motor as the load, the maximum stalled or locked rotor current can be calculated to be about 77 A as shown:

$$I_{max} = \frac{E_{max}}{R_{mtr}}$$

where E_{max} = maximum power supply voltage (14.4 V) and R_{mtr} = minimum motor resistance (0.18 Ω).

When the maximum motor current has been calculated, the specifications of the power transistor can be determined. In this case, the device needs an average current rating of at least 77 A. However, a further consideration for reliable power transistor operation is its worst case heat dissipation.

The worst case includes maximum values for the supply voltage, ambient temperature and motor current. A junction temperature of 150°C for the power transistors is used as a maximum point. The following equation calculates the transistor's maximum allowable heat dissipation for use in an 85°C environment using a 2.7°C/W heatsink and a 1°C/W junction to case power FET thermal resistance.

$$PD_{max} = \frac{TJ_{max} - TA_{max}}{R\varphi JC + R\varphi CS + R\varphi SA}$$

where TJ_{max} = maximum allowable junction temperature (150°C); TA_{max} = maximum ambient temperature (85°C); $R\varphi JC$ = junction to case thermal resistance (1°C/W); $R\varphi CS$ = case to heatsink interface thermal resistance (0.1°C/W);

$R\varphi SA$ = heatsink to ambient thermal resistance (2.7°C/W).

Using the given figure results in a value of about 17.1 W. This is considerably better than using a dropping resistor, but to achieve this, several power transistors would have to be connected in parallel. Significant heat sinking is also necessary.

This technique may become popular because of its significant improvement in efficiency over conventional methods and the possibilities for greater control over the speed of a PM motor.

12.7 New developments in auxiliary systems

12.7.1 Wiper blade pressure control

Bosch has a system of wiper pressure control, which can infinitely vary the pressure of the blade onto the screen, depending on vehicle speed. At high speeds the air stream can cause the blades to lift and judder. This seriously reduces the cleaning effectiveness. If the original pressure is set to compensate this, the pressure at rest could deform the arms and blades.

The pressure control system is shown in Fig-

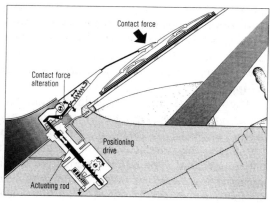

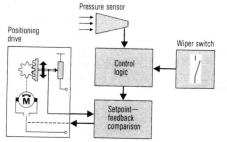

Figure 12.19 Wiper blade pressure control system

ure 12.19. Sensors are used to determine the air stream velocity and intensity of the rain. An ECU evaluates the data from these sensors and passes an appropriate signal to the servo motor. When the blades are in the rest position, pressure is very low to avoid damage. The pressure rises with increasing vehicle speed and heavy rain.

The system is able to respond very quickly such that, when overtaking, the deluge of spray is cleared by increased pressure and also, if the screen dries off, the pressure is reduced to prevent scraping.

12.7.2 Totally silent rear wiper module

Carmakers are constantly looking for ways to reduce the noise generated by wiper systems. The two main sources of noise are the wiper blade (particularly when it turns over at the end of each movement) and the wiper motor.

Valeo has produced a new rear wiper module offering an original solution to these problems in the form of a specific, integrated electronics control system. This system is designed around an H-bridge power stage, which has no relays. This eliminates all switching noise. The control algorithm provides pinpoint management of wiper speed; it slows the blades at the end of each cycle, thus cutting out turning noise.

Note: an H-bridge uses four power devices that are connected to reverse the voltage across both terminals of a load. This is used to control the direction of a motor.

12.7.3 Linear rear wiper concept

Current wiper systems that are based on an alternative rotary movement cover a wipe area of between 50 and 60% of the total surface area of the rear window. This limit is due to the height/width ratio and the curve of the window. Valeo's linear rear wiper concept ensures optimum visual comfort as it covers over 80% of the rear window surface; this is a visibility gain for the driver exceeding 60%.

This increase in the driver's field of vision enhances safety, especially during low-speed manoeuvres such as reversing or parking. The linear rear wiper concept is in keeping with the trend towards narrower, highly convex rear windows and can be fully integrated into vehicle design. Figure 12.20 shows this technique.

Figure 12.20 Linear wiper system

12.7.4 Silencio: a new wiper concept

The Silencio windshield wiper offers two major innovations to enhance passenger comfort and safety:

- A new extended-life rubber coating called 'Skin'.
- A wear indicator that tells the driver when to change the wiper.

External wear factors such as UV, ozone, pollution, windshield wiper fluid, etc damage the rubber blade and affect wiping quality. 'Skin' is a new coating that protects the blade.

This surface coating, composed of a slipping agent, a polymer bonding agent and an 'impermeability' agent, can be applied to natural or synthetic rubber. An innovative polymerization process ensures long-lasting adhesion to the blade. By protecting the blade from wear, 'Skin' maintains initial wiping quality longer and also eliminates rubber squeaking and friction noise on dry glass.

Silencio is also fitted with a wear indicator that tells the driver the state of wear of the wiper blade. The indicator – a round tab fixed to the wiper – degrades at the same speed as the rubber blade. External wear factors such as UV, ozone and pollution activate chemicals in the indicator which then gradually changes colour, going from black to yellow, as the wiper wears out.

12.7.5 Front end modules

Valeo has taken a new step in its modular approach by developing front end units. The main benefits for the customer are reduced cost

for the same function, reduced vehicle assembly time and space saving in plants.

The front end consists of a crash resistant structure to which are attached the mainframe carrier, the bumper beam, the engine cooling system (radiator, condenser, charge air cooler, oil cooler, fan motor system and shroud) and the lighting system (headlamps, turn signals, headlamp levelling actuators). Options such as the wiring harness or the bumper skin can also be supplied.

Valeo has been producing front end modules since 1997 and is developing a new generation of products that include a metal/plastic hybrid mainframe carrier. Component assembly has also been largely optimized through simplified assembly/dismantling operations – no screws, only clipping components – so that only the relevant parts have to be replaced in the event of a minor accident.

12.7.6 Electronic fan system control

The electronic control of the fan system is a further step in the drive to improve engine cooling management. Besides reducing electrical consumption, one of the main benefits of Valeo's concept is the reduction in noise levels thanks to continuous fan speed regulation, adjusted to the minimum air flow required for engine cooling and A/C management.

Valeo is due to start producing these variable speed fan/motor units in 2000. They have the following technical features.

- Electrical consumption reduced by half for an average usage profile.
- Noise level reduced by 15 dBa at half speed.
- Soft start of the fan, which removes peak starting currents and provides a better subjective sound level.

Electronic functions designed to improve the safety of the fan are possible; speed can be adapted to the minimum required, diagnostic functions are possible and self-protection in case of fan lock due to contamination is built in.

The fan electronic management unit can be easily installed in different places in the engine compartment to meet all types of customer specifications, even the most demanding ones in terms of high temperature. Valeo is currently developing a new concept that has a compact pulse width modulation (PWM) module integrated into the motor.

12.8 Self-assessment

12.8.1 Questions

1. State four electrical systems considered to be 'auxiliaries'.
2. Describe briefly how a flasher/indicator unit is rated.
3. Make a clearly labelled sketch to show a typical wiper motor linkage.
4. Draw a circuit diagram of an indicator circuit, and label each part.
5. List five requirements of a wiper system.
6. Explain how off-screen parking is achieved by some wiper systems.
7. Describe what is meant by the term 'stall protection' in relation to wiper motors.
8. Draw a clearly labelled brake light circuit. Include three 21 W bulbs, a relay and fuse as well as the brake light switch.
9. Calculate the rating of the fuse required in Question 8.
10. Explain with the aid of a sketch what is meant by 'windscreen zones'.

12.8.2 Assignment

Investigate a modern vehicle and produce a report of the efficiency and operation of the washer and wiper systems (front and rear).

Make a reasoned list of suggestions as to how improvements could be made. Consider for the purposes of lateral thinking that, in this case, money is not an issue!

13

Instrumentation

13.1 Gauges and sensors

13.1.1 Introduction

The topic of instrumentation has now reached such a level as to have become a subject in its own right. This chapter covers some of the basic principles of the science, with examples as to how it relates to automobile systems. By definition, an instrumentation system can be said to convert a 'variable', into a readable or usable display. For example, a fuel level instrument system will display, often by an analogue gauge, a representation of the fuel in the tank.

Instrumentation is not always associated with a gauge or a read-out type display. In many cases the whole system can be used just to operate a warning light. However, the system must still work to certain standards, for example if a low outside temperature warning light did not illuminate at the correct time, a dangerous situation could develop.

This chapter will cover vehicle instrumentation systems in use and examine in more detail the issues involved in choosing or designing an instrumentation system. Chapter 2 contains many details associated with sensors, an integral part of an instrumentation system, and it may be appropriate to refer back for some information related to this chapter.

13.1.2 Sensors

In order to put some limit on the size of this section, only electrical sensors associated with vehicle use will be considered. Sensors are used in vehicle applications for many purposes; for example, the coolant temperature thermistor is used to provide data to the engine management system as well as to the driver via a display. For the purpose of providing information to the driver, Table 13.1 gives a list of measurands (things that are measured) together with typical sensors, which is representative of today's vehicles.

Table 13.1 Measurements and sensors

Measurement required	Sensor example
Fuel level	Variable resistor
Temperatures	Thermistor
Bulb failure	Reed relay
Road speed	Inductive pulse generator
Engine speed	Hall effect
Fluid levels	Float and reed switch
Oil pressure	Diaphragm switch
Brake pad wear	Embedded contact wire
Lights in operation	Bulb and simple circuit
Battery charge rate	Bulb circuit/voltage monitor

Figure 13.1 shows some of the sensors listed above.

13.1.3 Thermal-type gauges

Thermal gauges, which are ideal for fuel and engine temperature indication, have been in use for many years. This will continue because of their simple design and inherent 'thermal' damping. The gauge works by utilizing the heating effect of electricity and the benefit of the widely adopted bimetal strip. As a current flows through a simple heating coil wound on a bimetal strip, heat causes the strip to bend. The bimetal strip is connected to a pointer on a suitable scale. The amount of bend is proportional to the heat, which in turn is proportional to the current flowing. Providing the sensor can vary its resistance in proportion to the measurand (e.g. fuel level), the gauge will indicate a suitable representation providing it has been calibrated for the particular task. Figure 13.2 shows a representation of a typical thermal gauge.

The inherent damping is due to the slow thermal effect on the bimetal strip. This causes the needle to move very slowly to its final position. It can be said to have a large time constant. This is a particular advantage for displaying fuel level, as the variable resistor in the tank will move, as the fuel moves, due to vehicle movement! If the gauge were able to react quickly it

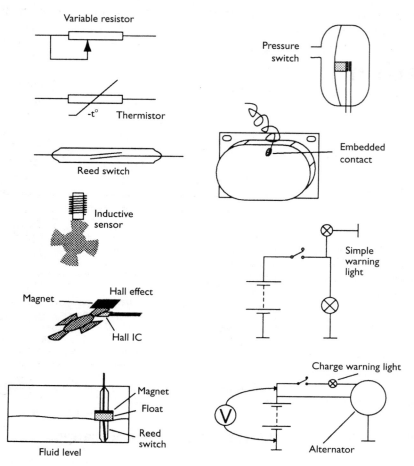

Figure 13.1 Sensors used for instrumentation

would be constantly moving. The movement of the fuel however is, in effect, averaged out and a relatively accurate display can be obtained. Some electronically driven thermal fuel gauges are damped even more by the control system.

Thermal-type gauges are used with a variable resistor and float in a fuel tank or with a thermistor in the engine water jacket. Figure 13.3 shows the circuit of these two together. The resistance of the fuel tank sender can be made non-linear to counteract any non-linear response of the gauge. The sender resistance is at a maximum when the tank is empty.

A constant voltage supply is required to prevent changes in the vehicle system voltage affecting the reading. This is because, if the system voltage increased, the current flowing would increase and hence the gauges would read higher. Most voltage stabilizers are simple Zener diode circuits, as shown in Figure 13.4.

13.1.4 Moving iron gauges

The moving iron gauge was in use earlier than the thermal type but is now gaining popularity

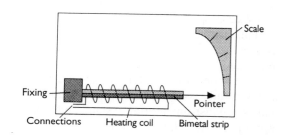

Figure 13.2 Bimetal strip operation in a thermal-type gauge

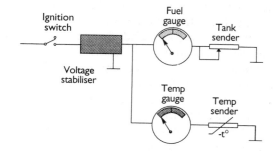

Figure 13.3 Bimetal fuel and temperature gauge circuit

for some applications. Figure 13.5 shows the circuit and principle of the moving iron gauge system. Two small electromagnets are used which act upon a small soft iron armature connected to a pointer. The armature will position itself between the cores of the electromagnets depending on the magnetic strength of each. The ratio of magnetism in each core is changed as the linear variable resistance sender changes and hence the needle is moved. This type of gauge reacts very quickly (it has a small time constant) and is prone to swing about with movement of the vehicle. Some form of external damping can be used to improve this problem. Resistor R_1 is used to balance out the resistance of the tank sender. A good way to visualize the operation of the circuit is to note that when the tank is half full, the resistance of the sender will be the same as the resistance of R_1. This makes the circuit balanced and the gauge will read half full. The

sender resistance is at a maximum when the tank is full.

13.1.5 Air-cored gauges

Air-cored gauges work on the same principle as a compass needle lining up with a magnetic field. The needle of the display is attached to a very small permanent magnet. Three coils of wire are used and each produces a magnetic field. The magnet will line up with the resultant of the three fields. The current flowing and the number of turns (ampere-turns) determine the strength of the magnetic flux produced by each coil. As the number of turns remains constant the current is the key factor. Figure 13.6 shows the principle of the air-cored gauge together with the circuit for use as a temperature indicator. The ballast resistor on the left is used to limit maximum current and the calibration resistor is used for calibration. The thermistor is the temperature sender. As the thermistor resistance is increased, the current in all three coils will change. Current through C will be increased but the current in coils A and B will decrease. The resultant magnetic fields are shown in Figure 13.6. This moves the magnetic armature accordingly.

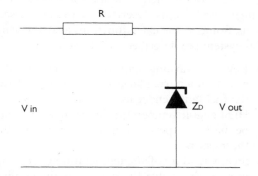

Figure 13.4 A voltage stabilizer

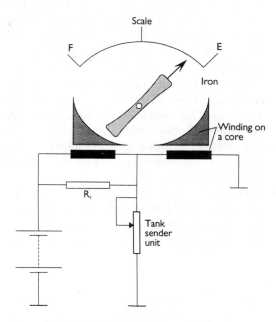

Figure 13.5 Circuit/principle of the moving iron gauge

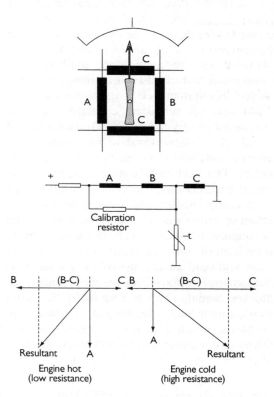

Figure 13.6 Principle of the air-cored gauge together with the circuit when used as a fuel level or temperature indicator and the resultant magnetic fields

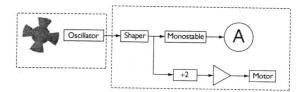

Figure 13.7 Block diagram of a speedometer system which uses a simple ammeter as the gauge

The air-cored gauge has a number of advantages. It has almost instant response and, as the needle is held in a magnetic field, it will not move as the vehicle changes position. The gauge can be arranged to continue to register the last position even when switched off or, if a small 'pull off' magnet is used, it will return to its zero position. As a system voltage change would affect the current flowing in all three coils variations are cancelled out, negating the need for voltage stabilization. Note that the operation is similar to the moving iron gauge.

13.1.6 Other types of gauges

A variation of any of the above types of gauge can be used to display other required outputs, such as voltage or oil pressure. Gauges to display road or engine speed, however, need to react very quickly to changes. Many systems now use stepper motors for this purpose although some retain the conventional cable driven speedometers.

Figure 13.7 shows a block diagram of a speedometer, which uses an ammeter as the gauge. This system uses a quenched oscillator sensor that will produce a constant amplitude signal even at very low speed. The frequency of the signal is proportional to road speed. The sensor is driven from the gearbox or a final drive output. The electronic control or signal conditioning circuit consists firstly of a Shmitt trigger, which shapes the signal and suppresses any noise picked up in the wiring. The monostable is used to produce uniform signals in proportion to those from the pulse generator. The moving coil gauge will read an average of the pulses. This average value is dependent on the frequency of the input signal, which in turn is dependent on vehicle speed. The odometer is driven by a stepper motor, which is driven by the output of a divider and a power amplifier. The divider is to

calibrate the action of the stepper motor to the distance covered. The actual speedometer gauge can be calibrated to any vehicle by changing the time delay of the monostable (see Chapter 2).

A system for driving a tachometer is similar to the speedometer system. Pulses from the ignition primary circuit are often used to drive this gauge. Figure 13.8 shows the block diagram of a typical system.

13.1.7 A digital instrumentation system

Figure 13.9 shows a typical digital instrumentation system. All signal conditioning and logic functions are carried out in the ECU. This will often form part of the dashboard assembly. Standard sensors provide information to the ECU, which in turn will drive suitable displays. The ECU contains a ROM section, which allows it to be programmed to a specific vehicle. The gauges used are as described in the above sections. Some of the extra functions available with this system are described briefly as follows.

- Low fuel warning light – can be made to illuminate at a particular resistance reading from the fuel tank sender unit.
- High engine temperature warning light – can be made to operate at a set resistance of the thermistor.
- Steady reading of the temperature gauge – to prevent the gauge fluctuating as the cooling system thermostat operates, the gauge can be made to read only at, say, five set figures. For example, if the input resistance varies from 240 to 200 Ω as the thermostat operates, the ECU will output just one reading, corresponding to 'normal' on the gauge. If the resistance is much higher or lower the gauge will read to one of the five higher or lower positions. This gives a low resolution but high readability for the driver.
- Oil pressure or other warning lights can be made to flash – this is more likely to catch the driver's attention.
- Service or inspection interval warning lights can be used – the warning lights are operated broadly as a function of time but, for example, the service interval is reduced if the

Figure 13.8 Block diagram of a tachometer which uses signals from the ignition coil

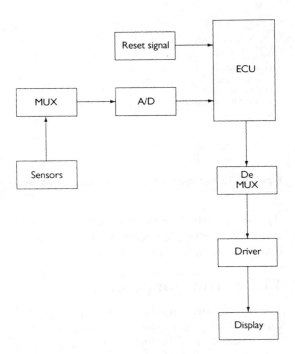

Figure 13.9 Digital instrumentation system

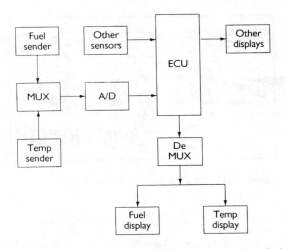

Figure 13.10 Block diagram of high temperature and low fuel warning lights. The A/D converter is time division multiplexed to various sensors

engine experiences high speeds and/or high temperatures. Oil condition sensors are also used to help determine service intervals.

- Alternator warning light – works as normal but the same or an extra light can be made to operate if the output is reduced or if the drive belt slips. This is achieved by a wire from one phase of the alternator providing a pulsed signal, which is compared to a pulsed signal from the ignition. If the ratio of the pulses changed this would indicate a slipping belt.

As an example of how some of this system works consider the high temperature and low fuel warning lights as examples. Figure 13.10 shows a block diagram of just this part of the overall system.

The analogue to digital converter is time division multiplexed to various sensors. The signals from the temperature and fuel level sensors will produce a certain digital representation of a numerical value when they reach say $180\,\Omega$ (about $105\,^\circ$C) and $200\,\Omega$ (10 litres left), respectively. These figures (assigned to variables 'temp_input' and 'fuel_input') can then be compared with those pre-programmed into memory, variables 'high_temp' and 'low_fuel'. The following simplified lines of computer program indicate the logical result.

    IF temp_input > high_temp THEN
    high_temp_light = on

    IF fuel_input > low_fuel THEN
    low_fuel_light = on

A whole program is built up which can be made suitable for any particular vehicle requirements.

13.2 Driver information

13.2.1 Vehicle condition monitoring

VCM or vehicle condition monitoring is a form of instrumentation. It has now become difficult to separate it from the more normal instrumentation system discussed in the first part of this chapter. The complete VCM system can include driver information relating to the following list of systems that can be monitored.

- High engine temperature.
- Low fuel.
- Low brake fluid.
- Worn brake pads.
- Low coolant level.
- Low oil level.
- Low screen washer fluid.
- Low outside temperature.
- Bulb failure.
- Doors, bonnet or boot open warning.

Figure 13.11 shows a trip computer display, which also incorporates the vehicle map (see next section).

The circuit shown in Figure 13.12 can be used to operate bulb failure warning lights for whatever particular circuit it is monitoring. The

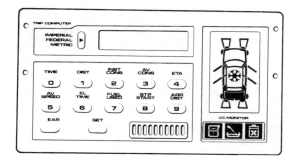

Figure 13.11 Trip computer display and a vehicle 'map'

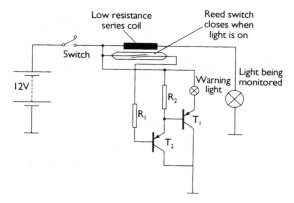

Figure 13.12 Bulb failure warning circuit

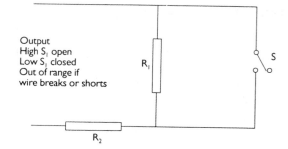

Figure 13.13 Equivalent circuit of a dual resistance self-testing system

simple principle is that the reed relay is only operated when the bulb being monitored is drawing current. The fluid and temperature level monitoring systems work in a similar way to the systems described earlier but in some cases the level of a fluid is monitored by a float and switch.

Oil level can be monitored by measuring the resistance of a heated wire on the end of the dipstick. A small current is passed through the wire to heat it. How much of the wire is covered by oil will determine its temperature and therefore its resistance.

Many of the circuits monitored use a dual resistance system so that the circuit itself is also checked. Figure 13.13 shows the equivalent circuit for this technique. In effect, it will produce one of three possible outputs: high resistance, low resistance or an out-of-range reading. The high or low resistance readings are used to indicate say correct fluid level and low fluid level. A figure outside these limits would indicate a circuit fault of either a short or open circuit connection.

The display is often just a collection of LEDs or a back lit LCD. These are arranged into suitable patterns and shapes such as to represent the circuit or system being monitored. An open door will illuminate a symbol that looks like the door of the vehicle map (plan view of the car) is open. Low outside temperature or ice warning is often a large snowflake.

13.2.2 Trip computer

The trip computer used on many top range vehicles is arguably an expensive novelty, but is popular nonetheless. The display and keypad of a typical trip computer are shown in Figure 13.11. The functions available on most systems are:

● Time and date.
● Elapsed time or a stop watch.
● Estimated time of arrival.
● Average fuel consumption.
● Range on remaining fuel.
● Trip distance.

The above details can usually be displayed in imperial, US or metric units as required. In order to calculate the above outputs the inputs to the system shown in Table 13.2 are required.

Figure 13.14 shows a block diagram of a trip computer system. Note that several systems use

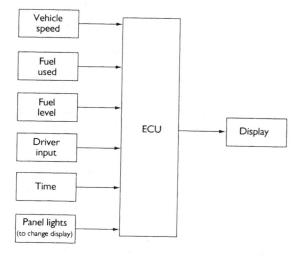

Figure 13.14 Display of a typical trip computer

Table 13.2 Input to the system

Input	Source
Clock signal	Crystal oscillator
Vehicle speed	Speed sensor or instruments ECU
Fuel being used	Injector open time or flow meter
Fuel in the tank	Tank sender unit
Mode/Set/Clear	Data input by the driver

the same inputs and that several systems 'communicate' with each other. This makes the overall wiring very bulky – if not complicated. This type of interaction and commonality between systems has been one of the reasons for the development of multiplexed wiring techniques (see Chapter 3).

13.2.3 Traffic information

Over 25 years have passed since we first watched James Bond use a tracking device, which showed a moving blip across a screen on the dashboard of his Aston Martin. Advances in computer technology and GPS systems have turned this into reality.

In California, many motor vehicles have been equipped with a gadget called the Navigator, which helps drivers get to a destination by displaying their vehicle's location on a glowing green map. The Navigator, introduced by a company known as Etak, is an electronic road map that calculates position by means of dead reckoning. Data from a solid-state compass installed in the vehicle's roof and from sensors mounted on its wheels are processed by a computer and displayed on a dashboard screen. The car's position is represented as a fixed triangle on a map, which scrolls down as the car moves forward and rotates sideways when it turns.

Toyota already offers a computerized dashboard map on an expensive model sold only in Japan, but many manufacturers are considering fitting these devices in the near future. Jaguar, as part of a project called 'Prometheus', in conjunction with other manufacturers, has developed a computerized system that picks up information from static transmitters. This system gives directions and advanced warning of road junctions, signposts and speed limits.

Other forms of driver information systems are being considered, such as one being developed in the USA. 'DriverGuide' is the electronic equivalent of winding down a window and asking for directions. By choosing from a variety of screen menus, the driver can specify where he or

she wants to go. Twenty seconds later a printed sheet of driving instructions constructed from a cartographic database will be printed. Computerized route finding software is already very popular. Its one problem is that the data on disk is out of date instantly due to roadworks and other restrictions. Transmitting live data to the vehicle is the answer.

13.3 Visual displays

13.3.1 Choosing the best display – readability

The function of any visual display is to communicate information to the desired level of accuracy. Most displays used in the vehicle must provide instant data but the accuracy is not always important. Analogue displays can provide almost instant feedback from one short glance. For example, if the needle of the temperature gauge is about in the middle then the driver can assume that the engine temperature is within suitable limits. A digital read-out of temperature such as 98°C would not be as easy to interpret. This is a good example as to why even when digital processing and display techniques are used, the actual read-out will still be in analogue form. Figure 13.15 shows a display using analogue gauges.

Figure 13.16 shows an instrument display using digital representation. Numerical and other forms of display are, however, used for many applications. Some of these are as follows:

- Vehicle map.
- Trip computer.
- Clock.
- Radio displays.
- Route finding displays.
- General instruments.

These displays can be created in a number of ways; the following sections examine each of these in more detail. To drive individual segments or parts of a complete display, a technique called *time division multiplexing* is often used.

13.3.2 Light-emitting diode displays

If the PN junction of a diode is manufactured from gallium arsenide phosphide (GaAsP), light

Figure 13.15 Analogue display

Figure 13.16 A display using LEDs

will be emitted from the junction when a current is made to pass in the forward-biased direction. This is a light-emitting diode (LED) and will produce red, yellow or green light with slight changes in the manufacturing process. LEDs are used extensively as indicators on electronic equipment and in digital displays. They last for a very long time (50 000 hours) and draw only a small current.

LED displays are tending to be replaced for automobile use by the liquid crystal type display, which can be backlit to make it easier to read in the daylight. However, LEDs are still popular for many applications.

The actual display will normally consist of a number of LEDs arranged into a suitable pattern for the required output. This can range from the standard seven-segment display to show numbers, to a custom-designed speedometer display. A small number of LED displays are shown in Figure 13.17.

13.3.3 Liquid crystal displays

Liquid crystals are substances that do not melt directly from a solid to the liquid phase, but first pass through a paracrystalline stage in which the molecules are partially ordered. In this stage, a liquid crystal is a cloudy or translucent fluid but still has some of the optical properties of a solid crystal.

The three main types of liquid crystals are smectic, nematic and cholesteric (twisted nematic), which are differentiated by the alignments of the rod-shaped molecules. Smectic liquid crystals have molecules parallel to one another, forming a layer, but within the layer no pattern exists. Nematic types have the rod-like molecules oriented parallel to one another but have no layer structure. The cholesteric types have parallel molecules, and the layers are arranged in a helical, or spiral, fashion.

Mechanical stress, electric and magnetic

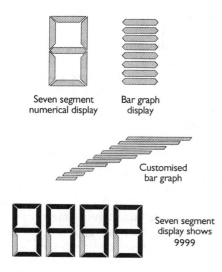

Seven segment numerical display

Bar graph display

Customised bar graph

Seven segment display shows 9999

Figure 13.17 LED displays

fields, pressure and temperature can alter the molecular structure of liquid crystals. A liquid crystal also scatters light that shines on it. Because of these properties, liquid crystals are used to display letters and numbers on calculators, digital watches and automobile instrument displays. LCDs are also used for portable computer screens and even television screens. The LCD has many more areas of potential use and developments are ongoing. In particular, this type of display is now good enough to reproduce pictures and text on computer screens.

One type of display uses the cholesteric type of liquid crystal. This display is achieved by only allowing polarized light to enter the liquid crystal which, as it passes through the crystal, is rotated by 90°. The light then passes through a second polarizer, which is set at 90° to the first. A mirror at the back of the arrangement reflects the light so that it returns through the polarizer, the crystal and the front polarizer again. The net result is that light is simply reflected, but only when the liquid crystal is in this one particular state.

When a voltage of about 10 V at 50 Hz is applied to the crystal, it becomes disorganized and the light passing through it is no longer twisted by 90°. This means that the light polarized by the first polarizer will not pass through the second, and will therefore not be reflected. This will show as a dark area on the display.

These areas are constructed into suitable segments in much the same way as with LEDs to provide whatever type of display is required. The size of each individual area can be very small, such as to form one pixel of a TV or computer screen if appropriate. Figure 13.18 shows a representation of how this liquid crystal display works.

LCDs use very low power but do require a source of light to operate. To be able to read the display in the dark some form of lighting for the display is required. Instead of using a reflecting mirror at the back of the display a source of light known as backlighting can be used. A condition known as DC electroluminescence is an ideal phenomenon. This uses a zinc-sulphide based compound, which is placed between two electrodes in much the same way as the liquid crystal, but it emits light when a voltage is applied. Figure 13.19 shows how this backlighting effect can be used to good effect for display purposes.

13.3.4 Vacuum fluorescent displays

A vacuum fluorescent display (VFD) works in much the same way as a television tube and screen. It is becoming increasingly popular for vehicle use because it produces a bright light (which is adjustable) and a wider choice of colours than LED or LCD displays. Figure 13.20 shows that the VFD system consists of three main components. These are the filament, the grid and the screen with segments placed appropriately for the intended use of the display. The filament forms the cathode and the segments the anode of the main circuit. The control grid is used to control brightness as the voltage is altered.

When a current is passed through the tungsten filaments they become red hot (several hundred degrees centigrade) and emit electrons. The whole unit is made to contain a good vacuum so that the electrons are not affected by any outside

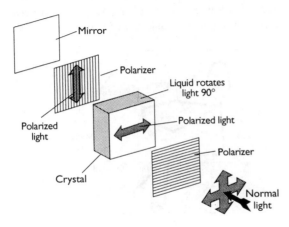

Figure 13.18 Principle of a liquid crystal display

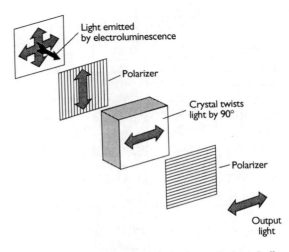

Figure 13.19 Back lighting effect can be used to good effect for display purposes

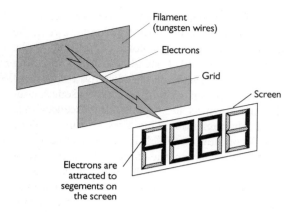

Figure 13.20 Vacuum fluorescent display

influence. The segments are coated with a fluorescent substance and connected to a control wire. The segments are given a positive potential to attract the electrons. When electrons strike the segments they fluoresce, emitting a yellow-green or a blue-green light depending on the type of phosphor used to coat the segments. If the potential of the grid is changed, the number of electrons striking the segments can be changed, thus affecting the brightness. If no segments are connected to a supply (often only about 5 V), then all the electrons emitted are stopped at the grid. The grid is also important in that it tends to organize the movement of electrons.

Figure 13.21 shows a circuit used to control a VFD. Note how the potential of the segments when activated is above that of the grid. The driver circuit for this system is much the same, in principle, as any other display, i.e. the electronic control will connect one or more of the appropriate segments to a supply to produce the desired output.

The glass front of the display can be coloured to improve the readability and aesthetic value. This type of display has many advantages but the main problem for automobile use is its susceptibility to shock and vibration. This can be overcome, however, with suitable mountings.

13.3.5 Head-up displays

One of the main problems to solve with any automobile instrument or monitoring display is that the driver has to look away from the road to see the information. Also, in many cases, the driver does not actually need to look at the display, and hence could miss an important warning such as low oil pressure. Many techniques can be used such as warning beepers or placing the instruments almost in view, but one of the most innovative is the head-up display (HUD). This was originally developed by the aircraft industry for fighter pilots; aircraft designers had similar problems in displaying up to 100 different warning devices in an aircraft cockpit. Figure 13.22 shows the principle of a head-up display. Information from a display device, which could be a CRT (cathode ray tube), is directed onto a partially reflecting mirror. The information displayed on the CRT would therefore have to be reversed for this system. Under normal circumstances the driver would be able to see the road through the mirror. The brightness of the display would, of course, have to be adjusted to suit ambient lighting conditions. A great deal of data could be presented when this system is computer controlled.

A problem, however, is which information to provide in this way. The speedometer could form part of a lower level display and a low oil pressure could cause a flash right in front of the dri-

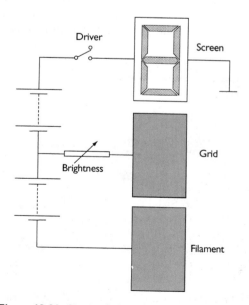

Figure 13.21 Circuit which could be used to control a VFD

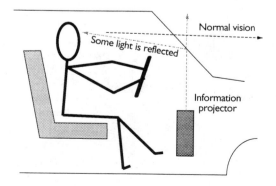

Figure 13.22 Head-up display

ver. A visual warning could also be displayed when a forward facing radar detects an impending collision. Current HUD systems are for straight-ahead vision, but liquid crystal rear view mirrors, used to dim and cut headlight glare automatically, can be used as an effective display screen for rear facing, blind spot detecting radar.

One of the most interesting studies is to determine exactly where the driver is looking at any point in time, which could be used to determine where the head-up display would be projected at any particular time. The technique involves tiny video cameras, coupled to a laser beam that reflects from the cornea of the driver's eye and can measure exactly where he or she is looking. Apart from its use in research, the eye motion detector is one of a series of tools used in bio-mechanical research that can directly monitor the physical well-being of the driver. Some of these tools could eventually be used actively to control the car or to wake up a driver who is at risk of falling asleep.

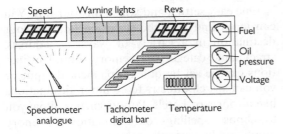

Figure 13.23 Displays which combine some of the devices discussed

13.3.6 Display techniques summary

Most of the discussion in previous sections has been related to the activation of an individual display device. The techniques used for – and the layout of – dashboard or display panels are very important. To a great extent this again comes back to readability. When so many techniques are available to the designer it is tempting to use the most technologically advanced. This, however, is not always the best. It is prudent to ask the one simple question: what is the most appropriate display technique for this application? Figure 13.23 shows a display that combines some of the devices discussed previously.

Many of the decisions regarding the display are going to be according to the preference of the designer. I find numerical display of vehicle speed or engine rev/min irritating. Even the bar graph displays are not as good as simple analogue needles (this, however, is only my opinion).

The layout and the way that instruments are combined is an area in which much research has been carried out. This relates to the time it takes the driver to gain the information required when looking away from the road to glance at the instrument pack. Figure 13.24 shows the instrument panel and other readout displays on the new 'S-type' Jaguar. Note how compact it is so that the information can be absorbed almost without the driver having to scan to each readout in turn. The aesthetic looks of the dashboard are an important selling point for a vehicle. This could be at odds with the best readability on some occasions.

Figure 13.24 Jaguar S-type instruments and displays

13.4 Case studies

13.4.1 Air-cored temperature gauge – Rover

Figure 13.25 shows the system used on some Rover vehicles for the temperature gauge. It is an air-cored device with fluid damping. The temperature gauge is fitted with a spiral pull off spring to make the gauge read 'cold' when the ignition is switched off. The fuel gauge is very similar but retains its position when the ignition is off.

When the system receives a supply from the ignition the resistance of the thermistor determines the current flowing through the coils. When engine coolant temperature is low, the resistance of the sender will be high. This will cause the voltage at point X to be higher than that at point Y. This will be above the Zener voltage and so the diode will conduct in its reverse direction. Current will flow through coil A and coil B directly but also a further path will exist through R and the diode, effectively bypassing coil A. This will cause the magnetism of coil B to be greater than coil A, deflecting the magnet and pointer towards the cold side. As the resistance of the sender falls with increasing temperature, the voltage at X will fall, reducing the

current through coil B, allowing the needle to rise.

At normal operating temperature, the voltage at X will be just under the Zener diode breakdown voltage. Current through each coil will now be the same and the gauge will read in the centre. If coolant temperature increases further, then current will flow through the diode in its forward direction, thus increasing the current through coil A, which will cause the needle to move to the hot side. Operation of the fuel gauge is similar but a resistor is used in place of the Zener diode. The diode is used to stabilize the gauge when reading 'normal' to reduce fluctuations due to thermostat operation.

13.4.2 Car navigation system – Alpine Electronics

The 'Alpine' navigation system is one of the most advanced systems in current use. It features very accurate maps, is easy to use and even offers some voice guidance. The system consists of the base unit, a monitor, an antenna, a remote control and CD-ROM disks. Figure 13.26 shows the system in a vehicle. The following features are highlighted by 'Alpine'.

One easy setting and you're on your way. You can input and have the system search for your destination in a variety of ways: by address, street name, category or memory point. Destinations can be set by quick alphabetical input or you can switch directly to common destinations like airports or hotels. Pop-up menus allow you to choose spellings of destinations, memory inputs, etc by using the remote control cursor.

Once inputting is done, the system calculates the best route to your destination according to your instructions. You can choose whether to go via motorway or normal streets, and also include local-points (like restaurants or fuel stations) or exclude avoid-points, which you set. If traffic flow is obstructed, use 'Alternate Route Setting' instantly to get a new route. Cross-border routes can also be specified.

Alpine gives 'Voice Guidance' to the destination, as well as a wide selection of display options. The 'Basic Direction Mode' displays only the most essential information, so as not to distract you from driving. It clearly shows the car's direction, distance to next junction, and time remaining to destination. The direction at the next junction is also shown – a big advantage in heavy traffic. 'Intersection Zoom' is a facility allowing a closer look, and any of several dis-

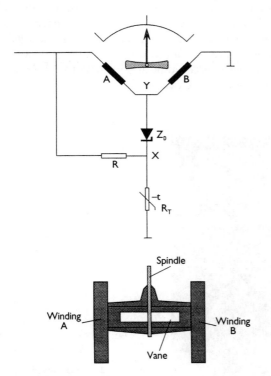

Figure 13.25 Air-cored gauge with fluid damping

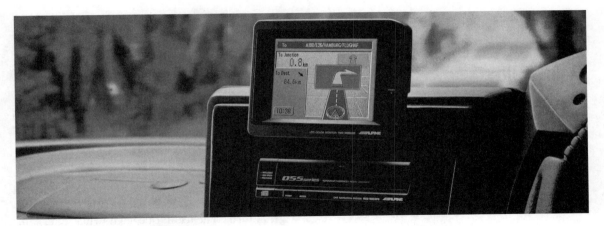

Figure 13.26 'Alpine' navigation system mounted in a vehicle

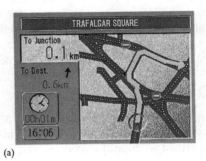

(a)

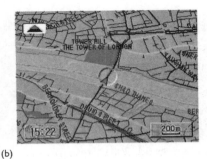

(b)

Figure 13.27 Screenshots from 'Alpine' showing (a) intersection zoom and (b) automatic rerouting

play modes, such as north-up or heading-up, are available. Figure 13.27 shows two screenshots from the system.

Intersection Zoom is an interesting feature of the Alpine system and the key to its easy to understand guidance. As you approach an intersection, the upcoming junction is enlarged so you know exactly what turns are required to stay on your route. If a junction is missed, the 'Auto Reroute' function calculates a new route within seconds. It works so smoothly and quickly you may not even realize you have missed your original way!

Alpine has become the most successful navigation systems in the world. This has been achieved by meeting present demands and also by anticipating future needs. For instance, if you change from summer to winter tyres, the system may have to be calibrated. The Alpine system auto-calibrates during the first few miles and software updates are easily downloaded from CD into the flash memory.

13.5 Diagnosing instrumentation system faults

13.5.1 Introduction

As with all systems the six stages of fault-finding should be followed.

1. Verify the fault.
2. Collect further information.
3. Evaluate the evidence.
4. Carry out further tests in a logical sequence.
5. Rectify the problem.
6. Check all systems.

The procedure outlined in the next section is related primarily to stage 4 of the process. Table 13.3 lists some common symptoms of an instrumentation system malfunction together with suggestions for the possible fault. The faults are very generic but will serve as a good reminder.

Table 13.3 Common symptoms and possible faults of an instrumentation system malfunction

Symptom	Possible fault
Fuel and temperature gauges both read high or low	• Voltage stabilizer.
Gauges read full/hot or empty/cold all the time Instruments do not work	• Short/open circuit sensors. • Short or open circuit wiring. • Loose or broken wiring/connections/fuse. • Inoperative instrument voltage stabilizer. • Sender units (sensor) faulty. • Gauge unit fault (not very common).

13.5.2 Testing procedure

The process of checking a thermal gauge fuel or temperature instrument system is broadly as follows.

1. Hand and eye checks (loose wires, loose switches and other obvious faults) – all connections clean and tight.
2. *Either* fit a known good $200\,\Omega$ resistor in place of the temperature sender – gauge should read full.
3. *Or* short fuel tank sender wire to earth – gauge should read full.
4. Check continuity of wire from gauge to sender – 0 to $0.5\,\Omega$.
5. Check supply voltage to gauge (pulsed 0–12 V on old systems) – 10 V stabilized on most.
6. If all above tests are OK the gauge head is at fault.

13.6 Advanced instrumentation technology

13.6.1 Multiplexed displays

In order to drive even a simple seven-segment display, at least eight wiring connections are required. This would be one supply and seven earths (one for each segment). This does not include auxiliary lines required for other purposes, such as backlighting or brightness. To display three seven-segment units, up to about 30 wires and connections would be needed.

To reduce the wiring, time division multiplexing is used. This means that the individual display unit will only be lit during its own small time slot. From Figure 13.28 it can be seen that, if the bottom connection is made at the same time as the appropriate data is present on the seven input lines, only one seven-segment display will be activated. This is carried out for each in turn, thousands of times a second and the human eye does not perceive a flicker.

The technique of multiplexing is taken a stage further by some systems, in that one digital controller carries out the whole of the data or signal processing. Figure 13.29 shows this in block diagram form. The technique is known as data sampling. The electronic control unit samples each input in turn in its own time slot, and outputs to the appropriate display again in a form suitable for the display device used. The electronics will contain a number of A/D and D/A converters and these will also be multiplexed where possible.

13.6.2 Quantization

When analogue signals are converted to digital, a process called quantization takes place. This could be described as digital encoding. Digital encoding breaks down all data into elementary binary digits (bits), which enable it to be processed, stored, transmitted and decoded as required by computer technologies.

The value of an analogue signal changes smoothly between zero and a maximum. This infinitely varying quantity is converted to a series of discrete values of 0 or 1 by a process known as quantization. The range of values from zero to the maximum possible is divided into a discrete number of steps or quantization levels. The number of steps possible depends on the bit size of the word the digital processors can deal with. For an 8-bit word, the range can be divided into 256 steps (2^8), i.e. from 00000000_2 to 11111111_2. These digital 'samples' should always

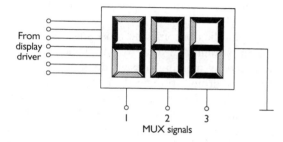

Figure 13.28 Time divisions multiplexing is used so the individual display unit will only be lit during its own small time slot

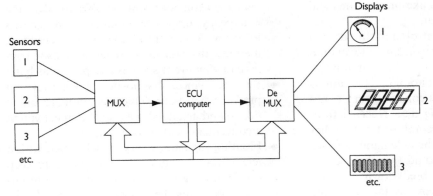

Figure 13.29 Block diagram showing how multiplexing is taken a stage further by some systems

be taken at more than twice the frequency of the analogue signal to ensure accurate reproduction.

Quantization introduces an error into the process, as each value is 'rounded' to the nearest quantization level. The greater the number of quantization levels the more accurate the process will be, but obviously, increased accuracy involves more bits being used to define the increased number of levels.

13.7 New developments in instrumentation systems

13.7.1 Introduction

As with many vehicle systems the future for instrumentation has much yet to offer. As the computerization of all vehicle systems continues to increase, more functions become possible. This is particularly so as, with the more powerful processors, some spare capacity is available. This will allow the systems to exhibit 'intelligence' so that only relevant information will be displayed at any particular time. For example, if the engine coolant temperature is correct it is not necessary to display a reading. Also, for example, if the range of the vehicle is displayed then the fuel quantity in the tank is not so important as far as the display is concerned.

The system could be programmed to interrupt at a suitable time to display a message. For example, '*Range now only 50 km*,' or '*Offside rear tyre pressure is falling!*' or '*You're lost aren't you?*' The limits are governed only by the imagination.

13.7.2 Holography

A holographic image is a three-dimensional representation of the original subject. It can be cre-

ated by splitting a laser beam into object and reference beams. These beams produce an interference pattern, which can be stored on a plate or projected on to a special screen. Some research is currently ongoing towards using holography to improve night driving safety. Information from infrared cameras can be processed, and then an enhanced holographic image can be projected onto a vehicle windscreen.

13.7.3 Telematics

The information provided here is taken from information provided by the Automobile Association (AA), a well-respected organization, in the UK. Similar developments are taking place across the world. It was difficult to know whether 'Telematics' should be included in the instrumentation section or elsewhere – but here it is anyway.

The car is a necessary component of our lives. Over the last 50 years the number of vehicles has grown 10-fold and, by 2030, traffic is expected to have increased by a further 60%. The cost of personal transport is high; we should be acting now to ease congestion, save fuel and protect the environment. The technology to create some of the solutions is already available.

First-generation telematics services are already available, or under development. They include:

- Voice based roadside assistance, emergency dispatch, traffic information services and route advice.
- Travel guidance, points of interest, touring and travel information.
- Stolen vehicle tracking by satellite.
- Radio Data System (RDS) built into most car radios and the recent launch of RDS-TMC (Traffic Message Channel).

A fifth of all driving time is spent getting lost on unfamiliar roads even though it is possible to pinpoint specific locations like fuel stations and then guide a vehicle to them.

A small telematics control unit fitted in a vehicle can open up a new world of information services, using a combination of communications and computing technology. The unit is connected to a receiver that constantly calculates the vehicle's position using data received from satellites. These data are combined with other information and fed to the telematics service centre. The information could then be used to guide a patrol vehicle to a breakdown. By linking the engine management system into the telematics unit, the service centre could even use a diagnostic program to identify the mechanical or electrical problem.

As Europe's largest traffic information broadcaster, the AA has taken a leading role in nine separate EC transport studies and has developed real-time traffic management systems that provide instant information about road problems and uncongested routes. When an on-board telematics unit is linked to a vehicle's engine management system it will be able to monitor vehicle performance and give advance warning of mechanical problems. In the near future, a wide range of vital new services may be on offer.

- Traffic information. To give drivers the best and quickest route destination given the road conditions at the time.
- Route guidance. The service centre will be able to calculate the best route to a nominated destination, taking into account traffic conditions along the way, and relay it to a visual and audible display in the vehicle.
- Radio Data System Traffic Message Channel (RDS-TMC). This is coded traffic information, broadcast continuously as a sub-carrier on a national radio channel, with updates made every 20 seconds. A driver can choose precisely when he or she receives the information, and can even specify particular roads that are relevant to their own journey.
- Vehicle tracking. This is tracking technology that can trace a stolen vehicle and identify its location.
- Remote services. To lock, unlock or immobilize a vehicle remotely. The operator will even be able to flash the vehicle's lights to help you locate it in a car park.
- Emergency dispatch. An in-vehicle emer-

gency button that will be able to alert the emergency services to an incident and give its location. Alternatively, the services could be alerted automatically by a vehicle sensor, triggered by an event such as a deployed airbag.
- Remote vehicle diagnostics. Telematics will predict when your vehicle is about to break down, and arrange for a patrol to meet you at a convenient nearby location.
- Floating car data. Every vehicle fitted with a telematics unit could eventually help to keep traffic moving by automatically and continuously providing the service centre with details of traffic flow in its immediate location. That traffic condition data can then be assessed and fed back out to other drivers who may be approaching the same area and possible congestion.

(Data from Automobile Association, 1998)

13.8 Self-assessment

13.8.1 Questions

1. State the main advantage of a thermal gauge.
2. Make a clearly labelled sketch of a thermal fuel gauge circuit.
3. Describe why moving iron and air-cored gauges do not need a voltage stabilizer.
4. Define the term, 'driver information'.
5. Explain why digital displays are multiplexed.
6. Draw the circuit of a bulb failure system and describe its operation.
7. List five typical outputs of a trip computer and the inputs required to calculate each of them.
8. Describe with the aid of a sketch how a head-up display (HUD) operates.
9. Explain the operation of an air-cored fuel gauge system.
10. Describe what is meant by 'Telematics'.

13.8.2 Assignment

Design an instrument display for a car. Choose whatever type of display techniques you want, but make a report justifying your choices. Some key issues to consider are readability, accuracy, cost and aesthetic appeal.

14
Air conditioning

14.1 Conventional heating and ventilation

14.1.1 Introduction

The earliest electrical heating I have come across was a pair of gloves with heating elements woven into the material (c. 1920). These were then connected to the vehicle electrical system and worked like little electric fires. The thought of what happened in the case of a short circuit is a little worrying!

The development of interior vehicle heating has been an incremental process and will continue to be so – the introduction of air conditioning being the largest step. The comfort we now take for granted had some very cold beginnings, but the technology in this area of the vehicle electrical system is still evolving. Systems now range from basic hot/cold air blowers to complex automatic temperature and climate control systems.

Any heating and ventilation system has a simple set of requirements, which are met to varying standards. These can be summarized as follows.

- Adjustable temperature in the vehicle cabin.
- Heat must be available as soon as possible.
- Distribute heat to various parts of the vehicle.
- Ventilate with fresh air with minimum noise.
- Facilitate the demisting of all windows.
- Ease of control operation.

The above list, whilst by no means definitive, gives an indication of what is required from a heating and ventilation system. As usual, the more complex the system the more the requirements are fulfilled. This is directly related to cost.

Some solutions to the above requirements are discussed below, starting with simple ventilation and leading on to full automatic temperature control. Figure 14.1 shows a representation of the perceived comfortable temperature in the vehicle compared with the outside temperature.

14.1.2 Ventilation

To allow fresh air from outside the vehicle to be circulated inside the cabin, a pressure difference must be created. This is achieved by using a plenum chamber. A plenum chamber by definition holds a gas (in this case air), at a pressure higher than the ambient pressure. The plenum chamber on a vehicle is usually situated just below the windscreen, behind the bonnet hood. When the vehicle is moving the air flow over the vehicle will cause a higher pressure in this area. Figure 14.2 shows an illustration of the plenum chamber effect. Suitable flaps and drains are utilized to prevent water entering the car through this opening.

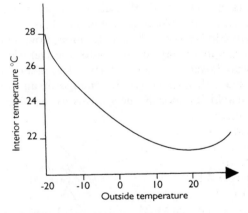

Figure 14.1 Representation of comfortable temperature

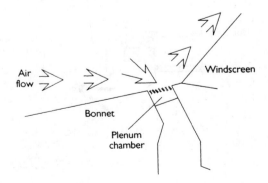

Figure 14.2 Plenum chamber effect

By means of distribution trunking, control flaps and suitable 'nozzles', the air can be directed as required. This system is enhanced with the addition of a variable speed blower motor. Figure 14.3 shows a typical ventilation and heating system layout.

When extra air is forced into a vehicle cabin the interior pressure would increase if no outlet was available. Most passenger cars have the outlet grills on each side of the vehicle above or near the rear quarter panels or doors.

14.1.3 Heating system – water-cooled engine

Heat from the engine is utilized to increase the temperature of the car interior. This is achieved by use of a heat exchanger, called the heater matrix. Due to the action of the thermostat in the engine cooling system the water temperature remains broadly constant. This allows for the air being passed over the heater matrix to be heated by a set amount depending on the outside air temperature and the rate of air flow. A source of hot air is therefore available for heating the vehicle interior. However, some form of control is required over how much heat (if any), is required. The method used on most modern vehicles is the blending technique. This is simply a control flap, which determines how much of the air being passed into the vehicle is directed over the heater matrix. The main drawback of this system is the change in air flow with vehicle speed. Some systems use a valve to control the hot coolant flowing to the heater matrix.

By a suitable arrangement of flaps it is possible to direct air of the chosen temperature to selected areas of the vehicle interior. In general, basic systems allow the warm air to be adjusted between the inside of the windscreen and the driver and passenger foot wells. Most vehicles also have small vents directing warm air at the drivers and front passenger's side windows. Fresh cool air outlets with directional nozzles are also fitted.

One final facility, which is available on many vehicles, is the choice between fresh or recirculated air. The main reason for this is to decrease the time it takes to demist or defrost the vehicle windows, and simply to heat the car interior more quickly to a higher temperature. The other reason is that, for example, in heavy congested traffic, the outside air may not be very clean.

14.1.4 Heater blower motors

The motors used to increase air flow are simple permanent magnet two-brush motors. The blower fan is often the centrifugal type and, in many cases, the blades are positioned asymmetrically to reduce resonant noise. Figure 14.4 shows a typical motor and fan arrangement. Varying the voltage supplied controls motor speed. This is achieved by using dropping resistors. The speed in some cases is made 'infinitely' variable by the use of a variable resistor. In most cases the motor is controlled to three or four set speeds.

Figure 14.5 shows a circuit diagram typical of a three-speed control system. The resistors are usually wire wound and are placed in the air stream to prevent overheating. These resistors will have low values in the region of $1\,\Omega$ or less.

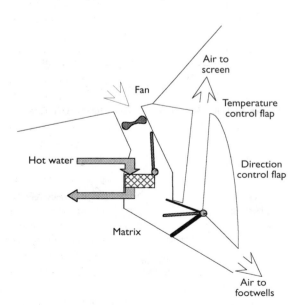

Figure 14.3 Ventilation and heating system

Figure 14.4 HVAC motor mounted in spiral housing

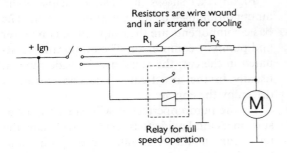

Figure 14.5 Circuit diagram of a three-speed control system

14.1.5 Electronic heating control

Most vehicles that have electronic control of the heating system also include air conditioning, which is covered in the next section. However, a short description at this stage will help to lead into the more complex systems. Figure 14.6 shows a block diagram representing an electronically controlled vehicle heating system.

This system requires control of the blower motor, blend flap, direction flaps and the fresh or recirculated air flap. The technique involves one or a number of temperature sensors suitably positioned in the vehicle interior, to provide information for the ECU. The ECU responds to information received from these sensors and sets the controls to their optimum positions. The whole arrangement is, in fact, a simple closed loop feedback system with the air temperature closing the loop. The ECU has to compare the position of the temperature control switch with the information that is supplied by the sensors and either cool or heat the car interior as required.

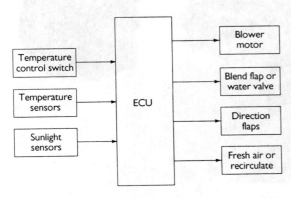

Figure 14.6 An electronically controlled vehicle heating system

14.2 Air conditioning

14.2.1 Introduction

A vehicle fitted with air conditioning allows the temperature of the cabin to be controlled to the ideal or most comfortable value determined by the ambient conditions. The system as a whole still utilizes the standard heating and ventilation components, but with the important addition of an evaporator, which both cools and dehumidifies the air.

Air conditioning can be manually controlled or, as is now often the case, combined with some form of electronic control. The system as a whole can be thought of as a type of refrigerator or heat exchanger. Heat is removed from the car interior and dispersed to the outside air.

14.2.2 Principle of refrigeration

To understand the principle of refrigeration the following terms and definitions will be useful.

- Heat is a form of energy.
- Temperature means the degree of heat of an object.
- Heat will only flow from a higher to a lower temperature.
- Heat quantity is measured in 'calories' (more often kcal).
- 1 kcal heat quantity, changes the temperature of 1 kg of liquid water by $1\,^{\circ}C$.
- Change of state, is a term used to describe the changing of a solid to a liquid, a liquid to a gas, a gas to a liquid or a liquid to a solid.
- Evaporation is used to describe the change of state from a liquid to a gas.
- Condensation is used to describe the change of state from gas to liquid.
- Latent heat describes the energy required to evaporate a liquid without changing its temperature (breaking of molecular bonds), or the amount of heat given off when a gas condenses back into a liquid without changing temperature (making of molecular bonds).

Latent heat in the change of state of a refrigerant is the key to air conditioning. A simple example of this is that if you put a liquid such as methylated spirits on your hand it feels cold. This is because it evaporates and the change of state (liquid to gas) uses heat from your body. This is why the process is often thought of as 'unheating' rather than cooling.

The refrigerant used in many air conditioning systems is known as R134A. This substance changes state from liquid to gas at –26.3°C. R134A is hydrofluorocarbon (HFC) rather than chlorofluorocarbon (CFC) based, due to the problems with atmospheric ozone depletion associated with CFC-based refrigerants. Note that this type of refrigerant is *not* compatible with older systems.

A key to understanding refrigeration is to remember that a low-pressure refrigerant will have low temperature, and a high-pressure refrigerant will have a high temperature.

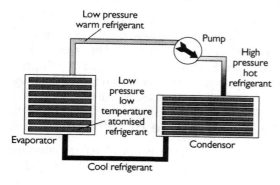

Figure 14.7 Basic principle of an air conditioning or refrigeration system

Figure 14.7 shows the basic principle of an air conditioning – or refrigeration – system. The basic components are the evaporator, condenser and pump or compressor. The evaporator is situated in the car; the condenser outside the car, usually in the air stream. The compressor is driven by the engine.

As the pump operates it will cause the pressure on its intake side to fall, which will allow the refrigerant in the evaporator to evaporate and draw heat from the vehicle interior. The high pressure or output of the pump is connected to the condenser. The pressure causes the refrigerant to condense (in the condenser); thus giving off heat outside the vehicle as it changes state.

Several further components are needed for efficient operation; these are explained over the next few sections. Figure 14.8 shows some typical components of an air conditioning system.

14.2.3 Air conditioning overview

The operation of the system is a continuous cycle. The compressor pumps low pressure but heat laden vapour from the evaporator, com-

Figure 14.8 Heating ventilation and air conditioning (HVAC) components

presses it and pumps it as a super-heated vapour under high pressure to the condenser. The temperature of the refrigerant at this stage is much higher than the outside air temperature, hence it gives up its heat via the fins on the condenser as is changes state back to a liquid.

This high-pressure liquid is then passed to the receiver-drier where any vapour which has not yet turned back to a liquid is stored, and a desiccant bag removes any moisture (water) that is contaminating the refrigerant. The high-pressure liquid is now passed through the thermostatic expansion valve and is converted back to a low-pressure liquid as it passes through a restriction in the valve into the evaporator. This valve is the element of the system that controls the refrigerant flow and hence the amount of cooling provided. As the liquid changes state to a gas in the evaporator, it takes up heat from its surroundings, thus cooling or 'unheating' the air that is forced over the fins. The low pressure vapour leaves the evaporator returning to the pump, thus completing the cycle. The cycle is represented in Figure 14.9.

If the temperature of the refrigerant increases beyond certain limits, condenser cooling fans can be switched in to supplement the ram air effect.

A safety switch is fitted in the high-pressure side of most systems. It is often known as a high–low pressure switch, as it will switch off the compressor if the pressure is too high due to a component fault, or if the pressure is too low due to a leakage, thus protecting the compressor.

14.2.4 Automatic temperature control

Full temperature control systems provide a comfortable interior temperature in line with the passenger controlled input. The electronic control unit has full control of fan speed, air distribution, air temperature, fresh or recirculated air and the air conditioning pump. These systems

will soon be able to control automatic demist or defrost, when reliable sensors are available. A single button will currently set the system to full defrost or demist.

A number of sensors are used to provide input to the ECU.

- An ambient temperature sensor mounted outside the vehicle will allow compensation for extreme temperature variations. This device is usually a thermistor.
- A solar light sensor can be mounted on the fascia panel. This device is a photodiode and allows a measurement of direct sunlight from which the ECU can determine whether to increase the air to the face vents.
- The in-car temperature sensors are simple thermistors but, to allow for an accurate reading, a small motor and fan can be used to take a sample of interior air and direct it over the sensing elements.
- A coolant temperature sensor is used to monitor the temperature of the coolant supplied to the heater matrix. This sensor is used to prevent operation of the system until coolant temperature is high enough to heat the vehicle interior.
- Driver input control switches.

The ECU takes information from all of the above sources and will set the system in the most appropriate manner as determined by the software. Control of the flaps can be either by solenoid controlled vacuum actuators or by small motors. The main blower motor is controlled by a heavy duty power transistor and is constantly variable. These systems are able to provide a comfortable interior temperature when exterior conditions range from –10 to +35°C even in extreme sunlight.

14.3 Other heating systems

14.3.1 Seat heating

The concept of seat heating is very simple. A heating element is placed in the seat, together with an on–off switch and a control to regulate the heat. However, the design of these heaters is more complex than first appears.

The heater must meet the following criteria.

- The heater must only supply the heat loss experienced by the person's body.

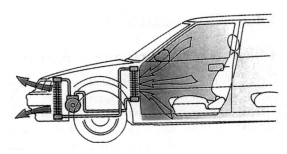

Figure 14.9 Air conditioning operation

- Heat to be supplied only at the major contact points.
- Leather and fabric seats require different systems due to their different thermal properties.
- Heating elements must fit the design of the seat.
- The elements must pass the same rigorous tests as the seat, such as squirm, jounce and bump tests.

Figure 14.10 shows a seat containing heating elements.

In order for the passengers (including the driver) to be comfortable, rigorous tests have been carried out to find the optimum heat settings and the best position for the heating elements. Many tests are carried out on new designs, using a manikin with sensors attached, to measure the temperature and heat flow.

The cable used for most heating elements is known as a Sine Cable and consists of multistrand alloyed copper. This cable may be coated with tin or insulated as the application demands. The heating element is laminated and bonded between layers of polyurethane foam.

The traditional method of control is a simple thermostat switch. Recent developments, however, tend to favour electronic control combined with a thermistor. A major supplier of seat

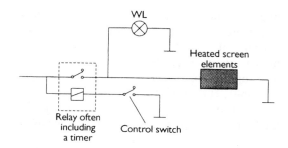

Figure 14.11 Screen heating circuit

heaters, Scandmec Ltd, supplies an electronic system that includes push button switches, potentiometers, timer function, short and open circuit detection. This is in addition to accurate control of the chosen temperature setting. These seat heaters will heat up to provide an initial sensation in 1 minute and to full regulated temperature in 3 minutes.

14.3.2 Screen heating

Heating of the rear screen involves a very simple circuit as shown in Figure 14.11. The heating elements consist of a thin metallic strip bonded to the glass. When a current is passed through the elements, heat is generated and the window will defrost or demist. This circuit can draw high current, 10–15 A being typical. Because of this, the circuit often contains a timer relay to prevent the heater being left on too long. The timer will switch off after 10–15 minutes. The elements are usually positioned to defrost the main area of the screen and the rest position of the rear wiper blade if fitted.

Front windscreen heating is being introduced on some vehicles. This of course presents more problems than the rear screen, as vision must not be obscured. The technology, drawn from the aircraft industry, involves very thin wires cast into the glass. As with the heated rear window, this device can consume a large current and is operated by a timer relay.

14.4 Case studies

14.4.1 Air conditioning – Rover

Figure 14.12 is the air conditioning system layout showing all the main components.

The compressor shown in Figure 14.13 is belt-driven from the engine crankshaft and it acts as a pump circulating refrigerant through the system. The compressor shown is a piston

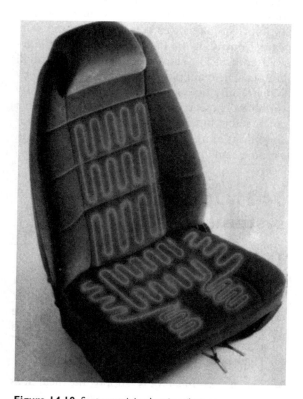

Figure 14.10 Seat containing heating element

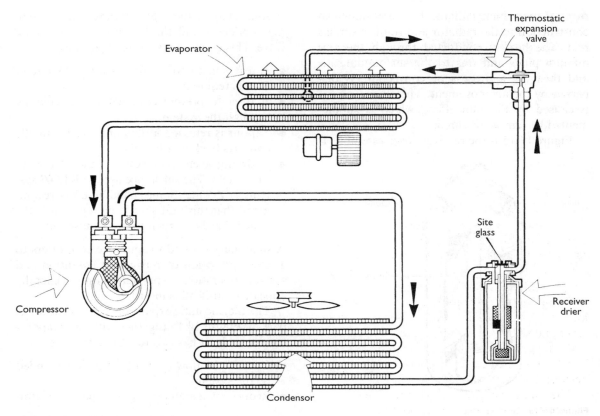

Figure 14.12 Air conditioning system layout

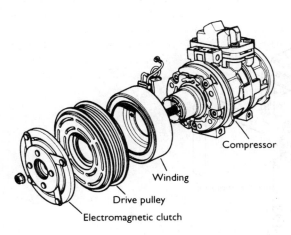

Figure 14.13 Air conditioning compressor

and reed valve type. As the refrigerant is drawn into the cylinder due to the action of the piston, the outlet valve is closed due to the pressure. When the piston begins its compression stroke the inlet reed valve closes and the outlet opens. This compressor is controlled by an electromagnetic clutch, which may be either under manual control or electronic control depending on the type of system.

Figure 14.14 shows the condenser fitted in

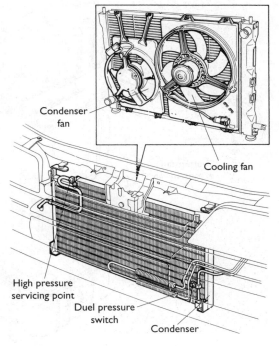

Figure 14.14 Air conditioing condenser

front of the vehicle radiator. It is very similar in construction to the radiator and fulfils a similar role. The heat is conducted through the aluminium pipes and fins to the surrounding air and then, by a process of convection, is dispersed by the air movement. The air movement is caused by the ram effect, which is supplemented by fans as required.

Figure 14.15 is the receiver–drier assembly. It

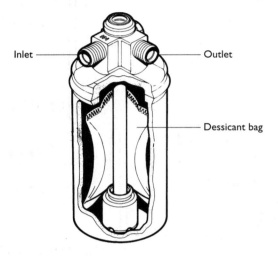

Figure 14.15 Receiver–drier

is connected in the high-pressure line between the condenser and the thermostatic expansion valve. This component has four features.

- A reservoir to hold refrigerant until a greater flow is required.
- A filter to prevent contaminants circulating through the system.
- Vapour is retained in this unit until it finally converts back to a liquid.
- A drying agent removes any moisture from the system. The substance used in R134A systems is Zeolite. Some manufacturers recommend that this unit should be replaced if the system has been open to the atmosphere.

A sight glass is fitted to some receiver–driers to give an indication of refrigerant condition and system operation. The refrigerant generally appears clear if all is in order.

The thermostatic expansion valve is shown as part of Figure 14.16 together with the evaporator assembly. It has two functions to fulfil:

- Control the flow of refrigerant as demanded by the system.
- Reduce refrigerant pressure in the evaporator.

The thermostatic expansion valve is a simple

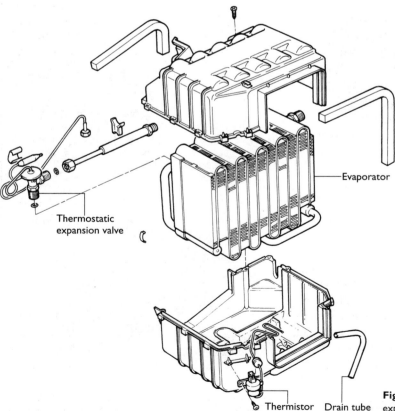

Figure 14.16 Evaporator and thermostatic expansion valve

spring-controlled ball valve, which has a diaphragm attached to a spring. A temperature sensitive gas such as carbon dioxide acts upon the diaphragm. The gas is in a closed system including a capillary tube and a sensing bulb. This sensing bulb is secured on the evaporator. If the temperature of the evaporator rises, the gas in the bulb expands and acts on the diaphragm such as to open the ball valve and allow a greater flow of refrigerant. If the evaporator were to become too cold, then the gas in the bulb will contract and the ball valve will close. In this way, the flow of refrigerant is controlled and the temperature of the evaporator is held fairly constant under varying air flow conditions.

The evaporator assembly is similar in construction to the condenser, consisting of fins to maximize heat transfer. It is mounted in the car under the dash panel, forming part of the overall heating and ventilation system. The refrigerant changes state in the evaporator from a liquid to a vapour. As well as cooling the air passed over it, the evaporator also removes moisture from the air. This is because the moisture in the air will condense on the fins and can be drained away. The action is much like breathing on a cold pane of glass. A thermistor is fitted to the evaporator on some systems to monitor temperature. The compressor is cycled if the temperature falls below about 3 or 4°C to prevent the chance of water freezing on the evaporator, which would restrict air flow.

The electrical circuit is shown in Figure 14.17. The following points are worthy of note.

- A connection exists between the air conditioning ECU and the engine management ECU. The reasons for this are so that the compressor can be switched off under very hard acceleration and to enable better control of engine idle.
- Twin cooling fans are used to cool the condenser. These can be run at two speeds using relays to connect them in series for slow operation, or in parallel for full speed.
- A number of safety features are included such as the high/low pressure switches.

14.5 Diagnosing air conditioning system faults

14.5.1 Introduction

As with all systems, the six stages of fault-finding should be followed.

1. Verify the fault.
2. Collect further information.
3. Evaluate the evidence.
4. Carry out further tests in a logical sequence.
5. Rectify the problem.
6. Check all systems.

Table 14.1 lists some common symptoms of an air conditioning system malfunction together with suggestions for the possible fault. The faults are generic but will serve as a good reminder. It is assumed an appropriate pressure gauge set has been connected.

Table 14.1 Symptoms and faults of an air conditioning system

Symptom	Possible fault
After stopping the compressor, pressure falls quickly to about 195 kPa and then falls gradually	Air in the system or, if no bubbles are seen in the sight glass as the condenser is cooled with water, excessive refrigerant may be the fault.
Discharge pressure low	Fault with the compressor or, if bubbles are seen, low refrigerant.
Discharge temperature is lower than normal	Frozen evaporator.
Suction pressure too high	High pressure valve fault, excessive refrigerant or expansion valve open too long.
Suction and discharge pressure too high	Excessive refrigerant in the system or condenser not working due to fan fault or clogged fins.
Suction and discharge pressure too low	Clogged or kinked pipes.
Refrigerant loss	Oily marks (from the lubricant in the refrigerant) near joints or seals indicate leaks.

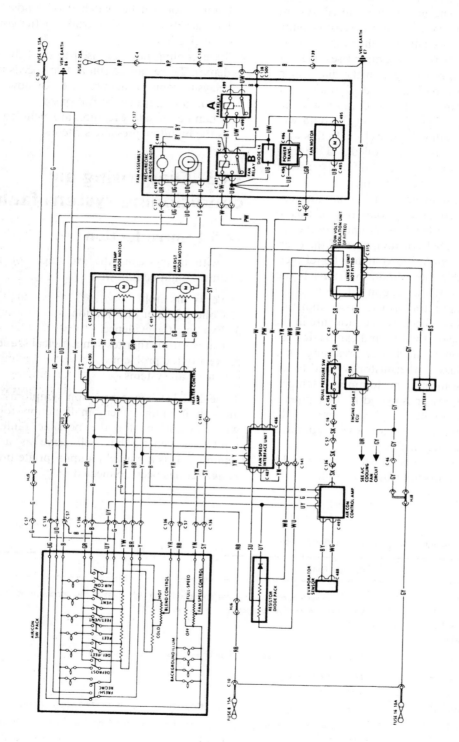

Figure 14.17 Air conditioning electrical circuit

14.5.2 Testing procedure

Do not work on the refrigerant side of air conditioning systems unless you have been trained and have access to suitable equipment.

The process of checking an air conditioning system is broadly as follows.

1. Hand and eye checks (loose wires, loose switches and other obvious faults) – all connections clean and tight.
2. Check system pressures.
3. Check discharge temperature.
4. Inspect receiver–drier sight glass.
5. Refer to the previous table.

14.6 Advanced temperature control technology

14.6.1 Heat transfer

Here is a reminder of the key terms associated with heat transfer.

Convection

Heat energy transfer that involves the movement of a fluid (gas or liquid). Fluid in contact with the source of heat expands and tends to rise within the bulk of the fluid. Cooler fluid sinks to take its place, setting up a convection current.

Conduction

Flow of heat energy through a material without the movement of any part of the material itself. Heat energy is present in all materials in the form of the kinetic energy of their vibrating molecules, and may be conducted from one molecule to the next in the form of this mechanical vibration. In the case of metals, which are particularly good conductors of heat, the free electrons within the material carry heat around very quickly.

Radiation

In physics, radiation is the emission of radiant energy as particles or waves; for example, heat, light, alpha particles and beta particles.

When designing a heating or air conditioning system, calculations can be used to determine the heating or cooling effect required. The following is the main heat current equation and can be used, for example, to help determine the heat loss through the windows.

$$\frac{\Delta Q}{\Delta t} = -kA\frac{\Delta T}{\Delta x} = -\frac{\Delta T}{\Delta x/kA}$$

where ΔQ = heat energy, ΔT = temperature, Δx = thickness/distance of material, Δt = time, k = thermal conductivity of the material ($\mathrm{W\,m^{-1}\,K^{-1}}$), A = cross-sectional area, $\Delta Q/\Delta t$ can be thought of as 'heat current'.

14.6.2 Armature reaction

Most heater motors, like many other motors, are unidirectional due to the positioning of the brushes. When a motor is running it also acts as a generator producing a back EMF. The brushes of a motor (or generator) must be placed around the commutator in such a way that, as the armature rotates and the brushes effectively short one commutator segment to the next, no EMF must be present in that associated armature winding.

If an EMF is present, then current will flow when the short is made. This creates sparks at the brushes and is known as armature reaction. To overcome this problem, the brushes are moved from the geometric neutral axis to the magnetic neutral axis of the motor fields. This is because, as armature current flows, the magnetism created around the armature windings interacts with the main magnetic field causing it to warp. Figure 14.18. shows this field warp diagrammatically. This phenomenon was used in some very early generators as a way of controlling output.

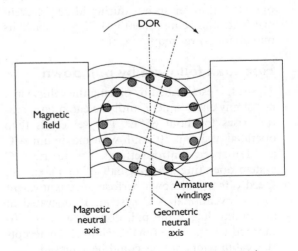

Figure 14.18 Field warp which causes armature reaction

14.7 New developments in temperature control systems

14.7.1 Electrically driven air conditioning

To drive the air conditioning pump electrically could provide the following benefits:

- Sealed motor and pump assembly.
- Smaller and less complex compressor.
- Flexible positioning (no drive belt).
- Full cooling capacity at any engine speed.
- Greater control is possible.

The motor output power necessary to drive an electric automotive air conditioning (A/C) system depends on the cooling capacity of the system, its efficiency and the boundary conditions (temperatures) it is operating against. All of these quantities are variable under normal vehicle operation. The use of a 'brushless' motor has been considered. The following 'standard' rating conditions are useful in assessing maximum power levels.

Stop-and-go driving in heavy traffic

Under these test conditions, high compressor discharge pressures will tend to overload the motor. To prevent this problem, fresh air must be restricted at idle to reduce evaporator load and, if possible, the condenser fan should operate at overspeed conditions. The motor must be operated at lower speeds during idle to prevent overload and, consequently, will not reach its maximum power requirement.

Hot soak followed by pull down

This test is established by placing the vehicle in a hot sunny environment until the cabin temperature rises to about 65°C. The vehicle is then operated at about 50 km/h with maximum A/C and fan speed control settings. An electric A/C system operating at about half of its maximum speed offers pull down performance equivalent to a conventional A/C system. If operated at maximum during the pull down test, a significant reduction in the time taken to reach acceptable cabin temperatures could be achieved.

Cruising with full fresh air intake

This operating condition requires the A/C system to maintain comfortable cabin temperatures

while processing significant quantities of outside air. This establishes a maximum capacity level, which in turn sets the size of the motor and its drive electronics. For conventional A/C systems, a 3.75 kW motor is a reasonable estimate for this condition. About 70% of the total load is used to condition the fresh outside air. Reducing or eliminating fresh air load at highway speeds has a direct influence on the size of the electric drive system.

A series of computer simulations was conducted to explore ways of reducing the motor power requirement. Using a two-stage cycle with 25% fresh air results in a 1.5 kW load on the motor. A conventional cycle using a high-efficiency compressor coupled with a 20% fresh air limitation also results in a 1.5 kW load. A 1.5 kW motor is a realistic option for automotive air conditioning.

The combined motor and electronics cost significantly affects the feasibility of electric automotive A/C systems. Cost increases significantly as the required motor power increases. Development of more efficient A/C systems is ongoing. Assuming these developments are successful, a 1.5 kW electrically driven A/C system will be possible and will be able to provide performance equal to or better than today's systems.

In this application, brushless DC motor systems are expected to achieve efficiencies of 85–90% when designed specifically for sealed automotive A/C applications. This translates into a maximum electrical demand from the vehicle power supply system of 1.7 kW when the A/C electrical drive operates under maximum cooling conditions.

Research is continuing in this area but, like many other developments, the current extra costs of the system may soon be outweighed by the benefits of extra control.

14.8 Self-assessment

14.8.1 Questions

1. State the meaning of 'plenum chamber'.
2. Make a clearly labelled sketch to show the main components of an air conditioning system.
3. Explain the principle of refrigeration.
4. Draw a circuit showing how 'dropping' resistors are used to control motor speed.
5. Describe the operation of an air conditioning system.

6. State three potential benefits of an electrically driven air conditioning compressor.
7. Define: heat flow, radiation, convection and conduction.
8. Describe the reason for *and* the operation of a thermostatic expansion valve.
9. Draw a circuit of a screen heater that includes a timer relay.
10. List four functional requirements of a seat heater.

14.8.2 Assignment

1. In relation to heating and air conditioning systems, discuss why the temperature and climate settings in a vehicle may need to be changed under different external conditions to achieve the same 'perceived or ideal' feeling of comfort. Draw a block diagram of the system and add appropriate comments as to how this 'ideal' effect could be achieved.

Figure 14.9 Chrysler cooling module

2. Figure 14.19 shows a cooling module for a Chrysler. Design a suitable electrical circuit to run the motors at two speeds.
3. Produce a report, following the standard format, about the operation of an air conditioning system fitted to a vehicle.

15
Chassis electrical systems

15.1 Anti-lock brakes

15.1.1 Introduction

The reason for the development of anti-lock brakes (ABS) is very simple. Under braking conditions, if one or more of the vehicle wheels locks (begins to skid), there are a number of consequences.

- Braking distance increases.
- Steering control is lost.
- Abnormal tyre wear.

The obvious result is that an accident is far more likely to occur. The maximum deceleration of a vehicle is achieved when maximum energy conversion is taking place in the brake system. This is the conversion of kinetic energy to heat energy at the discs and brake drums. The potential for this conversion process between a tyre skidding, even on a dry road, is far less. A good driver can pump the brakes on and off to prevent locking but electronic control can achieve even better results.

ABS is becoming more common on lower price vehicles, which should be a significant contribution to safety. It is important to remember, however, that for normal use, the system is not intended to allow faster driving and shorter braking distances. It should be viewed as operating in an emergency only. Figure 15.1 shows how ABS can help to maintain steering control even under very heavy braking conditions.

15.1.2 Requirements of ABS

A good way of considering the operation of a complicated system is to ask: 'what must the system be able to do?' In other words, what are the requirements? These can be considered for ABS under the following headings.

Fail-safe system

In the event of the ABS system failing the conventional brakes must still operate to their full potential. In addition, a warning must be given to the driver. This is normally in the form of a simple warning light.

Manoeuvrability must be maintained

Good steering and road holding must continue when the ABS system is operating. This is arguably the key issue, as being able to swerve around a hazard whilst still braking hard is often the best course of action.

Immediate response must be available

Even over a short distance the system must react such as to make use of the best grip on the road. The response must be appropriate whether the driver applies the brakes gently or slams them on hard.

Operational influences

Normal driving and manoeuvring should produce no reaction on the brake pedal. The stability and steering must be retained under all road

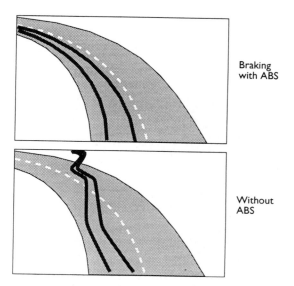

Braking with ABS

Without ABS

Figure 15.1 ABS can help maintain steering control

conditions. The system must also adapt to braking hysteresis when the brakes are applied, released and then re-applied. Even if the wheels on one side are on dry tarmac and the other side on ice, the yaw (rotation about the vertical axis of the vehicle) of the vehicle must be kept to a minimum and only increase slowly in order to allow the driver to compensate.

Controlled wheels

In its basic form, at least one wheel on each side of the vehicle should be controlled on a separate circuit. It is now general for all four wheels to be separately controlled on passenger vehicles.

Speed range of operation

The system must operate under all speed conditions down to walking pace. At this very slow speed even when the wheels lock the vehicle will come to rest very quickly. If the wheels did not lock then, in theory, the vehicle would never stop!

Other operating conditions

The system must be able to recognize aquaplaning and react accordingly. It must also still operate on an uneven road surface. The one area still not perfected is braking from slow speed on snow. The ABS will actually increase stopping distance in snow but steering will be maintained. This is considered to be a suitable trade-off.

A number of different types of anti-lock brake systems are in use, but all try to achieve the requirements as set out above.

15.1.3 General system description

As with other systems, ABS can be considered as a central control unit with a series of inputs and outputs. An ABS system is represented by the closed-loop system block diagram shown in Figure 15.2. The most important of the inputs are the wheel speed sensors, and the main output is some form of brake system pressure control.

The task of the control unit is to compare signals from each wheel sensor to measure the acceleration or deceleration of an individual wheel. From these data and pre-programmed look-up tables, brake pressure to one or more of the wheels can be regulated. Brake pressure can be reduced, held constant or allowed to increase.

The maximum pressure is determined by the driver's pressure on the brake pedal.

A number of variables are sensed, used or controlled by this system.

Pedal pressure

Determined by the driver.

Brake pressure

Under normal braking this is proportional to pedal pressure but under control of the ABS it can be reduced, held or allowed to increase.

Controlled variable

This is the actual result of changes in brake pressure, in other words the wheel speed, which then allows acceleration, deceleration and slip to be determined.

Road/vehicle conditions

Disturbances such as the vehicle load, the state of the road, tyre condition and brake system condition.

From the wheel speed sensors the ECU calculates the following.

Vehicle reference speed

Determined from the combination of two diagonal wheel sensor signals. After the start of braking the ECU uses this value as its reference.

Wheel acceleration or deceleration

This is a live measurement that is constantly changing.

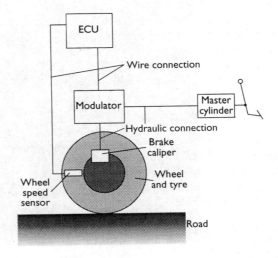

Figure 15.2 Anti-lock brake system

330 Automobile electrical and electronic systems

Brake slip

Although this cannot be measured directly, a value can be calculated from the vehicle reference speed. This figure is then used to determine when/if ABS should take control of the brake pressure.

Vehicle deceleration

During brake pressure control, the ECU uses the vehicle reference speed as the starting point and decreases it in a linear manner. The rate of decrease is determined by the evaluation of all signals received from the wheel sensors.

Driven and non-driven wheels on the vehicle must be treated in different ways as they behave differently when braking.

A logical combination of wheel deceleration/acceleration and slip is used as the controlled variable. The actual strategy used for ABS control varies with the operating conditions.

15.1.4 ABS components

There are a few variations between manufacturers involving a number of different components. For the majority of systems, however, there are three main components.

- Wheel speed sensors.
- Electronic control unit.
- Hydraulic modulator.

Wheel speed sensors

Most of these devices are simple inductance sensors and work in conjunction with a toothed wheel. They consist of a permanent magnet and a soft iron rod around which is wound a coil of wire. As the toothed wheel rotates, the changes in inductance of the magnetic circuit generate a signal; the frequency and voltage of which are proportional to wheel speed. The frequency is the signal used by the electronic control unit. The coil resistance is of the order of $1\,k\Omega$. Coaxial cable is used to prevent interference affecting the signal. Some systems now use 'Hall effect' sensors, as described in Chapter 2.

Electronic control unit

The function of the ECU (Figure 15.3 shows part of an ECU) is to take in information from the wheel sensors and calculate the best course of action for the hydraulic modulator. The heart of a modern ECU consists of two microprocessors such as the Motorola 68HC11, which run the same program independently of each

Figure 15.3 A microprocessor as used in an ABS ECU

other. This ensures greater security against any fault, which could adversely affect braking performance because the operation of each processor should be identical. If a fault is detected, the ABS disconnects itself and operates a warning light. Both processors have non-volatile memory into which fault codes can be written for later service and diagnostic access. The ECU also has suitable input signal processing stages and output or driver stages for actuator control.

The ECU performs a self-test after the ignition is switched on. A failure will result in disconnection of the system. The following list forms the self-test procedure.

- Current supply.
- Exterior and interior interfaces.
- Transmission of data.
- Communication between the two microprocessors.
- Operation of valves and relays.
- Operation of fault memory control.
- Reading and writing functions of the internal memory.

All this takes about 300 ms.

Hydraulic modulator

The hydraulic modulator as shown in Figure 15.4 has three operating positions.

- Pressure build-up – brake line open to the master cylinder.
- Pressure reducing – brake line open to the accumulator.
- Pressure holding – brake line closed.

The valves are controlled by electrical solenoids, which have a low inductance so they react very quickly. The motor only runs when ABS is activated.

15.1.5 Anti-lock brake system control

The control of ABS can be summarized under a number of headings as given below.

Brake pressure control commencement

The start of ABS engagement is known as 'first control cycle smoothing'. This smoothing stage is necessary in order not to react to minor disturbances such as an uneven road surface, which can cause changes in the wheel sensor signals. The threshold of engagement is critical as, if it were too soon, it would be distracting to the driver and cause unnecessary component wear; too late and steering/stability could be lost on the first control cycle.

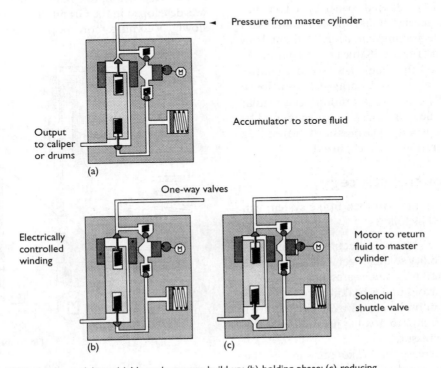

Figure 15.4 ABS hydraulic modulator. (a) Normal pressure build-up; (b) holding phase; (c) reducing

Even road surface regulation

Under these ideal circumstances adhesion is almost constant. ABS works at its best under these conditions, regulation frequency is relatively low with small changes in brake pressure.

Vehicle yaw (twist about the vertical axis, swerving moment)

When braking on a road surface with different adhesion under the left and right wheels, the vehicle will yaw or start to spin. The driver can control this with the steering if time is available. This can be achieved if when the front wheel with poor adhesion becomes unstable, the pressure to the other front wheel is reduced. This acts to reduce the vehicle yaw, which is particularly important, when the vehicle is cornering.

Axle vibration

Wheel speed instability occurs frequently and at random on rough roads. Due to this instability, brake pressure tends to be reduced more than it is increased, during ABS operation. This could lead to loss of braking under certain conditions. Adaptation to the conditions is therefore necessary to overcome this problem. An increase in brake pressure is made easier during hard re-acceleration of the wheel after an unstable instant. With modern soft suspension systems the axle may be subject to vibration. This can cause superimposed signals on the wheel speed sensors. The indicated accelerations can be the same as for actual unstable braking conditions. A slight delay in the reaction of the ABS due to the delay in signal smoothing – the time taken to move control valves and a time lag in the brake lines – helps to reduce the effect of axle vibration. The regular frequency of the vibrations can be recognized by the ECU. A constant brake pressure is introduced when axle vibrations are recognized.

15.1.6 Control strategy

The strategy of the anti-lock brake system can be summarized as follows.

- Rapid brake pressure reduction during wheel speed instability so the wheel will re-accelerate fast without too much pressure reduction, which will avoid under braking.
- Rapid rise in brake pressure during and after a re-acceleration to a value just less than the instability pressure.
- Discrete increase in brake pressure in the event of increased adhesion.

- Sensitivity suited to the prevalent conditions.
- Anti-lock braking must not be initiated during axle vibration.

The application of these five main requirements leads to the need for compromise between them. Optimum programming and prototype testing can reduce the level of compromise but some disadvantages have to be accepted. The best example of this is braking on uneven ground in deep snow, as deceleration is less effective unless the wheels are locked up. In this example, priority is given to stability rather than stopping distance, as directional control is favoured in these circumstances.

15.1.7 Variations of ABS

A novel approach to ABS has been developed which uses springs and a motor to produce the brake pressure conditions of reducing, holding or increasing. The potential advantage of this technique is that the response is smooth rather than pulsed. Figure 15.5 shows the layout of the motor and spring system.

15.2 Active suspension

15.2.1 Introduction

Active suspension, like many other innovations, was developed in the Grand Prix world. It is now slowly becoming more popular on production

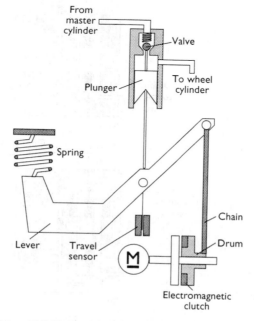

Figure 15.5 Motor and spring anti-lock brake system

vehicles. It is interesting to note that just as some Formula 1 teams perfected it, the rules changed (1993–94) to prevent its use!

Conventional suspension systems are always a compromise between soft springs for comfort and harder springing for better cornering ability. A suspension system has to fulfil four main functions.

- Absorb bumps.
- Manage nose dive when braking.
- Prevent roll when cornering.
- Control body movement.

This means that some functions have to be compromised in order to fulfil others to a greater extent.

15.2.2 Operation

Active suspension allows the best of both worlds. This is achieved by replacing the conventional springs with double-acting hydraulic units. These are controlled by an ECU, which receives signals from various sensors. Oil pressure in excess of 150 bar is supplied to the hydraulic units from a pump. A servo valve controls the oil, which is arguably the most critical component.

The main benefits of active suspension are as follows.

- Improvements in ride comfort, handling and safety.
- Predictable control of the vehicle under different conditions.
- No change in handling between laden and unladen.

15.2.3 Sensors, actuators and system operation

To control the hydraulic units to the best advantage, the ECU needs to 'know' certain information. This is determined from sensor readings from various parts of the vehicle. A number of sensors are used to provide information to the suspension ECU.

Load sensor

A load cell used to determine whether actual load is positioned on each hydraulic ram.

Displacement and vertical acceleration

This sensor can take a number of forms, as simple as a variable resistor or a more accurate and sensitive linear sensor such as the LVDT (see Chapter 2).

Lateral and longitudinal acceleration

Acceleration can be determined from a pendulum-type sensor using strain gauges linked to a mass, or devices similar to an engine knock sensor.

Yaw transducer

Yaw can be determined from lateral acceleration if the sensor is mounted at the front or rear of the vehicle.

Steering position

As well as steering position, rate of change of position is determined from a rotary position sensor. This device can be a light beam and detector type or similar. If the rate of change of steering position is beyond a threshold the system will switch to a harder suspension setting.

Vehicle speed

The speed of the vehicle is taken from a standard-type sensor as used for operating the speedometer.

Throttle position

Similar to the existing throttle potentiometers. This gives data on the driver's intention to accelerate or decelerate allowing the suspension to switch to a harder setting when appropriate.

Driver mode selection

A switch is provided allowing the driver to choose soft or hard settings. Even if the soft setting is selected, the system will switch to hard, under certain operational conditions.

The layout of the suspension system also shows a simplified view of the hydraulic unit. This is, in effect, a hydraulic ram and can have oil under very high pressure fed to the upper or lower chamber. The actual operation of the whole system is as follows. As a wheel meets a bump in the road there is increased upward acceleration and vertical load. This information is fed to the ECU, which calculates the ideal wheel displacement. A control signal is now sent to the servo valve(s), which control the position of the main hydraulic units. As this process can occur hundreds of times per second, the wheel can follow the contour of the road surface. This cushions the vehicle body from unwanted forces.

By considering information from other sensors, such as the lateral acceleration sensor, which gives data relating to cornering, and the longitudinal sensor, which gives data relating to braking or acceleration forwards, the actuators can be moved to provide maximum stability at all times.

Active suspension looks set to have an easy ride in the future. The benefits are considerable and, as component prices reduce, the system will become available on more vehicles. It is expected that even offroad vehicles may be fitted with active suspension in the near future. A representation of an electronically controlled suspension system is shown in Figure 15.6.

15.3 Traction control

15.3.1 Introduction

The steerability of a vehicle is not only lost when the wheels lock up on braking; the same effect arises if the wheels spin when driving off under severe acceleration. Electronic traction control has been developed as a supplement to ABS. This control system prevents the wheels from spinning when moving off or when accelerating sharply while on the move. In this way, an individual wheel, which is spinning is braked in a controlled manner. If both or all of the wheels are spinning, the drive torque is reduced by

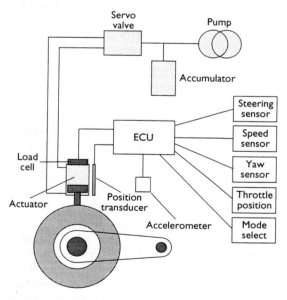

Figure 15.6 Sensors used to provide information to suspension ECU and general layout of an active suspension system

means of an engine control function. Traction control has become known as ASR or TCR.

Traction control is not normally available as an independent system, but in combination with ABS. This is because many of the components required are the same as for the ABS. Traction control only requires a change in logic control in the ECU and a few extra control elements such as control of the throttle. Figure 15.7 shows a block diagram of a traction control system. Note the links with ABS and the engine control system.

Traction control will intervene to achieve the following:

- Maintain stability.
- Reduction of yawing moment reactions.
- Provide optimum propulsion at all speeds.
- Reduce driver workload.

The following list of advantages can be claimed for a good traction control system.

- Improved tractive force.
- Better safety and stability on poor surfaces.
- Less driver stress.
- Longer tyre life.
- No wheel spin on turning and cornering.

An automatic control system can intervene in many cases more quickly and precisely than the driver of the vehicle. This allows stability to be maintained at time when the driver might not have been able to cope with the situation. Figure 15.8 shows an ABS and traction control modulator, complete with an ECU.

15.3.2 Control functions

Control of tractive force can be by a number of methods. Figure 15.9 shows a comparison of three techniques used to prevent wheel spin, throttle, ignition and brake control.

Throttle control

This can be via an actuator, which can move the throttle cable, or if the vehicle employs a

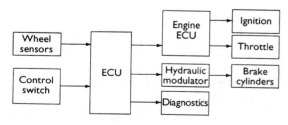

Figure 15.7 Traction control system

Figure 15.8 ABS and traction control ECU on the modulator

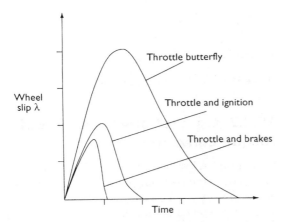

Figure 15.9 Comparison of three techniques used to prevent wheel spin: throttle, ignition and brake control

drive-by-wire accelerator, then control will be in conjunction with the engine management ECU. This throttle control will be independent of the driver's throttle pedal position. This method alone is relatively slow to control engine torque.

Ignition control

If ignition is retarded, the engine torque can be reduced by up to 50% in a very short space of time. The timing is adjusted by a set ramp from the ignition map value.

Braking effect

If the spinning wheel is restricted by brake pressure, the reduction in torque at the affected wheel is very fast. Maximum brake pressure is not used, to ensure passenger comfort is maintained.

15.3.3 **System operation**

The layout of a traction control system, which includes links with other vehicle control systems, is shown in Figure 15.10. The description that follows is for a vehicle with an electronic (drive-by-wire) accelerator.

A simple sensor determines the position of the accelerator and, taking into account other variables such as engine temperature and speed for example, the throttle is set at the optimum position by a servo motor. When accelerating, the increase in engine torque leads to an increase in driving torque at the wheels. In order for optimum acceleration, the maximum possible driving torque must be transferred to the road. If driving torque exceeds that which can be transferred, then wheel slip will occur, at least at one wheel. The result of this is that the vehicle becomes unstable.

When wheel spin is detected the throttle position and ignition timing are adjusted but the best results are gained when the brakes are applied to the spinning wheel. This not only prevents the wheel from spinning but acts such as to provide a limited slip differential action. This is particularly good when on a road with varying braking force coefficients. When the brakes are applied, a valve in the hydraulic modulator assembly moves over to allow traction control operation. This allows pressure from the pump to be applied to the brakes on the offending wheel. The valves – in the same way as with ABS – can provide pressure build-up, pressure hold and pressure reduction. This all takes place without the driver touching the brake pedal.

The summary of this is that the braking force must be applied to the slipping wheel, such as to equalize the combined braking coefficient for each driving wheel.

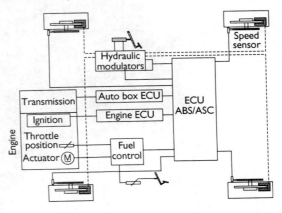

Figure 15.10 Layout of a traction control system which includes links with other vehicle control systems

15.4 Automatic transmission

15.4.1 Introduction

The main aim of electronically controlled automatic transmission (ECAT) is to improve conventional automatic transmission in the following ways.

- Gear changes should be smoother and quieter.
- Improved performance.
- Reduced fuel consumption.
- Reduction of characteristic changes over system life.
- Increased reliability.

The actual operation of an automatic gearbox is beyond the scope of this book. However, the important points to remember are that gear changes and lock-up of the torque converter are controlled by hydraulic pressure. In an ECAT system, electrically controlled solenoid valves can influence this hydraulic pressure. Figure 15.11 is a block diagram of an ECAT system.

Most ECAT systems now have a transmission ECU that is in communication with the engine control ECU (by a CAN – controller area network – databus in many cases). The system as a whole consists of a number of sensors providing data to the ECU, which in turn is able to control a number of actuators or output devices. Figure 15.12 shows a modern automatic gearbox as used by the Porsche Carrera.

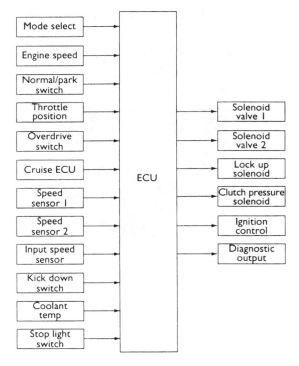

Figure 15.11 Block diagram of an ECAT system

15.4.2 Control of gear shift and torque converter

With an ECAT system, the actual point of gear shift is determined from pre-programmed memory within the ECU. Data from the sensors are used to reference a look-up table mainly as a function of engine speed and vehicle speed. Data from other sensors are also taken into consideration. Actual gear shifts are initiated by

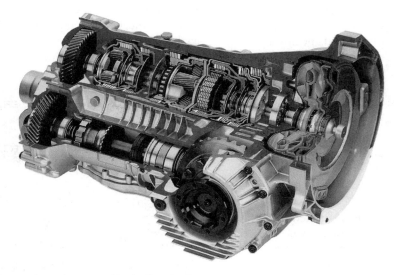

Figure 15.12 Automatic gearbox as used in the Porsche Carrera

changes in hydraulic pressure, which is controlled by solenoid valves.

The two main control functions of this system are hydraulic pressure and engine torque. The temporary reduction in engine torque during gear shifting (about 200 ms) allows smooth operation. This is because the peaks of gearbox output torque are suppressed, which causes the characteristic surge as the gears change on conventional automatics. Figure 15.13 shows a comparison of transmission output torque from systems with and without engine torque control. Also shown are the transmission speed and the timing control of the engine. Engine torque control can be by throttle movement, fuel cut-off or ignition timing retardation. The latter seems to have proved the most appropriate for modern systems.

Figure 15.14 shows how control of hydraulic pressure during gear up-shift again prevents a surge in transmission output torque. The hydraulic pressure control is in three stages as shown in the figure.

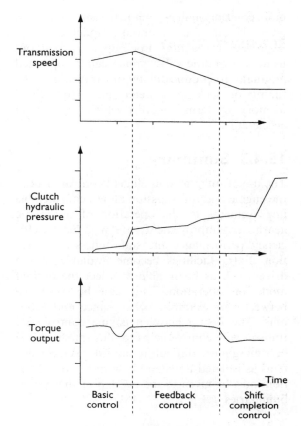

Figure 15.14 Control of hydraulic pressure

Basic pressure control

Pressure is set to an optimum value for speed of the gear shift. This can be adapted as the system learns the ideal pressure by monitoring shift time and changing the pressure accordingly.

Feedback control

The ECU detects the deviation of the rotational speed of the input shaft from a target value and adjusts pressure to maintain fine control.

Completion control

Torque converter hydraulic pressure is reduced momentarily so that as the engine torque output control is released, the potential surge is prevented. Because of these control functions, smooth gear shifts are possible and, due to the learning ability of some ECUs, the characteristics remain constant throughout the life of the system.

Torque converter lock-up

The ability to lock up the torque converter has been used for some time, even on vehicles with

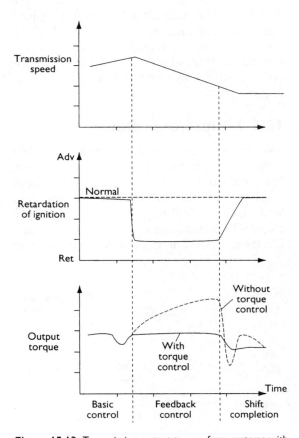

Figure 15.13 Transmission output torque from systems with and without engine torque control

more conventional automatic transmission. This gives better fuel economy, quietness and improved driveability. Lock-up is carried out using a hydraulic valve, which can be operated gradually to produce a smooth transition. The timing of lock-up is determined from ECU memory in terms of the vehicle speed and acceleration.

15.4.3 Summary

The use of integrated intelligent control of both the engine and transmission allows considerable improvements to the operation of automatic gearboxes. Improvements are possible to efficiency, performance and smoothness of operation. Extra facilities become available to the driver such as being able to select the desired mode of operation. This can be a choice between, for example, performance and economy. The tie up between engine control and transmission control helps to illustrate how it is becoming more difficult to consider vehicle systems as isolated units and how more consideration must be given to the overlap of the system boundaries.

15.5 Other chassis electrical systems

15.5.1 Electric power steering

There are three electric power steering techniques.

- Replacing the conventional system pump with an electric motor whilst the ram remains much the same.
- A drive motor, which directly assists with the steering and has no hydraulic components.
- Active steering in which the steering wheel is replaced with a joystick.

The first of these systems is popular, as the pump will only run when needed. This gives some savings in the fuel consumption and also allows the drive belt arrangement at the front of the engine to be simplified.

The second system listed is now becoming the most common. An electric motor acts directly on the steering via an epicyclic gear train. This completely replaces the hydraulic pump and servo cylinder. This eliminates the fuel penalty of the conventional pump and greatly simplifies the drive arrangements. Engine stall when the power

steering is operated at idle speed is also eliminated. An optical torque sensor is used to measure driver effort on the steering wheel. The sensor works by measuring light from an LED, which is shining through holes that are aligned in discs at either end of a 50 mm torsion bar fitted into the steering column. This system occupies little under-bonnet space (something which is at a premium these days), and the 400 W motor only averages about 2 A under urban driving conditions. The cost benefits over conventional hydraulic methods are considerable.

'Active steering' is the name given to a system developed by Saab from its experience in the aircraft industry. The technique is known as drive-by-wire. A joystick is used in place of the steering wheel and an array of sensors determines the required output and, via the control unit, operates two electro-hydraulic control valves. The ECU filters out spurious data from the sensors and provides a feedback to the joystick in order to maintain driver feel. As a safety feature, electronic circuits have built in self-test facilities and backup modules. Hydraulic fluid pressure is also held in reserve in an accumulator. Figure 15.15 shows an overview of the active steering system. Great benefit could be gained using this technique due to the removal of the steering column – although some opposition is expected to this radical approach! Disabled drivers, however, may consider this to be a major improvement.

15.5.2 Electronic clutch

The electronic clutch was developed for racing vehicles to improve the getaway performance. For production vehicles, a strategy has been developed to interpret the driver's intention. With greater throttle openings, the strategy changes to prevent abuse and drive line damage. Electrical control of the clutch release bearing

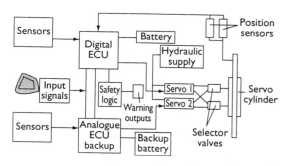

Figure 15.15 Overview of active steering system

position is by a solenoid actuator, which can be modulated by signals from the ECU. This allows the time to reach the ideal take-off position to be reduced and the ability of the clutch to transmit torque to be improved. The efficiency of the whole system can therefore be increased. Figure 15.16 shows the torque transmitted curve for an electronic clutch system. A switch could be provided to change between performance or economy mode.

15.5.3 Active roll reduction

The conventional anti-roll bar, as fitted to many vehicles, is replaced with a bar containing a rotary actuator. The actuators are hydraulically operated from a dedicated pump. Lateral acceleration is calculated by the ECU from steering angle and road speed. Hydraulic pressure is then regulated as required to the front and rear actuators such as to provide a force on the roll bar preventing the body of the vehicle from tilting. A good use for this system is on larger panel vans although it is being offered as an option to a range of vehicles. Figure 15.17 shows the positioning of one of the actuators for active roll reduction.

15.5.4 Electronic limited slip differential

Conventional limited slip differentials (LSDs) cannot be designed for optimum performance because of the effect on the vehicle when cornering. Their characteristics cannot be changed when driving. Front-wheel drive vehicles have even more problems due to the adverse effect on the steering. These issues have prompted the development of electronic control for the LSD.

The slip limiting action is controlled by a multi-disc clutch positioned between the crown

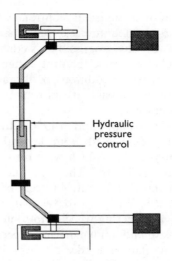

Figure 15.17 Positioning of one of the actuators for active roll reduction

wheel and the differential housing. It is able, if required, to lock the axle fully. The pressure on the clutch plates is controlled by hydraulic pressure, which in turn is controlled by a solenoid valve under the influence of an ECU. Data are provided to the ECU from standard ABS-type wheel sensors. Figure 15.18 shows a block diagram of a final drive and differential unit with electronic control.

15.6 Case studies

15.6.1 Tiptronic S – Porsche

Developed by Porsche, Tiptronic S is a fully 'intelligent' multi-programme automatic transmission

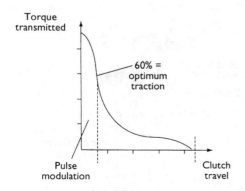

Figure 15.16 Torque transmitted curve for an electronic clutch system

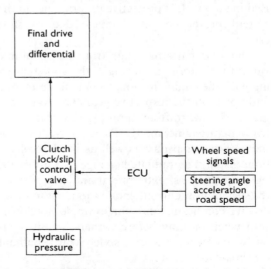

Figure 15.18 Final drive and differential unit electrical control

with additional fingertip control. The dual-function, 5-speed, automatic transmission, with active shift driving programmes is controlled by the 'Porsche Tiptronic' control system. As an alternative, and in addition to, the automatic mode, it is also possible to shift manually with fingertip controls.

Tiptronic first appeared in 1990 with technology directly descended from Fl and the 'Le Mans' Porsche 962s, which went on to win the 1994 'Le Mans 24 Hrs'. The Tiptronic S system features manual shift control integrated into the steering wheel. With two rocker switches in the spokes of the steering wheel, Tiptronic S offers an impressive and unique driving experience.

Once in manual mode, a driver can shift manually by pushing one of the two rocker switches. Slight pressure of the thumb is all it takes to shift up – tipping it downward will produce a downshift. The location and design of the rocker switches, as well as the distinctly perceptible pressure points, in combination with electronic transmission management, rule out any shift errors. The chosen gear is always indicated via a read-out located on the speedometer.

The quick responses of the transmission triggered by just a thumb push generate an absolutely spontaneous driving impression, gear changes being twice as quick as a manual gearbox. Since the hands remain at the wheel, Porsche, with its Tiptronic S, has extended levels of primary safety.

The Tiptronic S system 'learns' about a particular driving style by monitoring eight sensors around the car, which include throttle speed and position, road speed, engine speed and temperature, lateral acceleration and deceleration. Redline-controlled protective programmes in the system prevent engine damage due to shift errors.

The shift patterns range from an economic variant for smooth motoring to dynamic motoring with the engine revving to maximum torque and power in the respective gears before changing, and downshifting appropriately from relatively high engine speeds. Rapid movements of the accelerator pedal, as well as hard acceleration, result in a graduated change of shift maps, right up the most extreme variant. In addition, the system is intelligent enough to react to other driving conditions and, for example, to downshift when braking before corners, which obviously reflects driving style with manual gearboxes.

This 'intelligent shift program' – ISP for short

– of the Porsche Tiptronic S is characterized by the following special features in addition to the five automatic electronic shift maps.

- A warm-up program, which suppresses early up-shifting to ensure a rapid rise to the engine operating temperature to ensure clean emissions.
- Active shifting – when the accelerator pedal is depressed and released rapidly, the most 'dynamic' shift map is available instantaneously.
- Suppression of the overrun up-shift on a sudden lifting of the throttle – e.g. no gear change mid-bend.
- Brake-initiated downshift to the next lower gear for more efficient engine braking.
- Holding onto a gear in curves – i.e. no gear change whilst in mid-bend.
- Graduated up-shifting from lower gears to prevent immediate changeover to the top gear, especially after active downshifting.
- Identification of uphill stretches to stay in the lower gears as long as possible when driving up or downhill.
- Slip-induced up-shifting initiated by inertia forces when braking on slippery surfaces (rain, snow) to improve lateral guidance of the driving wheels and consequently driving stability.

15.6.2 Honda anti-lock brakes

A Honda anti-lock brake system is based on the plunger principle. Figure 15.19 shows the schematic diagram. When anti-lock is not operating, the chamber labelled W is connected to the reservoir via the outlet valve. The chamber is held at atmospheric pressure because the inlet

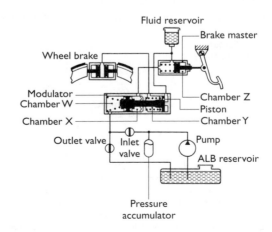

Figure 15.19 Honda ABS

valve blocks the line from the pressure accumulator. During braking, pressure is created in the master cylinder and fluid flows from chamber Z into chamber X, moving the piston and increasing pressure in chamber Y.

If a wheel threatens to lock, the outlet valve closes, pressure in chamber W rises and prevents further movement of the piston thus *holding* the pressure. If the risk of lock-up continues the inlet valve opens and allows fluid to flow from the accumulator into chamber W. This pressure moves the piston back, thus reducing the pressure to the wheel cylinder. When the risk of lock-up has gone, the inlet valve closes and the *hold* phase is restored.

The Honda system is a relatively simple ABS and has just two control channels. The front wheel which has the *higher* coefficient of friction determines the brake pressure for both front wheels. The result is that one front wheel may lock during extreme braking. The rear wheel with the *lower* coefficient of friction determines the rear brake pressure.

15.6.3 ABS – Chevrolet Corvette

The anti-lock braking system (ABS) was introduced on the Corvette in 1986 and is designed to maintain vehicle control even under severe braking conditions. The system does this by monitoring the speed of each wheel and then controlling the brake line pressure to prevent the wheels from locking.

Every time the vehicle is started, the anti-lock warning light illuminates for a short time and then goes out to indicate that the system is operating correctly. A test, which actually runs the

modulator valve, ensures that the system is fully functional. This test occurs when the vehicle is first started and when it reaches about 7 km/h (4 miles/h). During vehicle operation, the control module constantly monitors the system. If a fault occurs in any part of the system, the dashboard warning light will illuminate. If a fault in the ABS system occurs, the conventional brake system will remain fully functional.

The modulator valve is located in the compartment behind the driver's seat. The purpose of the modulator valve is to maintain or reduce the brake fluid pressure to the wheel callipers. It cannot increase the pressure above that transmitted to the master cylinder, and can never apply the brakes by itself! The modulator valve receives all its instructions from the control unit.

The control module is also located in the rear storage compartment behind the driver's seat. The function of the control module is to read and process information received from the wheel speed sensors. Acceleration, deceleration and slip values are calculated to produce control instructions for the modulator valve.

The lateral acceleration switch is located on the floor, just under the air conditioning control head. This switch is used to detect if the vehicle is cornering faster than a given curve speed. If so, a signal is sent to the control module indicating this hard cornering situation.

The wheel speed sensors are located at each wheel and send an electric signal to the control module indicating rotational speed. They are fitted in the knuckles and have toothed rings pressed onto the front hub and bearing assemblies, and the rear drive spindles. Figure 15.20 shows the location of the main ABS components.

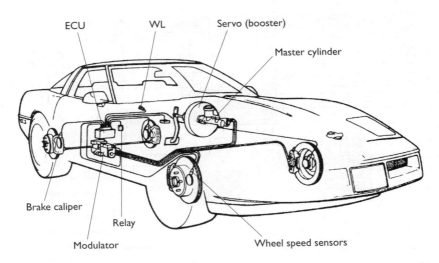

Figure 15.20 Chevrolet Corvette ABS components

15.6.4 'Jatco' automatic transmission

The 'Jatco SFPO' is the first electronically controlled automatic transmission (ECAT) fitted to a Rover Group vehicle. The new ECAT is part of a new series of transmissions with a standalone diagnostics system. It has five-speed adaptive control and is European on-board diagnostics (EOBD) compliant. Gear change and torque converter lock-up are determined by the throttle angle and the vehicle speed. Torque converter lock-up is available in third, fourth and fifth gears. Lock-up of the torque converter transmits maximum power from the engine to the wheels of the vehicle without slip occurring. The automatic transmission control unit (ATCU) is located in the passenger footwell.

Like all automatic transmission vehicles, the engine will only start in Park or Neutral. The EWS-3 (Elektronische Wegfahrsperre) immobilizer monitors the gear selector position transmitted by the ATCU on the Controller Area Network Bus (CAN-Bus). The start inhibitor switch is also hard-wired from the transmission to the EWS-3. Starting is allowed when the EWS-3 immobilizer receives a closed inhibitor switch signal or an appropriate CAN message transmitted from the ATCU. A further safety feature of the transmission system is 'reverse inhibition'. When the vehicle speed exceeds 10 km/h in the forward direction, the ATCU switches a solenoid, which drains the oil from the reverse clutch, thus preventing reverse selection and subsequent transmission damage.

The inhibitor switch consists of seven sets of contacts; the ATCU monitors these switches to determine the position of the gear range selector and choose the best shift pattern. It also transmits a signal relating to the selected gear on to the CAN-Bus. This CAN signal is used to illuminate the corresponding part of the 'PRND432' display on the instrument pack. A display also informs the driver which mode has been selected.

- 'D' means the transmission is in normal drive mode.
- 'S' means sport mode.
- A snowflake symbol indicates winter/snow mode.
- 'EP' means the ATCU has entered a fail-safe mode. Fault(s) are stored in the ATCU non-volatile memory.

Solenoid actuators that are controlled by the ATCU cause the automatic transmission gear changes. This is achieved using nine solenoids, which regulate the control valve operation.

The solenoid valve block is located inside the transmission system. The three shift solenoids, which engage the various gear ratios within the transmission, are called A, B and C in the block diagram of Figure 15.21 and there is a given combination of these solenoid states for the selection of each gear. Figure 15.21 is a block diagram of the system showing where the solenoids are used.

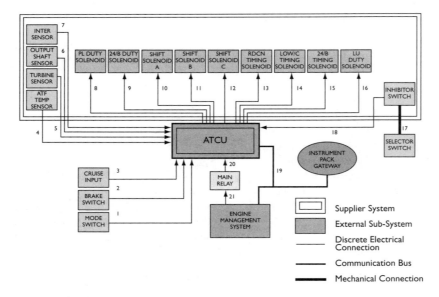

Figure 15.21 Jatco system fitted to a Rover car

The pressure of the transmission fluid must be regulated correctly. If pressure becomes too high, gear shifting will occur at high speeds, which is uncomfortable for passengers and can damage the transmission. If the line pressure becomes too low, gear shifting will take longer to complete and can shorten the life of the various clutches within the transmission. One solenoid control valve – called the line pressure duty solenoid (PL) – regulates line pressure. The required pressure is calculated by the ATCU from current engine speed, vehicle speed, current engine torque and throttle angle signals.

Being electronically controlled means it is possible to vary the characteristics of the shift maps. The shift maps can be selected manually by the driver mode options.

- Snow mode.
- Sport mode.
- Normal drive mode, 4, 3, 2.

Automatic intervention by the ATCU will occur if demanded by prevailing driving conditions. The shift map adaptations are called strategies; on power-up of the vehicle, the ATCU will default to normal drive mode. The system supports, and can automatically initiate, the following strategies.

1. Hill/trailer mode engagement. This is an adaptive mode with which the ATCU detects steep gradients and automatically enters this mode. ATCU detection is by monitoring of engine torque values, throttle angle and engine speed. Pulling a trailer has a similar effect on a vehicle in terms of torque requirements as a vehicle climbing a hill. This mode helps to prevent the gears shifting up and down in response to frequent throttle pedal adjustments that the driving conditions may require.
2. Downhill recognition. This strategy decreases the need for the application of the brakes when driving downhill. The system recognizes the decrease in throttle angle and the increase in speed as a slope. When the brakes are applied, the transmission changes down a gear. It stays in this mode until application of the throttle.
3. Cooling strategy engagement. Torque converter lock-up will not usually occur in second gear, and under high loading conditions the transmission can generate excessive heat. Locking the torque converter or changing gear can reduce the amount of heat gener-

ated. The ATCU recognizes that a low gear has been selected and that engine speed, engine torque and throttle angles are all high and it will engage the cooling strategy.
4. Cold start/climate strategy. This strategy holds onto the gears for longer than usual. It also prevents lock-up until the oil has reached a set temperature. This warms up the powertrain of the vehicle and it reaches its optimum performing temperature earlier. An improvement in vehicle emissions, fuel economy and driveability is the result.

The above strategies are controlled automatically by the ATCU. The driver can select various shift maps by choosing the sport mode or the snow/winter mode from the driver mode switch on the centre console. The transmission defaults to normal/drive mode on vehicle start up. In sport mode, the ATCU will hold on to the gears for longer than usual, improving acceleration performance, and will downshift more readily giving faster overall vehicle responsiveness. When snow/winter mode is selected, the ATCU limits the amount of wheel slip when the transmission is shifting between the gears by shifting gear at reduced engine torque loads. This mode is designed for use in icy and wet conditions.

15.6.5 Power steering – ZF Servoelectric

The ZF Servoelectric system is one of the most user-friendly Electric Power-Steering Systems available to date. It offers extensive economic and environmental improvements over hydraulic steering systems for a wide range of cars. In addition, ZF's new electric system is much easier to install by the original equipment (OE) vehicle manufacturers. Instead of a complex range of parts, ZF Servoelectric is offered as a modular kit, ensuring universal and cost-effective applications. Figure 15.22 shows a ZF steering system. The kit is available in three versions.

For small passenger cars

A Servoelectric system with an integral servo unit incorporated in the steering column. This is primarily suitable for small passenger cars with restrictions in engine compartment space. Maximum steering axle load is 600 kg.

For mid-range cars

A Servoelectric system designed for mid-range

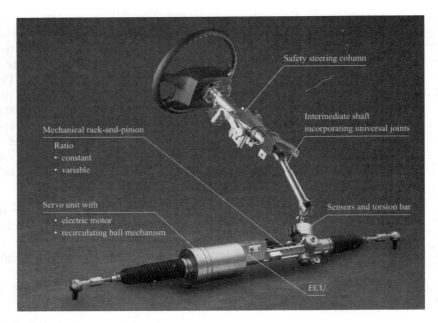

Figure 15.22 Electric power steering system

cars with the servo unit working on the pinion. Maximum steering load is 900 kg.

For upper mid-range cars

A Servoelectric system designed for upper mid-range passenger cars and light duty commercial vehicles where the steering rack itself is driven by an electric motor.

A reduction in energy consumption of up to 80% over hydraulic systems is possible. The average mid-sized car fitted with the Servoelectric system would experience reductions in fuel consumption of around 0.25 litres per 100 km. This is possible because the electric motor is operating only whilst the vehicle is being steered, unlike a continuously operating oil pump which is neither economical nor environmentally friendly. Electric steering also offers considerable benefits to vehicle manufacturers since the system is easier – not to mention a lot quicker – to install. An in-built ECU offers OE manufacturers the opportunity to adapt the steering system to their specific requirements, for instance, to the precise vehicle steering parameters, or to offer road-speed related servo-assistance. Integrated sensors housed in the steering system can transmit information about steering angles and speeds to chassis control units, or even to satellite-navigated driver information systems.

The Servoelectric system offers steering comfort levels equal to conventional hydraulic steering systems. In addition, driving on uneven road surfaces is made effortless, due to the system's programmable damping. There are many factors to bear in mind when selecting a steering system; these include performance, safety, strength, installation conditions and, of course, costs. ZF can now offer a wide range of possibilities for a particular solution from its range of hydraulic, electric or even electric–hydraulic power steering solutions – the latter supplies pressure by means of an electrically-driven oil pump. Estimates have shown that, by the year 2000, one-third of all power-steering systems manufactured in Western Europe will have electrical assistance. Furthermore, it has been predicted that eventually, the electric solution will completely replace hydraulic systems.

15.6.6 Porsche stability management

The new 911 Carrera 4 is the first Porsche to feature Porsche Stability Management (PSM), which is a combination of four-wheel drive designed for sports motoring and electronic suspension control, carefully geared to the character of the car. The result is not only a high standard of driving safety, but also that very special driving pleasure Porsche drivers have learnt to appreciate so much over the last 50 years. Figure 15.23 shows the layout of the PSM system.

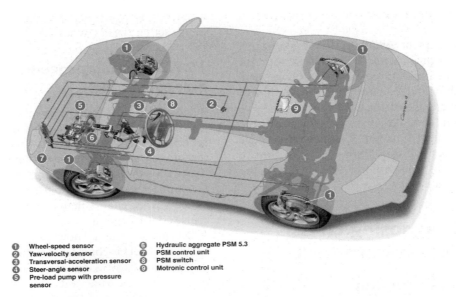

①	Wheel-speed sensor	⑥	Hydraulic aggregate PSM 5.3
②	Yaw-velocity sensor	⑦	PSM control unit
③	Transversal-acceleration sensor	⑧	PSM switch
④	Steer-angle sensor	⑨	Motronic control unit
⑤	Pre-load pump with pressure sensor		

Figure 15.23 Porsche stability management system

This objective calls for control and suspension management features different from those to be found in other cars incorporating similar systems. A Porsche will retain its agile, sporting and dynamic driving behaviour all the way to the most extreme limit. In addition, thanks to the high standard of safety reserves offered by the suspension, the driver only has to intervene in the car's behaviour on dry roads when driving under near-racing conditions. At the same time, PSM discreetly and almost unnoticeably corrects any minor deviations in directional stability attributable to load change or application of the brakes in a bend.

Porsche's engineers allow PSM to intervene more energetically at an even earlier point on wet or slippery roads and, in particular, on road surfaces with varying frictional coefficients. It is here, too, that PSM makes stopping distances much shorter while keeping the car stable and firmly on course when applying the brakes.

In its operation, PSM follows two fundamental control strategies. First, it offers the well-known concept of longitudinal control with ABS anti-lock brakes, anti-spin control and the Automatic Brake Differential, keeping the car smoothly on course when accelerating and applying the brakes on a straight or in bends.

Second, PSM also offers lateral or transverse control keeping the car reliably on course even when subject to substantial lateral forces in a bend. The corrections required for this purpose are provided by the specific, carefully controlled application of the brakes.

Any tendency to oversteer with the rear end of the car swerving around is counteracted by the exact, perfectly metered application of the brake on the outer front wheel in a bend. Understeering, in turn, is prevented by applying the brake on the rear inner wheel. Lengthwise dynamic control also comes in here to provide a supportive effect, with E-Gas technology in the Carrera 4 serving to adjust the position of the throttle butterfly according to specific requirements. On the road, this means much easier and smoother steering.

To ensure precise function at all times, PSM features a whole number of monitoring units. The wheel speed sensors introduced for the first time together with ABS not only provide information on the speed of the car, acceleration and deceleration, but are also able, by considering the difference in speed from left to right, to 'detect' bends and their radius. Further units are the steering angle sensor, a lateral acceleration sensor and a yaw sensor serving to detect any drift inclination of the car.

All data determined by the sensors are stored within the PSM computer, evaluated within fractions of a second and passed on as instructions to the E-Gas or brake system. As a result, PSM responds a lot faster in threatening situations than even the most experienced driver.

Really enthusiastic drivers wishing to try out the 'natural' dynamic behaviour of their Carrera

4 on the race track are able to deactivate the lateral dynamic control provided by Porsche Stability Management simply by flipping a switch on the instrument panel. Even then the risk involved when taking the car into a power slide is reasonably limited, since all the driver has to do when the angle of the car becomes excessive is to step on the brakes in order to reactivate the dynamic control function. Consequently under circumstances like this, PSM is able to 'bend', slightly but of course never fully override, the laws of physics.

15.7 Diagnosing chassis electrical system faults

15.7.1 Introduction

As with all systems, the six stages of fault-finding should be followed.

1. Verify the fault.
2. Collect further information.
3. Evaluate the evidence.
4. Carry out further tests in a logical sequence.
5. Rectify the problem.
6. Check all systems.

Table 15.1 lists some common symptoms of chassis electrical system malfunctions together with suggestions for the possible fault. The faults are generic but will serve as a good reminder. It assumed an appropriate pressure gauge set has been connected.

15.7.2 Testing procedure – black box technique

'Chassis electrical systems' covers a large area of the vehicle. The generic fault-finding lists presented in other chapters may be relevant but the technique that will be covered here is known as 'black box fault-finding'. This is an excellent technique and can be applied to many vehicle systems from engine management and ABS to cruise control and instrumentation.

As most systems now revolve around an ECU, the ECU is considered to be a 'black box', in other words we know what it should do but how it does it is irrelevant! 'Any colour, so long as it's black,' said Henry Ford in the 1920s. I doubt that he was referring to ECUs though.

Figure 15.24 shows a block diagram that could be used to represent any number of automobile electrical or electronic systems. In reality the arrows from the 'inputs' to the ECU and from the ECU to the 'outputs' are wires. Treating the ECU as a 'black box' allows us to ignore its complexity. The theory is that if all the sensors and associated wiring to the 'black box' are OK, all the output actuators and their wiring are OK and the supply/earth connections are OK, then the fault must be the 'black box'. Most ECUs are very reliable, however, and it is far more likely that the fault will be found in the

Table 15.1 Common symptoms and possible faults of a chassis electrical system malfunction

Symptom	Possible fault
ABS not working and/or warning light on	● Wheel sensor or associated wiring open circuit/high resistance. ● Wheel sensor air gap incorrect. ● Power supply/earth to ECU low or not present. ● Connections to modulator open circuit. ● No supply/earth connection to pump motor. ● Modulator windings open circuit or high resistance.
Traction control inoperative	● Wheel sensor or associated wiring open circuit/high resistance. ● Wheel sensor air gap incorrect. ● Power supply/earth to ECU low or not present. ● ABS system fault. ● Throttle actuator inoperative or open circuit connections. ● Communication link between ECUs open circuit.
ECAT system reduced performance or not working	● Communication link between engine and transmission ECUs open circuit. ● Power supply/earth to ECU low or not present. ● Transmission mechanical fault. ● Gear selector switch open/short circuit. ● Speed sensor inoperative.
Power steering assistance low or not working	● Power supply/earth to ECU low or not present. ● Mechanical fault. ● Power supply/earth to drive motor low or not present. ● Steering sensor inoperative.

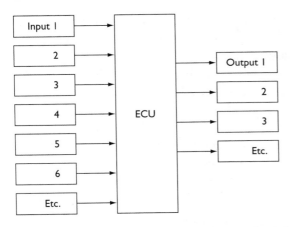

Figure 15.24 Block diagram representing many electrical systems

inputs or outputs.

Normal fault-finding or testing techniques can be applied to the sensors and actuators. For example, if an ABS system uses four inductive-type wheel speed sensors, then an easy test is to measure their resistance. Even if the correct value were not known, it would be very unlikely for all four to be wrong at the same time, so a comparison can be made. If the same resistance reading is obtained on the end of the sensor wires at the ECU, almost all of the 'inputs' have been tested with just a few ohmmeter readings.

The same technique will often work with 'outputs'. If the resistance of all the operating windings in, say, a hydraulic modulator were the same, then it would be reasonable to assume the figure was correct.

Sometimes, however, it is almost an advantage *not* to know the manufacturer's recommended readings. If the 'book' says the value should be between 800 and 900 Ω, what do you do when your ohmmeter reads 915 Ω? Answers on a postcard please...

Finally, don't forget that no matter how complex the electronics in an ECU, they will not work without a good power supply and an earth.

15.8 Advanced chassis systems technology

15.8.1 Road surface and tyre friction

The friction between the tyre and the road surface is a key issue when considering anti-lock brakes. Frictional forces must be transferred between the tyre contact patch and the road surface when the vehicle is accelerating or braking. The normal rules for friction between solid bodies have to be adapted because of the springy nature of rubber tyres. To get around this complicated problem, which involves molecular theory, the term 'slip' is used to describe the action of tyre and road.

Slip occurs when braking effort is applied to a rotating wheel. This can be defined as follows:

$$\lambda = \frac{\omega_0 - \omega}{\omega_0} \times 100\%$$

or

$$\lambda = \frac{V_v - V_r}{V_v} \times 100\%$$

0% is a free rolling wheel and 100% is a locked wheel where λ = slip; ω_0 = angular velocity of freely rotating wheel; ω = angular velocity of braked wheel; V_v = vehicle road speed = $\omega_0 r_d$; V_r = circumferential velocity of braked wheel = ωr_d; r_d = dynamic rolling radius of the wheel

The braking force or the adhesion coefficient of braking force (μ_F), measured in the direction the wheel is turning, is a function of slip. μ_F depends on a number of factors, the main ones being:

- Road surface material/condition.
- Tyre material, inflation pressure, tread depth, tread pattern and construction.
- Contact weight.

Figure 15.25 shows the relationship between the adhesion coefficient of braking effort and the amount of slip. Note that the graph is divided into two areas, stable and unstable. In the stable zone a balance exists between the braking effort

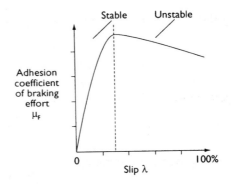

Figure 15.25 Relationship between adhesion coefficient of braking effort and amount of slip

applied and the adhesion of the road surface. Non-slip braking is therefore possible. In the unstable zone when the critical slip (l_c) is passed, no balance exists and the wheel will lock unless the braking force is reduced.

The value of critical slip (l_c) can vary between about 8 and 30% depending on the tyres and the road surface conditions. Figure 15.26 shows the difference between road surface conditions. This serves to highlight that a fixed slip threshold as a reference point, for when ABS should operate, would not make the best use of the available adhesion coefficient.

Lateral slip of the vehicle wheels must also be considered. This occurs when the wheel centre-line forms an angle of drift with the intended path of the wheel centre. The directional movement of the vehicle is defined as the correlation between the slip angle and the lateral force. This is shown in Figure 15.27, which is a graph of the coefficient of adhesion for lateral force, designated as μ_L, against slip angle (α). The critical slip angle (α_c) lies, in general, between 12 and 15°.

To regulate braking, it is essential that braking force and lateral guidance forces be considered. Figure 15.28 shows the combination of adhesion coefficient (μ_F), and the lateral adhe-

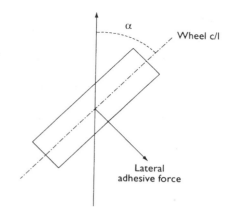

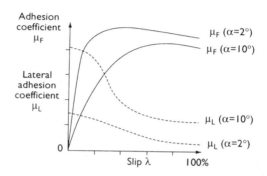

Figure 15.28 The combination of adhesion coefficient (μ_F) and lateral adhesion coefficient (μ_L) against braking (λ)

sion coefficient (μ_L) against braking slip (λ). The slip angle is shown at 2° and 10° and the test is on a dry road. Note the considerable reduction in lateral adhesion (μ_L) when the braking slip (λ) increases. When $\lambda = 28$, the value of μ_L is as a result of the steered angle of the wheel. This can be calculated as:

$$\mu_L(\text{min}) = \mu_F \sin\alpha$$

This serves to demonstrate how a locked wheel provides little steering effect. From Figure 15.27 it can be seen that ABS control must be extended for larger slip angles. If full braking occurs when the vehicle is experiencing high lateral acceleration (larger α) then ABS must intervene early and progressively allow greater slip as the vehicle speed decreases. These data are stored in look-up tables in a read only memory in the electronic control unit.

15.8.2 ABS control cycles

Figure 15.29 shows the braking control cycle for a high adhesion road (good grip). Figure 15.30 shows control cycles for a low adhesion surface

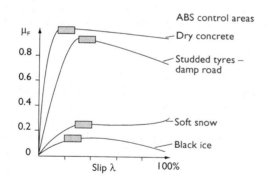

Figure 15.26 Difference between road surface conditions

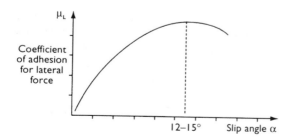

Figure 15.27 Graph of the coefficient of adhesion for lateral force μ_L, against slip angle (α)

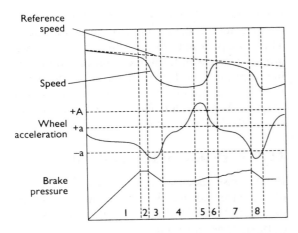

Figure 15.29 Braking control cycle for a high-adhesion surface

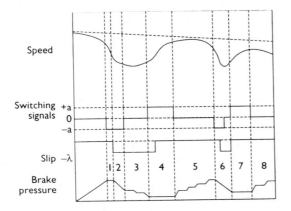

Figure 15.30 Braking control cycles for a low-adhesion surface

(slippery). Each figure is split into eight phases, which are described as follows.

High adhesion

1. Initial braking, ABS not yet activated.
2. Wheel speed exceeds the threshold calculated from the vehicle reference speed and brake pressure is held at a constant value.
3. Wheel deceleration falls below a threshold $(-a)$ and brake pressure is reduced.
4. Brake pressure holding is now occurring and wheel speed will increase.
5. Wheel acceleration exceeds the upper limit $(+A)$ so brake pressure is now allowed to increase.
6. Pressure is again held constant as the limit $(+a)$ is exceeded.
7. Brake pressure is now increased in stages until wheel speed threshold $(-a)$ is exceeded.

8. Brake pressure is decreased again and then held constant when $(-a)$ is reached.

The process as above continues until the brake pedal is released or the vehicle speed is less than a set minimum, at which time the wheels will lock to bring the vehicle finally to rest.

Low adhesion

1. Initial braking, ABS not yet activated.
2. Wheel speed exceeds the threshold calculated from the vehicle reference speed and brake pressure is held at a constant value.
3. During this phase a short holding time is followed by a reduction in brake pressure. The wheel speed is compared with, and found to be less than, the calculated slip threshold so pressure is reduced again followed by a second holding time. A second comparison takes place and the pressure is reduced again.
4. A brake pressure holding phase allows the wheel speed to increase.
5. There is a gradual introduction of increased brake pressure and holding pressure in steps until the wheel again slips.
6. Brake pressure is decreased allowing wheel speed to increase.
7. Pressure holding as the calculated slip value is reached.
8. Stepped increase in pressure with holding phases to keep high slip periods to a minimum. This ensures maximum stability.

The process again continues until the brake pedal is released or the vehicle comes to rest.

15.8.3 Traction control calculations

Figure 15.31 shows the forces acting on the wheels of a vehicle when accelerating on a non-homogeneous road surface. The maximum propulsion force can be calculated:

$$F = FH + FL = 2FL + FB$$

where F = total propulsion force; FH = force transmitted to μ_H part of road; FL = force transmitted to μ_L part of road; FB = braking force; μ_H = high braking force coefficient; μ_L = low braking force coefficient.

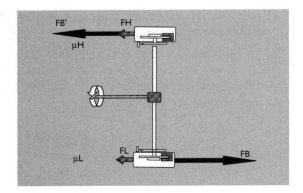

Figure 15.31 Forces acting on wheels of a vehicle when accelerating on a non-homogeneous road surface

15.9 New developments in chassis electrical systems

15.9.1 Brake by wire

Figure 15.32 shows a simple graph demonstrating the 10% improvement expected in stopping distances when electronically controlled brakes are used in comparison with a conventional system. The main advantages are that the electronics can continuously modify the braking effort to give the ideal distribution and also the time response of the system is greatly reduced. This is of particular importance for large articulated vehicles when it is vital that all the brakes react together and at a balanced effort. The area yet to be perfected is that of the load sensors; this is a question of cost versus reliability.

The main change in having electro-pneumatic control of the brakes is that the foot brake control is now situated next to the wheels. The con-trol is via electric relay valves, which have a progressive action due to modulation by the ECU. This close proximity to the actual air brake actu-ators is the reason for reduced propagation time in the air pipes. Anti-lock facilities are all part of the package. Figure 15.33 shows the layout of a brake-by-wire system. The wiring is shown con-nected to the CAN data bus.

15.9.2 Brake assist systems

Brake assist systems may be developed because of evidence showing that drivers are not realiz-ing the full benefit of anti-lock braking systems (ABSs). The introduction of ABS has not resulted in the reduction of accidents that had been hoped for. The reason for this is debatable; one view is that many drivers do not push hard enough on the brake pedal during an emergency stop, therefore the tyres do not slip sufficiently to engage the antilock system.

To counteract this problem, companies are developing brake assist systems that apply more hydraulic pressure than normal if an emergency condition is sensed. The system's ability to dis-cern whether a braking operation is an emer-gency, or not, is critical. Pedal force sensors only, as well as pedal force sensors in conjunc-tion with apply rate sensors, are under develop-ment, as are the control strategies. If field tests produce satisfactory results, brake assist sys-tems could be introduced relatively quickly into mass production.

Electric actuators may even begin to take the place of conventional wheel cylinders. Precisely controlled DC motors operating on drum brakes have the potential advantages of lower total sys-tem weight and cost. Developments are occur-ring in the area of magnetic braking, which has the potential to remove all wearing components from the vehicle!

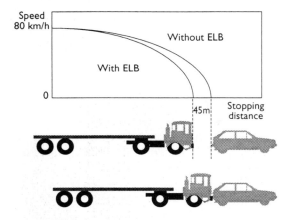

Figure 15.32 Graph demonstrating 10% improvement expected in stopping distances with electronically controlled brakes

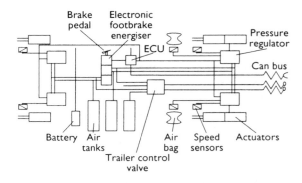

Figure 15.33 Layout of brake-by-wire system

15.9.3 Total vehicle dynamics

Throughout this chapter on chassis electronic systems, and in previous chapters on engine control, it may have become apparent that more and more electronic systems are required to be in communication to achieve optimum results. This is one of the driving forces behind data bus communications (Chapter 4), because many of the sensors used by various systems are common. Data are used by each in a slightly different way but many systems have an effect on others.

Systems, which are now quite common, that lend themselves to combined control, are as follows:

- Anti-lock brakes.
- Traction control.
- Active suspension.
- Four-wheel steering.
- Engine management.
- Automatic transmission.

Figure 15.34 is a block diagram showing how the above systems could be linked together. When these systems are working together, significant improvements in the operation of each can be produced. Research is still to be carried out in this area and further significant benefits are still possible in the future.

15.9.4 Automatic clutch

Valeo's Compact Automatic Clutch (CAC) eliminates the need for a clutch pedal. Cars equipped

with Valeo's CAC are therefore more comfortable and more fun to drive as the driver is freed from the tiresome effort of depressing and releasing the clutch pedal every time he or she changes gear. The gear lever remains, however, leaving active control of the car with the driver.

The Valeo CAC is an add-on system, which can be fitted to conventional manual transmissions. It consists of a clutch actuator, a powerful CPU and specific sensors driven by dedicated software that is optimized for each vehicle type. Figure 15.35 shows the clutch actuator.

The Valeo CAC uses electromechanical actuation. It is therefore more compact, weighs less and costs less compared with hydraulic systems. Its internal compensation spring allows for very fast response time (declutching time: 70–100 ms) combined with low electrical consumption (20 W). Its 16-bit electronic control unit and power electronics were developed and produced by Valeo Electronics.

Valeo has developed high performance computer simulation tools to design operating software that can make the car respond exactly in line with vehicle manufacturers' requirements.

The Valeo CAC is also designed to provide maximum safety and includes fault modes to minimize the impact of potential component failure. Figure 15.36 shows a 'robotized' manual gearbox.

15.9.5 Drive-by-wire – Delphi

Automotive technology has advanced gradually, and much of it has simply been a refinement of existing systems. Drive-by-wire technologies will change everything we know about designing, manufacturing and driving cars.

Drive-by-wire technology involves the replacement of traditional mechanical systems for steering, braking, throttle and suspension functions, with electronic controllers, actuators and

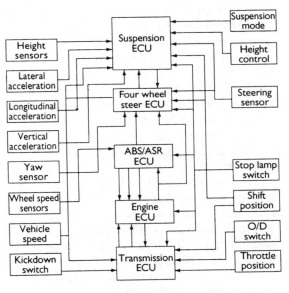

Figure 15.34 How several systems could be linked together: anti-lock brakes; traction control; active suspension; four-wheel steering; engine management; automatic transmission

Figure 15.35 Compact automatic clutch actuator

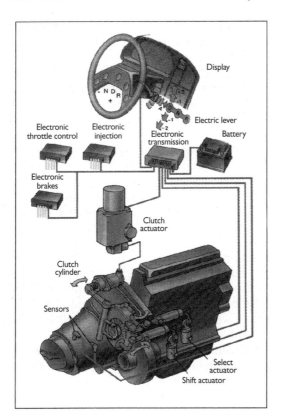

Figure 15.36 Robotized manual gearbox

sensors. For example, in Delphi's steer-by-wire system, the mechanical links between the steering wheel and the front wheels are replaced with two motorized assist mechanisms, a hand–wheel feedback unit and an electronic controller. As a result, the steering column, steering shaft, the pump and intermediate shaft and the hoses, fluids and belts associated with a traditional power steering system are completely eliminated. This enables improved system performance, simplified packaging and design flexibility.

Delphi's brake-by-wire system, known as the Galileo Intelligent Brake Control System, is already in use in General Motors' electric vehicle, the EV–1. The front hydraulic brakes are applied electrically, while the rear brakes are applied with a fully electric system. This eliminates the need for a vacuum booster, which gives increased packaging flexibility. Additional benefits include reduced mass for improved fuel efficiency in gas-powered vehicles, easier assembly and improved braking performance.

Electronic throttle control (ETC) replaces the throttle cables that run from the accelerator pedal to the engine. An electronic link replaces the traditional mechanical link, which commu-

nicates with an engine control module (ECM). The 1997 Chevrolet Corvette featured an ETC, developed by Delphi, and was the first gas-powered General Motors' passenger car to be so equipped. There are a number of benefits associated with ETC, including reduced mass, lower emissions, and increased throttle response.

Other drive-by-wire systems under development include damping by wire and roll by wire, where actuators and controllers replace conventional dampers and roll bars. Sensors measure vehicle yaw and levelling, as well as vehicle speed. Utilizing these data, a signal is sent to the proper actuators to damp the suspension actively. It is an infinitely variable system that can even compensate for the emptying of the fuel tank. Such a suspension reduces mass, which helps improve fuel economy, while improving ride and handling. It also reduces assembly time due to fewer parts and a simpler design.

Taken as a whole, these drive-by-wire systems allow for greatly increased modularity, which simplifies vehicle assembly, with the resultant, lower vehicle cost potential. In addition, they are environmentally friendly, since the number of hoses, pulleys and fluids are reduced or eliminated. More importantly, however, drive-by-wire systems bring a new freedom to how vehicles are designed, manufactured and ultimately used by the consumer. Prepare for a dramatic paradigm shift as entirely new design and assembly philosophies evolve.

Vehicle design will take different forms, allowing manufacturers to do things previously thought impossible with traditional technology and manufacturing processes. The vehicle's steering wheel is one example. Steer-by-wire technology will allow eventual replacement of the steering wheel and column, since the mechanical link between the steering wheel and the front wheels is no longer necessary. The space will be available for designers to do something completely different, such as incorporating new energy-absorbing systems in the body structure.

From an assembly standpoint, the entire system for steering, damping and braking can be contained in one module, which will arrive at the assembly point as a fully tested unit that is simply plugged into the vehicle. The simplicity of this module approach will greatly reduce assembly time, while at the same time increasing quality, since it will arrive fully tested. Manufacturers will find cost savings in a variety of areas.

15.10 Self-assessment

15.10.1 Questions

1. Describe the three main control phases of an ABS system.
2. Describe what is meant by 'black box fault-finding'.
3. Explain with the aid of a labelled sketch the operation of a wheel speed sensor.
4. State four advantage of electric power steering.
5. Draw a graph to show the effectiveness of traction control when only the throttle is controlled.
6. Make a simple sketch of a block diagram for an electronically controlled automatic transmission (ECAT) system and state the purpose of each part.
7. List eight chassis systems that can be controlled by electronics.
8. Define: 'Total vehicle dynamics'.
9. Describe the operation of an active suspension system.
10. State three possible disadvantages of an ABS system.

15.10.2 Assignment

Investigate the possibilities of producing a vehicle with a central control unit (CCU) that is able to control *all* operations of the vehicle from engine management to instrumentation and stability control.

Produce a report for the board of a major vehicle manufacturer showing the possible advantages and disadvantages of this approach. Make a clear recommendation to the board as to whether they should make this idea into a reality – or not. Justify your decision.

16

Comfort and safety

16.1 Seats, mirrors and sun-roofs

16.1.1 Introduction

Electrical movement of seats, mirrors and the sun-roof are included in one chapter as the operation of each system is quite similar. The operation of electric windows and central door locking is also much the same.

Fundamentally, all the above mentioned systems operate using one or several permanent magnet motors, together with a supply reversing circuit. A typical motor reverse circuit is shown in Figure 16.1. When the switch is moved, one of the relays will operate and this changes the polarity of the supply to *one* side of the motor. If the switch is moved the other way, then the polarity of the other side of the motor is changed. When at rest, both sides of the motor are at the same potential. This has the effect of regenerative braking so that when the motor stops it will do so instantly.

Further refinements are used to enhance the operation of these systems. Limit switches, position memories and force limitations are the most common.

16.1.2 Electric seat adjustment

Adjustment of the seat is achieved by using a number of motors to allow positioning of different parts of the seat. Movement is possible in the following ways.

- Front to rear.
- Cushion height rear.
- Cushion height front.
- Backrest tilt.
- Headrest height.
- Lumber support.

Figure 16.2 shows a typical electrically controlled seat. This system uses four positioning motors and one smaller motor to operate a pump, which controls the lumber support bag. Each motor can be considered to operate by a simple rocker-type switch that controls two relays as described above. Nine relays are required for this, two for each motor and one to control the main supply.

When the seat position is set, some vehicles have set position memories to allow automatic re-positioning if the seat has been moved. This is often combined with electric mirror adjustment. Figure 16.3 shows how the circuit is constructed

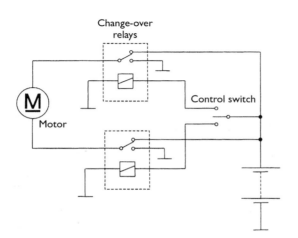

Figure 16.1 Typical motor reverse circuit

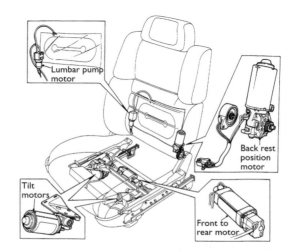

Figure 16.2 Electrically controlled seat

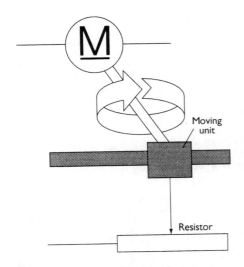

Figure 16.3 Position memory for electric seats

16.1.3 Electric mirrors

Many vehicles have electric adjustment of mirrors, particularly on the passenger side. The system used is much the same as has been discussed above in relation to seat movement. Two small motors are used to move the mirror vertically or horizontally. Many mirrors also contain a small heating element on the rear of the glass. This is operated for a few minutes when the ignition is first switched on and can also be linked to the heated rear window circuit. Figure 16.4 shows an electrically operated mirror circuit, which includes feedback resistors for positional memory.

16.1.4 Electric sun-roof operation

The operation of an electric sun-roof is similar to the motor reverse circuit discussed earlier in this chapter. However, further components and circuitry are needed to allow the roof to slide, tilt and stop in the closed position. The extra components used are a micro switch and a latching relay. A latching relay works in much the same way as a normal relay except that it locks into position each time it is energized. The mechanism used to achieve this is much like that used in ball-point pens that use a button on top.

The micro switch is mechanically positioned such as to operate when the roof is in its closed position. A rocker switch allows the driver to adjust the roof. The circuit for an electrically operated sun-roof is shown in Figure 16.5. The switch provides the supply to the motor to run it in the chosen direction. The roof will be caused to open or tilt. When the switch is operated to

to allow position memory. As the seat is moved a variable resistor, mechanically linked to the motor, is also moved. The resistance value provides feedback to an electronic control unit. This can be 'remembered' in a number of ways; the best technique is to supply the resistor with a fixed voltage such that the output relative to the seat position is proportional to position. This voltage can then be 'analogue-to-digital' converted, which produces a simple 'number' to store in a digital memory. When the driver presses a memory recall switch, the motor relays are activated by the ECU until the number in memory and the number fed back from the seat are equal. This facility is often isolated when the engine is running to prevent the seat moving into a dangerous position as the car is being driven. The position of the seats can still be adjusted by operating the switches as normal.

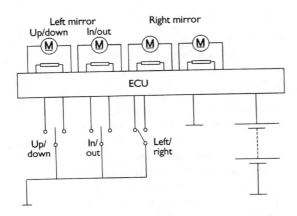

Figure 16.4 Feedback resistors for positional memory and the circuit

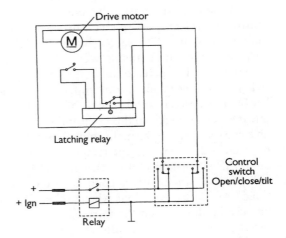

Figure 16.5 Sun-roof circuit

close the roof, the motor is run in the appropriate direction until the micro switch closes when the roof is in its closed position. This causes the latching relay to change over, which stops the motor. The control switch has now to be released. If the switch is pressed again, the latching relay will once more change over and the motor will be allowed to run.

16.2 Central locking and electric windows

16.2.1 Door locking circuit

When the key is turned in the driver's door-lock, all the other doors on the vehicle should also lock. Motors or solenoids in each door achieve this. If the system can only be operated from the driver's door key, then an actuator is not required in this door. If the system can be operated from either front door or by remote control, then all the doors need an actuator. Vehicles with sophisticated alarm systems often lock all the doors as the alarm is set.

Figure 16.6 shows a door locking circuit. The main control unit contains two change-over relays (as in Figure 16.1), which are actuated by either the door lock switch or, if fitted, the remote infrared key. The motors for each door-lock are simply wired in parallel and all operate at the same time.

Most door actuators are now small motors which, via suitable gear reduction, operate a linear rod in either direction to lock or unlock the doors. A simple motor reverse circuit is used to achieve the required action. Figure 16.7 shows a typical door-lock actuator.

Infrared central door locking is controlled by a small hand-held transmitter and an infrared sensor receiver unit as well as a decoder in the main control unit. This layout will vary slightly between different manufacturers. When the infrared key is operated by pressing a small switch, a complex code is transmitted. The number of codes used is well in excess of 50 000. The infrared sensor picks up this code and sends it in an electrical form to the main control unit. If the received code is correct, the relays are triggered and the door-locks are either locked or unlocked. If an incorrect code is received on three consecutive occasions when attempting to unlock the doors, then the infrared system will switch itself off until the door is opened by the key. This will also reset the system and allow the

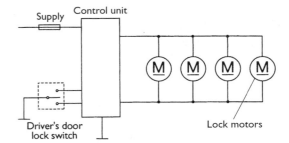

Figure 16.6 Door lock circuit

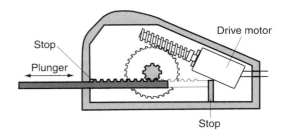

Figure 16.7 Door lock actuator

correct code to operate the locks again. This technique prevents a scanning type transmitter unit from being used to open the doors. Figure 16.8 shows a flow diagram representing the operation of a system that uses a 'rolling code' (MAC stands for Message Authentication Code).

16.2.2 Electric window operation

The basic form of electric window operation is similar to many of the systems discussed so far in this chapter; that is, a motor reversing system that is operated either by relays or directly by a switch.

More sophisticated systems are now becoming more popular for reasons of safety as well as improved comfort. The following features are now available from many manufacturers:

- One shot up or down.
- Inch up or down.
- Lazy lock.
- Back-off.

The complete system consists of an electronic control unit containing the window motor relays, switch packs and a link to the door lock and sunroof circuits. This is represented in the form of a block diagram in Figure 16.9.

When a window is operated in one-shot or one-touch mode the window is driven in the chosen direction until either the switch position is reversed, the motor stalls or the ECU receives a signal from the door-lock circuit. The problem with one-shot operation is that if a child, for example, should become trapped in the window there is a serious risk of injury. To prevent this, the back-off feature is used. An extra commutator is fitted to the motor armature and produces a signal via two brushes, proportional to the motor speed. If the rate of change of speed of the motor is detected as being below a certain threshold when closing, then the ECU will reverse the motor until the window is fully open.

By counting the number of pulses received, the ECU can also determine the window position. This is important, as the window must not reverse when it stalls in the closed position. In order for the ECU to know the window position it must be initialized. This is often done simply by operating the motor to drive the window first fully open, and then fully closed. If this is not done then the one-shot close will not operate.

On some systems, Hall effect sensors are used to detect motor speed. Other systems sense the current being drawn by the motor and use this as an indication of speed.

The lazy lock feature allows the car to be fully secured by one operation of a remote infrared key. This is done by the link between the door-lock ECU and the window and sun-roof ECUs. A signal is supplied and causes all the windows to close in turn, then the sun-roof, and finally it locks the doors. The alarm will also be set if required. The windows close in turn to prevent the excessive current demand that would occur if they all tried to operate at the same time.

A circuit for electric windows is shown in Figure 16.10. Note the connections to other systems such as door locking and the rear window isolation switch. This is commonly fitted to

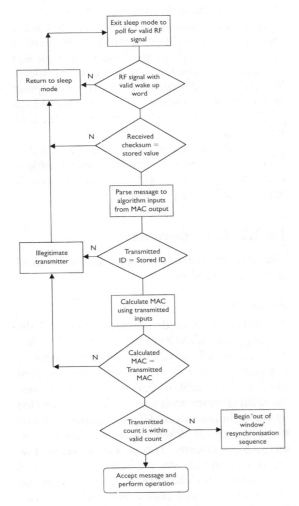

Figure 16.8 Flow diagram representing the 'Rolling code' system

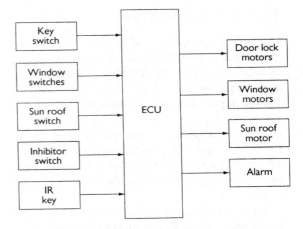

Figure 16.9 Block diagram showing links between door locks, windows and sun-roof – controlled by an infrared key

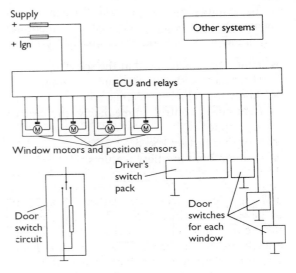

Figure 16.10 Electric window control circuit

Figure 16.11 Window lift motor for cable or arm-lift systems

allow the driver to prevent rear window operation for child safety, for example.

Figure 16.11 shows a typical window lift motor used for cable or arm-lift systems.

16.3 Cruise control

16.3.1 Introduction

Cruise control is the ideal example of a closed loop control system. Figure 16.12 illustrates this in the form of a block diagram. The purpose of cruise control is to allow the driver to set the vehicle speed and let the system maintain it automatically.

The system reacts to the measured speed of the vehicle and adjusts the throttle accordingly. The reaction time is important so that the vehicle's speed does not feel to be surging up and down.

Other facilities are included such as allowing the speed to be gradually increased or decreased at the touch of a button. Most systems also remember the last set speed and will resume this again at the touch of a button.

To summarize and to add further refinements, the following is the list of functional requirements for a good cruise control system.

- Hold the vehicle speed at the selected value.
- Hold the speed with minimum surging.
- Allow the vehicle to change speed.
- Relinquish control immediately the brakes are applied.
- Store the last set speed.
- Contain built in safety features.

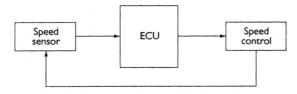

Figure 16.12 Cruise control closed loop system

16.3.2 System description

The main switch turns on the cruise control, this in turn is ignition controlled. Most systems do not retain the speed setting in memory when the main switch has been turned off. Operating the 'set' switch programs the memory but this normally will only work if conditions similar to the following are met.

- Vehicle speed is greater than 40 km/h.
- Vehicle speed is less than 12 km/h.
- Change of speed is less than 8 km/h/s.
- Automatics must be in 'drive'.
- Brakes or clutch are not being operated.
- Engine speed is stable.

Once the system is set, the speed is maintained to within about 3–4 km/h until it is deactivated by pressing the brake or clutch pedal, pressing the 'resume' switch or turning off the main control switch. The last 'set' speed is retained in memory except when the main switch is turned off.

If the cruise control system is required again then either the 'set' button will hold the vehicle at its current speed or the 'resume' button will accelerate the vehicle to the previous 'set' speed. When cruising at a set speed, the driver can press and hold the 'set' button to accelerate the vehicle until the desired speed is reached when the button is released.

If the driver accelerates from the set speed to overtake, for example, then when the throttle is released, the vehicle will slow down until it reaches the last set position.

16.3.3 Components

The main components of a typical cruise control system are as follows.

Actuator

A number of methods are used to control the throttle position. Vehicles fitted with driven by-wire systems allow the cruise control to operate the same actuator. A motor can be used to control the throttle cable or, in many cases, a vacuum-operated diaphragm is used which is controlled by three simple valves. This technique is shown in Figure 16.13. When the speed needs to be increased, valve 'x' is opened allowing low pressure from the inlet manifold to one side of the diaphragm. The atmospheric pressure on the other side will move the diaphragm and hence the throttle. To move the other way, valve 'x' is closed and valve 'y' is opened allowing

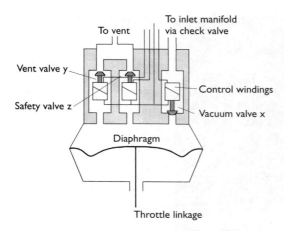

To vent
To inlet manifold via check valve
Vent valve y
Safety valve z
Control windings
Vacuum valve x
Diaphragm
Throttle linkage

Figure 16.13 Cruise control 'vacuum' actuator

atmospheric pressure to enter the chamber. The spring moves the diaphragm back. If both valves are closed then the throttle position is held. Valve 'x' is normally closed and valve 'y' normally open; thus, in the event of electrical failure cruise control will not remain engaged and the manifold vacuum is not disturbed. Valve 'z' provides extra safety and is controlled by the brake and clutch pedals.

Main switch and warning lamp

This is a simple on/off switch located within easy reach of the driver on the dashboard. The warning lamp can be part of this switch or part of the main instrument display as long as it is in the driver's field of vision.

Set and resume switches

These are fitted either on the steering wheel or on a stalk from the steering column. When the switches are part of the steering wheel, slip rings are needed to transfer the connection. The 'set' button programs the speed into memory and can also be used to increase the vehicle and memory speed. The 'resume' button allows the vehicle to reach its last set speed or temporarily to deactivate the control.

Brake switch

This switch is very important, as it would be dangerous braking if the cruise control system was trying to maintain the vehicle speed. This switch is normally of superior quality and is fitted in place or as a supplement to the brake light switch activated by the brake pedal. Adjustment of this switch is important.

Clutch or automatic gearbox switch

The clutch switch is fitted in a similar manner to the brake switch. It deactivates the cruise system to prevent the engine speed increasing if the clutch is pressed. The automatic gearbox switch will only allow the cruise to be engaged when it is in the 'drive' position. This is again to prevent the engine over-speeding if the cruise control tried to accelerate to a high road speed with the gear selector in the '1' or '2' position. The gearbox will still change gear if accelerating back up to a set speed as long as it 'knows' top gear is available.

Speed sensor

This will often be the same sensor that is used for the speedometer. If not, several types are available – the most common produces a pulsed signal, the frequency of which is proportional to the vehicle speed.

16.3.4 Adaptive cruise control

Conventional cruise control has now developed to a high degree of quality. It is, however, not always very practical on many European roads as the speed of the general traffic varies constantly and traffic is often very heavy. The driver has to take over from the cruise control system on many occasions to speed up or slow down. Adaptive cruise control can automatically adjust the vehicle speed to the current traffic situation. Figure 16.14 shows the operation of the system. The system has three main aims.

- Maintain a speed as set by the driver.
- Adapt this speed and maintain a safe distance from the vehicles in front.
- Provide a warning if there is a risk of collision.

The main components of basic and more complex adaptive cruise systems are shown in Figure 16.15. Note the main extra components are the 'headway' sensor and the steering angle sensor; the first of these is clearly the most important. Information on steering angle is used to enhance further the data from the headway sensor by allowing greater discrimination between hazards and spurious signals. Two types of the headway sensor are in use, the *radar* and the *lidar*. Both contain transmitter and receiver units. The radar system uses microwave signals at about 35 GHz, and the reflection time of these gives the distance to the object in front. Lidar uses a laser diode to produce infrared light signals, the

Figure **16.14** Adaptive cruise control operation

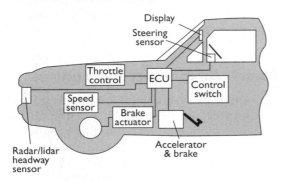

Figure 16.15 Adaptive cruise control

Figure 16.16 Headway sensor fitted at the front of a vehicle

reflections of which are detected by a photodiode.

These two types of sensors have advantages and disadvantages. The radar system is not affected by rain and fog but the lidar can be more selective by recognizing the standard reflectors on the rear of the vehicle in front. Radar can produce strong reflections from bridges, trees, posts and other normal roadside items. It can also suffer loss of signal return due to multipath reflections. Under ideal weather conditions, the lidar system appears to be the best but it becomes very unreliable when the weather changes. A beam divergence of about 2.5° vertically and horizontally has been found to be the most suitable whatever headway sensor is used. An important consideration is that signals from other vehicles fitted with this system must not produce erroneous results. Figure 16.16 shows a typical headway sensor. Fundamentally, the operation of an adaptive cruise system is the same as a conventional system except when a signal from the headway sensor

detects an obstruction, in which case the vehicle speed is decreased. If the optimum stopping distance cannot be achieved by just backing off the throttle, a warning is supplied to the driver. A more complex system can also take control of the vehicle transmission and brakes but this, while very promising, is further behind in development. It is important to note that adaptive cruise control is designed to relieve the burden on the driver, not take full control of the vehicle!

16.4 In-car multimedia

16.4.1 Introduction

These days it would be almost unthinkable not to have at least a radio cassette player in our vehicles. It does not seem too long ago, however, that these were an optional extra. Looking back just a little further, the in-car record player must

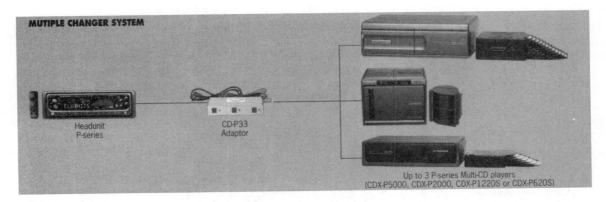

MUTIPLE CHANGER SYSTEM

Headunit
P-series

CD-P33
Adaptor

Up to 3 P-series Multi-CD players
(CDX-P5000, CDX-P2000, CDX-P1220S or CDX-P620S)

Figure 16.17 ICE system

have been interesting to operate – it was evidently quite successful in large American cars in the US but left a bit to be desired in British vehicles and on British roads. Figure 16.17 shows a typical high quality in-car entertainment (ICE) system with a multi-CD changer.

We now have ICE systems fitted to standard production cars, which are of good hi-fi quality. Facilities such as compact disc players and multiple compact disc changers together with automatic station search and re-tune are popular.

We have seen the rise and fall of the CB radio and the first car telephones – which were so large the main unit had to be fitted in the car boot. 'Hands-free' car telephones, which allow both hands to be kept free to control the car, are in common use and voice activation of other systems is developing.

The 'In-car PC' or the 'Auto PC' is an emerging technology that will soon become the 'norm'. The 'digital' automobile is here!

16.4.2 Speakers

Good ICE systems include at least six speakers, two larger speakers in the rear parcel shelf to produce good low frequency reproduction, two front door speakers for the mid-range and two front door tweeters for high frequency notes. Figure 16.18 shows a Pioneer sub-woofer speaker.

Speakers are a very important part of a sound system. No matter how good the receiver or CD player is, the sound quality will be reduced if inferior speakers are used. Equally, if the speakers are of a lower power output rating than the set, distortion will result at best, and damage to the speakers at worst. Speakers generally fall into the following categories.

- Tweeters – high frequency reproduction.

Figure 16.18 Pioneer sub-woofer

- Mid-range – middle range frequency reproduction (treble).
- Woofers – low frequency reproduction (bass).
- Sub-woofers – very low frequency reproduction.

Figure 16.19 shows the construction of a speaker.

16.4.3 ICE

Controls on most ICE sets will include volume, treble, bass, balance and fade. Cassette tape options will include Dolby filters to reduce hiss and other tape selections such as chrome or metal. A digital display, of course, will provide a visual output of the operating condition. This is also linked into the vehicle lighting to prevent

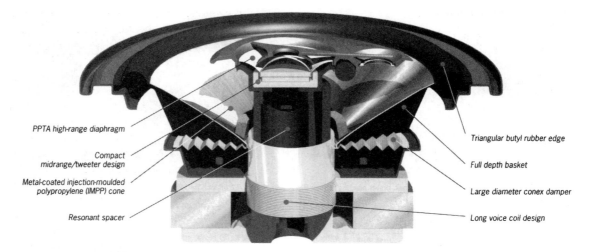

PPTA high-range diaphragm

Compact midrange/tweeter design

Metal-coated injection-moulded polypropylene (IMPP) cone

Resonant spacer

Triangular butyl rubber edge

Full depth basket

Large diameter conex damper

Long voice coil design

Figure 16.19 Speaker construction

glare at night. Track selection and programming for one or several compact discs is possible.

Many ICE systems are coded to deter theft. The code is activated if the main supply is disconnected and will not allow the set to work until the correct code has been re-entered. Some systems now include a plug-in electronic 'key card', which makes the set worthless when removed.

16.4.4 Radio data system (RDS)

RDS has become a standard on many radio sets. It is an extra inaudible digital signal, which is sent with FM broadcasts in a similar way to how teletext is sent with TV signals. RDS provides information so a receiver can appear to act intelligently. The possibilities available when RDS is used are as follows.

- The station name can be displayed in place of the frequency.
- Automatic tuning is possible to the best available signal for the chosen radio station. For example, in the UK, a journey from the south of England to Scotland would mean the radio would have to be re-tuned up to ten times. RDS will do this without the driver even knowing.
- Traffic information broadcasts can be identified and a setting made so that whatever you are listening to at the time can be interrupted.

RDS has six main features, which are listed here with a brief explanation.

1. Programme identification to allow the re-tune facility to follow the correct broadcasts.
2. Alternative frequencies, again to allow the receiver to try other signals for re-tuning as required.
3. Programme service name for displaying the name of the station on the radio set.
4. Traffic information, which provides for two codes to work in conjunction with route finding equipment.
5. Traffic programme, which allows the set to indicate that the station broadcasts traffic information.
6. A traffic announcement is transmitted when an announcement is being broadcast. This allows the receiver either to adjust the volume, switch over from the cassette during the announcement, lift an audio mute or, of course, if the driver wishes it, to do nothing.

16.4.5 Radio reception

There are two main types of radio signal transmitted; these are amplitude modulation (AM) and frequency modulation (FM). Figure 16.20 shows the difference between AM and FM signals.

Amplitude modulation is a technique for varying the height, or amplitude, of a wave in order to transmit information. Some radio broadcasts still use amplitude modulation. A convenient and efficient means of transmitting information is by the propagation of waves of electromagnetic radiation. Sound waves in the audible range, such as speech and music, have a frequency that is too low for efficient transmission through the air for significant distances. By the process of modulation, however, this low-frequency audio information can be impressed

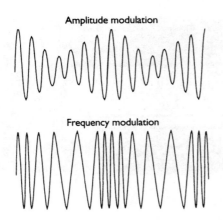

Amplitude modulation

Frequency modulation

Figure 16.20 Difference between AM and FM signals

on a carrier wave that has a much higher frequency and can propagate through space for great distances. The transmitter at a radio station generates a carrier wave having constant characteristics, such as amplitude and frequency. The signal containing the desired information is then used to modulate the carrier.

This new wave, called the modulated wave, will contain the information of the signal. In AM, it is the amplitude of the carrier wave that is made to vary so that it will contain the information of the signal. When the modulated wave reaches a radio receiver tuned to the proper frequency, it is demodulated, which is essentially the opposite of modulation. The set can then reproduce the desired sound via an amplifier and the loudspeakers. AM radio is still a popular form of radio broadcasting, but it does have a number of disadvantages. The quality of reproduction is relatively poor because of inherent limitations in the technique and because of interference from other stations and other electrical signals, such as those produced by lightning or by electronic devices – of which the car has more than its fair share. Some of these drawbacks can be overcome by using FM.

Frequency modulation is a method of modulation in which the frequency of a wave is varied in response to a modulating wave. The wave in which frequency is varied is called the carrier, and the modulating wave is called the signal. Frequency modulation requires a higher-frequency carrier wave and a more complex method for transmitting information than does AM; however, FM has an important advantage in that it has constant amplitude; it is therefore much less susceptible to interference from both natural and artificial sources. Such sources cause static in an amplitude-modulated radio.

Both types of modulation, however, are used in radio broadcasting. FM radio is a generally a far better source of high fidelity music. This is because the quality of AM reception, as well as the problems outlined above, is limited by the narrow bandwidth of the signal. During the winter months, reception of AM signals becomes worse due to changes in the atmosphere. FM does, however, present problems with reception when mobile. As most vehicles use a rod aerial, which is omni-directional, it will receive signals from all directions. Because of this, reflections from buildings, hills and other vehicles can reach the set all at the same time. This can distort the signal and is heard as a series of clicks or signal flutter as the signal is constantly enhanced or reduced. The best FM reception is considered to be line-of-sight from the transmitter. In general, the coverage or footprint of FM transmitters is quite extensive and, especially with the advent of RDS, the reception when mobile is quite acceptable.

16.4.6 Radio broadcast data system (RBDS)

The Radio Broadcast Data System is an extension of the Radio Data System (RDS), which has been in use in Europe since 1984. The system allows the broadcaster to transmit text information at the rate of about 1200 bits per second. The information is transmitted on a 57 kHz suppressed sub-carrier as part of the FM multiplexed (MPX) signal.

RBDS was developed for the North American market by the National Radio Systems Committee (NRSC), a joint committee composed of the Electronic Industries Association (EIA) and the National Association of Broadcasters (NAB). The applications for the transmission of text to the vehicle are interesting.

- Song title and artist.
- Traffic, accident and road hazard information.
- Stock information.
- Weather.

In emergency situations, the audio system can be enabled to interrupt the cassette, CD or normal radio broadcast to alert the user.

16.4.7 Digital audio broadcast (DAB)

Digital Audio Broadcasting is designed to provide high-quality, multiservice digital radio broadcasting for reception by stationary and

Figure 16.21 Clarion DAB receiver

mobile receivers. It is being designed to operate at any frequency up to 3 GHz. A system is being demonstrated and extensively tested in Europe, Canada and the United States. It is a rugged and also a very efficient sound and data broadcasting system.

The system uses digital techniques to remove redundancy and perceptually irrelevant information from the audio source signal. It then applies closely controlled redundancy to the transmitted signal for error correction. All transmitted information is then spread in both the frequency and the time domains (multiplexed) so a high quality signal is obtained in the receiver, even under poor conditions.

Frequency reallocation will permit broadcasters to extend services, virtually without limit, using additional transmitters, all operating on the same radiated frequency. A common worldwide frequency in the L band (around 1.5 GHz) is being considered, but some disagreement still exists. The possibilities make the implementation of DAB inevitable. Figure 16.21 shows the front panel of the Clarion system, capable of receiving digital broadcast signals.

16.4.8 Interference suppression

The process of interference suppression on a vehicle is aimed at reducing the amount of unwanted noise produced from the speakers of an ICE system. This, however, can be quite difficult. To aid the discussion, it is necessary first to understand the different types of interference. Figure 16.22 shows two signals, one clean and the other suffering from interference. The amount of interference can be stated as a signal-to-noise ratio. This is the useful field strength compared with the interference field strength at the receiver. This should be as high as possible but a value in excess of 22.1 for radio reception is accepted as a working figure. Interference is an

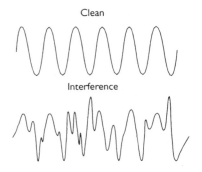

Figure 16.22 Two signals, one clean and the other suffering from interference

electromagnetic compatibility (EMC) issue and further details can be found in Chapter 4.

There are two overall issues to be considered relating to suppression of interference on a vehicle. These are as follows.

1. Short range – the effect of interference on the vehicle's radio system.
2. Long range – the effect of the vehicle on external receivers such as domestic televisions. This is covered by legislation making it illegal to cause disturbance to radios or televisions when using a vehicle.

Interference can propagate in one of four ways.

● Line borne, conducted through the wires.
● Air borne, radiated through the air to the aerial.
● Capacitive coupling by an electric field.
● Inductive coupling magnetic linking.

The sources of interference in the motor vehicle can be summarized quite simply as any circuit, which is switched or interrupted suddenly. This includes the action of a switch and the commutation process in a motor, both of which produce rapidly increasing signals. The secret of suppression is to slow down this increase. Interference is produced from four main areas of the vehicle.

- Ignition system.
- Charging system.
- Motors and switches.
- Static discharges.

The ignition system of a vehicle is the largest source of interference, particularly the high tension side. Voltages up to 30 kV are now common and the peak current for a fraction of a second when the spark plug fires can peak in excess of 100 A. The interference caused by the ignition system is mostly above 30 MHz and the energy can peak, for fractions of a second, of the order of 500 kW.

The charging system produces noise because of the sparking at the brushes. Electronic regulators produce little problems but regulators with vibrating contacts can cause trouble.

Any motor or switch, including relays, is likely to produce some interference. The most popular sources are the wiper motor and heater motor. The starter is not considered due to its short usage time.

The build-up of static electricity is due to friction between the vehicle and the air, and the tyres and the road. If the static on, say, the bonnet builds up more than the wing then a spark can be discharged. Using bonding straps to ensure all panels stay at the same potential easily prevents this. Due to the action of the tyres, a potential can build up between the wheel rims and the chassis unless suitable bonding straps are fitted. The arc to ground can be as much as 10 kV.

There are five main techniques for suppressing radio interference.

- Resistors.
- Bonding.
- Screening.
- Capacitors.
- Inductors.

Resistance is used exclusively in the ignition HT circuit, up to a maximum of about 20 kΩ per lead. This has the effect of limiting the peak current, which in turn limits the peak electromagnetic radiation. Providing excessive resistance is not used, the spark quality is not affected. These resistors effectively damp down the interference waves.

Bonding has been mentioned earlier, it is simply to ensure all parts of the vehicle are at the same electrical potential to prevent sparking due to the build-up of static.

Screening is generally only used for specialist applications such as emergency services and the military. It involves completely enclosing the ignition system and other major sources of noise, in a conductive screen, which is connected to the vehicle's chassis earth. This prevents interference waves escaping; it is a very effective technique but expensive. Often, a limited amount of screening – metal covers on the plugs for example – can be used to good effect.

Capacitors and inductors are used to act as filters. This is achieved by using the changing value of 'resistance' to alternating signals as the frequency increases. The correct term for this resistance is either capacitive or inductive reactance.

By choosing suitable values of a capacitor in parallel and or an inductor in series it is possible to filter out unwanted signals of certain frequencies.

The aerial is worth a mention at this stage. Several types are in use; the most popular still being the rod aerial, which is often telescopic. The advantage of a rod aerial is that it extends beyond the interference field of the vehicle. For reception in the AM bands the aerial represents a capacitance of 80 pF with a shunt resistance of about 1 MΩ. The set will often incorporate a trimmer to ensure the aerial is matched to the set. Contact resistance between all parts of the aerial should be less than 20 mΩ. This is particularly important for the earth connection.

When receiving in the FM range, the length of the aerial is very important. The ideal length of a rod aerial for FM reception is one quarter of the wavelength. In the middle of the FM band (94 MHz) this is about 80 cm. Due to the magnetic and electrical field of the vehicle and the effect of the coaxial cable, the most practical length is about 1 m. Some smaller aerials are available but whilst these may be more practical the signal strength is reduced. Aerials embedded into the vehicle windows or using the heated rear window element are good from the damage prevention aspect and insensitivity to moisture, but produce a weaker signal, often requiring an aerial amplifier to be included. Note that this will also amplify interference. Some top-range vehicles use a rod aerial and a screen aerial, the set being able to detect and use the strongest signal. This reduces the effect of reflected signals and causes less flutter.

Consideration must be given to the position of an external aerial. This has to be a compromise taking into account the following factors.

- Rod length – 1 m if possible.
- Coaxial cable length – longer cable reduces the signal strength.
- Position – as far away as reasonably possible from the ignition system.
- Potential for vandalism – out of easy reach.
- Aesthetic appearance – does it fit with the style of the vehicle?
- Angle of fitting – vertical is best for AM, horizontal for FM.

Most quality sets also include a system known as interference absorption. This is a circuit built into the set consisting of high quality filters.

Figure 16.23 shows a circuit of a typical ICE system. An electric aerial is included and also the connection to a multi compact disc unit via a data bus.

16.4.9 Mobile communications

If the success of the cellular industry is any indication of how much use we can make of the telephone, the future promises an even greater expansion. Cellular technology started to become useful in the 1980s and has continued to develop from then – very quickly!

The need and desire we perceive to keep in touch with each other is so great that an increasing number of business people now have up to five telephone numbers: home, office, pager, fax and cellular. But within the foreseeable future, high-tech digital radio technology and sophisticated telecommunications systems will enable all communications to be processed through a single number.

With personal numbering, a person carrying a pocket-size phone will need only one phone number. Instead of people calling places, people will call people – we will not be tied to any particular place. Personal numbering will make business people more productive because they will be able to reach, and be reached by, colleagues and clients, anywhere and anytime, indoors or outdoors. When travelling from home to office or from one meeting to the next, it will be possible to communicate with anyone, whenever the need arises.

But where does this leave communication systems relating to the vehicle? It is my opinion that 'in-vehicle' communication equipment for normal business and personal use will be by the simple pocket sized mobile phone and that there is no further market for the car telephone. Hands-free conversions will still be important.

CB radios and short-range two-way systems such as used by taxi firms and service industries will still have a place for the time being. However, even these may decline as the cellular network becomes cheaper and more convenient to use.

16.4.10 Auto PC

A revolution in the use of information technology in vehicles is taking place! Advanced computing, communications and positioning developments are being introduced in even the most basic vehicles. Figure 16.24 shows an Auto PC/Car Multimedia system. However, there were several barriers to the widespread use of such new technology.

- Not robust enough.
- Too costly.
- Difficult to install.
- Lack of common standards.
- Difficult to operate.

Most of these problems either have been resolved or are about to be, and other developments are also beneficial:

- Computers have become smaller.
- Prices have reduced.
- Performance has improved.
- Standards are being agreed.

Many leading computer companies, including Microsoft, IBM, and Intel have identified the vehicle as their next big marketplace. Plans have been announced for in-vehicle computers with a range of integrated functions. Microsoft's Auto PC, for example, uses the Windows CE operating system, a cut-down version of Windows 95/98/2000.

Many suppliers of Windows programs are

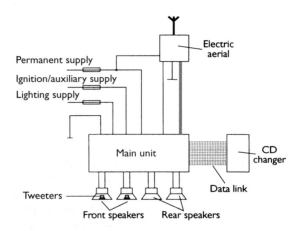

Figure 16.23 ICE system wiring

Figure 16.24 Car multimedia

now committed to offering Windows CE versions of their programs for use in car computers and hand-held personal computers (PCs). Just like a desktop PC, the car computer supports a range of programs. A car computer that will give the driver spoken directions while passengers browse the Internet or watch football will be a reality.

The Auto PC will be able to run familiar desktop programs whilst also offering the following.

- Spoken turn-by-turn navigation.
- Digital map database of useful sites, such as filling stations and cinemas.
- Voice memo system.
- Vehicle diagnostics program.
- Vehicle security and tracking system.
- Emergency roadside assistance service.

The unit could also be a high-performance stereo system capable of playing CDs and receiving FM radio. An optional communications interface will enable cellular phones to be controlled by spoken instructions, and traffic news received over a pager or cellular service. Intel, the largest computer chip manufacturer, envisages a car computer that is even more highly specified than Microsoft's Auto PC.

The Intel Connected Car PC has a full Windows operating system. As well as providing the driver with similar functions to the Auto PC, this also gives passengers access to a monitor for browsing the Internet or watching television programmes. IBM is working with car manufacturers to help them create networking capabilities in their vehicles.

Whether car computers ultimately succeed or not, there is little doubt that there will be much greater integration of all electronic systems in cars in the future. Efforts are underway in Europe, Japan and the United States to develop a standard data-bus system linking and powering non-safety related electronic systems in vehicles, such as CD players, positioning systems, air conditioning and electric windows. Adding electronic systems later would be by what is described as 'plug and play'.

Tying the computer in with the mobile communication system opens up even more possibilities. Cellular phone systems can provide an excellent means of tracking vehicles. Phone operators divide the country into separate cells and monitor phones as they move between them to ensure that each phone communicates through the best transmitter. Mobile communication systems will have a profound impact on how vehicles are used. Development work is underway on the exchange of information between vehicles and the road infrastructure.

(See also the section on 'Telematics' in Chapter 13).

16.5 Security

16.5.1 Introduction

Stolen cars and theft from cars account for about a quarter of all reported crime. A huge number of cars are reported missing each year and over 20% are never recovered. Even when returned many are damaged. Most car thieves are opportunists, so even a basic alarm system can serve as a deterrent.

Car and alarm manufacturers are constantly fighting to improve security. Building the alarm system as an integral part of the vehicle electronics has made significant improvements. Even so, retro-fit systems can still be very effective. Three main types of intruder alarm are used.

- Switch operated on all entry points.
- Battery voltage sensed.
- Volumetric sensing.

There are three main ways to disable the vehicle.

- Ignition circuit cut off.
- Starter circuit cut off.
- Engine ECU code lock.

A separate switch or IR transmitter can be used to set an alarm system. Often, they are set automatically when the doors are locked.

16.5.2 Basic security

To help introduce the principals of a vehicle alarm, this section will describe a very simple system, which can be built as a DIY retro fit. First, the requirements of this particular alarm system.

- It must activate when a door is opened.
- The ignition to be disabled.
- The existing horn is used as the warning.
- Once triggered, the horn must continue even when the door is closed.
- It must reset after 15 seconds.

The design will be based around a simple relay circuit. When a door is opened, the switches make an earth connection. This will be used to trigger the relay, which in turn will operate the horn. The delay must be built in using a capacitor, which will keep the relay energized even after the door closes, for a further 15 seconds. An external key switch is to be used to arm and disarm whilst isolating the ignition supply. Figure 16.25 shows a simple alarm circuit, which should achieve some of the aims. The delay is achieved by using a CR circuit; the 'R' is the resistance of the relay coil. Using the following data the capacitor value can be calculated.

- Time delay = 15 s.
- Relay coil = 120 Ω.
- Supply voltage = 12 V.
- Relay drop out = 8 V.

A capacitor will discharge to about 66% of its full value in CR seconds. The supply voltage is 12 V, so 66% of this is 8 V.

Therefore, if $CR = 15$, then, $C = 15/120$

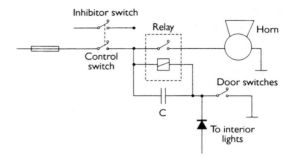

Figure 16.25 Simple alarm circuit the entry delay is made by using a CR circuit

$$C = 125\,\text{mF}$$

This seems an ideal simple solution – but it is not. As an assignment, find the problem and design a simple electronic circuit using a transistor, resistor and capacitor.

16.5.3 Top of the range security

The following is an overview of the good alarm systems now available either as a retro-fit or factory fitted. Most are made for 12 V, negative earth vehicles. They have electronic sirens and give an audible signal when arming and disarming. They are all triggered when the car door opens and will automatically reset after a period of time, often 1 or 2 minutes. The alarms are triggered instantly when an entry point is breached. Most systems can be considered as two pieces, with a separate control unit and siren; most will have the control unit in the passenger compartment and the siren under the bonnet.

Most systems now come with two infrared remote 'keys' that use small button-type batteries and have an LED that shows when the signal is being sent. They operate with one vehicle only. Intrusion sensors such as car movement and volumetric sensing can be adjusted for sensitivity.

When operating with flashing lights most systems draw about 5 A. Without flashing lights (siren only) the current drawn is less than 1 A. The sirens produce a sound level of about 95 dB, when measured 2 m in front of the vehicle.

Figure 16.26 shows a block diagram of a complex alarm system. The system, as is usual, can be considered as a series of inputs and outputs.

Inputs
- Ignition supply.
- Engine crank signal.
- Volumetric sensor.
- Bonnet switch.
- Trembler switch.
- IR/RF remote (Figure 16.27).

- Doors switches.
- Control switch.

Outputs

- Volumetric transmitter.
- System LED.
- Horn or siren.
- Hazard lights.
- Ignition immobilizer.
- Loop circuit.
- Electric windows, sun-roof and door locks.

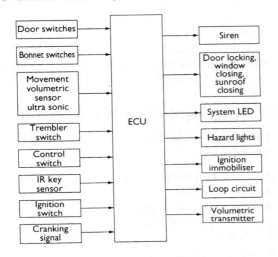

Figure 16.26 Block diagram of a complex alarm system

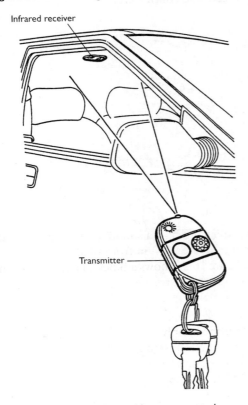

Figure 16.27 Alarm system with remote control

Some factory fitted alarms are combined with the central door locking system. This allows the facility mentioned in a previous section known as lazy lock. Pressing the button on the remote unit, and as well as setting the alarm, the windows and sun-roof close, and the doors lock.

16.5.4 Security coded ECUs

A security code in the engine electronic control unit is a powerful deterrent. This can only be 'unlocked' to allow the engine to start when it receives a coded signal. Ford and other manufacturers use a special ignition key that is programmed with the required information. Even the correct 'cut' key will not start the engine. Citroen, for example, have used a similar idea but the code has to be entered via a numerical keypad.

Of course nothing will stop the car being lifted on to a lorry and driven away, but this technique will mean a new engine control ECU will be needed by the thieves. The cost will be high and also questions may be asked as to why a new ECU is required.

16.6 Air bags and belt tensioners

16.6.1 Introduction

A seat-belt, seat-belt tensioner and an air bag are, at present, the most effective restraint system in the event of a serious accident. At speeds in excess of 40 km/h the seat-belt alone is no longer adequate. Research after a number of accidents has determined that in 68% of cases an air bag provides a significant improvement. It is suggested that if all cars in the world were fitted with an air bag then the number of fatalities annually would be reduced by well over 50 000. Some air bag safety issues have been apparent in the USA where air bags are larger and more powerful. This is because in many areas the wearing of seatbelts is less frequent.

The method becoming most popular for an air bag system is that of building most of the required components into one unit. This reduces the amount of wiring and connections, thus improving reliability. An important aspect is that some form of system monitoring must be built in, as the operation cannot be tested – it only ever works once. Figure 16.28 shows the air bags operating in a Peugeot.

Figure 16.28 Don't be a crash test dummy!

16.6.2 Operation of the system

The sequence of events in the case of a frontal impact at about 35 km/h, as shown in Figure 16.29, is as follows.

1. The driver is in the normal seating position prior to impact. About 15 ms after the impact the vehicle is strongly decelerated and the threshold for triggering the air bag is reached. The igniter ignites the fuel tablets in the inflater.
2. After about 30 ms the air bag unfolds and the driver will have moved forwards as the vehicle's crumple zones collapse. The seat-belt will have locked or been tensioned depending on the system.
3. At 40 ms after impact the air bag will be fully inflated and the driver's momentum will be absorbed by the air bag.
4. About 120 ms after impact the driver will be moved back into the seat and the air bag will have almost deflated through the side vents, allowing driver visibility.

Passenger air bag events are similar to the above description. A number of arrangements are used with the mounting of all components in the steering wheel centre becoming the most popular. Nonetheless, the basic principle of operation is the same.

16.6.3 Components and circuit

The main components of a basic air bag system are as follows.

- Driver and passenger air bags.
- Warning light.
- Passenger seat switches.
- Pyrotechnic inflater.
- Igniter.
- Crash sensor(s).
- Electronic control unit.

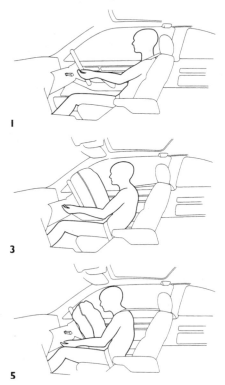

Figure 16.29 Airbag in action

The air bag is made of a nylon fabric with a coating on the inside. Prior to inflation the air bag is folded up under suitable padding that has specially designed break lines built in. Holes are provided in the side of the air bag to allow rapid deflation after deployment. The driver's air has a volume of about 60 litres and the passenger air bag about 160 litres.

A warning light is used as part of the system monitoring circuit. This gives an indication of a potential malfunction and is an important part of the circuit. Some manufacturers use two bulbs for added reliability.

Consideration is being given to the use of a seat switch on the passenger side to prevent deployment when not occupied. This may be more appropriate to side-impact air bags mentioned in the next section.

The pyrotechnic inflater and the igniter can be considered together. The inflater in the case of the driver is located in the centre of the steering wheel. It contains a number of fuel tablets in a combustion chamber. The igniter consists of charged capacitors, which produce the ignition spark. The fuel tablets burn very rapidly and produce a given quantity of nitrogen gas at a given pressure. This gas is forced into the air bag through a filter and the bag inflates breaking through the padding in the wheel centre. After deployment, a small amount of sodium hydroxide will be present in the air bag and vehicle interior. Personal protection equipment must be used when removing the old system and cleaning the vehicle interior.

The crash sensor can take a number of forms; these can be described as mechanical or electronic. The mechanical system (Figure 16.30) works by a spring holding a roller in a set position until an impact above a predetermined limit, provides enough force to overcome the spring and the roller moves, triggering a micro switch. The switch is normally open with a resis-

tor in parallel to allow the system to be monitored. Two switches similar to this may be used to ensure the bag is deployed only in the case of sufficient frontal impact. Note that the air bag is not deployed in the event of a roll over.

The other main type of crash sensor can be described as an accelerometer. This will sense deceleration, which is negative acceleration. Figures 16.31 is a sensor based on strain gauges.

Figure 16.32 shows two types of piezoelectric crystal accelerometers, one much like an engine knock sensor and the other using spring elements. A severe change in speed of the vehicle will cause an output from these sensors as the seismic mass moves or the springs bend. Suitable electronic circuits can monitor this and be preprogrammed to react further when a signal

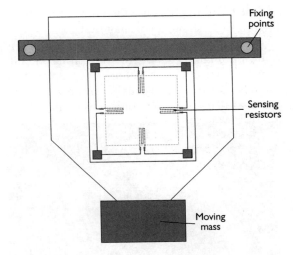

Figure 16.31 Strain gauges accelerometer

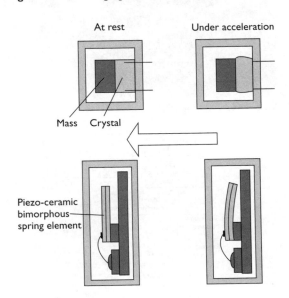

Figure 16.32 Piezoelectric crystal accelerometer

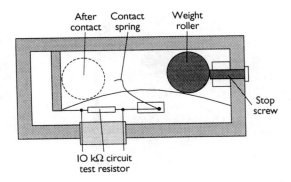

Figure 16.30 The mechanical impact sensor works by a spring holding a roller

beyond a set threshold is reached. The advantage of this technique is that the sensors do not have to be designed for specific vehicles, as the changes can be software based.

The final component to be considered is the electronic control unit or diagnostic control unit. When a mechanical-type crash sensor is used, in theory no electronic unit would be required. A simple circuit could be used to deploy the air bag when the sensor switch was operated. However, it is the system monitoring or diagnostic part of the ECU, that is most important. If a failure is detected in any part of the circuit then the warning light will be operated. Up to five or more faults can be stored in the ECU memory, which can be accessed by blink code or serial fault readers. Conventional testing of the system with a multimeter and jump wires is not to be recommended as it might cause the air bag to deploy! Figure 16.33 shows an air bag ECU.

A block diagram of an air bag circuit is shown in Figure 16.34. Note the 'safing' circuit, which is a crash sensor that prevents deployment in the event of a faulty main sensor. A digital-based system using electronic sensors has about

10 ms at a vehicle speed of 50 km/h, to decide if the restraint systems should be activated. In this time about 10 000 computing operations are necessary. Data for the development of these algorithms are based on computer simulations but digital systems can also remember the events during a crash, allowing real data to be collected.

16.6.4 Seat-belt tensioners

Taking the 'slack' out of a seat-belt in the event of an impact is a good contribution to vehicle passenger safety. The decision to take this action is the same as for the air bag inflation. The two main types of tensioners are:

- Spring tension.
- Pyrotechnic.

The mechanism used by one type of seat-belt tensioner is shown in Figure 16.35. When the explosive charge is fired, the cable pulls a lever on the seat-belt reel, which in turn tightens the belt. The unit must be replaced once deployed. This feature is sometimes described as anti-submarining.

16.6.5 Side air bags

Air bags working on the same techniques to those described previously are being used to protect against side impacts. In some cases bags are stowed in the door pillars or the edge of the roof. Figure 16.36 shows this system.

Figure 16.37 shows a full seat-belt and air bag system used by Ford.

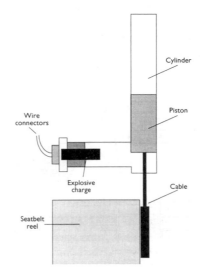

Figure 16.35 The mechanism used by one type of seat-belt tensioner

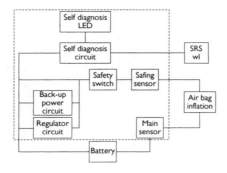

Figure 16.33 Airbag ECU

Figure 16.34 A block diagram of an airbag circuit

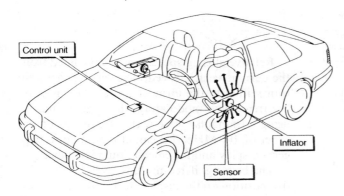

Figure 16.36 Side airbag system

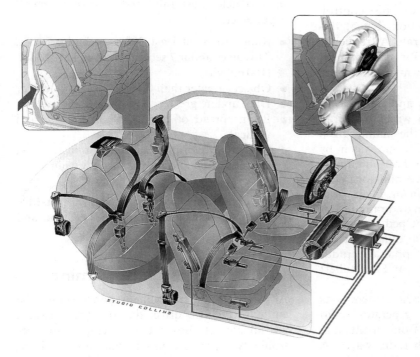

Figure 16.37 Seat belt and airbag operation

16.7 Other safety and comfort systems

16.7.1 Obstacle avoidance radar

This system, sometimes called collision avoidance radar, can be looked at in two ways. First, as an aid to reversing, which gives the driver some indication as to how much space is behind the car. Second, collision avoidance radar can be used as a vision enhancement system.

The principal of radar as a reversing aid is illustrated in Figure 16.38. This technique is, in effect, a range-finding system. The output can be audio or visual, the latter being perhaps most appropriate, as the driver is likely to be looking backwards. The audible signal is a 'pip pip pip' type sound, the repetition frequency of which increases as the car comes nearer to the obstruc-

tion, and becomes almost continuous as impact is imminent.

The technique is relatively simple as the level of discrimination required is fairly low and the radar only has to operate over short distances. The main problem is to ensure the whole width of the vehicle is protected.

Obstacle avoidance radar, when used as a vision enhancement system, is somewhat different. Figure 16.39 is a block diagram to demonstrate the principal of this system. In the future,

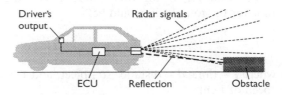

Figure 16.38 Obstacle avoidance radar

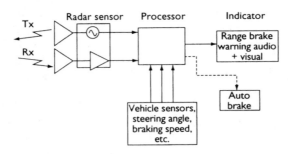

Figure 16.39 Block diagram of obstacle avoidance radar when used as a vision enhancement system

this may be linked with adaptive cruise control, as discussed in an earlier section, but at this stage the two systems are separate. A frequency of 94 GHz has been used for development work; this frequency is known as millimetre waves.

A short look at the history and principal of radar at this stage will help with an overall understanding. Radar was the name given during World War II to an electronic system by which radio waves were bounced off an aircraft in order to detect its presence and locate its position. The term is an acronym, made from the fuller term 'radio detection and ranging'. A large number of researchers helped to develop the devices and techniques of radar, but the development of the earliest practical radar system is usually credited to Sir Robert Watson-Watt.

The operation of a basic radar system is as follows: a radio transmitter generates radio waves, which are then radiated from an antenna, 'lighting up' the airspace with radio waves. A target, such as another vehicle that is in this space, scatters a small portion of the radio energy back to a receiving antenna. This weak signal is amplified by an electronic amplifier and displayed, often on a cathode ray tube. To determine its position, the distance (range) and bearing must be measured. Because radio waves travel at a known constant velocity, the speed of light, which is 3×10^8 m/s, the range may be found by measuring the time taken for a radio wave to travel from transmitter to obstacle and back to the receiver.

For example, if the range were 150 m, the time for the round trip would be:

$$t = \frac{2d}{C}$$

where t = time, d = distance to object, and C = speed of light.

In this example:

$$t = \frac{2 \times 150}{3 \times 10^8}$$

Relative closing speed can be calculated from the current vehicle speed. The radar is actually transmitted in the form of pulses. This is done by frequency modulating the signal, maybe using a triangular wave with a frequency of the order of 100 MHz: this can also be used to trigger a display and for calculation of distance.

The bearing, if required, is given by the relative position on the display device. Radar for use in a vehicle must fulfil the following general requirements.

- Range to be at least 300 m in bad weather. This gives about 7 seconds warning at 160 k/h (100 mile/h).
- Objects greater than 0.1 m^2 must be detected.
- Data update greater than one per second.
- Beam spread of about 15° horizontal and vertical.
- The driver's display should not intrude on concentration and only act as a warning.

The type of display or output that may be used on a motor vehicle will vary from an audible warning to a warning light or series of lights and possibly a display screen.

16.7.2 Tyre pressure warning

A glance at the instrument panel should be enough to tell the driver that the tyre pressures are all correct. Bosch has developed an electronic tyre pressure monitoring system. Each wheel has its own pilot lamp, which lights up if the pressure falls below a set value. Poorly inflated tyres cause loss of control and worse fuel consumption. The idea is to give the driver warning of reduced pressure – as an instant deflation is generally apparent to the driver!

There are three basic components to the system. Mounted in the wheel rim is a pressure operated switch, the contacts of which close when pressure falls. This is recognized by a high frequency sender which the switch passes but does not contact as the wheel rotates. The high frequency sender transmits an appropriate pulse to the electronic evaluator. If the pressure drops below the set value then the switch contacts open, causing the high frequency sender to interrupt its stream of pulses to the evaluation circuit and the warning lamp comes on. The system measures the tyre pressure with an accuracy of ±50 mbar. The design of the switch is such that changes in temperature of the air in the tyre will not cause false readings.

If the tyre pressure warning system is used in conjunction with wheels fitted with 'limp-home' tyres, it will provide a reminder that the limp-home mode is in use.

Bosch is also developing another tyre pressure warning system using active analogue sensors in the tyre and wireless transmission of the signal from the wheel to the body. The advantage is that absolute values of pressure and temperature are measured continuously, even when the car is at rest. Values such as vehicle speed and load are also included in the calculation.

16.7.3 Noise control

The principle of adaptive noise control is that of using sound, which is identical and 180° out of phase, or in anti-phase, to cancel out the original source of noise. Figure 16.40 shows three signals, the original noise, the anti-phase cancelling waveform and the residual noise.

A microphone picks up the original noise. It is then inverted and amplified, and then replayed by a suitably positioned speaker. This effectively cancels out the noise. Whilst the theory is relatively simple, until recently it has not been particularly suitable for motor vehicle use. This is due to the wide range of noise frequencies produced, and the fast response time, which is needed to give acceptable results. Low frequency noise (< 200 Hz), causes 'boom' in a vehicle, this is very difficult to reduce by conventional methods.

Much development time and money has been spent on reducing cabin noise levels. This can range from simple sound-deadening material to a special design of engine mountings, exhaust systems and using balance shafts on the engine. Even so, the demand still exists to reduce noise further and this is becoming ever more expensive.

Most vehicles today are susceptible to some low frequency boom in the passenger compartment, even when a large amount of sound deadening is used. The trend to produce lighter vehicles using thinner grade metal further exacerbates the problem. Conventional techniques solve the problem at certain frequencies, not all across the range.

To apply the adaptive noise control system to a car required the development of high-speed digital signal processors as well as a detailed understanding of noise generation dynamics in the vehicle. A typical four-cylinder engine running between 600 and 6000 rev/min has a firing frequency of about 20–200 Hz. There are several critical speeds at which the vehicle will display unpleasant boom. Low-profile tyres and harder suspension also generate considerable low frequency noise.

Lotus Engineering has developed a system which uses eight microphones embedded in the vehicle headlining to sample the noise. A digital signal processor measures the average sound pressure energy across the cabin and adjusts the phase and amplitude of the anti-noise signals. These are played through the in-car speaker system until, by measuring the error signal from the microphones, a minimum noise is achieved. The maximum active noise control can be achieved in about 70 ms. A quality loudspeaker system is needed which must be able to produce up to 40 W RMS per channel. This is not uncommon on many ICE systems. Figure 16.41 shows a typical layout of an adaptive noise control system. The greatest improvements are gained in small vehicles where the perceived reduction is as much as 80%.

16.8 Case studies

16.8.1 Volvo safety

The following information is extracted from information relating to features on the Volvo S80. It shows the clear commitment of manufacturers in general, and perhaps Volvo in particular, to safety developments.

Safety is very much part of Volvo's soul and, as a result, it is always present (claims the company). It is an integral part of the first design work and a vital part at every stage of the development process. Active safety can be summarized as active accident avoidance, passive safety can be summed up in three words: passenger protection priority. One of Volvo's prerequisites is that every new Volvo has to be safer than the previous one. Figure 16.42 shows the Volvo S80 air bags.

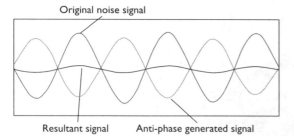

Original noise signal

Resultant signal   Anti-phase generated signal

Figure 16.40 Three signals; the original noise, the anti-phase cancelling waveform and the residual noise

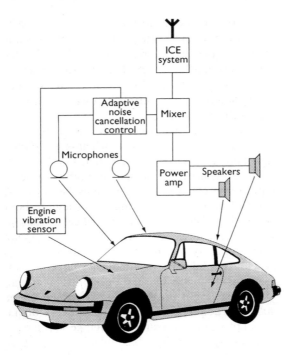

Figure 16.41 Layout of an adaptive noise control system and how it could be fitted

When it comes to the Volvo S80, this is very much the case. One of the objectives when designing the Volvo S80 was to strengthen further Volvo's position as the world leader in the field of passenger protection. This aim has been realized. With two new and important technical features, the level of passenger protection has taken yet another step forward. It would perhaps be no exaggeration to say that the Volvo S80 is the safest passenger car on the market at present. Although safety developments in the automotive industry have progressed by leaps and bounds in recent years, there is still some truth in the statement that a large car is safer than a small one. Size is related to safety. This is part of the laws of nature. A larger, heavier car suffers the least damage in a collision with a smaller, lighter car, thus providing better protection for

its occupants. Crumple zones and energy absorption are two vital parameters that can be more effectively designed if there is more space. A well-designed, rigid body structure is the perfect base on which to build.

Volvo has always claimed that the most important protective feature in a car is the seat-belt. The Volvo S80 has three-point belts on all five seating positions; all equipped with pyrotechnical pretensioners. The pretensioners automatically tighten the belts in a crash, eliminating the slack, which is normal in a belt. The front seat-belts are also equipped with force limiters, which control and regulate the roll speed of the belt webbing and provide more gentle restraint. The front seat-belts also have automatic belt height adjusters for optimum belt geometry. The belt system has been integrated with the airbag systems as these systems interact.

The passenger airbag is invisibly stored under the upper part of the dashboard and is designed to activate in a 'friendly' way in order to protect the passenger rather than being a risk. A belt sensor indicates whether or not the front seat passenger is wearing a seat-belt and adapts the airbag trigger level accordingly. This means that more crash energy is needed to trigger the bag when the passenger is wearing a seat-belt than when he is not.

In 1997, the Volvo Car Corporation presented the Whiplash Protection Study, WHIPS, which was an R&D project designed to produce a seat that would reduce the risk of whiplash injuries in rear-end collisions (Figure 16.43). Although they are most frequently caused at low speeds in relatively minor accidents, whiplash injuries are extremely painful, both physically and mentally, for the people who incur them, as well as being difficult to detect and define. They are also perhaps the single most expensive injury in insurance terms.

Since rear-end collisions often occur in city traffic, the WHIPS system is optimized to be

Figure 16.42 Volvo S80 airbags

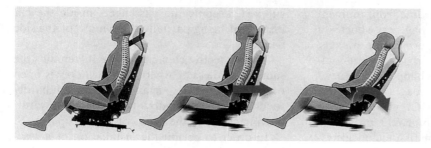

Figure 16.43 Volvo S80 'WHIPS'

most effective at speeds ranging from 15 to 30 km/h. The system consists of two elements. The first element of the WHIPS system is a brand new device that adjusts the angle between the seat cushion and the backrest. The system is activated in two phases.

1. The backrest of the seat is allowed to move backwards together with the occupant, reducing G-forces.
2. The angle of the backrest folds backwards by up to 15°, effectively catching the body and preventing a catapult effect.

The second element of WHIPS are six modified springs in the backrest with limiters that provide even support of the spine when pressed into the seat. The fixed head restraint, which remains close to the head, minimizes head movement and reduces forces on the neck. Consequently, the entire back is pressed against the backrest in a controlled manner. Tests conducted by Volvo during the development of the system reveal that the WHIPS system can reduce the acceleration forces in the neck by some 50%.

Passenger protection in side impacts is perhaps the most difficult area in terms of safety development, because of the lack of space and the minimal crumple zone, only 25–30 cm. Passengers sit very close to the point of impact. This must therefore be compensated for one way or another. The Side Impact Protection System (SIPS) structure has been extensively upgraded and its interacting components consist of the energy-absorbing elements in bottom rails, pillars, cross-members, roof and seats, plus energy-absorbing materials in the doors. This has been supplemented with more, further improved, padding in all the roof pillars and along the edges of the headliner. This material feels hard when it is touched, but it yields in a 'friendly' manner and absorbs energy when it is hit in an impact. The second step in the continued development of the SIPS system was the introduction of the SIPS bags in 1994 – now a standard item on all Volvo cars.

The Volvo side airbag (Figure 16.44) is located in the outer part of the backrest and is therefore always in the optimum protective position in relation to the occupant. The SIPS further reduces the risk of severe chest and pelvic injuries, as its function is to keep the occupant away from the side of the car. The side airbags are trigged by electronic sensors, one in the B pillar and one behind the rear door. Their

Figure 16.44 Volvo new SIPS airbag

position makes the reaction time from moment of impact to triggering the bag very short – a factor that is of vital importance in side impacts. However, padding and side airbags cannot completely make up for what can happen to the head when the car is hit from the side.

The Inflatable Curtain (IC) was presented together with WHIPS as an R&D project in 1997, and is claimed to be the first technical system for this type of protection. The purpose of the system (Figure 16.45) is to reduce further injuries in a side impact by protecting the head and neck of the occupants both in the front and rear seats. The curtains, one on each side, are woven in one piece and hidden inside the roof lining. They cover the upper part of the interior, from the 'A' pillar to the rear side pillar. The same sensors as used with the SIPS bags activate the IC. They are 'slave' sensors to a central sensor, which determines where the impact is and which bag should be triggered in order to protect the occupants.

If only the rear sensor is affected, the IC is activated but not the SIPS bag. The curtain is filled within 2.5 ms and stays inflated about three seconds in order to provide maximum protection in complicated collisions. The ducts do not cover the entire surface of the curtain. Instead, they are concentrated in the areas that are most likely to be hit by the occupants' heads. As a result, the need for gas is limited and the activation time is minimal. The ducts act as controlled head restraints and prevent the head from hitting the inside of the car. The curtain also prevents the head from impacting on collision obstacles, such as lampposts and similar objects. The size of the curtain also provides

support, keeping the passengers inside the car instead of being partially thrown out of the side windows.

The protective capacity of the IC remains the same, regardless of whether the window is open or closed. When the curtain is activated, it hardly touches the side window but expands inwards, moving closer to the heads of the occupants.

In order to permit the installation of a rear-facing child seat in the front passenger position, the passenger airbag can be switched on and off using a switch. This switch, which can be fitted only by a Volvo dealer, works via the ignition key. When the ignition is turned on, an indicator lamp on the switch comes on and shows whether or not the passenger airbag is activated. If the switch suffers electronic failure, the supplementary restraint system (SRS) lamp comes on, just as it does if any other defect occurs in the SRS system.

16.8.2 Rover electric windows

The circuit of the electric window system used by some Rover vehicles is shown in Figure 16.46. The windows will only operate when the ignition is switched on. When the ignition is switched on, the window lift relay is energized by the supply from fuse 18 in the passenger compartment fuse-box on the LG wire, which passes to earth on a B wire. With the relay energized, the battery supply from fusible link 4 on the N wire feeds the four window lift fuses on an N/U wire.

The driver's window can only be operated from the switchblock on the driver's door, which is supplied from fuse 30 in satellite fuse block 2, on an S/G wire. When the 'up' switch is pressed,

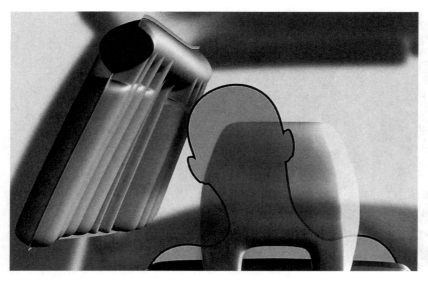

Figure 16.45 Volvo S80 inflatable curtain

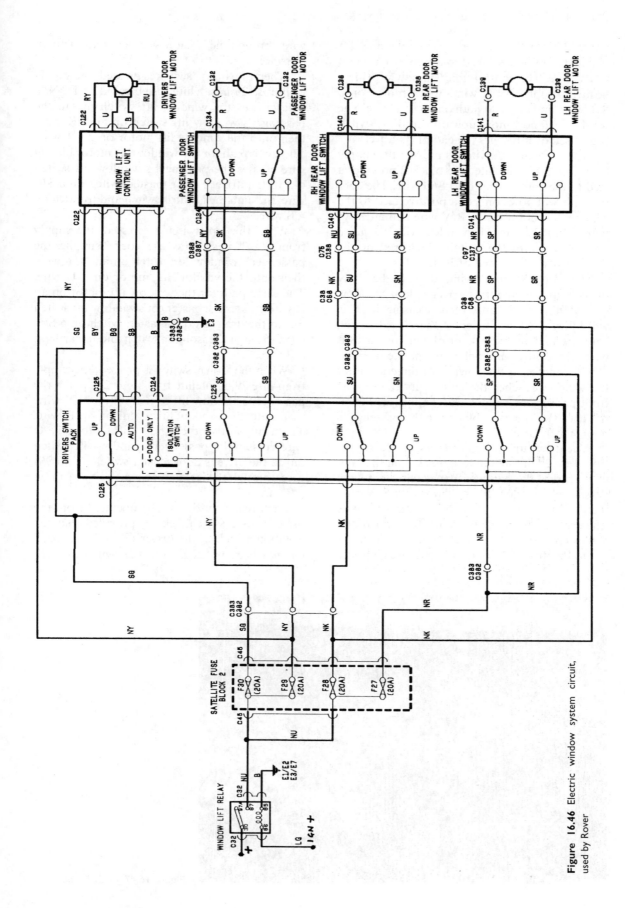

Figure 16.46 Electric window system circuit, used by Rover

the feed from the fuse crosses the window lift switch and provides a feed to the control unit on a B/Y wire. The control unit will now provide a positive supply to the window lift motor on a R/U wire and an earth path on an R/Y wire. The window will now move upwards until the switch is released or it reaches the end of its travel.

When the 'down' switch is pressed the supply from fuse 30 in satellite fuse block 2 provides a feed to the control unit on a S/G wire. The control unit will now connect a positive feed to the window lift motor on an R/Y wire and an earth path on a R/U wire. The window will now move downwards until the switch is released or the window reaches the end of its travel.

The driver's door window may be fully opened by moving the driver's door window switch fully downwards then releasing it. This will allow a supply to cross the closed switch contacts and feed the control unit on an S/B wire. The control unit will now operate the window lift motor in the downward direction until the window reaches the end of its travel. The front passenger's window can be operated from the driver's door switchback or the passenger's door switchback.

When the 'up' switch is pressed, the supply from fuse 29 in satellite fuse block 2 on the N/Y wire crosses the window lift switch out to the passenger's window lift switch on an S/B wire, then onto the window lift motor on a U wire. The earth path for the window lift motor on an R wire crosses the passenger's window lift switch out to the driver's door master switch on an S/K wire, through the isolator switch and to earth on a B wire.

When the 'down' switch is pressed, the supply from fuse 29 in satellite fuse block 2 on an N/Y wire crosses the window lift switch out to the passenger's window lift switch on an S/K wire, then onto the window lift motor on an R wire. The earth path for the window lift motor on a U wire crosses the passenger's window lift switch out to the driver's door master switch on a S/B wire, through the isolator switch and to earth on a B wire.

When the 'up' switch is pressed, the supply from fuse 29 in satellite fuse block 2 supplies the passenger's window lift switch on an N/Y wire, then onto the window lift motor on a U wire. The earth path for the window lift motor on an R wire crosses the passenger's window lift switch out to the driver's door master switch on a S/K wire, through the isolator switch and to earth on a B wire.

When the 'down' switch is pressed, the supply from fuse 29 in satellite fuse block 2 supplies the passenger's window lift switch on an N/Y wire, then onto the window lift motor on an R wire. The earth path for the window lift motor on a U wire crosses the passenger's window lift switch out to the driver's door master switch on an S/B wire, through the isolator switch and to earth on a B wire.

Each rear window can be operated from the driver's door switchback or, provided that the isolation switch in the driver's door switchback has not been pressed, from the switch on each

Figure 16.47 Jaguar S-type

rear door. The operation of the rear windows is similar in operation to the front passenger's window.

16.8.3 Jaguar 'S' type audio, communications and telematics

The following information is extracted from information relating to features on the Jaguar 'S' (Figure 16.47). It shows the general trend and developments relating to 'communication' systems.

For the first time on a production car (Jaguar claims), optional voice-activated controls for the audio (radio/cassette/CD), phone and climate control systems, responding to the spoken instructions of the driver, provide safe, hands-free operation. The system responds to a wide diversity of English and North American accents, but also provides for training to recognize a specific voice.

A first for Jaguar is the optional, fully integrated onboard satellite navigation system using multi-lingual, digitized map data on CD-ROM. The system can point out useful landmarks and points of interest and links with the UK's Trafficmaster system to provide real-time data on traffic delays.

The 175 W, 12-speaker, premium sound system, features two active 'centre fill' speakers, an active sub-woofer enclosure and 6-disc CD autochanger. Digital sound processing, working with Dolby, provides special audio effects and compensates for the number of vehicle occupants.

The premium specification Motorola portable GSM phone is a factory fit option, combining the advantages of vehicle integration, safety, convenience and performance with the versatility of a pocket phone.

16.9 Diagnosing comfort and safety system faults

16.9.1 Introduction

As with all systems the six stages of fault-finding should be followed.

1. Verify the fault.
2. Collect further information.
3. Evaluate the evidence.
4. Carry out further tests in a logical sequence.
5. Rectify the problem.
6. Check all systems.

The procedure outlined in the next section is related primarily to stage 4 of the process. Table 16.1 lists just a few faults as examples for this chapter.

16.9.2 Testing procedure

The following procedure is very generic but with a little adaptation can be applied to any

Table 16.1 Common symptoms and possible faults of comfort systems

Symptom	Possible fault
Radio interference	• Tracking HT components. • Static build-up on isolated body panels. • High resistance or open circuit aerial earth. • Suppression device open circuit.
Electric windows not operating	If all windows not operating: • Open circuit in main supply. • Main fuse blown. • Relay coil or contacts open circuit or high resistance. If one window is not operating: • Fuse blown. • Control switch open circuit. • Motor seized or open circuit. • Back-off safety circuit signal incorrect.
Cruise control will not set	• Brake switch sticking on. • Safety valve/circuit fault. • Diaphragm holed. • Actuating motor open circuit or seized. • Steering wheel slip ring open circuit. • Supply/earth/fuse open circuit.

electrical system. Refer to the manufacturer's recommendations if in any doubt. The process of checking any system circuit is broadly as follows.

1. Hand and eye checks (loose wires, loose switches and other obvious faults) – all connections clean and tight.
2. Check battery (see Chapter 5) – must be 70% charged.
3. Check motor/solenoid/linkage/bulbs/unit – visual check.
4. Fuse continuity – (do not trust your eyes) voltage at both sides with a meter or a test lamp.
5. If used, does the relay click (if yes, jump to stage 8) – this means the relay has operated, but it is not necessarily making contact.
6. Supply to switch – battery volts.
7. Supply from the switch – battery volts.
8. Supplies to relay – battery volts.
9. Feed out of the relay – battery volts.
10. Voltage supply to the motor – within 0.5 V of the battery.
11. Earth circuit (continuity or voltage) – $0\,\Omega$ or $0\,V$.

16.9.3 ECU auto-diagnostic function

Many ECUs are equipped to advise the driver of a fault in the system and to aid the repairer in detection of the problem. The detected fault is first notified to the driver by a dashboard warning light. A code giving the details is held in RAM within the ECU. The repairer, as an aid to fault-finding, can read this fault code.

Each fault detected is memorized as a numerical code and can only be erased by a voluntary action. Only serious faults will light the lamp but minor faults are still recorded in memory. The faults are memorized in the order of occurrence.

Faults can be read as two-digit numbers from the flashing warning light by shorting a diagnostic wire to earth for more than 2.5 seconds but less than 10 seconds. Earthing this wire for more than 10 seconds will erase the fault memory as does removing the ECU constant battery supply. Earthing a wire to read fault codes should only be carried out in accordance with the manufacturer's recommendations. The same coded signals can be more easily read on many after-sales service testers. On some systems it is not possible to read the fault codes without a code reader.

16.9.4 Fault-finding by luck

If four electric windows stopped working at the same time, it would be very unlikely that all four motors had burned out. On the other hand, if just one electric window stopped working, then it may be reasonable to suspect the motor. It is this type of reasoning that is necessary when fault-finding. However, be warned, it is theoretically possible for four motors to burn out apparently all at the same time!

Using this 'playing the odds' technique can save time when tracing a fault in a vehicle system. For example, if both stop lights do not work and everything else on the vehicle is OK, I would suspect the switch (stages 1 to 3 of the normal process). At this stage though, the fault could be anywhere – even two or three blown bulbs. Nonetheless a quick test at the switch with a voltmeter would prove the point. Now, let us assume the switch is OK and it produces an output when the brake pedal is pushed down. Testing the length of wire from the front to the back of the vehicle further illustrates how 'luck' comes into play.

Figure 16.48 represents the main supply wire from the brake switch to the point where the wire 'divides' to each individual stop light (the odds say the fault must be in this wire). For the purpose of this illustration we will assume the open circuit is just before point 'I'. The procedure continues in one of the two following ways.

One

- Guess that the fault is in the first half and test at point F.
- We were wrong! Guess that the fault is in the first half of the second half and test at point I.
- We were right! Check at H and we have the fault … On test number *three*.

Two

- Test from A to K in a logical sequence of tests.
- We would find the fault … On test number *nine*.

You may choose which method you prefer!

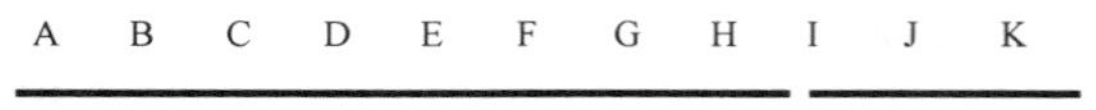

A   B   C   D   E   F   G   H   I   J   K

Figure 16.48 Representation of a wire with an open circuit between 'H' and 'I'

16.10 Advanced comfort and safety systems technology

16.10.1 Cruise control and system response

Figure 16.49 shows a block diagram of a cruise control ECU. Many cruise control systems work by the proportional-integral control technique. Proportional control means that an error signal is developed via the feedback loop, which is proportional to the difference between the required and actual outputs. The final output of a cruise control system is the vehicle speed but this depends on the throttle position, which is controlled by the actuator. The system electronics must take into account the lag between throttle movement and the required change in vehicle speed.

If the system overreacts, then the vehicle speed would become too high and then an over-reaction would cause the speed to become too low and so on. In other words, the system is not damped correctly (under damped) and will oscillate, much like a suspension spring without a damper. Proportional control alone is prone to this problem because of steady-state errors in the system. To improve on this, good system design will also include integral control. Thus, the final signal will be the sum of proportional and integral control signals. An integral controller produces a signal, which is a ramp, increasing or decreasing, proportional to the original error signal.

The use of integral control causes the final error signal to tend towards zero. The combination therefore of these two forms of control in the weighting given to each determines the damping factor of the control electronics. Figure 16.50 shows the effect on vehicle speed of different damping factors. These four responses are well known in engineering and electronics and can be modelled by mathematics to calculate the response of a system.

The above technique can be based on analogue or digital electronics. The principal is much the same in that for any system the proportional and integral control can be used. The theoretical values can be calculated prior to circuit design as follows:

$$G_i = \omega_n^2 M$$
$$G_p = (2d\omega_n M) - C$$

where G_i = integral gain, G_p = proportional gain, ω_n = natural frequency of the system $(2\pi f_n)$, M = mass of the vehicle, C = experimentally determined frictional factor (mechanical), and d = damping coefficient.

16.10.2 Radio suppresser calculations

Capacitors and inductors are used to act as filters. This is achieved by using the changing value of 'resistance' to alternating signals as the frequency increases. The correct term for this resistance is either capacitive or inductive reactance. These can be calculated as follows:

$$X_C = \frac{1}{2\pi f C}$$
$$X_L = 2\pi f L$$

where X_c = capacitive reactance (ohms), X_L = inductive reactance (ohms), C = capacitance

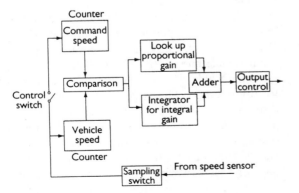

Figure 16.49 Cruise control system – detailed block diagram

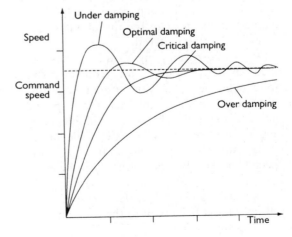

Figure 16.50 Damping factors

(farads), L = inductance (henrys), f = frequency of the interference (hertz).

Using the above formulae gives the following results with a 0.1 mF capacitor and a 300 mH inductor, first at 50 Hz and then at 1 MHz.

Frequency	100 Hz	1 MHz
Capacitive reactance	15.5 kΩ	1.6 Ω
Inductive reactance	0.18 Ω	1.9 kΩ

By choosing suitable values of a capacitor in parallel and or an inductor in series it is possible to filter out unwanted signals of certain frequencies. To home in on a specific or resonant frequency a combination of a capacitor and inductor can be used. The resonant frequency of this combination can be calculated:

$$f = \frac{1}{2\pi \sqrt{LC}}$$

When the range of the interference frequency is known, suitable values of components can be determined to filter out its effect.

16.11 New developments in comfort and safety systems

16.11.1 Noise control developments

A hydraulic engine mount, which is electronically controlled in response to the engine vibration, can significantly reduce noise. Some manufacturers, however, are now using a much simpler version, which can switch between hard and soft settings. A system developed by Lotus is claimed to be as effective as about 45 kg of sound deadening material.

An exhaust company, 'Walker', has developed an active muffler for reducing exhaust noise. The heart of this system is a digital processor. Two inputs are used, a microphone to measure the noise from the tailpipe and an engine speed sensor. The system calculates the correct anti-noise and delivers this by means of special speaker drivers mounted on the exhaust system. The residual noise is measured and adjustments can be made. Because the system is self-learning it will adapt to the changing noises of an ageing engine.

The active muffler allows straight gas flow from the exhaust after the catalytic converter. This allows improved engine performance that can mean less fuel is used. An average reduction in fuel consumption of 5% is possible. Future EC directives are expected relating to exhaust noise, which are currently set at 77 dB (A) in Germany. Larger mufflers will be needed to comply, which means this system may well become quite popular.

16.11.2 Alarming developments!

Professional car thieves will always find ways around the latest alarm systems. However, the vehicle manufacturers strive to stay one jump ahead. Tracking devices can be built in to an unknown part of the vehicle's chassis. This can be activated in the event of the car being stolen, allowing the police to trace the vehicle. A system popular in the UK is 'Tracker' and this works as follows.

1. The car is stolen.
2. Depending on the product, the owner tells 'Tracker' or 'Tracker' tell the customer.
3. The 'Tracker' unit in the car is activated by powerful transmitters.
4. Police with tracking computers detect the silent homing signal.
5. The police recover the car.

The 'Tracker' unit is a radio transponder. When the vehicle is reported stolen the police are informed and the 'Tracker' unit is activated. The unit then broadcasts a unique reply code, which can be detected and decoded by police tracking computers, which are fitted in police cars, helicopters and fixed land sites. The police then track the vehicle, taking appropriate action. Figure 16.51 shows a stolen car recovery in action.

A 'Tracker' unit can be fitted to any self-propelled road vehicle that has a suitable location where the unit can be hidden. The system currently only operates in mainland Great Britain. It constantly draws power from the main vehicle battery but if this is disconnected, a re-chargeable back-up battery provides power for up to 2 days. The presence of the unit is not disclosed to the thief, which means there is a greater likelihood of rapid recovery and minimal damage. The unit is not transferable from one vehicle to another but the new owner need only pay for the network subscription. Most insurance companies offer additional discounts of up to 20% if this system is fitted.

Figure 16.51 Since 1993 'Tracker' has helped police forces throughout the UK recover more than £35 million worth of stolen vehicles

16.11.3 ICE warning

The following is a description of a Blaupunkt 'New York RDA 127' ICE system.

This is a purely high-end system thanks to DSA, which is an automatic calibration program for linear frequency response in the car and the 'psycho-acoustic masking' of driving noises (DNC). An integrated high-end CD drive is included with an optional opto-changer.

The 'Sub-Out' and the many equalizer functions demonstrate its serious claim of sophistication among the high-end car hi-fi systems of today. The whole spectrum of new car audio technology is covered along with some fascinating options:

- FM, MW, LW
- TIM (Traffic Memo)
- Dual-tuner RDS
- RDS-EON-PTY
- Radiotext
- Travelstore
- CD 1 bit/8× Over-sampling
- Disc Management System (DMS)
- Digital Signal Adaptation (DSA)
- Dynamic Noise Covering (DNC)
- Self-adjusting equalizer
- 4 × 23 watts RMS power
- 4 × 35 watts maximum power
- Digital-in
- Four-channel pre-amp output
- Sub-Out

This type of mobile multimedia seems to have everything! In spite of high-end performance, it all remains uncomplicated. Good sized, easily readable displays, menu-controlled operator prompting and an ergonomic, award-winning design make an important contribution to driving pleasure.

High-end sound technology automatically perfects the acoustics in the vehicle interior, masks undesirable driving noises and uses incredible dynamics and 'spatiality' to make listening to the audio system on the road a real experience.

16.11.4 Intelligent airbag sensing system

Bosch has developed an 'Intelligent Airbag Sensing System' which can determine the right reaction for a specific accident situation. The system can control a one- or two-stage airbag inflation process via a two-stage gas generator. Acting on signals from vehicle acceleration and belt buckle sensors, which vary according to the severity of the accident, the gas generator receives different control pulses, firing off one airbag stage (de-powering), both stages (full inflation), or staged inflation with a time interval.

Future developments will lead to capabilities for multistage inflation or a controllable sequence of inflation following a pattern determined by the type of accident and the position of the vehicle occupants. The introduction of an automotive occupancy sensing (AOS) unit that uses ultrasonic and infrared sensors will provide further enhancements. This additional module will detect seat and child occupancy and will be

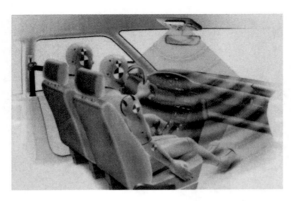

Figure 16.52 Intelligent airbag system

capable of assessing whether a passenger is in a particular position, such as feet on the dashboard!

Bosch hopes that the latest radar technology will assist the design of a pre-crash sensor capable of detecting an estimated impact speed prior to collision, and activating individual restraint systems, such as seatbelt pre-tensioners. Or, if necessary, all available restraint systems. Figure 16.52 shows a representation of this system.

16.12 Self-assessment

16.12.1 Questions

1. State what is meant by active and passive safety.
2. Draw a simple motor reverse circuit and explain its operation.
3. Describe briefly six features of a high-end ICE system.
4. State five sources of radio interference.
5. Explain why fault-finding sometimes involves 'playing the odds'.
6. Describe the operating sequence of a driver's air bag.
7. Define 'Latching relay'.
8. Describe, with the aid of a block diagram, the operation of a cruise control system.
9. State four advantages of an intelligent air bag.
10. Explain the key features of a top-end alarm system.

16.12.2 Assignment

Investigate the development of the 'Auto PC' with particular reference to:

- Digital map databases.
- Vehicle diagnostics programs.

Produce a report on some of the issues connected with these developments. A good technique for starting on this type of assignment is to ask the question: 'Who gains and who loses?'

Consider also issues of updating and cost.

17

Electric vehicles

17.1 Electric traction

17.1.1 Introduction

The pressure to produce a non-fossil-fuel vehicle is increasing. Indeed, recent legislation has set the requirement for the production of zero emission vehicles (ZEVs). The development of the electric vehicle is still in a state of flux (pun intended), but some major manufacturers now have electric vehicles available for sale to the general public.

In 1990, General Motors announced that its EV, the 'Impact', could accelerate to 100 km/h in just 8 s, had a top speed of 160 km/h (100 mile/h) and had a range of 240 km between charges. Running costs were about double the fossil-fuel equivalent but this cost was falling. The car was a totally new design with drag-reducing tyres and brakes which, when engaged, act as generators (regenerative braking). The car was powered by a 397 kg array of advanced gel electrolyte lead-acid batteries (32 at 10 V) and two small AC electric motors to drive the front wheels. The recharging time was about 2 hours but this could be reduced to 1 hour in an emergency. This was very impressive, but things have moved on still.

The following sections look at some of the issues in more detail, but the subject of 'electric vehicles' could (and does) fill many books in its own right. This chapter is presented as an introduction to a technology that is certain to become a major part of the general motor trade. The 'Case Studies' section looks, amongst other things, at two EVs in current use.

17.1.2 Electric drive vehicle layout

Figure 17.1 shows the general layout in block diagram form of an electric vehicle (EV). Note that the drive batteries are often a few hundred volts, so a lower 12/24 V system is still required for 'normal' lighting and other systems. Some of the components shown are optional.

17.1.3 EV batteries

A number of options are available when designing the electric car but, at the risk of oversimplification, the most important choice is the type of batteries.

Table 17.1 summarizes the current choice relating to batteries and will allow some comparisons to be made. Further details relating to some of these and other battery developments can be found in Chapter 5.

Currently the main advantage of lead-acid batteries is the existing mature technology, which is accepted by the motor industry. The disadvantage is their relatively low specific power. The sodium-sulphur battery is a good contender but has a far greater cost and new technologies are needed to cope with the operating conditions such as the high temperatures.

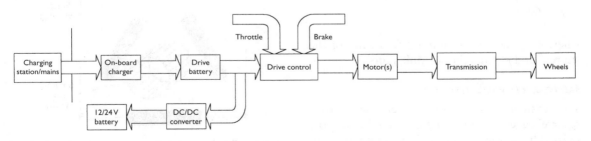

Figure 17.1 General electric vehicle (EV) layout

Table 17.1 Factors relating to batteries

Battery	Symbols	Specific energy, Wh/kg (Watt hours per kilogram)	Relative cost per kW/h (average est. in 1994)	Operating temperature range, °C	Cycle life, 80% depth of discharge (DOD)
Lead-acid	Pb-Acid	27–33	1	0–60	450–600
Nickel-cadmium	NiCd	35–64	10	–20–60	2000–500
Nickel-metal-hydride	NiMH	50–51	8	–20–60	500
Nickel-iron	NiFe	51	8	–20–60	1000
Zinc-bromine	ZnBr	56	5		500
Nickel-zinc	NiZn	73–79	3.5	–20–60	600
Lithium-ion/polymer	Li-ion	90	–	–20–60	1200–2000
Sodium-sulphur	NaS	79–81	6.5	300–380	1000
Silver-zinc	AgZn	117–139	15		100
Zinc air	Zn-Air	144–161	15	–20–40	150

Significant developments are occurring in relation to lithium-based batteries. However, most batteries in general use are lead-acid or nickel-based.

17.1.4 Drive motors

There are several choices of the type of drive motor. The basic choice is between an AC and a DC motor. The AC motor offers many control advantages but requires the DC produced by the batteries to be converted using an inverter. A DC shunt wound motor rated at about 50 kW is a popular choice for the smaller vehicles but AC motors are likely to become the most popular. The drive motors can be classed as AC or DC but it becomes difficult to describe the distinctions between an AC motor and a brushless DC motor.

AC motors

In general, all AC motors work on the same principle. A three-phase winding is distributed round a laminated stator and sets up a rotating magnetic field that the rotor 'follows'. The speed of this rotating field and hence the rotor can be calculated:

$$n = 60\,\frac{f}{p}$$

where n = speed in rev/min; f = frequency of the supply; and p = number of pole pairs.

Asynchronous motor

The asynchronous motor is often used with a squirrel cage rotor made up of a number of pole pairs. The stator is usually three-phase and can be star or delta wound. This is shown in Figure 17.2. The rotating magnetic field in the stator induces an EMF in the rotor which, because it is a complete circuit, causes current to flow. This creates magnetism, which reacts to the original field caused by the stator, and hence the rotor rotates. The amount of slip (difference in rotor and field speed) is about 5% when the motor is at its most efficient.

Synchronous with permanent excitation

This motor has a wound rotor known as the inductor, which is a winding magnetized by a DC supply, via two slip rings. The magnetism 'locks on' to the rotating magnetic field and produces a constant torque. If the speed is less than n (see above), fluctuating torque occurs and high current can flow. This motor needs special arrangements for starting rotation. An advan-

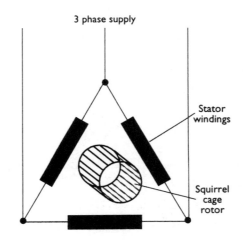

Figure 17.2 An asynchronous motor is used with a squirrel cage rotor made up of a number of pole pairs

tage, however, is that it makes an ideal generator. The normal vehicle alternator is very similar. Figure 17.3. shows a representation of the synchronous motor.

EC motors (electronically controlled)

The EC motor is, in effect, half way between an AC and a DC motor. Figure 17.4 shows a representation of this system. Its principle is very similar to the synchronous motor above except the rotor contains permanent magnets and hence no slip rings. It is sometimes known as a brushless motor. The rotor operates a sensor, which provides feedback to the control and power electronics. This control system produces a rotating field, the frequency of which determines motor speed. When used as a drive motor, a gearbox is

needed to ensure sufficient speed of the motor is maintained because of its particular torque characteristics. Some schools of thought suggest that if the motor is supplied with square-wave pulses it is DC, and if supplied with sine wave pulses then it is AC. This leaves a problem describing motors supplied with trapezoidal signals!

DC motor – series wound

The DC motor is a well proven device and has been used for many years on electric vehicles such as milk floats and fork lift trucks. Its main disadvantage is that the high current has to flow through the brushes and commutator.

The DC series wound motor has well known properties of high torque at low speeds. Figure 17.5 shows how a series wound motor can be controlled using a thyristor and also provide simple regenerative braking.

DC motor – separately excited shunt wound

The fields can be controlled either by adding a resistance or using chopper control in order to vary the speed. Start-up torque can be a problem but, with a suitable controller, can be overcome. This motor is also suitable for regenerative braking by increasing field strength at the appropriate time. Some EV drive systems only vary the field power for normal driving and this can be a problem at slow speeds due to high current.

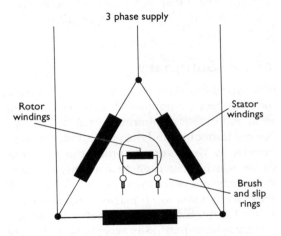

Figure 17.3 Representation of the synchronous motor

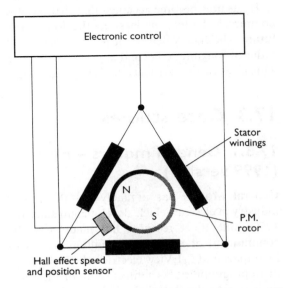

Figure 17.4 The EC motor is, in effect, halfway between an AC and a DC motor

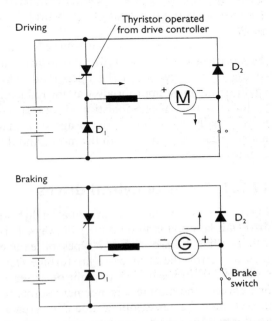

Figure 17.5 A series wound motor can be controlled using a thyristor and can also provide simple regenerative braking

17.1.5 EV summary

The concept of the electric vehicle is not new, the essential battery technology was developed in the late 19th century and many such cars were being manufactured by the year 1900. Although some models achieved high speeds at that time, the electric car was generally slow and expensive to operate. Its range was also limited by its dependence on facilities to recharge the battery. Many of these problems have been overcome, but not all of them. Cost is still an issue, but 'cost' is a relative value and when the consequences of pollution are considered the 'cost' may not be as high as it appears.

Although advances in battery technology have increased the range of the EV, the maximum cruising speed is also limited, as is the number of accessories that can be placed on the car. On the other hand, the electric car is expected to be mechanically more dependable and durable than its fossil-fuelled equivalent.

17.2 Hybrid vehicles

17.2.1 Introduction

The concept of a combined power source vehicle is simple. Internal combustion (IC) engines produce dangerous emissions and have poor efficiency at part load. Electric drives produce '*no*' emissions but have a limited range. The solution is to combine the best aspects and minimize the worst. Such is the principal of the hybrid drive system.

One way of using this type of vehicle is to use the electric drive in slow traffic and towns, and to use the IC engine on the open road. This could be the most appropriate way for reducing pollution in the towns. Sophisticated control systems actually allow even better usage such that under certain conditions both the motor and the engine can be used.

17.2.2 Types of hybrid drives

Figure 17.6. shows how the principal of hybrid drive can be applied in a number of ways. It is also possible to use different types of engine such as petrol, diesel or even gas turbine. The layout of the drives can be thought of as series or parallel. The parallel arrangement seems to be proving to be more popular due to its greater flexibility. The series arrangement, however, allows the fossil-fuel engine to run at a constant

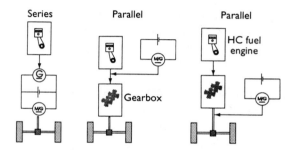

Figure 17.6 The hybrid drive principle can be applied in a number of ways

speed driving the generator. This makes use of the combustion engine in its own right more efficient, but the double energy conversion process (mechanical to electrical to mechanical) is less efficient than driving the vehicle transmission directly. The other advantage of series connection is that a transmission (gearbox) is not essential.

17.2.3 Summary

The hybrid or combined power source vehicle is likely to become popular. It appears to be the ideal and obvious compromise whilst drive and battery technology is developing. It may become possible in the future to produce a fossil-fuel engine which, when running at a constant speed, will produce a level of emissions that, if not zero, is very close to zero. This, when combined with a highly efficient electric motor and battery storage system, may be an acceptable ZEV (zero emission vehicle).

It has now become accepted that there will be no miracle battery, at least in the foreseeable future. The energy density of fossil fuels is of an order of magnitude beyond any type of battery. This gives further credence to the hybrid design.

17.3 Case studies

17.3.1 General motors – EV-1 (1999 version)

General Motors has arguably led the motor industry in electric vehicle development since the 1960s and, most recently, has made a major commitment of nearly half a billion dollars to its Impact and PrEView electric vehicle development programmes. As a direct result of these initiatives, GM developed the EV-1 electric car as the world's first specifically designed production electric vehicle, and became the first to go on sale (in the USA) in 1996. The EV-1 is shown in Figure 17.7.

Marketed as a stylish two-passenger coupe, the EV-1 has a drag coefficient of just 0.19 and an aluminium spaceframe chassis (40% lighter than steel) with composite body panels. Weighing just 1350 kg in total, the car has an electronically regulated top speed of 128 km/h (80 mile/h) – although a prototype EV-1 actually holds the world land-speed record for electric vehicles at 293 km/h (183 mile/h)! It can reach 96 km/h (60 mile/h) from a standing start in less than 9 s. The key to the success of the EV-1 is its electrical powertrain, based on a 103 kW (137 HP) three-phase AC induction motor with an integral, single-speed, dual-reduction gear-set driving the front wheels. The unit requires no routine maintenance for over 160 000 km (100 000 miles).

The battery pack uses 26, 12 V maintenance-free lead-acid batteries, giving a total voltage of 312 V and a range of 112 km (70 miles) per charge in urban conditions and 144 km (90 miles) on the open road. However, new nickel-metal-hydride (NiMH) batteries were phased into production during 1998, almost doubling the EV-1's range to 224 km (140 miles) in the city and 252 km (160 miles) on highways. An innovative regenerative braking system helps to extend that range still further by converting the energy used when braking back into electricity in order to recharge the battery pack partially.

Full recharging can be carried out safely in all weather conditions and takes 3–4 hours using a 220 V standard charger or 15 hours using the on-board 110 V convenience charger. Compared with normal fossil fuels, the lower cost of domestic electricity means operating costs are relatively low.

Regenerative braking is accomplished by using a blended combination of front hydraulic disc and rear electrically applied drum brakes and the electric propulsion motor. During braking, the electric motor generates electricity (regenerative) which is then used to partially recharge the battery pack.

The EV-1 comes with traction control, cruise control, anti-lock brakes, dual airbags, power windows, door locks and outside mirrors, AM/FM CD/cassette, tyre inflation monitor system and numerous other features.

17.3.2 Nissan – Altra

Nissan recently confirmed pricing for its Altra EV following the success of initial trials in the US during 1998. The Altra is an estate built for the US market and the EV version is the first zero-emission Nissan to go on sale outside Japan. The Altra EV is shown in Figure 17.8.

The Altra has a water-cooled, permanent magnet, synchronous electric motor, which is the first to use the highly efficient neodymium-iron-boron alloy (Nd-Fe-B). The alloy was discovered by accident, when an order for materials was misinterpreted! The Hitachi motor is one of the most powerful in the world, developing 62 kW (84 PS) and 159 Nm with a maximum rotor speed of 13 000 rev/min. Average motor speed is

Figure 17.7 General Motors EV-1

Figure 17.8 Nissan Altra EV

8000–9000 rev/min and the power-to-weight ratio of the 39 kg motor is 1.6 kW/kg – one of the best in the EV field.

A lithium-ion battery pack, developed by the Sony Corporation in a deal that is so far unique to Nissan, provides the power. It delivers a nominal output of 345 V from 12 modules of 8 cells, each producing 36 V when fully charged and 20 V when discharged. The gross weight of the battery pack is 350 kg and it has an energy density of 90 Wh/kg across the normal temperature range. Battery life is rated at 1200 cycles (to a 5% drop in efficiency) but Nissan claims batteries have endured in excess of 2000 cycles without significant further loss. The battery pack is mounted in a double-walled aluminium tray bolted to the centre of the platform between the front and rear axles beneath a flat floor; a dedicated ventilation system and fan keep it cool.

A vector controller developed by Nissan features twin fully redundant CPUs. The controller is water-cooled and has an input range of 216–400 V. Data are gathered on the state of charge, driving strategy, history, use of auxiliary systems and the function of the regenerative braking system to make accurate range predictions. It also performs relay control for battery cooling, provides the communication between the power supply and the Li-ion cell controller and determines the charging strategy based on the data it has collected.

Batteries are charged using an external inductive charger, which consists of a paddle inserted into a charging port in the front of the car. A fast charge takes 5 hours and provides a claimed range of 193 km, although on busy roads 135 km is more realistic.

The Altra has hydraulic power steering driven by an electric, rather than mechanical, hydraulic pump, which operates only when power assistance is required. A standard 12 V lead-acid battery, charged via a water-cooled DC/DC transformer from the main battery, powers auxiliary systems. Heating, ventilation and air conditioning consumes 50% of the energy of a conventional system in air conditioning mode and 66% when heating the cabin. R134a refrigerant serves both purposes and the system, like the power steering, uses an electric pressurization pump operating on demand.

The regenerative braking system operates on two levels.

- First stage – triggered when the driver lifts off the throttle and provides 'a similar feel to that of a conventional car,
- Second stage – is much more substantial and occurs when the driver applies moderate braking effort. The braking system itself has standard four-channel ABS.

Passive and active safety is unaffected by the extra weight compared with the standard vehicle; there are the standard front airbags, door beams and 8 km/h (5 mile/h) front and rear impact bumpers.

The instrument panel is digital with a large tachometer. Seven warning lamps alert the driver of 50 potentially dangerous situations with the battery or drive systems. Should critical problems arise, the systems can be shut down automatically to avoid damage.

17.3.3 'Nelco' – hybrid drive

A company called 'Nelco' has developed an interesting idea in hybrid EV drive technology.

The system is based around a drive package that could potentially be used to power existing internal combustion engined cars. The claimed performance is equivalent to a conventional front-wheel drive car, with two-thirds of the fuel consumption and just one-third of the noxious emissions. Figure 17.9. shows the parallel layout used for this system. It is hoped that the vehicle could have a range of 800 km (500 miles) and a top speed of 160 km/h (100 mile/h). The main components used are a deep discharge-tolerant lead-acid battery, a permanent magnet brushless DC motor and a 'Norton' rotary engine.

The special battery uses lead tin foil plate construction, which was developed for the aircraft industry. This allows deep cycling and long life as high internal pressures prevent loss of active material during deep discharge. Tests have shown that 18 batteries rated at 30 Ah and 12 V, can provide 50 kW for 5 minutes. Hawker Siddeley has developed a flat array of cells that can be placed under the passenger compartment of the vehicle. The pack measures $120 \times 120 \times 4\,cm^2$, weighs 170 kg and can supply 7.5 kWh. The battery can withstand 1100 discharges to 80% depth of discharge (DOD) and 11 000 cycles to 20% DOD. This is expected to last the life of the vehicle. The reason for this long life is a battery thermal management system, which keeps the lead-acid cells at a constant 30–40°C which is the most efficient operating temperature.

Norton rotary engines achieved fame by winning major awards in the motorcycle racing world. This engine has a fast warm-up and only an 8 Nm starting torque. Two electrically pre-heated catalytic converters are used and the injection system operates the engine on a lean burn setting at high load. The engine supplies a constant output with the electric motor adding power for transient loads.

Figure 17.10 shows a sectional representation of the permanent magnet brushless DC motor. The actual motor used weighs 45 kg and is liquid cooled; oil is used as the coolant to prevent

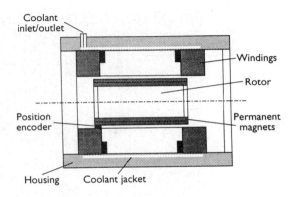

Figure 17.10 Permanent magnet brushless DC motor

freezing. A sophisticated inverter and control circuit controls the motor. The voltage supply to the motor is converted from the 216 V DC of the batteries to a 300 V DC stabilized rail. The motor is supplied with three-phase power as either trapezoidal or square waves, the phase of which can be altered to control braking or acceleration. The accelerator position provides an input to the control module and a Hall effect rotor position sensor provides a feedback signal. The position feedback is to ensure the three phases of the motor are energized in the correct order.

The whole power unit weighs about 100 kg compared with 200 kg for a conventional system. The batteries, however, add a further 130 kg above the normal, but allow a 48 km (30 mile) range without running the engine.

17.3.4 A sodium-sulphur battery EV system

The layout or interconnection of components on an EV depends on the type of battery and drive motor. Figure 17.11 represents a system using sodium-sulphur (NaS) batteries, and a shunt wound DC motor using conventional brushes.

Altering the field current and/or the armature current changes the speed and torque of this type of motor. The control characteristics used on this type of drive system are shown in Figure 17.12. The vehicle starts accelerating at time = zero. In the early stages of acceleration the field is held constant and the armature current is limited so as to match the demand.

As speed increases, the field current is decreased which will weaken the main fields so reducing the back EMF from the armature. The armature current demand can be met allowing increased speed. A motor such as this is likely to

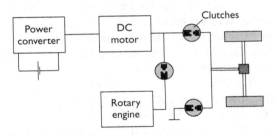

Figure 17.9 Parallel layout used for the 'Nelco' system

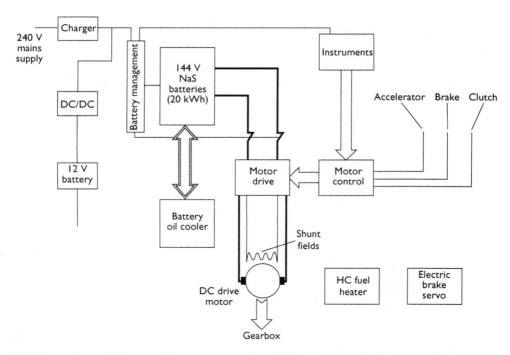

Figure 17.11 Layout that could be typical of a system using sodium-sulphur batteries and a shunt wound DC motor

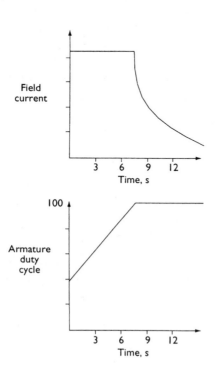

Figure 17.12 The control characteristics that can be used on this type of drive system

be air cooled. Some systems do, however, use liquid coolant. A variable regenerative braking system is used to maximize the efficiency of the system. This allows the batteries to be recharged during braking.

Batteries are often connected in series to increase the voltage. Motor design is easier for higher voltages mainly due to less current being required for the same power transfer. A battery management system is used to ensure the battery charge and discharge rates are controlled to the optimum value. A number of warning functions can be built in to indicate an abnormality and a warning about the remaining range of the vehicle is also possible. This information is displayed on the instrument pack.

The drive controller is made using existing power transistor technology. The transistors are controlled by a microprocessor, which in turn has its characteristics set by software. The controller receives input signals from the brake and accelerator pedals by using simple potentiometers. Signals from the other controls are from basic switches.

A simple method of controlling the rest of the vehicle electrical system is by fitting a conventional 12 V lead-acid battery. This can be charged when required from the drive batteries via a DC/DC converter.

17.4 Advanced electric vehicle technology

17.4.1 Motor torque and power characteristics

The torque and power characteristics of four types of drive motors are represented in Figure 17.13. The four graphs show torque and power as functions of rotational speed.

A significant part of the choice when designing an EV is the drive motor(s), and how this will perform in conjunction with the batteries and the mass of the vehicle.

17.4.2 Optimization techniques – mathematical modelling

The effects of design parameters on the performance of an EV can be modelled mathematically. This section presents some of the basic techniques. Refer to Figure 17.14 and Table 17.2 for an explanation of the symbols.

Aerodynamic drag force:

$$F_a = \frac{\rho C_d A_f}{2}(V_v \pm V_{wind})^2$$

Rolling resistive force:

$$F_r = \mu_r mg \cos(\theta)$$

Climbing resistive force:

$$F_c = mg \sin(\theta)$$

Therefore the total resistive force is:

$$F_{resistive} = F_a + F_r + F_c$$

Force developed at the wheels:

$$F_{dw} = F_{motor}\,\eta_e\,\eta_m$$

The tractive effort therefore is:

$$F_{tractive} = F_{dw} - F_{resistive}$$

The maximum tractive force that can be developed:

$$F_{dwmax} = \frac{\alpha\mu_a W/L}{1 + \mu_a\,h_{cg}/L}$$

The effective mass of a vehicle is:

$$m_{eff} = m + \frac{J_{eff}}{r^2}$$

Acceleration time can now be shown to be:

$$t = m_{eff}\int_{V_1}^{V_2}\frac{dV}{F_{tractive}}$$

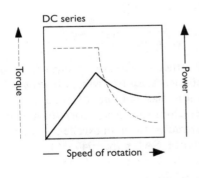

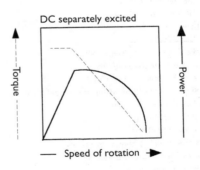

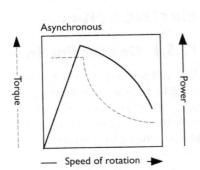

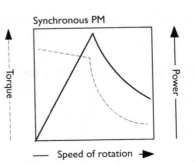

Figure 17.13 Motor torque and power characteristics

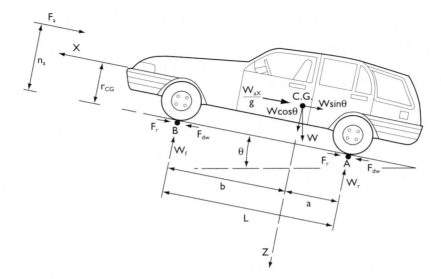

Figure 17.14 Mathematical modelling – values used

Table 17.2 Explanation of symbols

F_a	Aerodynamic drag force
ρ	Density of air
C_d	Coefficient of drag, e.g. 0.3 to 0.4
A_f	Area of the vehicle front
V_v	Velocity of the vehicle
V_{wind}	Velocity of the wind
F_r	Rolling resistive force
μ_r	Road coefficient of friction
μ_{const}	Tyre rolling coefficient of friction
F_c	Climbing resistive force
m	Mass of the vehicle (total)
g	Acceleration due to gravity
θ	Angle of the hill
$F_{resistive}$	Total resistive force
F_{dw}	Force developed at the driving wheels
η_e	Efficiency of the electric motor
η_m	Efficiency of the mechanical transmission
a	Centre of gravity position within the wheel base
μ_a	Coefficient of road adhesion
W	Weight of the vehicle (mg)
L	Length of the wheel base
h_{cg}	Height of the vehicle's centre of gravity
J_{eff}	Total effective inertia of the vehicle
M_B	Mass of the battery
Y	Power density of the battery (see Table 17.1)
x_i	Correlation between energy density as a function of power density

Power required to hold the vehicle at a constant speed:

$$Power = \frac{V_v \, F_{resistive}}{\eta_e \, \eta_m}$$

Power density of the batteries:

$$y = \frac{Power}{M_s}$$

The correlation between energy density as a function of power density can be calculated:

$$x_i = ay^5 + by^4 + cy^3 + dy^2 + ey + f$$

The range of the vehicle from fully charged batteries can be calculated from:

$$Hours = \frac{x_i}{y}$$

$$Range = V_v \times Hours$$

Further calculations are possible to allow modelling – a subject which, if grasped, can save an enormous amount of time and money during development. The information presented here is extracted from an excellent research paper. (Reference: SAE paper 940336).

17.5 New developments in electric vehicles

17.5.1 Gas turbine hybrid

The state-of-the-art gas turbine engine is very attractive to the automotive industry and is in line with environmental pressures towards low emissions and low fuel consumption. The turbine engine has a number of useful features:

- Good thermal efficiency.
- Clean combustion.
- High power-to-weight ratio.
- Multifuel capability.
- Smoothness of operation.

These advantages could make it a natural successor to the reciprocating engine. The automotive gas turbine is still in its infancy, despite many technical achievements made since the world's first gas turbine car – the Rover 'Jet 1'. The technical challenge posed by the automotive gas turbine remains considerable and is, in many ways, even greater today. This is due mainly to the challenge created by advancing combustion, mechanical, aerodynamics, material and electrical technologies.

A further factor that has been added, and which was described earlier, is the hybrid electrical vehicle system, for which the gas turbine engine is most suitable. The design scope of hybrid systems is also very wide. The gas turbine engine has many features that suit automotive applications. For example, it is compact and light, which allows flexibility in power train layout. This reduces the vehicle weight, which results in better vehicle performance and economy.

Modern combustion chamber design makes the engine produce very low emissions of all pollutants, even when burning diesel. This can be achieved without having to use catalytic converters. These advantages are becoming increasingly important in the current market place.

Compared with an equivalent reciprocating engine, gas turbines are smooth and quiet in operation, can run on various types of hydrocarbon fuel, and their inherent mechanical simplicity will result in improved reliability and increased servicing intervals.

Compared with the conventional reciprocating engine, the turbine has, until recently, had poor transient power response and part load fuel economy. There has also been a natural resistance to change by the automotive industry because of its huge investments in the infrastructure of the existing engine's manufacture and service. When the advantages of the turbine are combined with the advances in hybrid electrical systems, an exciting combination offers great potential for the future of the hybrid technique.

17.5.2 Inductive charging

The Nissan Altra, as described earlier in this chapter, uses inductive charging. In this case a 'paddle' connected to an external power source, is used to plug into a 'socket' in the car. The risk of electric shock and the possibility of overheating due to 'loose' connections are almost eliminated.

'Drive in' inductive charging is a possible development to help the advance of the electric vehicle. The principle is shown in Figure 17.15. A coil, which forms the secondary winding of a transformer, is positioned on the car in a suitable position. The primary winding of the transformer could be placed on a movable core which, when the vehicle is parked, could automatically lift into position and allow a magnetic link with the secondary winding.

17.5.3 ZOXY battery system – 'chemTEK'

The ZOXY zinc-air battery is not really a battery in the traditional sense. The core of the ZOXY 'P280' is a single, easy-to-handle and flexible unit. The battery dimensions are $220 \times 135 \times 39 \, \text{mm}^3$ and it weighs only 2 kg. Its energy density of 150 Wh/kg is five times the amount of lead-acid batteries. The ZOXY battery will keep its charge for very long periods; its typical energy discharge is under 1%. If the air supply is interrupted, the self-discharge falls well below 1%. Another advantage of the ZOXY system is that it works in a wide temperature range (–20 °C to +40 °C).

Although production levels currently are relatively small, the costs of the ZOXY are equal to the costs of ordinary lead-acid batteries per unit of energy. With economies of scale, the cost level of a ZOXY battery would fall considerably below that of its lead-acid counterpart.

While the ZOXY zinc-air battery is a 'high-energy reservoir', an additional 'booster' may be used for high acceleration vehicles to ensure optimum performance. The booster battery provides the power necessary for the acceleration demands of the vehicle. Higher speed driving

Figure 17.15 Inductive charging could help the development of the electric vehicle

may require an ability to merge into traffic relatively quickly. On the other hand, city driving is slower-paced and acceleration requirements are reduced. The cascade system, using two batteries, is an excellent way to provide for these and other driving needs.

An electronic control system manages the use of both battery types without the driver of the vehicle ever being aware of the changes. The driving characteristics of such a system are similar to petrol/gasoline and diesel-powered vehicles.

P280 Specifications

- Extended operating time (150 Wh/kg).
- Volume ($33.0 \times 13.5 \times 3.9\,cm^3$).
- Weight (1.25/2.00 kg dry/filled).
- Stable discharge curve, low self-discharge.
- Low-sensitivity to temperature changes ($-20\,°C$ to $+40\,°C$).
- Environmentally friendly.
- Voltages nominal/shut-off 1.1/0.6 V.
- Nominal current/peak current 0–30/40 A.
- Capacities at 10/20A 320/280 Ah.
- Energy content at 10/20 A 300/250 Wh.

17.6 Self-assessment

17.6.1 Questions

1. State what is meant by ZEV.
2. Describe briefly the term 'Hybrid'.
3. Explain what is meant by, and the advantages of, inductive charging.
4. Describe with the aid of sketches the different ways in which a hybrid vehicle can be laid out.
5. Explain the term 'Power density'.
6. List five types of EV batteries.
7. The GM EV-1 uses lead-acid or alkaline batteries. State three reasons for this.
8. Describe with the aid of a sketch the operation of a synchronous motor.
9. State four types of EV drive motor.
10. Describe how the Nissan Altra calculates the current range of the car.

17.6.2 Assignment

A question often posed about so-called ZEVs: as the electricity has to be generated at some point, often from burning fossil fuels, then how can they be said to produce no emissions?

The answer, in my opinion, is that at the point of use the vehicles are ZEVs. The production of the electricity for recharging, which will mostly be during the night, allows power stations to run at optimum efficiency and hence overall emissions are reduced.

Research and comment on this issue.

18
World Wide Web

18.1 Useful contacts

18.1.1 Introduction

If you have access to a computer and modem then you are no doubt already interested in the Internet and the World Wide Web (WWW). In this short chapter I want to highlight some of the resources available to you on the net and on disk.

All the major vehicle manufacturers have web pages, as do most of the parts suppliers. There are user groups, forums for discussions and many other sources of information and lets admit it, it's just good fun. So go ahead and surf the net. I would be most pleased to meet you at my WWW site where you will find up-to-date information about my books, new technologies and links to just about everywhere. Just point your browser to:

http://www.automotive-technology.co.uk

All the links to sites in the next section can be accessed from here, as well as many more. It will also be kept up to date as I find more interesting places to visit. Please let me know if you have any suggestions for useful links – I am mostly interested in sites that provide good technical information, not just advertising material!

tom_denton@compuserve.com

Figure 18.1 Information is often provided on a CD

18.1.2 Web site addresses

Here are some site addresses that I have found useful and interesting. I can't guarantee that the addresses will stay current so check my web site for updates.

Table 18.1 Automotive component or equipment suppliers

http://www.blaupunkt.de/	Blaupunkt ICE
http://www.bosch.de/	Bosch
http://www.delphiauto.com/	Delphi-Electrical & Electronic
http://www.eur.lighting.philips.com/automotive/	Philips Lighting
http://www.genrad.com/	Diagnostics & Equipment
http://www.golucas.com/	LucasVarity
http://www.hella.de/	Hella – Lighting
http://www.siemensauto.com/	Siemens Automotive
http://www.valeo.com/	Valeo – Electrical & Electronic

General

http://mot-sps.com/automotive/	Motorola Automotive
http://www.calstart.org/	Transport news and information
http://www.detr.gov.uk/	Department of the Environment, Transport & Regions (UK)
http://www.emclab.umr.edu/	Electromagnetic Compatibility
http://www.fuelinjection.com/accel.html	Performance tuning and 'Hot chipping'
http://www.njautosite.com/	New Jersey Automotive Site + 'Virtual Mechanic!'
http://www.nrtt.demon.co.uk/can.html	Controller Area Networks
http://www.sws.co.jp/	Sumitomo Wiring Systems
http://www.uta.com/	United Technologies Automotive
http://www.via.gov.uk/	Vehicle Inspectorate – UK

General automotive directories

http://www.autodirectory.com/	Automotive directory
http://www.findlinks.com/autolinks.html/	Automotive related links

Vehicle manufacturers

http://www.alfaromeo.com/	Alfa Romeo
http://www.audi.com/	Audi
http://www.bmw.com/	BMW
http://www.citroen.com/eng/	Citroen
http://www.dm.co.kr/	Daewoo
http://www.ferrari.it/	Ferrari
http://www.fiat.com/	Fiat
http://www.ford.com/	Ford
http://www.honda.com/	Honda
http://www.jaguarcars.com/	Jaguar
http://www.lamborghini.com/	Lamborghini
http://www.landrover.com/	Landrover
http://www.lotuscars.com/	Lotus
http://www.mazda.com/	Mazda
http://www.mercedes-benz.com/	Mercedes Benz
http://www.mitsubishi-motors.co.jp/	Mitsubishi
http://www.nissan.com/	Nissan
http://www.peugeot.com/	Peugeot
http://www.porsche.com/	Porsche
http://www.renault.com/	Renault
http://www.rover.co.uk/	Rover Cars
http://www.seat.com/	Seat
http://www.subaru.com/	Subaru
http://www.suzuki.com/	Suzuki
http://www.toyota.com/	Toyota
http://www.vauxhall.co.uk/	Vauxhall
http://www.volkswagen.com	Volkswagen
http://www.volvo.com/	Volvo

Research

http://www.mira.co.uk/	Motor Industry Research Association (MIRA)

Electronics companies

http://www.mot.com/	Motorola
http://www.national.com/	National Semiconductors
http://www.ti.com/	Texas Instruments

Electric and hybrid vehicles

http://www.ev.hawaii.edu/	Electric Hybrid Vehicles
http://www.gmev.com/	General Motors EV1

Automotive and engineering societies

http://www.iee.org.uk/	Institute of Electrical Engineers (IEE)
http://www.imeche.org.uk/	Institute of Mechanical Engineers (ImechE)
http://www.sae.org/	Society of Automotive Engineers (SAE)
http://www.irte.org/	Institute of Road Transport Engineers (IRTE)

Publishers and media

http://www.arnoldpublishers.co.uk/	Arnold Publishers
http://www.autodata.ltd.uk/autoweb4/english/	Autodata Reference Books
http://autologic.nmaster.com/index.html	Autologic Reference and CDs
http://www.macmillan-press.co.uk/	Macmillan Press
http://www.newscientist.com/	New Scientist Magazine
http://www.sae.org/automag/	Automotive International magazine
http://www.topgear.com/	Top Gear

Equipment suppliers

http://www.fluke.com/	Fluke Instruments
http://www.snapon.com/	Snap-on Tools

Oil companies

http://www.bpamoco.com/	BP/Amoco Oil
http://www.shell.com/home/	Shell Oil

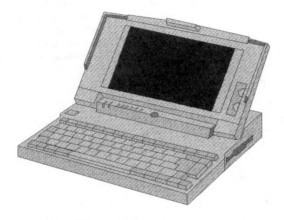

Figure 18.2 The WWW is a good source of information

18.2 Programs and information to download

18.2.1 Introduction

Many of the web sites listed above have information you can download. This is in addition to the information on the normal web pages. In many cases the information is in the form of a '.pdf' (Portable Document Format) file. To read this you will need 'Acrobat Reader', which is available free from http://www.adobe.com/acrobat/ or on many magazine CDs.

Use pictures from other sites to illustrate your work; it is OK to copy these if they are only for your own personal use but make sure you reference the source.

18.2.2 Searching the web

My favourite search engine is 'Yahoo!' at http://www.yahoo.com/. This is because a lot of work has already been done for you and it is easy to use. The following is a short extract from guidance given by Yahoo! on using their search site:

Yahoo!, like the Web itself, is too large to be explored entirely link by link. However, with over half a million sites divided into more than 25 000 categories, Yahoo! is both browseable and searchable. Use these two features and you will almost always find something to match your interests.

You can browse Yahoo! by simply clicking on the various categories listed on each page. Search Yahoo! by entering a word (or, a few words) into the seach box that appears on every page in the directory. Combine the two strategies and you can 'browse and then search' or 'search and then browse.'

If you put quotation marks around your search terms, you'll get search results that only contain that exact phrasing. If you leave off the quotation marks, you'll get

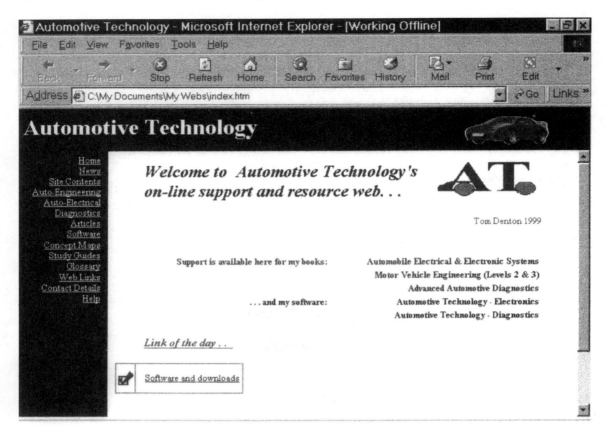

Figure 18.3 A screen shot from my web site

results that contain any of the words that you typed in.

If you know exactly what you're looking for or even have a general idea, try using Yahoo! search. Specify a keyword or set of keywords, and Yahoo! will search its entire database to find listings that match the keywords you provide.

After you have specified keyword(s) inside the query box and clicked on the search button, Yahoo! will search through the five areas of its database for keyword matches. The five areas are:

- Yahoo! Categories
- Yahoo! Web Sites
- Web Pages
- News Stories
- Yahoo! Net Events

The first page returned to you will be a list of matching Yahoo! Categories followed by a list of matching Yahoo! Sites. If no matching Yahoo! Categories and Sites are found, Yahoo! will automatically perform a Web-wide, full-text document search using the 'Inktomi' search engine.

You are able to navigate through matches from Yahoo! Categories and Yahoo! Sites, Web Sites, Yahoo! News and Yahoo! Net Events by clicking on the links in the bar at the top and bottom of the page.

18.2.3 Automotive Technology – electronics

Available as shareware from:
http://www.automotive-technology.co.uk
From the 'Help File':

Click on the key button in the toolbar to open the 'Controls' window if it is not already visible. Now click on the key button in this window to start the engine. You can also right click the screen and choose to start the engine from there. Don't forget to check that the car is not in gear – or else it won't start!

Next choose from the 'Run' menu the electronic system you would like to operate or work with. I suggest 'Engine Management' is the best place to start.

In common with all of the other simulation windows you will use, you can set or control the operating inputs to the system. For engine management control, these are engine speed, engine load, temperature and so on. The system will react and control the outputs in just the same way as a real vehicle. Be warned the unregistered version runs out of fuel!"

'An ideal way of learning about automotive electronic systems'.

18.3 Self-assessment

18.3.1 Questions

1. State the web address of General Motors in the UK, USA and Australia.
2. Check and comment on the latest news from: http://www.automotive-technology.co.uk
3. Determine a typical peak current figure (amps) for a modern ignition system and send me the answer via e-mail.
4. Describe four advantages and disadvantages of research via the WWW.
5. Describe briefly why this chapter only has five questions!

18.3.2 Assignment

Look back at any of the other assignments in this book and choose one for further study. Your task is to use the Internet/WWW as your research tool. Produce a report on the latest technology developments in your chosen subject area. E-mail it to me if you wish and I may be able to use it in the next edition – with acknowledgement to you, of course.

Good luck with your future studies and work – keep in touch . . .

Figure 18.4 A screen shot of the electronics simulation program

Appendix

This program runs under the Microsoft® QBASIC interpreter. It is also available to download free of charge from my web site.

```
Color 7, 9
CLS: Clear

load1: Data 14, 30, 40, 53:       Rem Timing lookup table
load2: Data 13, 28, 37, 49
load3: Data 11, 20, 30, 42
load4: Data 10, 20, 28, 40

volts14: Data 20, 30, 40, 50:     Rem Dwell lookup table
volts13: Data 21, 31, 41, 51
volts12: Data 23, 33, 43, 53
volts0: Data 24, 34, 44, 54

temp1: Data -5, -4, -3, -2:       Rem Timing correction temperature lookup table
temp2: Data -4, -3, -2, -1
temp3: Data -3, -2, -1, -1
temp4: Data -1, 0, 0, 0

loadi1: Data 0.2, 0.7, 1.2, 1.7: Rem Injection period lookup table
loadi2: Data 0.5, 1, 1.5, 2
loadi3: Data 1, 1.5, 2, 2.5
loadi4: Data 1.5, 2, 2.5, 3

Print
Print' Engine Management Simulation'
Print

Rem section to input engine speed and convert to a lookup value
start:
lastspeed = speed: lastload = Load
INPUT 'Enter engine speed ...'; speed
If speed = 0 Then Go To cutoff
If speed < 150 Then Go To cranking:    Rem engine is cranking
If speed > 500 Then speedlookup = 1:   Rem sets column in data lookup
If speed > 1500 Then speedlookup = 2
If speed > 2500 Then speedlookup = 3
If speed > 3500 Then speedlookup = 4

Rem section to input load and convert to a lookup value
INPUT 'Enter engine load [min] 1 2 3 4 [max] ...'; load
If Load = 1 Then RESTORE load1: Rem sets line in data lookup
If Load = 2 Then RESTORE load2
If Load = 3 Then RESTORE load3
If Load = 4 Then RESTORE load4
```

```
Rem section to lookup timing value from table
READ timing 1, timing2, timing3, timing4
If speedlookup = 1 Then timing = timing1
If speedlookup = 2 Then timing = timing2
If speedlookup = 3 Then timing = timing3
If speedlookup = 4 Then timing = timing4

timinglookup = timing

Rem section to enter temperature and lookup timing correction
INPUT 'Enter coolant temperature (-20 to 100) ...'; engtemp
If engtemp > -21 Then RESTORE temp1
If engtemp > -1 Then RESTORE temp2
If engtemp > 20 Then RESTORE temp3
If engtemp > 50 Then RESTORE temp4
READ tempcor1, tempcor 2, tempcor3, tempcor4
If speedlookup = 1 Then timing = timing + tempcor1
If speedlookup = 2 Then timing = timing + tempcor2
If speedlookup = 3 Then timing = timing + tempcor3
If speedlookup = 4 Then timing = timing + tempcor4

Rem section to enter battery voltage and lookup dwell
INPUT 'Enter battery voltage ...'; volts
If volts > 0 Then RESTORE volts0
If volts > 11.9 Then RESTORE volts12
If volts > 12.9 Then RESTORE volts13
If volts > 13.9 Then RESTORE volts 14
READ dwell1, dwell2, dwell3, dwell4
If speedlookup = 1 Then dwell = dwell1
If speedlookup = 2 Then dwell = dwell2
If speedlookup = 3 Then dwell = dwell3
If speedlookup = 4 Then dwell = dwell4

Rem section to lookup and correct injection time
If Load = 1 Then RESTORE loadi1: Rem sets line in data lookup
If Load = 2 Then RESTORE loadi2
If Load = 3 Then RESTORE loadi3
If Load = 4 Then RESTORE loadi4
READ injection1, injection2, injection3, injection4
If speedlookup = 1 Then injection = injection1
If speedlookup = 2 Then injection = injection2
If speedlookup = 3 Then injection = injection3
If speedlookup = 4 Then injection = injection4

Rem section to correct injection period for temperature and voltage
If engtemp < -10 Then injection = injection * 2
If engtemp < 0 Then injection = injection * 1.5
If engtemp < 20 Then injection = injection * 1.3
If engtemp < 50 Then injection = injection * 1.1
If volts < 12 Then injection = injection * 1.1
If volts < 13 Then injection = injection * 1.05
If volts < 14 Then injection = injection * 1.02

Rem section to check throttle position and implement fuel variation
INPUT 'Throttle position 0 normal, -1 over run, 1 full load ...'; throttle
If throttle = 1 And speed > 2000 And engtemp > 50 Then injection = 0
If throttle = 1 And engtemp > 50 And Load = 4 Then injection = 5

Rem section to adjust timing if knocking has started or stopped
Print
```

```
Print  'Timing lookup value ='; timinglookup; 'deg BTDC'
Print
If speed < lastspeed Or speed > lastspeed Then GoTo printout
If Load < lastload Or Load > lastload Then Go To printout
INPUT 'Is engine knocking [1=yes 0=no] ...'; knock
timinglookup = timing: Rem to be used as a test figure
If knock -0 Then timing = lastiming +2
If timing > timinglookup Then timing = timinglookup
If knock = 1 Then timing = lasttiming -2
If timing <8 Then timing = 8: Rem sets a bottom limit
GoTo printout

cutoff:
CLS
Print
Print 'Engine not running or cranking - coil not switched on and no fuel
injected'
Print
GoTo start

cranking:
CLS
timing = 10: dwell = 20: injection = 3.5: Rem crank timing dwell injection
Print
Print 'Basic settings for cranking condition have been set'

printout:
Print
Print  'Actual timing ='; timing; 'deg BTDC'
Print
Print  'Dwell ='; dwell; '%'
Print
Print  'Injection ='; injection; 'mS'
lasttiming = timing

GoTo start
```

Index

ABS control cycles 345
ABS 330, 335
AC motors 388
Accelerometer 41
Accuracy 35, 57
Active muffler 384
Active roll reduction 339
Active steering 338
Active suspension 332
Actuator thermal 51
Actuator, rotary 48, 53
Actuators 36, 47
Adaptive cruise control 359, 360
Adaptive noise control 375
Advance angle 154
Aerial, electric 366
Afterburning 223
Aiming board 268
Air bags 369, 373, 377
    sequence 370
    triggering unit 372
Air conditioning 315, 317, 318, 320, 324
    components 301, 310, 321
Air cored gauge 301, 310
Air shrouded injectors 219, 220
Alarms 384
Alkaline battery 111
Alternative fuels 45
Alternator 119, 125, 127, 129
    characteristics 131
    circuit 122
Ampere-hour capacity 103
Amplifier 20, 21
    differential 21
    operational 22
Amplitude modulation 363
Analogue to digital conversion 25
Antilock brakes 328, 331, 340
Anti-phase noise 375
Arithmetic logic unit 32
Armature reaction 325
Armature windings 325
Artificial intelligence (AI) 254
Asynchronous motor 388
Atom 13
Audio 381
Auto PC 366
Automatic clutch 351
Automatic temperature control (ATC) 319
Automatic transmission 336, 342
Auxiliaries 290

Auxiliary air device 51
Auxiliary circuits 284

Batteries 101, 338
    acid 11
    charging 104
    choosing 101
    construction 102
    faults 105
    lead-acid 108, 110
    maintenance 105
    testing 106
Beam setter 268
Bifocal reflector 266
Black box techniques 346
Bosch jetronic variations 201
Brake assist systems 350
Brake by wire 350
Brake lights 290
Bridge circuits 22
Bulb failure circuit 304
Bulbs 260, 261
Buses 31

Cables 78
    size 79
    volt drop 79
Capacitance 14, 40
Capacitor discharge ignition 163
Capacitor 17, 19, 54, 365
Carbon monoxide measurement 63
Carbon monoxide 87
Carburation 189
Carburation, electronic control 189
Catalyst 107, 212
Catalytic converter, electrically heated 224
Catalytic converters 223, 226
CD-ROM 68, 399
Central control unit 226
Central locking 356
Charge balance calculation 130
Charging 117
    circuits 125
    system 117, 128, 132
    voltages 119
Circuit 13
    breakers 83
    diagram 4, 93
    diagrams 93
    symbols 18
Claw pole rotor 119

Clock 29
Closed loop 162
Coded electronic control units 369
Cold cranking amps 103
Cold start injector 197
Colour codes 79
Combination relay 198
Combinational logic 26
Combustion process 180
Combustion 180, 186
Common rail diesel injection 214, 215
Communications 381
Complex shape reflectors 280
Component power consumption 118
Compression ignition 184
Conductors 13
Connectors 84
Constant dwell 157
Constant energy 157
Contact breaker ignition 174
Controller area network (can) 89, 91, 227, 342
Conventional flow 12
Coolant sensor 165
Cooling fan motors 290
Counters 28
Crankshaft sensor 164, 238
Cruise control 383
Cruise control, actuator 359
Cruise control, adaptive 359
Current flow diagrams 93, 97
Current flow, effects of 12
Current limiting 162

Dab 363
Damping factor 383
Darlington pair 24
Data bus 88
Day running lights 264
DC motor 138
    characteristics 138
    circuits 139, 140
    series wound 389
Detonation 182
Diesel emissions 199
Diesel knock sensor 52
Digital circuits 25
Digital instrumentation 303
Digital to analogue conversion 25
Dim dip 268, 272
Diodes 19
Dipped beam 272
Direct ignition 164, 169
Direct injection 228
Direct mixture injection 256
Display techniques 305
Distributor injection pump 199
Distributorless ignition (DIS) 168, 176
Door lock actuators 365
Door lock circuit 365
Door locking 356

Drive by wire 351
Drive motors 388
Driver information 303
Dual rail power supply 129
Dwell 161
    angle 161
    calculation 250
    map 239
Dynalto system 151

ECU auto-diagnostics 245, 382
Electric
    headlamp actuators 379
    mirrors 355
    power steering 338
    seats 354
    sunroof operation 355
    traction 386
    vehicles 386
    windows 356, 378
Electrical
    loads 117
    symbols 18
Electrochemistry 107
Electrode gap 172
Electroluminescence 307
Electrolyte 102, 106
Electromagnetic compatibility (EMC) 98
Electromagnetism 15
Electron flow 12
Electron 12
Electronic
    clutch 338
    components 17
    control of automatic transmission (ECAT) 336
    control of wipers 286
    control unit (ECU) 90, 166, 198, 239, 253, 330,
        338
    diesel control 200
    fan 297
    heating control 317
    ignition 155
    limited slip differential 339
    spark advance (ESA) 168
    starter control 151
Emissions 187, 199, 233
    regulations 189
    comparisons 200
Engine analysers 60
Engine cooling 290
Engine management 178, 217, 256, 403
EPROM 29
Equivalent circuit 304
Evaporator 322
Exhaust emissions 186
    control 154, 221
    test 62
    diesel 225
    regulations 189
Exhaust gas 186, 187

measurement 44, 62
  recirculation (EGR) 222
  recirculation valve 53
Expert lighting 277

Faraday Michael 1, 4
Fetch execute sequence 31
Fibre optics 98
Field windings 138
Filters 23
Flap type air flow sensor 41
Flasher unit 289
Flip-flops 27
Flowchart diagrams 167, 252, 274
Fog lights 264
Frequency modulation 383
Frequency response 22
Fuel cell 112, 113
Fuel injection 191
  diesel 198
  fuel supply 240
  injectors 194, 196, 205, 240
  mixture calculation 241
  multi/single point 191, 206
  petrol 191, 193, 237, 240
  pressure regulator 194, 197
  pump 194, 197, 208
Fuses 83, 84

Gas analyser 62
Gas discharge lamps (GDL) 269, 270
Gas turbine 396
Gasoline direct injection (GDI) 228, 236
Gauges 299
General motors EV1 390
Glow plug 201

Hall effect 38, 52, 159
  pulse generator 159
Hand tools 56
Hazard circuit 291
Head up display (HUD) 308
Headlights 264, 265
  beam patterns 265, 267
  cleaners 291
  circuits 269
  lenses 267
  levelling 267
  wipers 290
Heater blower motor 316
Heating system 315
High resistance 13
High tension 153
High voltage systems 150
History 1
Holography 313
Homifocal reflector 266
Horns 290
Hot chipping 253
Hot wire air flow sensor 42

Hot wire injection 203, 207
Hybrid drive 392
Hybrid vehicles 390
Hydraulic modulator 331
Hydrocarbons 187

Idle control 194
Ignition 2, 153
  amplifier 156
  calculations 248
  coil 155, 177, 178
  control principle 302
  conventional 154, 156
  electronic 155, 156, 158
  energy 157
  leads 155, 157
  module 156
  timing 166
Immobilizer 369
Impact sensor 371
In car entertainment (ICE) 360, 361, 385
In-car multimedia 360
Indicator circuits 292
Indicators 291
Inductance 14, 16, 17
Inductive charging 397
Inductive pulse generator 160
Inductive sensor 38
Inductor 19
Inertia engagement 140
Inertia starter 140
Inflatable curtain 378
Infra-red lights 280
Infra-red remote 369
Injection cut-off 244
Injection duration calculation 250
Injector 47, 208
  resistors 194
  signals 247, 248
Instrumentation 299, 302
Insulator 13
Integrated circuit 19
Integrated starters 151
Intelligent air bag sensing system 385
Intelligent front lighting 277
Interference suppression 364
Intermediate transmission starter 144
Intermittent wipe 288
Internal resistance 109

Jewel aspect signal lamps 278

Knock protection 244
Knock sensor 41, 165

Lambda control effects 225
Lambda 224
  control system 225
  fuel map 193
  sensor, wide range 219

Layout wiring diagrams 93, 95
Lead-acid batteries 102
Leaded fuel 188
Lean burn 212, 230, 232, 256
Led display 305
Led lighting 271
Lenses 265
Light sensors 45
Lighting circuit 268
Lighting 260, 278
Linear lighting 278
Linear rear wiper 296
Linear Variable Differential Transducer (LVDT) 41, 42
Liquid crystal display (LCD) 306
Local intelligence 91
Logic gates 26

Magnetic field 138
Magnetism 15
Magneto 3, 6
Main beam 272
Manifold absolute pressure (MAP) sensor 164
Measurement 35
Memory 28, 29, 31
Methanol sensor 45
Microcontroller 33
Microprocessor 30, 32, 288
Mirrors, electrical 354
Mobile communications 366
MOT regulations (UK) 189
Motor 48, 139
    characteristics 139
    reverse circuit 354
    actuators 48
Motronic 237
Moving iron gauge 300
Multimeter 57, 59, 88
Multiplex 88
    display 313
    wiring 88, 95, 98, 99, 312

Neon 279
Neural computing 255
Night vision 280, 281
Nissan Altra 391
Nitrogen oxides 187
Noise control 375, 384

Obstacle avoidance 373
Ohm's law 13
Oil condition sensor 52
On board diagnostics (OBD) 66, 95, 219
    standards (OBD2) 64, 220
One way clutch 142
Open circuit 13
Optical pulse generator 43
OPUS 158
Oscilloscope 56, 59, 60, 72
Otto 4

Over-run fuel cut-off 190, 206
Oxygen sensor 44

Parabolic reflector 264
Particulate emissions 187
Permanent magnet (PM) starter/motor 142, 295
Pitot tube 43
Poly ellipsoidal headlight 266
Ports 30
Position memory 355
Potentiometer 40
Power management 130
Power requirements 118
Power steering, electrical 343
Power 14
Pre-engaged starters 141, 144
Pre-engagement 141
Pre-ignition 182
Printed circuits 83
Programmed ignition 163, 165
Programmes 251, 402, 405
Programming 34
Pulse generators 159, 160
Pyrotechnic inflator 372

Quantization 312
Quenched oscillator 38

Radar 359, 360
Radio 360
    interference 381
    reception 362
    suppressors 383
Rain sensor 46
Random access memory (RAM) 29, 32
RBDS 363
RDS 362, 385
Read only memory (ROM) 29, 32, 192, 241
Rear lights 264
Rear wash/wipe 287
Rear wipers 287, 296
Receiver drier 322
Rectification 120
Rectifier pack 121
Rectifier 120
Reed switch 304
Reflector 264, 265
Refrigerant 317
Regenerative braking 391
Regulator 122
    electronic 124
    mechanical 123
Reserve capacity 103
Resistance 14, 40, 58
Resistor 14, 17, 54, 365
Resolution 35
Reverse lights 264
RGB lights 281
Rotor 119

Safety 11, 107
Scanner 56, 64
Schmitt trigger 22
Screen heaters 320
Seat 354
    belt tensioner 372
    electrical 354
    heaters 319
Security codes ECUs 369
Security 367, 368
Semiconductor 13
Sensors 36, 46, 299
Sensors, testing 53
Sequential injection 196
Sequential logic 27
Serial port communications 64, 65
Shareware 403
Short circuit 13
Side impact air bag 372
Side lights 272
Signalling circuits 289
Signalling 278
Single source lighting 281
Sodium sulphur batteries 114, 393
Solenoid actuators 47
Spark plugs 154, 170, 171, 172, 178
    diagnostics 176
    'V' groove 172
    heat range 171
    temperature 173
Speakers 361, 362
Speed control 295
Speed sensor 359
Speedometer 302
Spot lights 264
Stability management 344
Starter motor 136, 140
    heavy vehicle 144
    installation 145
    intermediate transmission 144
    internal circuit 138
    system circuits 136, 142, 146
    system 134
    torque 148
Stator 119, 120
Stepper motor 48, 49, 50
Stepper motor driver 24
Strain gauge 39, 371
Stratification 183, 222
Sub-woofers 361
Sun roof 354
Swing battery 115
Switches 78, 85, 86, 87
Symbols 93
Synchronous motor 50
Systems, open/closed loop 77, 78

Tachometer 302
Telematics 313, 381
Temperature control systems 326

Terminal designation 79
Terminal diagrams 93, 96
Terminals 78, 84, 85
Thermal actuator 51
Thermal gauge 299
Thermistor 36, 37, 195, 300
Thermocouple 37, 173
Thermostatic expansion valve 322
Thick film air flow sensor 45
Thin film air flow sensor 42
Third brush dynamo 3
Third harmonic 122
Throttle control 334
Throttle potentiometer 40, 195, 242
Thrust SSC 8
Timers 23, 28, 286
Timing map 239
Timing 154
Traction control (TCR) 334, 335
Traffic information 305
Transistor 19, 54, 55
Trip computer 304
Turbine flow sensor 43
Tweeters 361
Two stroke 257
Tyre pressure warning 374

Ultra capacitor 112
Ultra-violet headlights 271
Unit injection 213, 270, 271
Unleaded fuel 188

Vacuum fluorescent display (VFD) 307, 308
Variable inlet tract 217
Variable valve timing 218, 219
Vehicle condition monitoring (VCM) 303
Vehicle map 304
Vehicle position sensor 46
Ventilation 315
Video diagnostics 259
Visual display 305
Volta Alessandro 1, 4
Voltage 12, 14
    correction 244
    regulation 122
    stabilizer 300
Vortex flow sensor 43

Washer/wiper circuits 286
Water cooled alternators 133
Waveforms 61, 71
Web site addresses 399
Wheatstone bridge 22
Wheel speed sensor 330
Windscreen heaters 320
Windscreen washers 284, 286
Windscreen wipers 284
Wiper blades 284, 296
Wiper circuit 292
Wiper linkages 284

Wiper motors 284, 295
Wipers 284
Wiring 78
  diagrams 93
  harness 81, 82
  loom 81, 82
  terminals 78, 84, 85
Woofer 361

World wide web (WWW) 399

Xenon headlamps 282
Xenon lighting 272, 275

Zebra battery 112
Zener diode 19, 124
Zoxy battery system 397